P9-CJY-960

# THE MYCOPLASMAS

## VOLUME I

# THE MYCOPLASMAS

*EDITORS*

*M. F. Barile*

Mycoplasma Branch
Bureau of Biologics
Food and Drug Administration
Bethesda, Maryland

*S. Razin*

Biomembrane Research Laboratory
Department of Clinical Microbiology
The Hebrew University–Hadassah Medical School
Jerusalem, Israel

*J. G. Tully*

Mycoplasma Section
Laboratory of Infectious Diseases
National Institute of Allergy and Infectious Diseases
National Institutes of Health
Bethesda, Maryland

*R. F. Whitcomb*

Plant Protection Institute
Federal Research, Science and Education Administration
U.S. Department of Agriculture
Beltsville, Maryland

# THE MYCOPLASMAS

## VOLUME I

## Cell Biology

Edited by

*M. F. BARILE*

*Mycoplasma Branch*
*Bureau of Biologics*
*Food and Drug Administration*
*Bethesda, Maryland*

and

*S. RAZIN*

*Biomembrane Research Laboratory*
*Department of Clinical Microbiology*
*The Hebrew University–Hadassah Medical School*
*Jerusalem, Israel*

**ACADEMIC PRESS** New York San Francisco London 1979

A Subsidiary of Harcourt Brace Jovanovich, Publishers

ACADEMIC PRESS, INC.
111 Fifth Avenue, New York, New York 10003

*United Kingdom Edition published by*
ACADEMIC PRESS, INC. (LONDON) LTD.
24/28 Oval Road, London NW1 7DX

**Library of Congress Cataloging in Publication Data**
Main entry under title:

The Mycoplasmas.

Includes bibliographies.
CONTENTS: v. 1. Cell biology.
1. Mycoplasmatales. 2. Mycoplasma diseases. I. Barile, Michael Frederick, 1924– [DNLM: 1. Mycoplasma. QW143 M9973]
QR352.M89 589.9 78–20895
ISBN 0–12–078401–7

PRINTED IN THE UNITED STATES OF AMERICA

*79 80 81 82 83 84 9 8 7 6 5 4 3 2 1*

# CONTENTS

# LIST OF CONTRIBUTORS

Numbers in parentheses indicate the pages on which the authors' contributions begin.


***Edwin S. Boatman*** (63), Departments of Environmental Health and Pathobiology, School of Public Health, University of Washington, Seattle, Washington 98195

***Wolfgang Bredt*** (141), Institut für Allgemeine Hygiene und Bakteriologie, Zentrum für Hygiene der Albert-Ludwigs-Universität, D-7800 Freiburg, Germany

***Vincent P. Cirillo***[1] (323), Department of Biochemistry, State University of New York, Stony Brook, New York 11794

***Roger M. Cole*** (385), Laboratory of Streptococcal Diseases, National Institute of Allergy and Infectious Diseases, National Institutes of Health, Bethesda, Maryland 20014

***Jyotirmoy Das***[2] (411), Department of Microbiology, University of Rochester, School of Medicine and Dentistry, Rochester, New York 14642

***D. G. ff. Edward***[3] (1), Charlewood Cottage, Bickley, Kent, England

***E. A. Freundt*** (1), FAO/WHO Collaborating Centre for Animal Mycoplasmas, Institute of Medical Microbiology, University of Aarhus, DK-8000 Aarhus, Denmark

***George E. Kenny*** (351), Department of Pathobiology, SC-38, University of Washington, Seattle, Washington 98195

***Thomas A. Langworthy*** (495), Department of Microbiology, University of South Dakota, Vermillion, South Dakota 57069

***Jack Maniloff*** (411), Departments of Microbiology and of Radiation Biology and Biophysics, University of Rochester, School of Medicine and Dentistry, Rochester, New York 14642

***G. K. Masover*** (451), Department of Surgery, Division of Urology, Stanford University School of Medicine, Stanford, California

***Alana Mitchell*** (103), Russell Grimwade School of Biochemistry, University of Melbourne, Parkville, Victoria, 3052, Australia

***Harold Neimark*** (43), Department of Microbiology and Immunology, College of Medicine, State University of New York at New York City, Brooklyn, New York 11203


[1] Present address: Department of Biochemistry, The Weizmann Institute of Science, Rehovoth, Israel.

[2] Present address: Department of Microbiology, Bose Institute, Calcutta 700009, India.

[3] Deceased.

**Jan A. Nowak** (411), Department of Radiation Biology and Biophysics, University of Rochester, School of Medicine and Dentistry, Rochester, New York 14642

**J. D. Pollack** (187), Department of Medical Microbiology, College of Medicine, Ohio State University, Columbus, Ohio 43210

**Resha M. Putzrath**[4] (411), Department of Radiation Biology and Biophysics, University of Rochester, School of Medicine and Dentistry, Rochester, New York 14642

**Shmuel Razin** (213, 289), Biomembrane Research Laboratory, Department of Clinical Microbiology, The Hebrew University–Hadassah Medical School, Jerusalem, Israel

**Mitchell E. Reff** (157), Department of Microbiology and Molecular Genetics, Harvard Medical School and Department of Clinical Genetics, The Children's Hospital Medical Center, Boston, Massachusetts 02115

**I. M. Robinson** (515), National Animal Disease Center, Science and Education Administration, U.S. Department of Agriculture, Ames, Iowa 50010

**A. W. Rodwell** (103), CSIRO Division of Animal Health, Animal Health Research Laboratory, Parkville, Victoria, 3052, Australia

**Shlomo Rottem** (259), Biomembrane Research Laboratory, Department of Clinical Microbiology, The Hebrew University–Hadassah Medical School, Jerusalem, Israel

**M. C. Shepard** (451), Occupational and Preventive Medicine Service, Naval Regional Medical Center, Camp Lejeune, North Carolina

**Paul F. Smith** (231), Department of Microbiology, University of South Dakota, Vermillion, South Dakota

**Eric J. Stanbridge** (157), Department of Medical Microbiology, California College of Medicine, University of California at Irvine, Irvine, California 92717

**Joseph G. Tully** (431), Mycoplasma Section, Laboratory of Infectious Diseases, National Institute of Allergy and Infectious Diseases, National Institutes of Health, Bethesda, Maryland 20014

[4] Present address: Department of Physiology, Harvard Medical School, Boston, Massachusetts 02115.

# FOREWORD

It must be wondered why mycoplasmology has lagged so far behind other disciplines in microbiology. I think the explanation might be the uncertain relationship to bacteria that existed for so long. When I first took up the study of mycoplasmas, there were doubts among scientists as to whether they even existed, some claiming they were merely forms of bacteria. In fact, the field was regarded as slightly disreputable. There was disagreement about the morphology and the method of replication. Large bodies are characteristic of the pleuropneumonia-like morphology of mycoplasmas and of L-phase organisms, and there was much dispute about the significance of these and whether the large bodies were viable. In the older literature there are many descriptions of life cycles, but the morphology of these organisms was studied to the relative exclusion of all else. Today, there is agreement that mycoplasmas divide by binary fission, so far as the genome is concerned, with filaments being formed when there is delay in cell division. Rodwell and Mitchell discuss replication in Chapter 4.

Classification and nomenclature of the various members of the group lagged behind; we did not know how to refer to the organisms, except to say that they were organisms of the pleuropneumonia group—named after the first organisms isolated from bovine contagious pleuropneumonia. We used to speak of pleuropneumonia-like organisms, abominably abbreviated to P.P.L.O. Classification and nomenclature is discussed in Chapter 1 by Freundt and Edward.

Confusion also arose from the fact that bacteria, under the effect, for instance, of penicillin, appeared the same as a pleuropneumonia-like phase, so-called L forms, as regards morphology and colonial appearances under certain conditions of growth. Klieneberger-Nobel's series of isolates included one L-phase organism, the L1 organism, which actually belongs among the true mycoplasmas. Additional confusion was caused by the claims that mycoplasmas had reverted to bacteria. The reversion theory is postulated in at least one comparatively recent textbook, which does not consider the alternative hypothesis: that mixed cultures were being used, i.e., mycoplasmas being contaminated with bacteria. This is not to say I disbelieve that in the course of evolution mycoplasmas were descended from bacteria; Neimark in Chapter 2 provides evidence that this is so. My disbelief only extends to this being a contemporary occurrence. I have always regarded mycoplasmas as a distinct class of organisms.

In 1969 a comprehensive textbook, edited by Hayflick, with contributions from a number of experts in the fields of mycoplasmology and L-phase organisms was published. In the ten years that have elapsed since then, so much information has been collected by workers that a three-volume work is needed to present it, even though consideration of the L-phases of bacteria has been omitted. Klieneberger-Nobel wrote a Foreword to the Hayflick book, wherein she traced the early history of the mycoplasmas, dating from the initial isolation of the organism of contagious pleuropneumonia, and emphasized the original contributions she made at a time when there was little interest in the subject. I have no need to repeat this early history.

Since the earlier volume, much information has become available about the T-strains, now renamed ureaplasmas, which have been put into a new genus, *Ureaplasma*. They are reviewed in Chapter 17 by Shepard and Masover; the former is the discoverer of T-strains. The first acholeplasma was isolated in 1936 from sewage by Laidlaw and Elford; it is of interest that non-sterol-requiring mycoplasmas, with varying properties including serological ones, are at present being isolated from a wide variety of sources including plants. Accordingly, a chapter (Chapter 16) is included in this volume, where Tully discusses acholeplasmas and hypothesizes as to whether the effect of viruses can alter antigenicity. Darland and others in 1970 isolated from a coal refuse pile an organism they recognized as a mycoplasma, in spite of its needing for growth an extremely acid pH and a high temperature. A new genus, *Thermoplasma*, has been established for it, and information is given about this new genus by Langworthy in Chapter 18. Mycoplasmas have been isolated from another unlikely habitat, namely, the rumen of cows and sheep. These mycoplasmas are strict anaerobes, and an account of them is given by Robinson in Chapter 19; the problem of classification is dealt with in the chapter of taxonomy.

The most exciting development has been the visualization in the electron microscope of mycoplasma-like bodies in plants suffering from certain diseases, together with their insect vectors. Intensive efforts were made to cultivate the bodies, using established media successful with mycoplasmas from animal sources and many modifications of these media. However, all attempts were unsuccessful until Bové and collaborators succeeded in isolating an organism from citrus plants affected by stubborn disease and established Koch's postulates with the organism. A similar mycoplasma has been isolated from corn stunt. Morphologically these mycoplasmas are characterized by motile spirals, so the mycoplasma has been named *Spiroplasma*. These mycoplasmas will be dealt with in Volume III, and only classification will be considered in this volume, in the chapter on taxonomy.

Bacteria are coated with a cell wall, but only the membrane separates the mycoplasmas from their environment; it is therefore important to study the composition of the membrane and to know its function. This is dealt with comprehensively in Chapters 8 through 11 by three authors.

As far back as 1946 Andrewes and Welch demonstrated motility in *Mycoplasmel pulmonis*. Bredt (Chapter 5), finding other species to be motile, has reinvestigated the phenomenon of motility.

The study of mycoplasmas has now branched out to include the investigation of the viruses associated with them. Gourlay was the first to isolate a "lytic" virus from *Acholeplasma laidlawii*, and viruses have now been found associated with spiroplasmas. They are discussed in Chapters 14 and 15: Cole deals with the morphological aspects and Maniloff and colleagues with the molecular biology.

So, at last, there is substantial progress in the newest branch of microbiology. One speaks of the "golden age" of bacteriology; it seems that we are now going through a similar "golden age" in mycoplasmology. Recently, a link has been discovered between animal and insect mycoplasmas. Two agents from rabbit ticks have been identified as spiroplasmas; these agents are pathogenic for suckling mice and rats, the so-called suckling mouse cataract agent. They have been cultivated on special media and serologically are similar to other plant spiroplasmas involved in stubborn disease and corn stunt. There is, therefore, an indication to try out the special media to search for other mysterious infections.

These volumes constitute a reference work written by experts in their various fields. Each chapter is furnished with an extensive bibliography to provide further reading matter and to cater to mycoplasmologists with diverse interests. This work should provide an impetus for progress in the future.

D. G. ff. Edward

# PREFACE

"The Mycoplasmas," a comprehensive three-volume series, encompasses the various facets of mycoplasmology, emphasizing outstanding developments made in the field during the past decade. The pronounced information explosion in mycoplasmology was prompted primarily by the discovery of insect and plant mycoplasmas and mycoplasma viruses in the early 1970s, which attracted many new workers from different disciplines. During this period significant progress in the field of animal and human mycoplasmas was also made, providing important new insights into the nature of host–parasite relationships and into the mechanisms by which mycoplasmas infect and cause disease in man and animals.

Mycoplasmas are the smallest and simplest self-replicating microorganisms, and their use as models for the study of general biological problems has contributed considerably to our understanding of cell biology, particularly in the field of biological membranes. Volume I deals with the cell biology of the mycoplasmas, largely concentrating on problems regarding their classification, phylogenetics, and relatedness to wall-covered bacteria; their unique molecular biology, energy metabolism, transport mechanisms, antigenic structure, and membrane biochemistry. The characterization, ultrastructure, and molecular biology of the mycoplasma viruses, as well as the special properties of several groups of mycoplasmas, are also included.

Volume II is concerned with host–parasite relationships of mycoplasmas in man and animals. In part, emphasis is placed on recent developments in the study of classical mycoplasmal diseases of animals, such as cattle, sheep, goats, swine, and chickens. On the other hand, new information on the host range of mycoplasmas made it necessary to describe the mycoplasma flora of hosts not previously known to harbor mycoplasmas (for example, equines) or to document the increasing number of new mycoplasmas found in some other animal hosts (as observed in canines, felines, and nonhuman primates). This volume also offered the opportunity to record current knowledge about mycoplasmal diseases of man, including those involving the respiratory and genitourinary tracts. Humoral and cellular immune responses to mycoplasmas, which are assuming an ever-increasing significance in our understanding of the pathogenesis of human and animal mycoplasmal diseases, are covered in detail. The volume closes with reviews on mycoplasmas as arthritogenic agents and the interaction of mycoplasmas with cell and organ cultures.

Volume III represents the first serious attempt not only to present an extensive and critical review of the rapidly expanding field of plant and insect mycoplasmas but to integrate these important new subdisciplines into the total field of mycoplasmology. Many of the contributions involve current information on an entirely new group of helical mycoplasmas (spiroplasmas), stressing their part in plant and insect diseases. Tick-borne spiroplasmas and their possible role in vertebrate disease are also discussed here. Additional coverage in this volume updates our knowledge of other suspected mycoplasmal plant diseases, as well as vector transmission of mycoplasmas and spiroplasmas, and discusses the chemotherapy of mycoplasmal plant diseases.

Thus, this three-volume series provides a standard reference work for every mycoplasmologist and a current exhaustive treatment of recent advances in mycoplasmology for other interested microbiologists, cellular and molecular biologists, membrane biochemists, clinicians, veterinarians, plant pathologists, and entomologists.

M. F. Barile
S. Razin
J. G. Tully
R. F. Whitcomb

# 1 / CLASSIFICATION AND TAXONOMY

*E. A. Freundt and D. G. ff. Edward*[1]



## I. INTRODUCTION

The history of the classification and nomenclature of the mycoplasmas prior to the late 1960s was recorded (Edward and Freundt, 1969) in the predecessor of this book, the "Mycoplasmatales and the L-Phase of

[1] Deceased September 11, 1978.



THE MYCOPLASMAS, VOL. I

ISBN 0-12-078401-7

Bacteria," edited by Leonard Hayflick. During the subsequent short span of years we have witnessed an impressive and significant development of mycoplasma taxonomy. This development parallels a rapidly expanding knowledge of these organisms that is now more extensive than at any time during the preceding 70 years of mycoplasmology. In 1969 the proposal to assign the order Mycoplasmatales to a separate class, the Mollicutes, was several years old. Yet, only one family, the Mycoplasmataceae, with the genus *Mycoplasma,* was recognized under the order. Today, less than a decade later, a second genus, *Ureaplasma,* is recognized under the Mycoplasmataceae, and two new families with one genus each, the Acholeplasmataceae and the Spiroplasmataceae, have been established. In addition, two new genera, *Thermoplasma* and *Anaeroplasma*, have been tentatively recognized as members of the Mollicutes, although their exact position in the taxonomic hierarchy has not yet been defined. The total number of species classified within the Mollicutes has more than doubled during the same period from about 30 to 64, and—what is more important—the general standard of the descriptions of the new species has highly improved. In 1973, Edward and Freundt reviewed and clarified the status of type cultures for mycoplasma species that had been named up to and including 1972. In this study, type or neotype strains were designated for a number of species for which no strain had been previously designated. The importance of designating type strains for species described in the future, and of depositing these strains in one of the permanently established culture collections, was emphasized. Also, the general principles of mycoplasma classification have been repeatedly discussed (Freundt, 1973; Edward, 1974).

The Subcommittee on the Taxonomy of Mycoplasmatales that was established in 1966 under the International Committee on Nomenclature of Bacteria (now International Committee on Systematic Bacteriology, ICSB) has played a very active role in developing the taxonomy of mycoplasmas to its present stage. At a series of meetings held at intervals of a few years, the Subcommittee has discussed current specific and general problems in mycoplasma taxonomy and issued recommendations, as reported in the Minutes of the meetings (Subcommittee on The Taxonomy of Mycoplasmatales, 1967a, 1971, 1974, 1975, 1977). As one of the very first ICSB Subcommittees it has published, moreover, a proposal for minimal standards for descriptions of new species of the taxa within its responsibility (Subcommittee on the Taxonomy of Mycoplasmatales, 1972). This proposal has recently appeared in a revised edition to bring it up to date (Subcommittee on The Taxonomy of Mycoplasmatales, 1979).

It is the purpose of this chapter to describe and discuss the developments of mycoplasma taxonomy of the past decade, with special emphasis on emerging principles of classification and on newly emergent taxa whose existence has only recently been discovered.

## II. TAXONOMIC POSITION OF THE CLASS AND ITS RELATIONSHIP TO BACTERIA

The proposal made in 1967 (Subcommittee on The Taxonomy of Mycoplasmatales, 1967b; Edward and Freundt, 1967) to assign the mycoplasmas to a separate new class, the Mollicutes, has remained essentially uncontested and seems, in fact, to have received general acceptance. For example, whereas in the seventh edition of "Bergey's Manual of Determinative Bacteriology" (Breed *et al.*, 1957) Mycoplasmatales was placed as order X of the class Schizomycetes, the editors of the eighth edition of the Manual (Buchanan and Gibbons, 1974) *de facto* recognized the status of the mycoplasmas as a distinct class. Murray (1974) in his thought-provoking essay, "A Place for Bacteria in the Living World," suggested a provisional subdivision of the Kingdom Procaryotae into two divisions with three classes each. In Division I are placed the phototrophic prokaryotes ("Photobacteria") and in Division II the prokaryotes which are "indifferent" to light ("Scotobacteria"). The proposed grouping within the latter category that is obviously particularly pertinent to our theme is as follows:

Kingdom Procaryotae
  Division II: "Scotobacteria"
    Class I: The bacteria
    Class II: Obligate intracellular Scotobacteria in eukaryotic cells—Rickettsias
    Class III: Scotobacteria without cell walls—Mollicutes

In the proposed grouping of higher taxa within the Kingdom Procaryotae, a grouping stated to be constructed "without hierarchial prejudice," the term "bacteria" is used in two senses: (1) in the sense of all prokaryotic organisms ("Photobacteria" and "Scotobacteria") and (2) to denote Division II, Class I (formerly the Schizomycetes) of the Kingdom Procaryotae. When we compare Mollicutes with "bacteria," of course, we are thinking of bacteria in the latter sense.[1]

[1] Very recently, a proposal was made by Gibbons and Murray (1978) to divide the kingdom Procaryotae on the basis of cell wall types into three divisions: (i) Gracilicutes (with two classes: Photobacteria and Scotobacteria), which possesses the gram-negative type of cell wall; (ii) Firmacutes, which possesses the gram-positive type of cell wall; and (iii) Mollicutes. Thereby, the type of the cell boundary is recognized as a major criterion for the classification of prokaryotes at a high taxonomic level.

As pointed out earlier (Freundt, 1973) there should be very weighty reasons for creating a new class within the prokaryotes. The arguments originally adduced by the Subcommittee in support of its proposal (Subcommittee on the Taxonomy of Mycoplasmatales, 1967b) are too well known to need more than a brief recapitulation. Much emphasis was placed on the absence of a cell wall, together with the organisms' inability to synthesize the mucopeptide polymer and its precursors. Also, the significance of the requirement for sterols as a distinctive character shared by the vast majority of the mycoplasmas was stressed. Although our knowledge of the mycoplasmal genome was admittedly sparse in those days, the Subcommmittee (1967b) also included nucleic acid data in its considerations. Thus, importance was attached to the apparent fundamental genetic heterogeneity within the group as reflected by the apparent wide range of guanine + cytosine (G + C) ratios recorded for different mycoplasma species and to the fact that the G + C content of some species were among the lowest reported for any microorganism and lower than any known for "eubacteria."

It is interesting to note that the arguments that may be mobilized today to give additional support for the distinctness of the mycoplasmas at a high taxonomic level are primarily derived from further results obtained in the field of nucleic acid studies. The demonstration of a genome size of about $5 \times 10^8$ daltons for members of the family Mycoplasmataceae (Bak *et al.*, 1969; Black *et al.*, 1972), i.e., half the size of that of the smallest bacteria, and $1 \times 10^9$ daltons for the families Acholeplasmataceae (Bak *et al.*, 1969) and Spiroplasmataceae (Saglio *et al.*, 1973) and the genus *Thermoplasma* (Christiansen *et al.*, 1975) does confer special status upon the whole group of organisms. Pertinent, moreover, in this context is the very recent study by Reff *et al.* (1977) who compared the electrophoretic mobilities of ribosomal ribonucleic acids (rRNAs) from two or three *Mycoplasma* species, four *Acholeplasma* species, and four bacterial species (together with the L-phase variant of one of the bacteria) under nondenaturing and denaturing conditions. With one minor exception, the rRNAs of the bacterial species studied comigrated in nondenaturing gels. The *Mycoplasma* and *Acholeplasma* species, on the other hand, migrated in a pattern distinct from each other and from the bacteria, the differences observed being ascribed to conformational differences in the molecules. When subjected to electrophoresis in a denaturing formamide gel, the 23 S RNAs of all organisms comigrated. So did the mycoplasmal and acholeplasmal 16 S rRNAs whereas they differed in their mobility from that of bacterial 16 S RNAs. These studies, which are described in greater detail in Chapter 6 of this volume, confirm and extend earlier observations indicating differences between mycoplasmal and bacterial rRNA in base composition (Kirk and Morowitz, 1969; Johnson and Horowitz, 1971), sedimentation properties in sucrose gradient (Reich, 1967; Kirk and Morowitz, 1969; Johnson and Horowitz, 1971), and electrophoretic mobility in polyacrylamide gels (Harley *et al.*, 1973). The taxonomic significance of these observations lies in the growing recognition of prokaryotic rRNAs as evolutionarily highly conserved molecules (Pace, 1973) as borne out by a number of studies (Taylor *et al.*, 1967; Loening, 1968; Pace and Campbell, 1971; Woese *et al.*, 1974, 1975). The differences found between the RNAs of mycoplasmas and bacteria were interpreted by Reff *et al.* (1977) to reflect the existence of a significant evolutionary gap between mycoplasmas and bacteria. Although still incomplete, certain data on the characterization of DNA polymerase (E. J. Stanbridge, personal communication, 1977) and DNA-dependent RNA polymerases (Skripal, 1978, also personal communication, 1978) of mycoplasmas suggest another avenue in the field of nucleic acid research that may eventually lead to the detection of further fundamental differences between the mycoplasmas and other prokaryotes.

Up till now we have been considering only such arguments that may strengthen conservation of the Mollicutes as a separate class. Due allowance should also be made, of course, for observations that tend to provide contrary evidence. In this connection, rather considerable weight must be attributed to the studies by Neimark suggesting the existence of a phylogenetic relationship between the members of one taxon within the Mollicutes, viz., the family Acholeplasmataceae, and the lactic acid bacteria (Neimark and Lemcke, 1972; Neimark and Tung, 1973; Neimark, 1973, 1974). Evidence to that effect was obtained from the demonstration in *Acholeplasma* species, but not in other mycoplasmas, of lactate dehydrogenases (LDHs) that are specifically activated by fructose 1,6-diphosphate, a regulatory mechanism previously known to occur only among the Lactobacillaceae (Neimark, 1974). Comparison of the acholeplasmas and selected streptococci by a number of other methods, including polyacrylamide gel electrophoresis of cellular proteins and immunologic analysis of aldolase enzyme systems common to both groups of organisms—in the opinion of Neimark—provided proof of their phylogenetic relationship. For further details of this most interesting theory, first advanced in 1974 (Neimark, 1974), the reader is referred to Chapter 2 of this volume.

If accepted, the theory of Neimark does not necessarily invalidate the appropriateness of assigning the mycoplasmas to a separate class within the prokaryotes. After alı, the criteria used to define a "class" must of necessity be arbitrary. Hence the possible demonstration of a remote evolutionary relationship between two major groups of organisms does not exclude *a priori* their separation even at a very high taxonomic level, should such a separation be convenient and reasonable on the grounds of other criteria. It will be pertinent, moreover, to keep in mind that the alleged phylogenetic relationship between bacteria and mycoplasmas other than members of the family Acholeplasmataceae remains purely speculative so far. Anyway, it is clear from what has been said already in this section that a number of the properties distinguishing the Acholeplasmataceae do collectively suggest a closer relationship to bacteria than is apparent for the rest of the class. Thereby the Acholeplasmataceae obviously holds a key position as a possible intermediate between the Mollicutes and bacteria of Class I of the Scotobacteria (*sensu* Murray, 1974).

Our deliberate avoidance in this section of a lengthy discussion of the possible relationship of the mycoplasmas to L-phase variants and other cell wall-deficient bacteria is in fact in full agreement with a statement made by the Subcommittee in connection with its proposal, in 1967, to establish a separate class for mycoplasmas (Subcomittee, 1967b). In doing so, the Subcommittee expressed its disbelief in the view that the mycoplasmas in general be regarded as fixed L-phase variants of bacteria. Similarities and dissimilarities existing between L-phase variants and mycoplasmas were later again subjected to critical review (Edward and Freundt, 1969). The aspects discussed included colonial appearance, cellular morphology and ultrastructure, mode of reproduction, biochemistry and physiology, penicillin resistance, pathogenicity, and nucleic acid composition. From this thorough reconsideration of all available data it was concluded once again that mycoplasmas could not be regarded as stable L-phase variants. Since then, no observations have been made that call for a revival of the discussion of this topic. Altogether, if a phylogenetic relationship does exist between the mycoplasmas and bacteria we feel inclined to agree with Neimark (1974) in suggesting a more complex evolutionary process for the mycoplasmas than the developments leading to the simple L-phase variants we know.

## III. TAXONOMIC UNITS

### A. Order

At present only one order, Mycoplasmatales, is accepted in the class Mollicutes. In consequence, the description of the order is the same as for the class. When formally established in 1955 (Freundt, 1955) the order covered a relatively homogeneous group of no more than a dozen species. Serologically distinct but otherwise apparently very similar—except for a remarkable difference in the nutritional requirement for certain sterols—they were all placed in one family, Mycoplasmataceae, and one genus, *Mycoplasma*. As will appear from Table I the group of organisms that is today classified under the Mycoplasmatales is considerably more heterogeneous, comprising as it does three different families with one or two genera each. Even excluding the two genera whose exact affiliation is presently regarded as uncertain, *Thermoplasma* and *Anaeroplasma*, the order contains a multitude of organisms (about 60 different species) that differ widely in their morphology and ultrastructure, growth requirements, metabolic patterns, nucleic acid compositions, and genome size, as well as in their ecology and host relationships. There would seem to be very good reasons, therefore, for a future recognition of at least a second order within the class, the family Acholeplasmataceae being the most obvious candidate for elevation to the next higher rank in the hierarchy. Actually, almost a decade ago, at the time that the species then known as *Mycoplasma laidlawii* was assigned to a separate genus and family, we ventured the suggestion that this species might even deserve classification in a separate order. It may not be improper to quote here *in extenso* the considerations arguing for the said suggestion:

> Since nutritional dependence on sterols is accepted as a major characteristic of the mycoplasmas, it would be only logical and reasonable to assign to a separate family, or for that matter even to a separate order, a species which—in spite of lack of that fundamental property—must still be regarded as belonging within the Mollicutes, because of other essential points of similarity with the members of that class (Edward and Freundt, 1969).

If it is accepted that in 1969 there was a case for suggesting, although in vague terms only, the establishment of a separate order for the acholeplasmas, then the diversity of additional arguments that may be adduced today to that effect would certainly seem to be worth consideration. Some of the properties that distinguish members of the family Acholeplasmataceae as compared to Mycoplasmataceae have already been mentioned in the preceding section, and others are described below. Several of these properties undoubtedly give the acholeplasmas a unique position within the Mollicutes and we believe that their assignment to a separate order should be subjected to serious consideration.

## B. Families

The subdivision in 1970 of the order Mycoplasmatales into two families, Mycoplasmataceae and Acholeplasmataceae (Edward and Freundt, 1970), was followed by the proposal of Skripal (1974) to establish a third family, Spiroplasmataceae. This proposal received the endorsement of

TABLE I. **Taxonomy of the Class Mollicutes**[a]

Class: Mollicutes
Order: Mycoplasmatales
Family I: Mycoplasmataceae
1. Sterol required for growth
2. Genome size about $5.0 \times 10^8$ daltons
3. NADH oxidase localized in cytoplasm
Genus I: *Mycoplasma* (about 50 species current)
Do not hydrolyze urea
Genus II: *Ureaplasma* (single species with serotypes)
Hydrolyzes urea

Family II: Acholeplasmataceae
1. Sterol not required for growth
2. Genome size about $1.0 \times 10^9$ daltons
3. NADH oxidase localized in membrane
Genus I: *Acholeplasma* (seven species current)

Family III: Spiroplasmataceae
1. Helical organisms during some phase of growth
2. Sterol required for growth
3. Genome size about $1.0 \times 10^9$ daltons
4. NADH oxidase localized in cytoplasm
Genus I: *Spiroplasma* (one species current)

Genera of uncertain taxonomic position
*Thermoplasma* (single species)
*Anaeroplasma* (two species)

[a] From Tully (1978).

the Subcommittee on the Taxonomy of Mycoplasmatales (1977). The subdivision of the Mycoplasmatales into these three families is primarily based on nutritional and morphologic criteria. Whereas members of the family Acholeplasmataceae resemble bacteria in their nutritional independence of sterols, members of the families Mycoplasmataceae and Spiroplasmataceae differ from all other prokaryotes in their requirements of sterols for growth (Edward and Freundt, 1970; Saglio *et al.*, 1973). Members of the Spiroplasmataceae differ, as indicated by the name, from organisms in the other two families in exhibiting helical morphology of the cells together with typical rotatory and undulating motility (Davis and Worley, 1973; Cole *et al.*, 1973). There are other significant differences between the three families. Thus, the separation of the Acholeplasmataceae from Mycoplasmataceae (Edward and Freundt, 1970) is supported by the demonstration of the difference in the genome size (Bak *et al.*, 1969) as well as the more recent observation of presumably essential differences in the rRNAs of the two families (Reff *et al.*, 1977). Equally important is the presence, in all of the five *Acholeplasma* species hitherto examined, of lactate dehydrogenases not found in members of the family Mycoplasmataceae (Neimark, 1974). The ability of acholeplasmas to incorporate significant amounts of acetate into their lipids seems to be another useful criterion for differentiating them from *Mycoplasma* species. For example, in a comparative study of three *Acholeplasma* (15 strains) and five *Mycoplasma* species the acholeplasmas were shown by Herring and Pollack (1974) to synthesize fatty acids from acetate, while *Mycoplasma* species possessed no or very limited ability to utilize acetate for lipid synthesis. Mention may be made of a number of other properties which, although examined less systematically, collectively tend to strengthen the impression of a fundamental difference between the two families. These include differences with respect to osmotic fragility (Razin, 1964), localization of enzymatic activities (Pollack *et al.*, 1965; Rodwell, 1967; Low and Zimkus, 1973; Pollack, 1975; Larraga and Razin, 1976; Vinther and Freundt, 1977), and susceptibility to as yet undefined inhibitory substances released from certain bacteria (Kunze, 1973).

### C. Genera

The genera established within the three families are listed in Table I. The Mycoplasmataceae is the only family that has been subdivided. The interesting etymology of the name of the type genus, *Mycoplasma,* a term that dates as far back as 1889, has been described by Krass and Gardner (1973). *Ureaplasma* differs from this genus, and from all other mycoplasmas, in possessing urease. Only single genera (viz. *Acholeplasma* and *Spiroplasma*) are recognized for each of the other two families. Consequently, these genera are both defined by the distinguishing characters of their respective families.

Two genera recognized as members of the class Mollicutes, *Thermoplasma* and *Anaeroplasma*, are provisionally placed under the heading of "genera of uncertain taxonomic position." The reasons for postponing assignment of these two genera to families are essentially different.

In the case of *Thermoplasma*, a group of truly saprophytic organisms characterized by an optimum temperature for growth of about 56°–60°C and a pH optimum of about 1–2, the continued discussion of the taxonomic relationships of the genus is concerned above all with the problem of whether or not it does belong to the Mollicutes. The tentative classification of *Thermoplasma* as a member of the order Mycoplasmatales by the authors who described it (Darland *et al.*, 1970) was based on the absence of a cell wall and the apparent lack of wall precursors. Biochemical analyses of the cell membrane with respect to its composition of lipids

(Langworthy *et al.*, 1972) and amino acids (Smith *et al.*, 1973) provided additional support for the proposed classification of *Thermoplasma*. Some very recent studies seem, however, to cast doubt upon the alleged affiliation with the Mollicutes, suggesting as they do a closer relationship to the bacteria. For example, *Thermoplasma* has been found to differ from all other mycoplasmas examined in its capability of synthesizing lipoquinones, in which respect it resembles respiring bacteria (Holländer *et al.*, 1977). Also, as yet unpublished studies by Reff and his co-workers indicate that *Thermoplasma* is unlike any mycoplasma in the electrophoretic mobility of its rRNA; instead it bears a strong resemblance to a thermophilic acidophilic bacterium, *Sulfolobus acidocaldarius* (Eric Stanbridge, personal communication, 1977). A possible relationship between *Thermoplasma* and *Sulfolobus* and *Sulfolobus*-like organisms was in fact suggested on an earlier occasion by de Rosa *et al.* (1975), who proposed that they be classified together in a "form/habitat" group, Caldariella. Unfortunately, the available reports of the guanine plus cytosine contents of the individual members of this group of organisms are somewhat conflicting. Thus, the values of 24–29% reported by Darland *et al.* (1970) and Belly *et al.* (1973) for *Thermoplasma acidophilum* could not be confirmed by Christiansen *et al.* (1975) or by Searcy and Doyle (1975), who independently found a G + C content of 46% for this organism. This latter figure corresponds well with the range of 39–45% as determined for the *Sulfolobus*-like organisms examined by de Rosa *et al.* (1975) but is significantly lower than the values of 60–68% reported by Brock *et al.* (1972) for *Sulfolobus acidocaldarius*. Among the differences and similarities found to exist between *Thermoplasma* and *Sulfolobus*, those pertaining to the cell boundaries and extracellular appendages deserve particular attention. The cells of *Sulfolobus acidocaldarius* and the *Sulfolobus*-like strains are surrounded by an "envelope" (Brock *et al.*, 1972) or "coat" (Millonig *et al.*, 1975) which is different from both gram-positive and gram-negative cell walls, but apparently different also from the amorphous layer observed in *Thermoplasma* (Black *et al.*, 1979) in exhibiting a subunit structure. While lacking muramic acid, the *Sulfolobus* envelope contains hexosamine (Brock *et al.*, 1972) which could not be detected in *Thermoplasma* (Belly *et al.*, 1973). *Sulfolobus* and the *Sulfolobus*-like strains possess multiple extracellular appendages regarded as pili (Weiss, 1973; Millonig *et al.*, 1975), whereas *Thermoplasma* is equipped with a single polar flagellum responsible for the swimming motility exhibited by this organism (Black *et al.*, 1979). Although there are other apparent differences between *Sulfolobus* and *Thermoplasma*, final decision of the possible relationship between these two genera and definition of the taxonomic position of *Thermoplasma* in relation to the *Molli-*

*cutes* must await future direct comparison. Very obviously, determination of the relationship of *Thermoplasma* to *Sulfolobus* or to some other thermophilic acidophilic bacterium, with particular emphasis on the possibility that *Thermoplasma* may in fact be an L-phase variant, has decisive implications in the classification of *Thermoplasma*. Therefore, such comparative studies should be given high priority.

Quite different reasons have caused the delay in the allocation to a family of the genus *Anaeroplasma,* whose identity as a mycoplasma is beyond doubt. The difficulties involved in defining the taxonomic status of *Anaeroplasma* primarily stem from the proposal of Robinson *et al.* (1975) to include in one genus or even in a single species, *Anaeroplasma abactoclasticum,* strains that either require or do not require sterols. The possible acceptance of this proposal would, in the words of the Subcommittee on the Taxonomy of Mycoplasmatales, "necessitate a major revision of taxonomic concepts within Mollicutes because sterol dependence or independence is now used to separate families of the order Mycoplasmatales" (Subcommittee, 1977). It was only logical, therefore, that the Subcommittee urged great restraint in the inclusion of sterol-requiring and sterol-nonrequiring strains in a single genus. The Subcommittee did not, on the other hand, propose a classification of *Anaeroplasma* with respect to family, and the ultimate decision will probably depend on clarification of the degree of relatedness between the various anaerobic strains, and on their possible relatedness to other known mycoplasmas. An especially important datum would be the genome size of representative sterol-requiring and sterol-independent strains.

### D. Species

The binomially named species with its type strain is the cornerstone of any biologic classification system, the basic unit on which all of the other taxa rest. No wonder, therefore, that very considerable efforts are devoted by microbiologists in general, and by mycoplasmologists in particular, to define the species concept and to lay down general standards for the description and characterization of the individual species.

As stated by the Subcommittee on the Taxonomy of Mycoplasmatales (1979) "the species concept in Mollicutes, as in all prokaryotic taxa, is arbitrary and is designed as a convenience for designation of isolates." Perhaps the most simple and clear-cut definition of the species concept in microbiology is the one formulated by Mandel (1969) who refers to a species as "a type culture and those cultures resembling it." Provocative though this definition may be in its deliberate oversimplification of the problem, it has the particular advantage of emphasizing the paramount importance of the type culture as the prototype of the species it represents. The species is defined, in fact, by the properties of its type strain, and that is why the designation of a type strain is made a condition for valid publication of a new name [Lapage *et al.*, 1975, Rule 27 (3)] and deposition of that strain in a permanent Type Culture Collection strongly recommended (Lapage *et al.*, 1975, Recommendation 30a). The real problem is to define the degree of relatedness or "resemblance" with the type strain that may be acceptable for inclusion of other strains in the same species. This is an area that often requires arbitrary judgments. In this respect, the definition formulated by the Subcommittee (1979), when referring to mycoplasma species as "clusters of morphologically identical isolates whose genomes exhibit a high degree of relatedness," does not differ too much from that of Mandel (1969). However, it is more specific in that it suggests a possible means (comparison of genomes) of determining the degree of relatedness.

In commenting further on the methods to be used for classification of the species, it is appreciated by the Subcommittee (1979) that extensive DNA hybridization studies between isolates may not be feasible in practice. Among the alternative methods that may be used to establish patterns of relationship, mention is made of biochemical tests, gel electrophoresis of cellular proteins, and serology. It may be inferred here that the problem of arbitrariness is equally pertinent whether the genetic relationship is determined directly in terms of nucleic acid homologies or indirectly by recording and comparing phenotypic characters.

What observable properties of mycoplasmas should be used for species distinctions? In organisms that reproduce sexually, definition of the species depends on the ability of its members to interbreed. Sexuality is unknown in the mycoplasmas. Determination of the extent to which the genomes of different strains of mycoplasmas are able to hybridize is a somewhat analogous procedure for determining genetic relatedness. In practice, nucleic acid hybridization studies have obtained limited usage. Because they present complex technical difficulties, they have been generally used either to check the validity of less direct means of species classification or to solve particularly complicated problems. If nucleic acid hybridization is not feasible as a general technique, what other methods are available? A less direct method of description of genomes is the guanine + cytosine (G + C) ratio of the DNA. This is now recognized (Subcommittee, 1979) as an important property of mycoplasma species.

In the absence of definitive information on the genome, or to supplement such information, one must turn to phenotypic characters. Characters that lend themselves as a basis for species classification include morphologic and biochemical properties, patterns of cellular protein migration in gel electrophoresis, and antigenicity. Since, unfortunately, few mycoplasma species possess distinctive shapes, the use of morphology in species classification is rather limited. Biochemical pathways often prove not to be distinctive at the species level. The gel electrophoretic pattern of the cell proteins has been referred to as "the fingerprint" of the microorganisms. Because the diversity of proteins presumably reflects a number of genes, gel electrophoresis should theoretically provide an ideal indirect method of assessing the degree of genetic relatedness. In the hands of experienced workers, this method has in fact contributed rather significantly to solving a number of taxonomic problems, but it has not yet fulfilled the expectations for general applicability with which it was originally received.

In the search for a suitable species marker we then end up with the antigenic pattern, which has in fact been used most extensively in mycoplasmology to distinguish between species. This is true at least for species of the genera *Mycoplasma* and *Acholeplasma,* i.e., the vast majority of presently recognized mycoplasma species, which may in fact be defined as groups of strains showing consistent and significant serologic differences from other strains (Subcommittee, 1979). Formerly, direct agglutination and/or complement fixation were used most frequently to serve this purpose. Of the tests now available, the combination of growth inhibition, metabolism inhibition, and immunofluorescence tests best meet the requirements with respect to specificity and sensitivity that are needed to distinguish between most mycoplasma strains at the species level.

In 1974 a step was taken by the Subcommittee on the Taxonomy of Mycoplasmatales which implied a decisive break with the principle of defining the mycoplasma species as a serologically distinct group of strains. It was recommended by the Subcommittee (1974) that serologically distinct subdivisions of the genus *Ureaplasma* that were not otherwise significantly different should in the future be designated as "serotypes" rather than be assigned to new species. The unspoken purpose of this recommendation, which was carried into effect by Shepard *et al.* (1974), was to avoid an unnecessary increase in the number of named species that might eventually, in the words of Edward and Freundt (1969), "lead to a situation like the one experienced in the classification of salmonellas." The introduction of the new principle of subgeneric classification in *Ureaplasma* of course implies that serologic reactions are not applicable as criteria for species distinction within this genus. Or, in the wording of the Subcommittee on the Taxonomy of Mycoplasmatales, "serological tests which differentiate between species within the genera *Mycoplasma* and *Acholeplasma* are sufficient only for separation of serotypes within the genus *Ureaplasma*" (Subcommittee, 1979). Such conservative approaches are, of course, not to be construed as final judgments. It may be that eventually, with adequate study, the serotypes can be promoted to species rank with complete assurance of taxonomic stability. For example, it remains to be determined, by hybridization experiments, whether the eight *Ureaplasma urealyticum* serotypes are as different in their nucleic acid compositions as are the various species of *Mycoplasma* and *Acholeplasma*. Such experiments are in progress in the laboratory of one of us (E.A.F.). In any event, criteria for establishment of different species of the genus *Ureaplasma* have not yet been defined, and at present *U. urealyticum* remains the only species recognized for that genus.

No definition has been proposed, so far, for species of the genus *Spiroplasma,* which is represented to date only by the type species, *S. citri*. Another spiroplasma, the "corn stunt organism" (CSO), is serologically related to *S. citri* (Tully *et al.*, 1973). Also, the DNAs of the two organisms have identical G + C ratios, about 25–26%, but hybridize at the 30% level only (Christiansen *et al.*, 1979, in preparation). The suckling mouse cataract agent (SMCA) is serologically considerably less related to *S. citri* (Tully *et al.*, 1977); it has a somewhat higher G + C ratio, 30–31%, and the homology values between SMCA and *S. citri* are less than 10% (Christiansen *et al.*, 1979, in preparation).

In Table II are listed all of the 64 species, arranged according to family and genus, that are presently (July, 1978) recognized within the Mollicutes. The number of species has been steadily growing since Edward and Freundt (1956) published their first tentative list of binomially named species of the genus *Mycoplasma*. More importantly, there has been a conspicuous overall improvement of the general standard of descriptions of new species of the Mollicutes during the last decade. This may to a great extent—directly and indirectly—be ascribed to the activities displayed in that field by the Subcommittee on the Taxonomy of Mycoplasmatales and by its individual members. Very soon after its establishment in 1966 the Subcommittee took action by issuing a set of "Recommendations on nomenclature of the Order Mycoplasmatales" (Subcommittee, 1967b). In this document, mycoplasmologists intending to publish new specific names were urged to satisfy basic requirements that may be briefly summarized as follows: (1) to provide an adequate description that will allow laboratory identification of the proposed new species and its differentiation from other mycoplasma species, (2) to designate and supply to a National Type Culture Collection a type culture, and (3) to publish new specific names in journals of a known wide circulation among microbiologists in general. In contrast with the International Code of

TABLE II. **List of Named Species**

| Species | References | Type or neotype strain | ATCC[a] No. | NCTC[b] No. | % G + C of DNA |
|---|---|---|---|---|---|
| 1. *Acholeplasma axanthum* | Tully and Razin (1970) | S-743 | 25176 | 10138 | 31.3 |
| 2. *A. equifetale* | Kirchhoff (1974, 1978) | C112 | 29724 | 10171 | NT[c] |
| 3. *A. granularum* | Edward and Freundt (1970) | BTS-39 | 19168 | 10128 | 30.5–32.4 |
| 4. *A. hippikon* | Kirchhoff (1978) | Cl | 29725 | 10172 | NT |
| 5. *A. laidlawii* | Edward and Freundt (1970) | PG8 (Sewage A) | 23206 | 10116 | 31.7–35.7 |
| 6. *A. modicum* | Leach (1973) | Squire (PG49) | 29102 | 10134 | 29.3 |
| 7. *A. oculi* | Al-Aubaidi *et al.* (1973)<br>Al-Aubaidi (1975) | 19L | 27350 | 10150 | NT |
| 8. *Anaeroplasma abactoclasticum* | Robinson *et al.* (1975) | 6-1 | 27879 | | 29.3–29.5 |
| 9. *A. bactoclasticum* | Robinson and Allison (1975) | ATCC 27112 | 27112 | | 32.5–33.7 |
| 10. *Mycoplasma agalactiae* | Freundt (1955) | PG2[e] | | 10123 | 33.5–34.2 |
| 11. *M. alkalescens* | Leach (1973) | D12 | | 10135 | 25.9 |
| 12. *M. alvi* | Gourlay *et al.* (1977) | Ilsley | | 10157 | 26.4 |
| 13. *M. anatis* | Roberts (1964) | 1340 | 25524 | 10156 | NT |
| 14. *M. arginini* | Barile *et al.* (1968) | G230 | 23838 | 10129 | 27.6–28.6 |
| 15. *M. arthritidis* | Freundt (1955) | PG6 (Preston)[e] | 19611 | 10162 | 30.0–33.7 |
| 16. *M. bovigenitalium* | Freundt (1955) | PG11 (B2) | 19852 | 10122 | 28.0–32.0 |
| 17. *M. bovirhinis* | Leach (1967) | PG43 (5M331) | 27748 | 10118 | 24.5–25.7 |
| 18. *M. bovis* | Askaa and Ernø (1976) | Donetta (PG45) | 25523 | 10131 | 32.7–32.9 |
| 19. *M. bovoculi* | Langford and Leach (1973) | M165/69 | | 10141 | 29.0 |
| 20. *M. buccale* | Freundt *et al.* (1974) | CH20247 | 23636 | 10136 | 25–26.4 |
| 21. *M. canadense* | Langford *et al.* (1976) | 275C | | 10152 | 29 |
| 22. *M. canis* | Edward (1955) | PG14 (C55) | 19525 | 10146 | 28.5–29.1 |
| 23. *M. capricolum* | Tully *et al.* (1974) | California kid | 27343 | 10154 | 25.5 |
| 24. *M. caviae* | Hill (1971) | G122 | 27108 | 10126 | NT |

*(continued)*

TABLE II (*Continued*)

| Species | References | Type or neotype strain | ATCC[a] No. | NCTC[b] No. | % G + C of DNA |
|---|---|---|---|---|---|
| 25. *M. conjunctivae* | Barile *et al*. (1972) | HRC581 | 25834 | 10147 | NT |
| 26. *M. cynos* | Rosendal (1973a) | H831 | 27544 | 10142 | NT |
| 27. *M. dispar* | Gourlay and Leach (1970) | 462/2 | 27140 | 10125 | 28.5–29.3 |
| 28. *M. edwardii* | Tully *et al*. (1970) | PG24 (C21) | 23462 | 10132 | 29.2 |
| 29. *M. equirhinis* | Allam and Lemcke (1975) | M432/72 | 25966 | 10148 | NT |
| 30. *M. faucium* | Freundt *et al*. (1974) | DC-333 | 25293 | 10174 | NT |
| 31. *M. feliminutum* | Heyward *et al*. (1969) | Ben | 25749 | 10159 | NT |
| 32. *M. felis* | Cole *et al*. (1967) | CO | 23391 | 10160 | 25.0–25.4 |
| 33. *M. fermentans* | Edward (1955) | PG18 (G) | 19989 | 10117 | 27.5–29.1 |
| 34. *M. flocculare* | Meyling and Friis (1972) | Ms42 | 27399 | 10143 | NT |
| 35. *M. gallinarum* | Freundt (1955) | PG16 (Fowl) | 19708 | 10120 | 26.3–28.0 |
| 36. *M. gallisepticum* | Edward and Kanarek (1960) | PG31 (X95) | 19610 | 10115 | 31.6–35.7 |
| 37. *M. gateae* | Cole *et al*. (1967) | CS | 23392 | 10161 | 28.4–28.6 |
| 38. *M. hominis* | Edward (1955) | PG21 (H50) | 23114 | 10111 | 27.3–29.3 |
| 39. *M. hyopneumoniea*[d] | Maré and Switzer (1965) | J[e] | 25934 | 10110 | NT |
| 40. *M. hyorhinis* | Switzer (1955) | BTS-7 | 17981 | 10130 | 27–28 |
| 41. *M. hyosynoviae* | Ross and Karmon (1970) | S16 | 25591 | 10167 | NT |
| 42. *M. iners* | Edward and Kanarek (1960) | PG30 (M) | 19705 | 10165 | 28.9–29.5 |
| 43. *M. lipophilum* | Del Giudice *et al*. (1974) | MaBy | 27104 | 10173 | NT |
| 44. *M. maculosum* | Edward (1955) | PG15 (C27) | 19327 | 10168 | 26.5–29.6 |
| 45. *M. meleagridis* | Yamamoto *et al*. (1965) | 17529 | 25294 | 10153 | 28.0–28.5 |
| 46. *M. moatsii* | Madden *et al*. (1974) | MK 405 | 27625 | 10158 | NT |
| 47. *M. molare* | Rosendal (1974) | H542 | 27746 | 10144 | NT |

| | | | | | |
|---|---|---|---|---|---|
| 48. *M. mycoides* | | | | | |
| 48a. subsp. *mycoides* | Freundt (1955) | PG1[e] | | 10114 | 26.1–27.1 |
| 48b. subsp. *capri* | Freundt (1955) | PG3 | | 10137 | 23.6–26.3 |
| 49. *M. neurolyticum* | Freundt (1955) | Type A | 19988 | 10166 | 22.8–26.5 |
| 50. *M. opalescens* | Rosendal (1975) | MH5408 | 27921 | 10149 | NT |
| 51. *M. orale* | Taylor-Robinson *et al*. (1964) | CH19299 | 23714 | 10112 | 24.0–28.2 |
| 52. *M. ovipneumoniae* | Carmichael *et al*. (1972) | Y98 | | 10151 | NT |
| 53. *M. pneumoniae* | Somerson *et al*. (1963) | FH | 15531 | 10119 | 38.6–40.8 |
| 54. *M. primatum* | Del Giudice *et al*. (1971) | HRC292 | 25948 | 10163 | 28.6 |
| 55. *M. pulmonis* | Freundt (1955) | Ash (PG34) | 19612 | 10139 | 27.5–28.3 |
| 56. *M. putrefaciens* | Tully *et al*. (1974) | KS1 | 15718 | 10155 | 28.9 |
| 57. *M. salivarium* | Edward (1955) | PG20 (H110) | 23064 | 10113 | 27.0–31.5 |
| 58. *M. spumans* | Edward (1955) | PG13 (C48) | 19526 | 10169 | 28.4–29.1 |
| 59. *M. sualvi* | Gourlay *et al*. (1978) | Mayfield (clone B) | | 10170 | 23.7 |
| 60. *M. synoviae* | Olson *et al*. (1964) | WVU 1853 | 25204 | 10124 | 34.2 |
| 61. *M. verecundum* | Gourlay *et al*. (1974) | 107 | 27862 | 10145 | 27.0–29.2 |
| 62. *Spiroplasma citri* | Saglio *et al*. (1973) | Morocco (R8-A2) | 27556 | 10164 | 26.0–26.3 |
| 63. *Thermoplasma acidophilum* | Darland *et al*. (1970) | 122-1B2 | 25905 | | 46.0 |
| 64. *Ureaplasma urealyticum* | Shepard *et al*. (1974) | 960-(CX8) | 27618 | 10177 | 27.7–28.5 |

[a] ATCC: American Type Culture Collection, Rockville, Maryland, U.S.A.
[b] NCTC: National Collection of Type Cultures, London, England.
[c] NT: Not tested.
[d] *M. hyopneumoniae*: cf. Subcommittee (1974).
[e] Neotype.

Nomenclature of Bacteria and Viruses (Editorial Board of The International Committee on Bacteriological Nomenclature, 1958), which failed to specify criteria for the description of new species, the Subcommittee provided certain guidelines to that effect. An adequate description would, in the opinion of the Subcommittee, include serologic as well as biologic characters. The statements and recommendations made in this publication, which undoubtedly contributed much in themselves to improved descriptions of new mycoplasma species, were further elaborated and greatly extended a few years later in the "Proposal for minimal standards for descriptions of new species of the order Mycoplasmatales" (Subcommittee, 1972), prepared in part at the instigation of the Judicial Commission of the International Commission on Systematic Bacteriology.

Based on the recommendations made in 1967, guidelines were laid down in this new document for descriptions of new mycoplasma species within the frames of order, family, and genus. The obvious importance of basing the description on filter-cloned cultures was emphasized, Mention was made also of the kind of basic information that should be provided for a proposed new species: the origin and natural ecology of the strain(s), pathogenicity data, nutritional requirements, and the dependence of growth on pH, temperature, and atmospheric conditions. The biochemical and serologic tests recommended for establishing differences from existing species were divided into two categories: those tests that were considered particularly important for adequate species descriptions and a series of optional tests that might provide additional useful information. The tests recommended were described only in broad outline and in terms of general principles, reference being made to pertinent papers for technical details. In doing so, the Subcommittee abstained from following a request made by the Judicial Commission to provide exact details of the methods used to assess the biologic and antigenic characters of the mycoplasma species. Although the proposed "Minimal Standards" could not be made obligatory (Subcommittee, 1975), there are many indications to show that they have, nevertheless, served as a challenge and a model for authors publishing descriptions of new species in recent years. To assure that this document meets contemporary needs, the Subcommittee has now prepared a revised and extended second edition of the Minimal Standards (Subcommittee, 1979). The general principles set forth in the first version are followed rather rigidly in the new edition and the recommendations made are, with a few notable exceptions, essentially the same. For example, in the first proposal, the requirements for serologic comparison of a proposed new species with other named species were intentionally phrased in relatively vague terms. In the new edition they have been strengthened, and the recommendations for demonstration of serologic relatedness have been made more specific. It is now required that an organism which cannot be classified within one of the already established genera should be compared with all previously named species of Mollicutes. If it can be assigned to an existing genus it should at least be compared with all species of that genus. The serologic examination should, as a minimum, be based on two methods, the growth inhibition test and direct and indirect fluorescent antibody tests being recommended as particularly suitable for the purpose. References are given also for alternative methods (metabolism inhibition and indirect agglutination tests) and for methods that are likely to reveal the presence of internal antigens not measured by other tests (complement fixation and double immunodiffusion tests). Of the tests that were previously regarded as optional, determination of the guanine plus cytosine content of the DNA by $T_m$ or buoyant density, to define the genetic characters of a species, is listed in the revised version as a mandatory test.

Returning then to the list of named species presented in Table II it should be stated that this is essentially identical with the approved list of names of species of the class Mollicutes submitted in 1976 by the Taxonomy Subcommittee to the Ad Hoc Committee of the Judicial Commission of the ICSB, with the addition, *inter alia,* of a few names proposed since then. The names have been published in the First Draft of Approved Lists of Bacterial Names (Ad Hoc Committee of the Judicial Commission of the ICSB, 1976) intended to serve as a basis for a second draft to be published in June 1978. A third draft of the lists in final format will be produced for submission to the ICSB at the International Congress of Bacteriology in Munich, September 1978. Amendments to the Approved List will be accepted until June 30, 1979, whereupon the lists will be prepared for publication by January 1, 1980. Our reasons for quoting here in some length the proposed schedule for producing a final Approved List of Bacterial Names should be seen in light of the highly important and far-reaching implications of the policy adopted by the ICSB. The decision has been made that January 1, 1980, replace 1973 as the date for recognition of priority of new names. In consequence, after January 1, 1980, names not included in the Approved Lists, or names that are not validated by publication in the *International Journal of Systematic Bacteriology,* will no longer have any standing in nomenclature. This has the further implication that names previously proposed but invalidated by the new rules will be available for reuse (Ad Hoc Committee, 1976).

When submitting its lists of approved names to the Ad Hoc Committee of the Judicial Commission the position was taken by the Subcommittee that the approval of a name did not necessarily imply that the species concerned be a bona fide species. The status of the species listed in Table

II may, nevertheless, on the whole be regarded as well established with respect to adequate description. It is true, of course, that some species—and especially some of those proposed in the earliest years of mycoplasmology—do not fulfill all of the requirements laid down in the Minimal Standards. A survey of the species descriptions provided in the eighth edition of "Bergey's Manual of Determinative Bacteriology" (Freundt, 1974a) will show, for example, that there are some blanks with respect to adequate biochemical characterization and determination of the G + C ratio. As to the serologic distinctness of the species, we are certainly in a much better position today than we were a decade ago. Quite a considerable proportion of the species have been subjected, in connection with an evaluation of reference reagents for mycoplasmas (Freundt *et al.*, 1973b), to an extensive serologic cross-testing, using six different techniques. The number of different species included in the serologic comparison was further extended by as yet unpublished work, carried out on a collaborative basis by a number of working teams established under the FAO/WHO Programme on Comparative Mycoplasmology (1974). From these studies, and from the additional extensive experience gained from the work carried out at the FAO/WHO Collaborating Centre of Animal Mycoplasmas, it may be concluded with a great deal of confidence that there is a very minimal risk that a major serologic relatedness will be discovered in the future between any species listed in Table II.

The very essential requirement for designation of a type strain, and its deposition in one of the recognized culture collections, has been satisfied for every species listed. The status of species of the Order Mycoplasmatales, with respect to the existence of type strains, was reviewed a few years ago by Edward and Freundt (1973). Before that time the situation was less clear. Only a minor proportion of the original descriptions of species had been accompanied by an unequivocal formal designation of a type strain. In some instances, in which no formal designation was provided, only a single strain was described. In those cases, the single described strain became the type by monotypy, according to Rule 18c of the 1976 Revision of the International Code of Nomenclature of Bacteria (Lapage *et al.*, 1975). A total of 24 species (or subspecies) remained for which type strains had not been clearly designated by the authors that had named the species. In our subsequent designation of type strains for these species, the choice of the type could in most cases be based on available evidence as to the intentions of the original authors. In the case of three species (subspecies) it was necessary to propose neotype strains as provided by Rule 18c (Lapage *et al.*, 1975), viz. *M. mycoides* subsp. *mycoides, M. agalactia* subsp. *agalactiae,* and *M. arthritidis,* because the strains upon which the original descriptions were based no longer existed. Since 1973, type strains have been designated for all of the new species that have been proposed within the class Mollicutes by the authors who described them. Also, type cultures have been supplied to either the American Type Culture Collection or the National Collection of Type Cultures, or both. The fact that most proposals for new mycoplasma species that have appeared in recent years have been published in the *International Journal of Systematic Bacteriology* is in itself a guarantee that the requirements for an adequate description, as well as for designation and deposition of a type strain, be satisfied. The expected effectiveness of making publication in the IJSB mandatory for valid publication after a certain date (January 1, 1980) would thus seem to be indisputable.

## E. Subspecies

Among the new additions to the forthcoming revised version of the Minimal Standards document (Subcommittee, 1979) is a section reflecting on the definition of the subspecies. It is suggested that "the subspecies concept in Mollicutes should be reserved for designation of important strains that demonstrate consistent differences in a number of properties, but which nevertheless prove to be too closely related by serological, or especially by nucleic acid hybridization tests, to warrant species designation." The Subcommittee did not attempt to define standards for serologic and nucleic acid hybridization relatedness required for establishment of a subspecies. On the other hand, it did strongly discourage subspecies designation on the basis of insufficient relatedness data. Historically, there is one extremely negative example in which a subspecies (viz. *M. agalactiae* subsp. *bovis;* Hale *et al.*, 1962) was established for a mycoplasma isolate in the complete absence of evidence for relationship of the isolate with *M. agalactiae*. Nor was the confused situation arising from such a misnaming resolved by a subsequent proposal—on equally poor grounds—for the recognition of a distinct species, *M. bovimastitidis* (Jain *et al.*, 1967), to replace *M. agalactiae* subsp. *bovis*. A resolution of the problems in this case was proposed by Askaa and Ernø (1976), who elevated *M. agalactiae* subsp. *bovis* to the rank of species under the name *M. bovis* on the basis of renewed serology and nucleic acid homology studies.

Today, only one mycoplasma species, *M. mycoides,* is subdivided into subspecies. The rationale for maintaining the subdivision of *M. mycoides* into two subspecies, subsp. *mycoides* and subsp. *capri,* was reconsidered in another very recent publication by Askaa *et al.* (1978), who compared PG1 and PG3, the type strains of the two subspecies, serologically and by nucleic acid hybridization, as described in a following section (page 33).

Although Askaa *et al.* (1978) did not find the combined results of their study inconsistent with the continued classification of the two strains as subspecies, they preferred to submit their data for the consideration of the Taxonomy Subcommittee and to leave the final decision to the Subcommittee.

In their comparative study Askaa *et al.* (1978) included strains PG50 and B144P, representing "group 7" of Leach (1973) and "group L" of Al-Aubaidi and Fabricant (1971), respectively. They came to the conclusion that these two strains, which they found serologically closely related to each other, might together comprise a third subspecies of *M. mycoides,* although they did not propose formal subspecies designation.

## F. Infrasubspecific Subdivision

The Code of Nomenclature of Bacteria does not provide rules for the possible designation of taxa at the infrasubspecific level. With the purpose, however, of encouraging conformity and clarifying the application of such designations, this item is dealt with in Appendix 10 of the Code (Lapage *et al.*, 1975). According to the definition given in the Appendix, an infrasubspecific taxon is "one strain or a set of strains showing the same or similar properties, and treated as a taxonomic group." Infrasubspecific taxa may be referred to by terms such as *serovar, chemovar, phagovar, pathovar,* or *forma specialis,* the latter term being defined as "a parasitic, symbiotic, or commensal microorganism distinguished primarily by adaptation to a particular host or habitat." These terms should be preferred for isolates or groups of isolates previously termed serotype, chemotype, phagotype, etc. The term "type" again, according to Appendix 10 of the Code, should be used strictly for a nomenclatural type (e.g., a "type strain" or a "type species") and "should not be used to designate a division of a species nor to designate taxa based on antigenic characters." It is stated, moreover, that infrasubspecific terms may be applied to subdivisions within a species, even if the species has not been divided into subspecies. This implies that "the subdivisions so named would still be infrasubspecific subdivisions for nomenclatural purposes until such time as they may be raised to subspecific or specific rank." Finally the "group" is regarded as informal and of no nomenclatural standing.

In mycoplasmology, "type," "serotype," "group" and similar terms have been used quite extensively and rather indiscriminately to denote sets of strains exhibiting certain common characteristics, irrespective of the possible relationship of such strains to existing species. In addition, the terms "type" and "serotype" have been used for subdivisions of *Mycoplasma orale* ("types" 1, 2, and 3) and *U. urealyticum* ("serotypes"

I–VIII). In the case of *M. orale*, the authors who described types 2 and 3 actually regarded them as distinct new species rather than subdivisions of *M. orale*. In consequence, *M. orale* types 2 and 3 were later raised to species rank under the names of *M. buccale* and *M. faucium* (Freundt *et al.*, 1974). As discussed earlier (page 13), the eight "serotypes" of *U. urealyticum* may eventually be regarded as equivalents of the species of the genera *Mycoplasma* and *Acholeplasma,* although, according to the definitions given in Appendix 10 of the Code, they should be formally regarded today—under the name of *serovats*—as subdivisions of *U. urealyticum* at the infrasubspecific level.

The possibility provided by the Code of Nomenclature of Bacteria of establishing taxa at the infrasubspecific level, in the strict sense of the term, might serve a practical purpose if applied to *M. mycoides* subsp. *mycoides*. Strains that are serologically closely related to, or even indistinguishable from *M. mycoides* subsp. *mycoides,* the etiologic agent of contagious bovine pleuropneumonia, have been isolated quite frequently from diseased goats and sheep (Al-Aubaidi *et al.*, 1972; Ojo, 1976). Also, strains related to *M. mycoides* subsp. *mycoides* and/or subsp. *capri* have occasionally been recovered, in countries known to be free of contagious bovine pleuropneumonia, from other aberrant hosts, such as horses (R. M. Lemcke, U. Gupta, and H. Ernø, in preparation) and dogs (Rosendal, 1975). There is reason to believe, from experimental and epidemiologic evidence, that the goat and sheep strains of *M. mycoides* subsp. *mycoides* are nonpathogenic to cattle, and this may well be true also for the horse and dog strains. Since reports of the isolation of *M. mycoides* subsp. *mycoides* and related organisms in countries thought to be free of bovine pleuropneumonia would be a matter of great concern to veterinary health authorities in those countries, there is an obvious need for distinguishing, in the nomenclature, such strains from strains pathogenic to cattle. This might conveniently be done by assigning them to separate infrasubspecific taxa, under the names of, for example, *M. mycoides* subsp. *mycoides, pathovar capri,* and *M. mycoides* subsp. *mycoides, forma specialis equi,* etc. It is uncertain, however, that formal taxonomic recognition of a pathovar recognizable only by nonpathogenicity to a large animal host would in fact contribute substantially to mycoplasmology or to veterinary research.

## IV. PROPERTIES USED IN CLASSIFICATION

The criteria used for classification of the mycoplasmas at different taxonomic levels have been mentioned already in connection with descriptions of the individual taxa, and they are reviewed also in the Minimal Standards documents (Subcommittee, 1972, 1979). It may still be useful, however, to attempt—as we did in an earlier publication (Edward and Freundt, 1969)—a comprehensive evaluation and assessment of the relative importance of the properties that provide the basis for mycoplasma taxonomy.

## A. Ultrastructure and Morphology

The absence of a cell wall is obviously the most prominent single criterion distinguishing the Mollicutes. The demonstration by electron microscopy of sectioned material that the cells lack a wall and are bounded only by a single membrane is a condition, therefore, for the classification of an organism as a member of the Mollicutes.

The plasticity and the size (about 300 nm) of the smallest viable units, as determined by filterability through membrane filters of 220–450 nm pore diameter, are less remarkable, but significant characteristics of all members of the Mollicutes.

Members of the families Mycoplasmataceae and Acholeplasmataceae are usually described as "pleomorphic," the cells varying in shape from small coccoid bodies to fine branched filaments of variable length. In recent years, there has been a tendency to question the significance, in the logarithmic growth phase at least, of many of the cells that show greatest deviation from spherical shape (Robertson *et al.*, 1975).

Gliding motility has been described for three *Mycoplasma* species, *M. pneumoniae, M. pulmonis,* and *M. gallisepticum,* all of which possess specialized tiplike structures believed to have some bearing upon locomotion, possibly as attachment organelles (cf. Chapter 5 of this volume). Although gliding motility associated with specialized structures is typical of certain species, these characteristics do not yet play any practical role as an aid to classification.

Helical morphology, and rotatory and undulating motility of helical filaments are major criteria by which members of the family Spiroplasmataceae differ significantly from other families of the class (Cole *et al.*, 1973; Davis and Worley, 1973). These properties are best observed by phase contrast or dark-field microscopy of liquid cultures in the exponential growth phase but may be found with more difficulty by examination of agar cultures. Helical morphology may also be demonstrated by electron microscopy of material fixed with formalin or glutaraldehyde and negatively stained by ammonium molybdate (see Chapter 3). It should be noted that the helical appearance of the filaments may be lost on aging of the culture and under suboptimal growth conditions. Considerably more important, from a taxonomic point of view, is the occasional occurrence of nonhelical, nonmotile spiroplasma strains (Townsend *et al.*, 1977). It follows that the absence of helical morphology does not necessarily exclude classification in the family Spiroplasmataceae which must be based in such cases on other criteria.

The possession of monotrichous flagella and swimming motility by *Thermoplasma acidophilum* is unique among the mycoplasmas (Black *et al.*, 1979).

## B. Colonial Appearance

The potential of most, if not all, mycoplasmas to produce on solid medium colonies of a typical "fried egg" appearance is a cultural property which indicates that a candidate organism belongs to the class Mollicutes. This property is shared, in general, only by bacterial L-phase variants, with which new mycoplasma isolates may be confused in the earliest phase of research after isolation. On the other hand, colonial appearance usually does not play a significant role in species classification, although the colonies of different species may sometimes exhibit prominent differences. Colonial appearances depend, to a wide extent, upon the composition of the growth medium, which may determine *inter alia* the size of the colonies, the ability of their central areas to grow down into the agar medium, and the extent to which the peripheral surface growth develops. For example, *Ureaplasma urealyticum* isolates were originally defined in part by the tiny atypical colonies appearing on conventional media and were thus originally named T mycoplasmas. However, larger colonies of classical fried-egg appearance may develop on optimal media (Shepard *et al.*, 1974). The colonial appearance may also be greatly modified by phage infection (Milne *et al.*, 1972).

## C. Growth Requirements

The nutritional dependence on cholesterol or certain other sterols, which characterizes the great majority of mycoplasmas, is a property that is completely unknown in any species of bacteria.

Of the three families presently established within the Mollicutes, the Acholeplasmataceae is the only one whose members do not require sterols for growth. Since the classification of Spiroplasmataceae primarily rests on the demonstration of helical morphology, dependence versus nondependence on sterols for growth thus becomes a major criterion for the differentiation only between Mycoplasmataceae and Acholeplasmataceae. The fact that other significant differences, such as genome size (Bak *et al.*, 1969), separate the two families does not, of course, detract from the importance of sterol dependence as a practical means of distinguishing Mycoplasmataceae and Acholeplasmataceae. The taxonomic implications of the lack of dependence on sterols in the case of *Thermoplasma* and some isolates of *Anaeroplasma* (see also p. 11) remain to be defined.

The dependence on sterols for growth may be determined either indirectly by the digitonin test (Freundt *et al.*, 1973a), or directly, by comparing the growth on a series of different media whose composition varies with respect to cholesterol contents. The digitonin test depends on the ability of digitonin to inhibit the growth of members of the families Mycoplasmataceae and Spiroplasmataceae, possibly by interacting with cholesterol incorporated in the cell membrane, whereas the growth of members of the family Acholeplasmataceae is not influenced, or only insignificantly so, by the concentration of digitonin employed. Preliminary results obtained by digitonin tests should be confirmed by determination of the growth response to cholesterol (Subcommittee, 1972, 1979). Two alternative methods are available for this purpose (Razin and Tully, 1970; Edward, 1971).

Determination of the ability to utilize urea for growth is essential, of course, for the classification of an organism in the genus *Ureaplasma*. Among the methods described for the demonstration of urease activity, a test carried out on solid medium devised by Shepard and Howard (1970) may be particularly recommended.

In general, very little use can be made of specific nutritional requirements in classification of mycoplasma species. There are, however, two notable exceptions: *M. synoviae* which is unique in its requirement of $\beta$-nicotinamide dinucleotide (Frey *et al.*, 1968), and *A. bactoclasticum*, which is distinguished from *A. abactoclasticum* in requiring an as yet undefined factor(s) in bacterial lipopolysaccharide for growth (Robinson and Allison, 1975; Robinson *et al.*, 1975). Specific nutritional requirements may prove useful also in the taxonomy of members of the family Spiroplasmataceae (Jones *et al.*, 1977).

The dependence of the growth of mycoplasmas on physical conditions, such as the temperature, pH of the growth medium, and composition of the atmosphere, is of some taxonomic importance. *Thermoplasma acidophilum*, with its pH optimum at about 2–3 and temperature optimum at 56°–60°C, is of course the most outstanding example. The pH optimum of 5.5–6.5 for *Ureaplasma urealyticum*, about 0.5 lower than that of most other mycoplasmas, is considerably less conspicuous but still very significant. Members of the family Acholeplasmataceae generally grow over a wider temperature range than members of the Mycoplasmataceae; some of them at least are able to grow at room temperature. Some members of the family Spiroplasmataceae are characterized by a narrow temperature range and sharp temperatures for optimal growth. The optimum for *S. citri,* for example, is 32°C, whereas for the corn stunt agent it is 29°C. The suckling mouse mouse cataract agent, on the other hand, apparently grows equally well at 30° and 37°C. However, because physical growth requirements reflect the environment in which mycoplasmas occur more than the taxonomic origin of the organisms, the use that can be made of such differences in taxonomy is usually rather limited. As one exception, mention may be made of *Anaeroplasma,* whose dependence on strictly anaerobic conditions for growth is evidently an outstanding distinguishing characteristic of the genus.

## D. Biochemical Properties

Minimal requirements for proper species description include determination of the following properties: fermentation of glucose (preferably under both aerobic and anaerobic conditions of growth), hydrolysis of arginine, hydrolysis of urea, and—for species of the family Acholeplasmataceae—production of carotenoids (Razin and Cleverdon, 1965; Tully and Razin, 1968). Although the great majority of *Mycoplasma* species are either glucose-positive or arginine-positive, a few attack neither substrate and some are capable of metabolizing both. It is interesting to note that some serologic relatedness has been demonstrated between species within, but not between, the glucose-positive and arginine-positive groups (Taylor-Robinson *et al.*, 1963; Lemcke, 1965; Fox *et al.*, 1969; Kenny, 1969, 1972, 1973; Ernø and Jurmanova, 1973; Thirkill and Kenny, 1974). Moreover, the patterns obtained by isoelectric focusing of mycoplasma proteins were found to differ in a characteristic way for, *inter alia,* glucose-fermenting and arginine-hydrolyzing *Mycoplasma* species (Sayed and Hatten, 1976). The correlation thus demonstrated between biochemical properties, on the one hand, and antigenicity plus electrophoretic pattern of the cell proteins, on the other, attributes particular taxonomic significance to a grouping of *Mycoplasma* species according to their ability to catabolize glucose and arginine, and may well serve in the future as a basis for dividing the genus *Mycoplasma* into two distinct genera. The heterogeneity characterizing the Mycoplasmataceae with respect to glucose/arginine catabolism is in sharp contrast to the pattern exhibited by species of the family Acholeplasmataceae, all of which catabolize glucose but not arginine. Thereby the rationale for utilizing these properties as possible criteria for classification at the generic level would seem to become even more evident.

In addition to the relatively few biochemical tests that are considered of primary importance for taxonomic purposes, a number of optional tests that may provide additional useful information are listed in the Minimal Standards document (Subcommittee, 1979). These include fermentation of a number of carbohydrates and related compounds other than glucose, phosphatase activity, production of film and spots on agar plate medium containing egg yolk emulsion, proteolytic activity, tetrazolium reduction under aerobic and anaerobic conditions, and hemolysis of sheep and guinea-pig red blood cells together with tests for the possible identification of the hemolysin as peroxide. Most of these properties are shared by a great many different species, and although they add to the general characterization of the organisms they are of limited value as distinctive markers.

The biochemical tests that have been described herein are relatively simple tests that involve the recording of end products of metabolism or certain enzyme activities and which have the merit of convenience. The results of more detailed biochemical and physiologic studies, involving (to mention just a few) the composition of structural components of the cell, the properties and distribution of enzymes, transport mechanisms, metabolic pathways, and energy-yielding mechanisms, may in fact, in the future, provide a more rational and informative background for taxonomy. Mention has been made already, in another context (page 5), of the demonstration in species of Acholeplasmataceae, but not in Mycoplasmataceae, of lactate dehydrogenases specifically activated by fructose 1,6-diphosphate (Neimark, 1974). Significant differences between members of Acholeplasmataceae and Mycoplasmataceae have been found, moreover, in the utilization of acetate in the synthesis of lipids (Herring and Pollack, 1974) and in the cellular localization of oxidoreductase activity that is associated with the membrane in *Acholeplasma* species and localized in the cytoplasm of *Mycoplasma* species (for references, see page 9). The results of biochemical analyses of the cell membrane of *Thermoplasma acidophilum* with respect to its composition of lipids (Langworthy *et al.*, 1972) and amino acids (Smith *et al.*, 1973) were interpreted as additional evidence in support of the proposed classification of this organism as a member of the Mollicutes. In general, although a number of complex biochemical studies of great potential interest and utility have been carried out, the results cannot yet be utilized in taxonomy, mostly because only a few strains have been examined by each method.

## E. Serologic Properties

In an earlier publication (Edward and Freundt, 1969), we compared a number of different serologic methods and evaluated their relative merits with respect to technical complexity, specificity, and sensitivity. Some methods (e.g., direct agglutination, indirect hemagglutination, latex agglutination, growth inhibition, metabolism inhibition, and immunofluorescence tests) were considered best for distinctions between species. In contrast, minor antigenic overlapping was best detected by such tests as complement fixation and double immunodiffusion tests. Very little can be added today to our earlier discussion. However, considerable technical improvements have been made in certain of the tests, especially those that are used for determination of serologic distinctness or relatedness. It is interesting to note, for example, that although we pointed out the potential usefulness of immunofluorescence (Edward and Freundt, 1969), the techniques that were available at that time were considered too laborious and difficult to perform. However, as a direct consequence of the introduction in mycoplasmology of the epiimmunofluorescence test applied to agar colonies (Del Giudice *et al.*, 1971), this test is now recommended as a serologic method of primary significance—equal in importance to growth inhibition and/or metabolism inhibition—for species classification.

Historically, much of the earlier classification at the species level was based on direct agglutination and complement fixation tests that were later supplemented with the growth inhibition test. With the introduction of new serologic techniques, or the development of appropriate modifications of older ones, these were of course included in the armamentarium of serologic methods available for characterization of the mycoplasma species or "serotype." The selection, in the most recent version of the Minimal Standards document (Subcommittee, 1979), of the growth inhibition and epiimmunofluorescence tests as the methods of choice for differentiation between species in the case of *Mycoplasma* and *Acholeplasma,* and between serotypes (or *serovars)* in the case of *Ureaplasma,* was made on empirical grounds, taking into consideration requirements for practicability, specificity, and sensitivity. Because the growth inhibition test is often insensitive, even with otherwise highly potent sera, the Subcommittee (1979) recommended the metabolism inhibition test as a possible alternative "when the growth inhibition test is not technically feasible." The additional recommendation of the complement fixation or double-immunodiffusion tests for "resolution of internal antigens not measured by other tests" was made with the purpose of facilitating the demonstration of group antigens common to different generic or subgeneric groups of mycoplasma species. The particular suitability of the double-immunodiffusion test for that purpose has been repeatedly confirmed.

Although we have been considering in the present section serology almost exclusively as a means of species classification, the obvious importance of serology as an aid to classification at the subspecific or infrasubspecific level should be kept in mind as well. A description of the underlying principles of the serologic methods used in mycoplasma taxonomy, as well as of technical details, may be found in a publication by Freundt *et al.* (1979).

## F. Electrophoretic Patterns of Cell Proteins

The fact that relatively little emphasis is put in the Minimal Standards documents on determination of the electrophoretic patterns of cell proteins is a reflection of an existing divergency of opinions as to the practical usefulness of this method as an aid to taxonomy. From theoretical considerations, the composition of the electrophoretic pattern of the cell proteins would indeed seem to provide a rational basis for species differentiation, inasmuch as the synthesis of the proteins is genetically directed and should therefore represent a spectrum of the assemblage of genes that comprises the genomes of mycoplasma species. The experience gained with this method since its introduction, in our opinion, has sustained its practical applicability to species differentiation, with the provision, of course, that it be performed under well-defined standard conditions. Thus, although the electrophoretic pattern was originally thought to be independent of the composition of the medium and of the age of the culture, more systematic studies soon demonstrated the influence of variation of these parameters on test results. Since horse serum proteins tend to precipitate during prolonged incubation of the growth medium, it is recommended that PPLO Serum Fraction (Difco) (1–2% final concentration) is used to replace horse serum in the test medium. Although horse serum may be used with the fast- and heavy-growing mycoplasmas, which require only short incubation periods, it is still advisable that the serum supplement of the growth medium be reduced to 3–5%. A number of technical improvements developed during recent years have added further to the reproducibility of the method and facilitated reading and interpretation of test results. For example, release of the cell proteins by sodium dodecyl sulfate (SDS) treatment has definite advantages over the phenolacetic acid solution used in the original version of the method. In addition to ensuring complete solubilization and extraction of the proteins, separation in the SDS system depends only on the molecular weight of the proteins, whereas the rate of migration in the denaturing acidic gel is dependent on both molecular weight and electrical charge. The use of flat gel slabs instead of gel rods has the advantages of permitting several samples to be run under identical conditions and of allowing direct comparison of the patterns. This is also greatly facilitated by the technique of densitometric tracing of the proteins, which should probably be used much more extensively. These and other aspects of the polyacrylamide gel electrophoresis technique, including literature review, are discussed in more detail in a document prepared by a working group of the FAO/WHO Programme on Comparative Mycoplasmology (1975). The progress made since then in this field and the development of related electrophoretic techniques will undoubtedly further improve the results in the future, both with respect to high-resolution separation of the proteins and as regards reproducibility. Mention may be made, for example, of polyacrylamide gel isoelectric focusing of proteins (Sayed and Hatten, 1976), the use of slab gels containing a gradient of polyacrylamide from 7 to 15% (Dahl *et al.*, 1978), and two-dimensional gel electrophoresis (Rodwell *et al.*, 1978).

The taxonomic use that can be made of the gel electrophoresis technique as a possible aid to species classification depends, first and foremost, upon the distinctness of the electrophoretic pattern of proteins shown by the species. From the rather numerous studies that have been published on this subject it appears that the protein pattern is generally sufficiently distinct to allow differentiation between species. Perhaps one of the most informative studies in this area is one carried out by Rosendal (1973b) with the purpose of evaluating the reliability of the polyacrylamide gel electrophoresis technique as a means of species identification. Whereas in most other studies reported only very few strains of each species have been compared, Rosendal analyzed—in a blind trial—a total of 50 wild strains that had been serologically classified as *M. canis, M. edwardii,* or *M. spumans,* together with the type strains of these canine species. Although at first the patterns of *M. canis* and *M. edwardii* appeared difficult to distinguish from each other, they did differ in a few characteristic bands, and altogether no mistake was made in the classification of any of the 53 coded strains. As might be expected, some minor variations might occur in the pattern shown by different strains within a species, but these were mostly quite negligible. There was some evidence, however, of more constant differences between laboratory strains that had been maintained in numerous culture passages on artificial medium and recently isolated wild strains of the same species. Intraspecies variations in the gel electrophoretic pattern of the proteins, reflecting perhaps a genetically determined heterogeneity, have been reported for some other species, as for example *M. hominis* (Razin, 1968), *M. pulmonis* (Forshaw, 1972), and *M. gallisepticum* (Müllegger *et al.*, 1973; Rhoades *et al.*, 1974).

It is not clear to what extent these variations correlated with the serologic heterogeneity known to exist for the first two species.

### G. Nucleic Acid Composition

Whereas the gel electrophoretic patterns of cell proteins may be assumed to reflect, in an indirect way, genetic identity or nonidentity of microorganisms, comparison of the composition of their nucleic acids offers a most direct means of determining genetic relatedness. No wonder, therefore, that the introduction in the early 1960s of the new techniques in molecular biology, that made it possible to analyze the genetic makeup of microorganisms directly, received considerable attention from taxonomists. The achievements that have been made in this field at different levels of taxonomy are indeed quite significant.

Nucleic acid studies relevant to taxonomy have followed four major trends: estimates of the genome size, determination of nucleic acid base ratios, determination of nucleic acid homologies by DNA–DNA hybridization, and determination of the electrophoretic mobility of rRNAs. All of these aspects have been considered already, more or less thoroughly, in earlier sections of this chapter. Hence, only a few supplementary or summarizing remarks are needed here.

The data obtained from investigations of the nucleic acids of mycoplasmas contribute to the understanding of their taxonomy along quite a wide frontier, ranging from the status of the class itself to classification of the species and subspecies. For example, the justification for the assignment of the mycoplasmas to a separate class is supported by the information obtained from the study of the electrophoretic mobility of the rRNAs and, to some extent, by the demonstration of a genome size in a considerable number of species included in the class that is only half of that found in the smallest bacteria. It is a characteristic of the class also, that the guanine plus cytosine ratio of the great majority of mycoplasma species is within the lower range of those found for any microorganisms. Moreover, the genome size is a prominent characteristic and—in part—a distinguishing marker of the three families of the order Mycoplasmatales and their genera (Bak *et al.*, 1969; Black *et al.*, 1972; Saglio *et al.*, 1973).

In classification of the species, determination of the G + C ratio and nucleic acid hybridization may both be involved. In particular, determination of the guanine plus cytosine content of the DNA by $T_m$ or by buoyant density is now recommended as a minimum requirement. In fact, the G + C value (Table II) has become an almost obligatory part of the description of the species, as is reflected in Bergey's Manual (Freundt, 1974a).

The extent to which the DNA strands of two organisms will hybridize is a very clear and direct expression of their relationship. Although determination of genetic relatedness by nucleic acid homology is considered to be an optional test, it was recognized that it "may be necessary for resolution of closely related species." (Subcommittee, 1979). This test was listed as optional in appreciation of the fact that hybridization experiments must, for technical reasons, be performed in a few specialized laboratories. Fortunately, there is remarkable agreement between species classifications based on other, more readily performed methods, such as serology, and the results obtained by nucleic acid hybridization. Thus, it is justifiable to reserve nucleic acid homology studies for taxonomic problems that are particularly controversial and that are not easily solved by other methods.

An example of the specialized application of nucleic acid homology is provided by the recent elevation of *M. agalactiae* subsp. *bovis* to the rank of species, under the name of *M. bovis*. Earlier serologic comparisons between "PG2" and "Donetta," the type strains of *M. agalactiae* subsp. *agalactiae* and subsp. *bovis,* carried out in different laboratories were somewhat conflicting and did not provide conclusive evidence in support of a proper classification. Neither did comparisons of the gel electrophoretic patterns of the two strains. The Subcommittee on the Taxonomy of Mycoplasmatales (1975) therefore encouraged attempts at solving the problem by nucleic acid studies. The G + C ratio of the DNA had previously been shown to be almost identical for the two strains, viz. about 33–34 mol%. However, hybridization experiments carried out by Askaa and Ernø (1976) demonstrated homology values between PG2 and Donetta at about 40%, as further confirmed by competition experiments between DNA from the two strains. The results of renewed serologic comparisons carried out in the same study showed the two type strains to be distinct by growth inhibition, metabolism inhibition, and immunofluorescence tests and thereby provided additional support for the proposal of Askaa and Ernø (1976) to raise *M. agalactiae* subsp. *bovis* to the rank of species.

Another example is the case of *M. mycoides* subsp. *mycoides* and subsp. *capri*. It has been repeatedly debated (e.g., subsp. *mycoides* Subcommittee on the Taxonomy of Mycoplasmatales, 1975), whether the subspecies status of these two taxa should be maintained or whether they should be elevated to separate species. The problem was reconsidered by Askaa *et al*. (1978) in the light of renewed serologic studies and nucleic acid hybridization. In accordance with several earlier observations PG1 and PG3, the type strains of the two *M. mycoides* subspecies, were found to be distinct by growth inhibition, metabolism inhibition, and immunofluorescence tests, although common antigens had previously been demonstrated by other authors using complement fixation, direct agglutination, double-immunodiffusion, and growth precipitation tests. The nucleic acid hybridization experiments resulted, on an average, in about 80% hybridization, corresponding to a relatedness value of 0.70. The relatively close relationship between PG1 and PG3 supports the continued recognition of the two subspecies of *M. mycoides*.

It is interesting to compare closely the two cases just described. The relationship between PG2 and Donetta, on the one hand, and between PG1 and PG3, on the other, was studied by serology and nucleic acid hybridization by identical techniques by the same team of workers. The two strains of each pair are in both cases serologically distinct by the growth inhibition, metabolism inhibition, and immunofluorescence tests, but the relatedness in terms of nucleic acid homologies is 40% in the case of PG2 and Donetta and 80% for PG1 and PG3. Thus, the different conclusions arrived at with respect to the status of the taxa concerned depend primarily on the hybridization data.

It should be pointed out, however, that there is no agreement about the homology levels that might be used to define the species and the subspecies, although in the case of bacteria 60–70% homology has been proposed to indicate relatedness at the subspecies level (Johnson, 1973). This gives us the occasion to emphasize, as we have done before, that any approach to a definition of taxa is a matter of convenience and agreement between taxonomists. Even when endeavoring to define genetic relatedness by nucleic acid hybridization, the values chosen to circumscribe the taxonomic categories of species and subspecies must of necessity be arbitrary and dependent on agreements. This does not, of course, detract from the fact that the determination of nucleic acid homologies is still the most exact and accurate method of assessing the degree of genetic relatedness, expressing this as it does, so to speak, in plain figures.

## H. Habitat and Pathogenicity

The origin of the strain(s) of a proposed new species and their possible pathogenicity for the natural host(s) and for laboratory animals are considered to be among the properties to be determined for proper species description (Subcommittee, 1979).

During recent years, our understanding of the host relationships of mycoplasmas has markedly changed. Whereas previously the mycoplasmas were generally regarded as highly host specific, an increasing number of species have now been found to occur, more or less frequently, in more than one host. Several examples of animal mycoplasmas naturally associated with different vertebrate hosts were given in a recent review (Freundt, 1974b). Perhaps the most striking examples are to be found, however, among the spiroplasmas. Not only is one and the same *Spiroplasma* species capable of infecting and producing symptoms in a diversity of unrelated plants, but it is able to multiply also in a number of insects of wide phylogenetic position (Whitcomb and Williamson, 1975). The natural association of spiroplasmas with both plant and arthropod hosts is a most important feature, moreover, of their natural ecology and even a condition for their maintenance in nature. Mention may further be made of the suckling mouse cataract spiroplasma that is pathogenic to vertebrates, although it was first recovered from rabbit ticks (Tully *et al.*, 1977).

Recognition of the complex host relationships characterizing some mycoplasma species has at least one major taxonomic implication. The relatively moderate requirement made in the first edition of the Minimal Standards document (Subcommittee, 1972) that an organism thought to represent a new species should (although as a minimum) "be shown to differ antigenically from all species having the same habitat and/or sharing the same general properties," might be justified as long as the traditional belief in the overall host specificity of the mycoplasma species seemed to be relatively valid. However, the use of such a minimal guideline for serologic comparison has led to some mistakes in classification. Therefore, in consequence of the changed situation, the revised Proposal for Minimal Standards (Subcommittee, 1979) categorically states that an organism that cannot be classified within one of the already established genera "should be compared with *all* previously named species," and if it can be assigned to an existing genus "it should as a minimum be compared with all species of that genus."

In the past, the specific epithet of the name of a new mycoplasma species has been derived very frequently from the name of a host, the name of a disease it is causing, or from a combination of both. In the future, it may of course still be convenient and fully justifiable to base the name of a proposed new species on the name of a principal host, even if it does occur, now and again, in other hosts as well. One should obviously be cautious if a description of a new species is to be based on a limited number of isolates, since the first isolates of a given species may not be derived from the principal host. This and other aspects of the nomenclatural implications of the fact that several mycoplasmas are less host specific than hitherto believed are discussed in more detail elsewhere (Freundt, 1974b).

## V. TAXONOMIC STATUS OF NONCULTIVABLE MYCOPLASMAS

It is a well-known phenomenon experienced in many laboratories that mycoplasmas contaminating tissue cultures may become increasingly difficult to grow on conventional cell-free media following adaptation to the tissue culture system. Strains of *M. hyorhinis* have been found particularly frequently to become cell-dependent and apparently noncultivable (Zgorniak-Nowosielska *et al.*, 1967; Hopps *et al.*, 1973, 1976). The organisms grow to high titers in the tissue culture, produce typical cytopathology, and can be demonstrated also by immunofluorescence techniques; but they can be cultivated in mycoplasma media only with great difficulty. The hypothesis has been advanced that the cell-dependent strains may "turn off" essential metabolic systems which permit these organisms to grow on artificial broth and agar media (Hopps *et al.*, 1976). Whatever the mechanism may be, the phenomenon is of obvious importance, *inter alia*, as an *in vitro* model of noncultivable mycoplasmas possibly involved in chronic infections of animals, humans, and plants. Such infections are suspected relatively frequently on morphologic or histopathologic grounds.

If mycoplasmas exist that are noncultivable in the sense described above, or, in fact, if fastidious organisms exist that require special growth factors not included in conventional mycoplasma media, they would—as pointed out by Hopps *et al.* (1976)—raise a taxonomic problem. According to the existing proposals for Minimal Standards (Subcommittee, 1972, 1979) a "mycoplasma-like organism" cannot be formally assigned to a taxon unless it has been shown to produce typical agar colonies and unless it can be properly characterized by biochemical tests, etc. Thus, the corn stunt organism could not be classified at the time when it was still "noncultivable," even though its relatedness to the genus *Spiroplasma* was evident from morphologic and serologic observations (Davis and Worley, 1973; Tully *et al.*, 1973). In appreciation of the fact that there might under such circumstances be a case for deviating from strict adherence to the Minimal Standards, the Subcommittee on the Taxonomy of Mycoplasmatales (1977) recently considered the taxonomic status of noncultivable mycoplasmas. In concluding its discussion, the Subcommittee agreed "that the requirement for growth on agar was necessary at present to assure and encourage proper and careful studies for the characterization of candidate mycoplasmas." The comment was made, though, that "retention of this requirement does not preclude eventual classification in Mollicutes of noncultivable wall-free procaryotes, especially if biochemical deletions could be demonstrated to explain their noncultivability."

However, the need for "extensive and careful documentation" as provided, for example, for *Chlamydia psittaci* (Page, 1968) was emphasized.

## REFERENCES

Ad Hoc Committee of the Judicial Commission of the ICSB (1976). *Int. J. Syst. Bacteriol.* **26,** 563–599.

Al-Aubaidi, J. M. (1975). *Int. J. Syst. Bacteriol.* **25,** 221.

Al-Aubaidi, J. M., and Fabricant, J. (1971). *Cornell Vet.* **61,** 490–518.

Al-Aubaidi, J. M., Dardiri, A. H., and Fabricant, J. (1972). *Int. J. Syst. Bacteriol.* **22,** 155–164.

Al-Aubaidi, J. M., Dardiri, A. H., Muscoplatt, C. C., and McCauley, E. H. (1973). *Cornell Vet.* **63,** 117–129.

Allam, N. M., and Lemcke, R. M. (1975). *J. Hyg.* **74,** 385–408.

Askaa, G., and Ernø, H. (1976). *Int. J. Syst. Bacteriol.* **26,** 323–325.

Askaa, G., Ernø, H., and Ojo, M. O. (1978). *Acta Vet. Scand.* **19,** 166–178.

Bak, A. L., Black, F. T., Christiansen, C., and Freundt, E. A. (1969). *Nature* (London) **224,** 1209–1210.

Barile, M. F., Del Giudice, R. A., Carski, T. R., Gibbs, C. J., and Morris, J. A. (1968). *Proc. Soc. Exp. Biol. Med.* **129,** 489–494.

Barile, M. F., Del Giudice, R. A., and Tully, J. G. (1972). *Infect. Immun.* **5,** 70–76.

Belly, R. T., Bohlool, B. B., and Brock, T. D. (1973). *Ann. N.Y. Acad. Sci.* **225,** 94–107.

Black, F. T., Christiansen, C., and Askaa, G. (1972). *Int. J. Syst. Bacteriol.* **22,** 241–242.

Black, F. T., Freundt, E. A., Vinther, O., and Christiansen, C. (1979). *J. Bacteriol.* **137** (in press).

Breed, R. S., Murray, E. G. D., and Smith, N. R., eds. (1957). "Bergey's Manual of Determinative Bacteriology," 7th ed., pp. 914–926. Williams & Wilkins, Baltimore, Maryland.

Brock, T. D., Brock, K. M., Belly, R. T., and Weiss, R. L. (1972). *Arch. Mikrobiol.* **84,** 54–68.

Buchanan, R. E., and Gibbons, N. E., eds. (1974). "Bergey's Manual of Determinative Bacteriology," 8th ed., p. 929. Williams & Wilkins, Baltimore, Maryland.

Carmichael, L. E., St. George, T. D., Sullivan, N. D., and Horsfall, N. (1972). *Cornell Vet.* **62,** 654–679.

Christiansen, C., Freundt, E. A., and Black, F. T. (1975). *Int. J. Syst. Bacteriol.* **25,** 99–101.

Christiansen, C., Askaa, G., Freundt, E. A., Tully, J. G., and Whitcomb, R. (1979). In preparation.

Cole, B. C., Golightly, L., and Ward, J. R. (1967). *J. Bacteriol.* **94,** 1451–1458.

Cole, R. M., Tully, J. G., Popkin, T. J., and Bové, J. M. (1973). *J. Bacteriol.* **115,** 367–386.

Dahl, J. S., Hellewell, S. B., and Levine, R. P. (1977). *J. Immunol.* **119,** 1419–1426.

Darland, G., Brock, T. D., Samsonoff, W., and Conti, S. F. (1970). *Science* **170,** 1416–1418.

Davis, R. E., and Worley, J. F. (1973). *Phytopathology* **63,** 403–408.

Del Giudice, R. A., Carski, T. R., Barile, M. F., Lemcke, R. M., and Tully, J. G. (1971). *J. Bacteriol.* **108,** 439–445.

Del Giudice, R. A., Purcell, R. H., Carski, T. R., and Chanock, R. M. (1974). *Int. J. Syst. Bacteriol.* **24,** 147–153.

de Rosa, M., and Gambacorta, A. (1975). *J. Gen. Microbiol.* **86,** 156–164.

Editorial Board of the International Committee on Bacteriological Nomenclature (1958). "International Code of Nomenclature of Bacteria and Viruses. Bacteriological Code." Iowa State College Press, Ames, Iowa.
Edward, D. G. ff. (1955). *Int. Bull. Bacteriol. Nomencl. Taxon.* **5,** 85–93.
Edward, D. G. ff. (1971). *J. Gen. Microbiol.* **69,** 205–210.
Edward, D. G. ff. (1974). *Colloq. Inst. Natl. Santé Rech. Méd.* **33,** 13–18.
Edward, D. G. ff., and Freundt, E. A. (1956). *J. Gen. Microbiol.* **14,** 197–207.
Edward, D. G. ff., and Freundt, E. A. (1967). *Int. J. Syst. Bacteriol.* **17,** 267–268.
Edward, D. G. ff., and Freundt, E. A. (1969). *In* "The Mycoplasmatales and the L-Phase of Bacteria" (L. Hayflick, ed.), pp. 147–200. Appleton, New York.
Edward, D. G. ff., and Freundt, E. A. (1970). *J. Gen. Microbiol.* **62,** 1–2.
Edward, D. G. ff., and Freundt, E. A. (1973). *Int. J. Syst. Bacteriol.* **23,** 55–61.
Edward, D. G. ff., and Kanarek, A. D. (1960). *Ann. N.Y. Acad. Sci.* **79,** 696–702.
Ernø, H., and Jurmanová, K. (1973). *Acta Vet. Scand.* **14,** 524–537.
FAO/WHO Programme on Comparative Mycoplasmology (1974). *Vet. Rec.* **95,** 457–461.
FAO/WHO Programme on Comparative Mycoplasmology (1975). Working Document, VPH/MIC 75.3. World Health Organ., Geneva.
Forshaw, K. A. (1972). *J. Gen. Microbiol.* **72,** 493–499.
Fox, H., Purcell, R. H., and Chanock, R. M. (1969). *J. Bacteriol.* **98,** 36–43.
Freundt, E. A. (1955). *Int. Bull. Bacteriol. Nomencl. Taxon.* **5,** 67–78.
Freundt, E. A. (1973). *Ann. N.Y. Acad. Sci.* **225,** 7–13.
Freundt, E. A. (1974a). *In* "Bergey's Manual of Determinative Bacteriology" (R. E. Buchanan and N. E. Gibbons, eds.), 8th ed., Part 19, pp. 929–954. Williams & Wilkins, Baltimore, Maryland.
Freundt, E. A. (1974b). Colloq. *Inst. Natl. Santé Rech. Méd.* **33,** 19–25.
Freundt, E. A., Andrews, B. E., Ernø, H., Kunze, M., and Black, F. T. (1973a). *Zentralbl. Bakteriol., Parasitenkd., Infektionskr. Hyg., Abt. I: Orig., Reihe A* **225,** 104–112.
Freundt, E. A., Ernø, H., Black, F. T., Krogsgaard-Jensen, A., and Rosendal, S. (1973b). *Ann. N.Y. Acad. Sci.* **225,** 161–171.
Freundt, E. A., Taylor-Robinson, D., Purcell, R. H., Chanock, R. M., and Black, F. T. (1974). *Int. J. Syst. Bacteriol.* **24,** 252–255.
Freundt, E. A., Ernø, H., and Lemcke, R. M. (1979). *Methods Microbiol.* **13** (in press).
Frey, M. L., Hanson, R. P., and Anderson, D. P. (1968). *Am. J. Vet. Res.* **29,** 2163–2171.
Gibbons, N. E., and Murray, R. G. E. (1978). *Int. J. Syst. Bacteriol.* **28,** 1–6.
Gourlay, R. N., and Leach, R. H. (1970). *J. Med. Microbiol.* **3,** 111–123.
Gourlay, R. N., Leach, R. H., and Howard, C. J. (1974). *J. Gen Microbiol.* **81,** 475–484.
Gourlay, R. N., Wyld, S. G., and Leach, R. H. (1977). *Int. J. Syst. Bacteriol.* **27,** 86–96.
Gourlay, R. N. Wyld, S. G., and Leach, R. H. (1978). *Int. J. Syst. Bacteriol.* **28,** 289–292.
Hale, H. H., Helmboldt, C. F., Plastridge, W. N., and Stula, E. F. (1962). *Cornell Vet.* **52,** 582–591.
Harley, E. H., White, J. S., and Rees, K. R. (1973). *Biochim. Biophys. Acta* **299,** 253–263.
Herring, P. K., and Pollack, J. D. (1974). *Int. J. Syst. Bacteriol.* **24,** 73–78.
Heyward, J. T., Sabry, M. Z., and Dowdle, W. R. (1969). *Am. J. Vet. Res.* **30,** 615–622.
Hill, A. (1971). *J. Gen. Microbiol.* **65,** 109–113.
Holländer, R., Wolf, G., and Mannheim, W. (1977). *Antonie van Leeuwenhoek* **43,** 177–185.
Hopps, H. E., Meyer, B. C., and Barile, M. F. (1973). *Ann. N. Y. Acad. Sci.* **225,** 265–276.
Hopps, H. E., Del Giudice, R. A., and Barile, M. F. (1976). *Proc. Soc. Gen. Microbiol.* **3,** 143.
Jain, N. C., Jasper, D. E., and Dellinger, J. D. (1967). *J. Gen. Microbiol.* **49,** 401–410.
Johnson, J. D., and Horowitz, J. (1971). *Biochim. Biophys. Acta* **247,** 262–279.

Johnson, J. L. (1973). *Int. J. Syst. Bacteriol.* **23,** 308–315.
Jones, A. L., Whitcomb, R. F., Williamson, D. L., and Coan, M. E. (1977). *Phytopathology* **67,** 738–746.
Kenny, G. E. (1969). *J. Bacteriol.* **98,** 1044–1055.
Kenny, G. E. (1972). *Med. Microbiol. Immunol.* **157,** 174.
Kenny, G. E. (1973). *J. Infect. Dis.* **127,** Suppl. 2–5.
Kirchhoff, H. (1974). *Zentralbl. Veterinaermed. Reihe B* **21,** 207–210.
Kirchhoff, H. (1978). *Int. J. Syst. Bacteriol.* **28,** 76–81.
Kirk, R. G., and Morowitz, H. J. (1969). *Am. J. Vet. Res.* **30,** 287–293.
Krass, C. J., and Gardner, M. W. (1973). *Int. J. Syst. Bacteriol.* **23,** 62–64.
Kunze, M. (1973). *Zentralbl. Bakteriol., Parasitenkd., Infektionskr. Hyg., Abt. 1: Orig., Reihe A* **223,** 197–204.
Langford, E. V., and Leach, R. H. (1973). *Can. J. Microbiol.* **19,** 1435–1444.
Langford, E. V., Ruhnke, H. L., and Onoviran, O. (1976). *Int. J. Syst. Bacteriol.* **26,** 212–219.
Langworthy, T. A., Smith, P. F., and Mayberry, W. R. (1972). *J. Bacteriol.* **112,** 1193–1200.
Lapage, S. P., Sneath, P. H. A., Lessel, E. F., Skerman, V. B. D., Seeliger, H. P. R., and Clark, W. A., eds. (1975). "International Code of Nomenclature of Bacteria." Am. Soc. Microbiol., Washington, D.C.
Larraga, V. and Razin, S. (1976). *J. Bacteriol.* **128,** 827–833.
Leach, R. H. (1967). *Ann. N.Y. Acad. Sci.* **143,** 305–316.
Leach, R. H. (1973). *J. Gen. Microbiol.* **75,** 135–153.
Lemcke, R. M. (1965). *J. Gen. Microbiol.* **38,** 91–100.
Loening, U. (1968). *J. Mol. Biol.* **38,** 355–365.
Low, J. E., and Zimkus, S. M. (1973). *J. Bacteriol.* **116,** 346–354.
Madden, D. L., Moats, K. E., London, W. T., Matthew, E. B., and Sever, J. L. (1974). *Int. J. Syst. Bacteriol.* **24,** 459–464.
Mandel, M. (1969). *Ann. Rev. Microbiol.* **23,** 239–274.
Maré, C. J., and Switzer, W. P. (1965). *Vet. Med. (Kansas City, Mo.)* **60,** 841–846.
Meyling, A., and Friis, N. F. (1972). *Acta Vet. Scand.* **13,** 287–289.
Millonig, G., de Rosa, M., Gambacorta, A., and Bu'lock, J. D. (1975). *J. Gen. Microbiol.* **86,** 165–173.
Milne, R. G., Thompson, G. W., and Taylor-Robinson, D. (1972). *Arch. Gesamte Virusforsch.* **37,** 378–385.
Müllegger, P.-H., Gerlach, H., and Schellner, H.-P. (1973). *Zentralbl. Bakteriol., Parasitenkd., Infektionskr. Hyg., Abt. 1: Orig., Reihe A* **223,** 372–380.
Murray, R. G. E. (1974). *In* "Bergey's Manual of Determinative Bacteriology" (R. E. Buchanan and N. E. Gibbons, eds.), 8th ed., pp. 4–9. Williams & Wilkins, Baltimore, Maryland.
Neimark, H. (1973). *Ann. N.Y. Acad. Sci.* **225,** 14–21.
Neimark, H. (1974). *Colloq. Inst. Natl. Santé Rech. Méd.* **33,** 71–78.
Neimark, H., and Lemcke, R. M. (1972). *J. Bacteriol.* **111,** 633–640.
Neimark, H., and Tung, M. C. (1973). *J. Bacteriol.* **114,** 1025–1033.
Ojo, O. M. (1976). *Trop. Anim. Health Prod.* **8,** 137–146.
Olson, N. O., Kerr, K. M., and Campbell, A. (1964). *Avian Dis.* **8,** 209–214.
Pace, B., and Campbell, L. L. (1971). *J. Bacteriol.* **107,** 543–547.
Pace, N. R. (1973). *Bacteriol. Rev.* **37,** 562–603.
Page, L. A. (1968). *Int. J. Syst. Bacteriol.* **18,** 51–66.
Pollack, J. D. (1975). *Int. J. Syst. Bacteriol.* **25,** 108–113.
Pollack, J. D., Razin, S., and Cleverdon, R. C. (1965). *J. Bacteriol.* **90,** 617–622.

Razin, S. (1964). *J. Gen. Microbiol.* **36,** 451–459.
Razin, S. (1968). *J. Bacteriol.* **96,** 687–694.
Razin, S., and Cleverdon, R. C. (1965). *J. Gen. Microbiol.* **41,** 409–415.
Razin, S., and Tully, J. G. (1970). *J. Bacteriol.* **102,** 306–310.
Reff, M. E., Stanbridge, E. J., and Schneider, E. L. (1977). *Int. J. Syst. Bacteriol.* **27,** 185–193.
Reich, P. R. (1967). *Ann. N.Y. Acad. Sci.* **143,** 113.
Rhoades, K. R., Phillips, M., and Yoder, H. W. (1974). *Avian Dis.* **18,** 91–96.
Roberts, D. H. (1964). *Vet. Rec.* **76,** 470–473.
Robertson, J., Gomersall, M., and Gill, P. (1975). *J. Bacteriol.* **124,** 1007–1018.
Robinson, I. M., and Allison, M. J. (1975). *Int. J. Syst. Bacteriol.* **25,** 182–186.
Robinson, I. M., Allison, M. J., and Hartman, P. A. (1975). *Int. J. Syst. Bacteriol.* **25,** 173–181.
Rodwell, A. W. (1967). *Ann. N.Y. Acad. Sci.* **143,** 88–109.
Rodwell, A. W., Archer, D. B., Peterson, J. E., and Rodwell, E. S. (1978). *Zentralbl. Bakteriol., Parasitenkd., Infektionskr. Hyg., Abt. 1: Orig., Reihe A* **241,** 178–179.
Rosendal, S. (1973a). *Int. J. Syst. Bacteriol.* **23,** 49–54.
Rosendal, S. (1973b). *Acta Pathol. Microbiol. Scand., Sect. B* **81,** 273–281.
Rosendal, S. (1974). *Int. J. Syst. Bacteriol.* **24,** 125–130.
Rosendal, S. (1975). *Acta Pathol. Microbiol. Scand., Sect. B* **83,** 463–470.
Ross, R. F., and Karmon, J. A. (1970). *J. Bacteriol.* **103,** 707–713.
Saglio, P., LHospital, M., Laflèche, D., Dupont, G., Bové, J. M., Tully, J. G., and Freundt, E. A. (1973). *Int. J. Syst. Bacteriol.* **23,** 191–204.
Sayed, I. A., and Hatten, B. A. (1976). *Appl. Environ. Microbiol.* **32,** 603–609.
Searcy, D. G., and Doyle, E. K. (1975). *Int. J. Syst. Bacteriol.* **25,** 286–289.
Shepard, M. C., and Howard, D. R. (1970). *Ann. N.Y. Acad. Sci.* **174,** 809–819.
Shepard, M. C., Lunceford, C. D., Ford, D. K., Purcell, R. H., Taylor-Robinson, D., Razin, S., and Black, F. T. (1974). *Int. J. Syst. Bacteriol.* **24,** 160–171.
Skripal, I. G. (1974). *Mikrobiol. Zh. (Kiev)* **36,** 462–467.
Skripal, I. G. (1978). *Mikrobiol. Zh. (Kiev)* **39,** 98–103.
Smith, P. F., Langworthy, T. A., Mayberry, W. R., and Hougland, H. E. (1973). *J. Bacteriol.* **116,** 1019–1028.
Somerson, N. L., Taylor-Robinson, D., and Chanock, R. M. (1963). *Am. J. Hyg.* **17,** 122–128.
Subcommittee on the Taxonomy of Mycoplasmatales (1967a). *Int. J. Syst. Bacteriol.* **17,** 105–109.
Subcommittee on the Taxonomy of Mycoplasmatales (1967b). *Science* **155,** 1694–1696.
Subcommittee on the Taxonomy of Mycoplasmatales (1971). *Int. J. Syst. Bacteriol.* **21,** 151–153.
Subcommittee on the Taxonomy of Mycoplasmatales (1972). *Int. J. Syst. Bacteriol.* **22,** 184–188.
Subcommittee on the Taxonomy of Mycoplasmatales (1974). *Int. J. Syst. Bacteriol.* **24,** 390–392.
Subcommittee on the Taxonomy of Mycoplasmatales (1975). *Int. J. Syst. Bacteriol.* **25,** 237–239.
Subcommittee on the Taxonomy of Mycoplasmatales (1977). *Int. J. Syst. Bacteriol.* **27,** 392–394.
Subcommittee on the Taxonomy of Mycoplasmatales (1979). *Int. J. Syst. Bacteriol.* **29,** (in press).
Switzer, W. P. (1955). *Am. J. Vet. Res.* **16,** 540–544.

Taylor, M. M., Glasgow, J. E., and Storck, R. (1967). *Proc. Natl. Acad. Sci. U.S.A.* **57,** 164–169.
Taylor-Robinson, D., Somerson, N. L., Turner, H. C., and Chanock, R. M. (1963). *J. Bacteriol.* **85,** 1261–1273.
Taylor-Robinson, D., Canchola, J., Fox, H., and Chanock, R. M. (1964). *Am. J. Hyg.* **80,** 135–148.
Thirkill, C. E., and Kenny, G. E. (1974). *Infect. Immun.* **10,** 624–632.
Townsend, R., Markham, P. G., Plaskitt, K. A., and Daniels, M. J. (1977). *J. Gen. Microbiol.* **100,** 15–21.
Tully, J. G. (1978). *In* "Mycoplasma Infection of Cell Cultures" (G. J. McGarrity, D. G. Murphy, and W. W. Nichols, eds.), pp. 7–33, Plenum, New York.
Tully, J. G., and Razin, S. (1968). *J. Bacteriol.* **95,** 1504–1512.
Tully, J. G., and Razin, S. (1970). *J. Bacteriol.* **103,** 751–754.
Tully, J. G., Barile, M. F., Del Giudice, R. A., Carski, T. R., Armstrong, D., and Razin, S. (1970). *J. Bacteriol.* **101,** 346–349.
Tully, J. G., Whitcomb, R. F., Bové, J. M., and Saglio, P. (1973). *Science* **182,** 827–829.
Tully, J. G., Barile, M. F., Edward, D. G. ff., Theodore, T. S., and Ernø, H. (1974). *J. Gen. Microbiol.* **85,** 102–120.
Tully, J. G., Whitcomb, R. F., Clark, H. F., and Williamson, D. L. (1977). *Science* **195,** 892–894.
Vinther, O., and Freundt, E. A. (1977). *Acta Pathol. Microbiol. Scand., Sect. B* **85,** 184–188.
Weiss, R. L. (1973). *J. Gen. Microbiol.* **77,** 501–507.
Whitcomb, R. F., and Williamson, D. L. (1975). *Ann. N.Y. Acad. Sci.* **266,** 260–275.
Woese, C. R., Sogin, M. L., and Sutton, L. A. (1974). *J. Mol. Evol.* **3,** 293–299.
Woese, C. R., Fox, G. E., Zablen, L., Uchida, T., Boner, L., Peckham, K., Lewis, B. L., and Stahl, D. (1975). *Nature (London)* **254,** 83–86.
Yamamoto, R., Bigland, C. H., and Ortmayer, H. B. (1965). *J. Bacteriol.* **90,** 47–49.
Zgorniak-Nowosielska, I., Sedwick, W. D., Hummeler, K., and Koprowski, H. (1967). *J. Virol.* **1,** 1227–1237.

# 2 / PHYLOGENETIC RELATIONSHIPS BETWEEN MYCOPLASMAS AND OTHER PROKARYOTES

*Harold Neimark*



## I. INTRODUCTION

Mycoplasmas are prokaryotes with the general characteristics and cellular organization of bacteria, but they share several distinctive properties that set them apart from bacteria, rickettsias, chlamydias, and other microorganisms. (See Chapter 1 for a detailed description of the Mycoplasmatales.) Like bacteria, they possess a triple-layer "unit membrane," 70 S ribosomes, and a typical prokaryotic circular chromosome which is



THE MYCOPLASMAS, VOL. 1

ISBN 0-12-078401-7

"folded" (Bode and Morowitz, 1967; Teplitz, 1977). In contrast to bacteria, however, mycoplasmas are extremely small—they are the smallest organisms known to be capable of growth in cell-free media—and lack a cell wall. Indeed, it is this absence of a cell wall that appears to be responsible for a number of their interesting properties, such as their characteristic colonial morphology, filterability, staining properties, growth inhibition by specific antibody, and resistance to a variety of antibiotics that affect cell wall biosynthesis.

Underlying all these unique properties of mycoplasmas are their small genomes (Bak *et al.*, 1969; Morowitz *et al.*, 1967). *Mycoplasma* and *Ureaplasma* species genomes have molecular weights of approximately $5 \times 10^8$ daltons, a value smaller than almost any known for bacteria, and just one-fifth that of *Escherichia coli.* Only the DNAs from prokaryotes that are unable to grow independently on cell-free media, such as *Chlamydia,* are as small (Sarov and Becker, 1969; Kingsbury, 1969). *Acholeplasma, Thermoplasma, and Spiroplasma* species have genome sizes of about $1.0 \times 10^9$ daltons, slightly more than twice that of *Mycoplasma* species but still smaller than those of most bacteria (for comparison, *Hemophilus influenzae* with a genome size of approximately $1.0 \times 10^9$ daltons can be cited as a bacterium with a relatively small genome).

The question of the nature of mycoplasmas, their pathogenicity, and their relationships to bacteria was raised with the realization that virtually all bacteria can produce wall-deficient growth forms (so-called L forms) that closely resemble the naturally occurring mycoplasmas (Dienes and Weinberger, 1951). Two general hypotheses have been advanced to explain the relationship of mycoplasmas to other microorganisms (Dienes, 1963; Edward, 1960; Klieneberger-Nobel, 1960): One holds that mycoplasmas are a true biologic class whose members are related to one another through evolution; the other, that mycoplasmas are an assemblage of wall-deficient forms derived from various bacteria.

If the former view is correct, the Mycoplasmatales must represent the surviving descendants of exceedingly primitive bacteria—those which must have existed before the development of bacterial cell wall peptidoglycan synthesis. Such primitive bacteria also may have had smaller genomes than contemporary bacteria.

## II. HETEROGENEITY AMONG THE MYCOPLASMAS

Genetic heterogeneity is a major feature of the mycoplasmas (Neimark and Pickett, 1960; Neimark, 1967). This heterogeneity became discernible after the group was first assembled and when it was composed of relatively few organisms (Edward, 1954; Freundt, 1957); and this heterogeneity is still evident at present after approximately 60 distinct organisms have been isolated and identified. Clearly, the group is made up of microorganisms with diverse morphologic, nutritional, physiologic, and antigenic properties.

## A. Heterogeneity in DNA Base Composition

At a fundamental level we investigated relationships among the various mycoplasmas as well as possible relations to other prokaryotes by determining DNA base compositions (Neimark and Pene, 1965a,b). Similar DNA base compositions would be expected where a taxonomic relationship exists between organisms, although similarity in base composition is not necessarily an indication of a relation (since unrelated organisms may occasionally display the same base composition).

We found, even in this initial work involving only fermentative species, that there was a marked heterogeneity in base composition with the guanine plus cytosine (G + C) contents of these mycoplasma DNAs falling in a wide range from approximately 23 to 40 mol-%. This study thus established absolutely the fundamental nature of the heterogeneity of mycoplasmas.

In addition, the existence of a cluster group of mycoplasmas with very low G + C contents was revealed. The G + C contents found for the calf strain, *Mycoplasma capricolum* (California goat, kid strain), and *Mycoplasma neurolyticum* are among the lowest reported for any microorganisms and coincide with a value reported for the agent of caprine pleuropneumonia, *Mycoplasma mycoides* var. *capri* (Jones and Walker, 1963). No aerobic bacteria have been shown to have such low G + C contents. Indeed, these values are among the lowest known for any organisms. Other organisms known to have very low G + C contents are the ciliated protozoan, *Tetrahymena,* and the slime mold, *Dictyostelium discoideum*. A plausible conclusion was that the values establish these organisms as unique, and consequently they could not be stable L forms of any known bacteria (Jones and Walker, 1963).

Studies from many laboratories rapidly contributed additional DNA base composition values for most of the recognized species and many strains of mycoplasmas (Jones *et al.*, 1965; Rogul *et al.*, 1965; Somerson *et al.*, 1966; Reich *et al.*, 1966; McGee *et al.*, 1967; Morowitz *et al.*, 1967; Neimark, 1967, 1970; Bak and Black, 1968; Chelton *et al.*, 1968; Lynn and Haller, 1968; Brennan *et al.*, 1969; Kelton and Mandel, 1969; Peterson and Pollock, 1969; Williams *et al.*, 1969; Gourlay and Leach, 1970; Allen, 1971; Askaa *et al.*, 1973). Subsequently, most newly described type species have been characterized with respect to DNA G + C content (see Neimark, 1970, and Chapter 1 for a compilation of DNA base compositions). This rapid progress was motivated by the recognition of the usefulness of DNA base composition values in bacterial taxonomy and was paralleled by similar studies for most other groups of prokaryotes.

All of these studies confirm that there is a fundamental and widespread heterogeneity among the mycoplasmas which extends through both fermentative and nonfermentative species. Clearly, these diverse organisms are incompatible in a single genus.

### B. Divisions among the Mycoplasmas

The mycoplasmas can be arranged into sets based first on their fermentative capacity and then on the G + C content of their DNA (Neimark, 1967, 1970). The division of organisms by their pattern of energy metabolism reflects an evolutionary separation within the mycoplasmas. The span of G + C contents for each division is well over 10%. Since bacteria with G + C contents differing by more than 10% would be expected to have little genetic relatedness (Sueoka, 1961), each division probably contains at least two genetically unrelated subgroups. To provide "working" units for studying relations, the mycoplasmas were divided into six subgroups (Neimark, 1970):

1. *Mycoplasma pneumoniae*. G + C content ca. 40%. Set apart from all other mycoplasmas
2. Sterol-nonrequiring mycoplasmas (acholeplasmas). G + C ca. 29–35%
3. High G + C fermentative organisms. G + C ca. 35% and less
4. Low G + C fermentative organisms. G + C ca. 23% and greater
5. High G + C nonfermentative organisms. G + C ca. 34% and less
6. Low G + C nonfermentative organisms. G + C ca. 24% and greater

These subgroups are meant to serve as guides or basic starting points for searching for relationships and "affinity groups" among the mycoplasmas. Even these subgroups are heterogeneous and, excepting the acholeplasma biotype, the organisms within each group may not necessarily be related to one another.

If additional characters, such as basic biochemical characters or serologic properties, are considered, the subgroups can be further divided. For example, a cluster of three serologically related fermentative organisms (Kenny, 1969) also produce D(−)-lactic acid (Neimark, 1973); these organisms may represent a new affinity group or biotype. In addition, new, recently discovered organisms with distinctive properties such as the sterol-requiring, large genome spiroplasmas, the thermoplasmas (Chapter 18), and the anaeroplasmas (Chapter 19) constitute obvious new subgroups.

### C. Nucleic Acid Hybridization Studies

It was immediately recognized that DNA base composition values could be combined with prominent characters, such as ability to ferment glucose, to rationally choose mycoplasmas and bacteria (or pairs of mycoplasmas) for nucleic acid hybridization studies, thus allowing direct examination of the hypothesis that mycoplasmas are related to bacteria.

Hybridization studies have been carried out between selected mycoplasmas and bacteria, but none of these has provided any evidence of nucleic acid hybridization between a mycoplasma and a bacterium or its L-form derivative (Neimark and Pene, 1965b; Rogul *et al.*, 1965; McGee *et al.*, 1965, 1967; Somerson *et al.*, 1967; Neimark, 1967; Brennan *et al.*, 1969; Haller and Lynn, 1969). On the other hand, hybridizations between bacteria and their stable L forms have shown very close nucleic acid homology (Dowell *et al.*, 1964; Panos, 1965; McGee *et al.*, 1965; Somerson *et al.*, 1967), thus indicating that most known L forms are very little changed genetically from their bacterial parents. These results coincide with observations that L forms retain most of the properties of their parents except those associated with the cell wall. Consequently, there is no evidence available from homology studies to support the hypothesis that mycoplasmas are derived from bacteria or their L forms.

However, in considering these results it should be recognized that nucleic acid hybridization studies comparing whole genomes can detect only close relations, such as those existing at the level of related species, and more distant relations may go undetected.

Few hybridization studies have been carried out among the mycoplasmas themselves, and most of these have revealed only some low-level cross-reactions of uncertain significance (McGee *et al.*, 1967); a moderate level cross-reaction was found between *Acholeplasma laidlawii* and *Acholeplasma granularum* (Neimark, 1970).

### D. Divisions in Genome Size

If the mycoplasmas are a true biologic group, it should be possible to demonstrate some common fundamental properties in addition to the absence of a cell wall. One property that has been examined is genome size. Values for seven strains were determined from contour length measurements of DNA and found to fall in a range from $4.4 \times 10^8$ to $1.2 \times 10^9$ daltons (Bode and Morowitz, 1967; Morowitz *et al.*, 1967; Ryan and Morowitz, 1969). Bak *et al.* (1969) determined genome sizes in 12 mycoplasmas by studying renaturation rates of DNA. Their results were partly in agreement with those of Morowitz and co-workers. However, the mycoplasmas fell into only two groups, each composed of strains with nearly identical genome size. The first group, containing the sterol-nonrequiring acholeplasmas, had genome sizes of about $1.0 \times 10^9$ daltons. All the serum-requiring strains examined were in the second group, with genome sizes of about $4.6 \times 10^8$ daltons, a value approximately half that for the acholeplasmas and smaller than that for any known bacteria.

For comparison, the lowest values observed for bacteria are $1.0 \times 10^9$ and $0.99 \times 10^9$ for *Hemophilus influenzae* and *Neisseria gonorrhoeae,* respectively (Bak *et al.*, 1969, 1970). Additional species were examined by Askaa *et al.* (1973), who employed the same procedure and obtained the same results.

This finding of similar small genomes among mycoplasmas may reflect a common phylogenetic origin (Bak *et al.*, 1969). The *Acholeplasma* species, with genome sizes approximately twice as large as the mycoplasmas, constitute a distinct group, and the discontinuity in genome size between the two groups is remarkable. Subsequently, spiroplasmas and thermoplasmas were found to have genome sizes of $1 \times 10^9$ daltons.

Wallace and Morowitz (1973) have proposed a theory that attempts to explain the relationship between mycoplasmas and acholeplasmas. They noted that large increases or doublings of genome size had been widely observed in animals, plants, viruses, mitochondria, and bacterial plasmids (reviewed by Sparrow and Nauman, 1976) and hypothesized that doubling of the mycoplasma $5 \times 10^8$ dalton genome during evolution could have resulted in the $1 \times 10^9$ dalton genome of the acholeplasmas. (Genome size values for spiroplasmas and thermoplasmas of $1 \times 10^9$ daltons were reported after their paper was prepared and their hypothesis does not consider the occurrence of these other large genome organisms.)

## III. RELATIONSHIPS BETWEEN MYCOPLASMAS AND BACTERIA

The Mycoplasmatales include many physiologic and morphologic types that resemble, at least coincidentally, specific bacteria. Obvious similarities occur, for example, between thermoplasmas and certain thermophilic bacteria, between anaeroplasmas and other strict anaerobes isolated from ruminants, and *Mycoplasma lipophilum* resembles lipophilic corynebacteria found on human skin. Also, certain mycoplasmas capable of gliding movement (Bredt, 1973) share a general method of locomotion with various gliding bacteria. Any or all of these similarities could be examples of convergent evolution of the sort that is well known in animals and plants, or they could result from as yet unrecognized relationships.

The fermentative mycoplasmas are another prominent segment of the Mycoplasmatales that bears similarities to known bacteria. Approximately half of the 60 species of mycoplasmas known at present ferment glucose and accumulate acid end products, and at least 16 fermentative organisms so far examined produce lactic acid as a major product of glucose fermentation (Rodwell and Rodwell, 1954; Neimark and Pickett, 1960; Tourtellotte and Jacobs, 1960). Both homofermentative and heterofermentive mycoplasmas are recognized (Neimark and Pickett, 1960). Various workers (Neimark and Pickett, 1960; Rodwell, 1960; Smith *et al.*, 1963; Van Demark, 1967) have commented that certain of these mycoplasmas resemble lactic acid bacteria. Organisms in both groups ferment glucose by the Embden–Meyerhof pathway and accumulate lactate, usually do not synthesize heme enzymes, and have a flavin-terminated respiratory system.

### A. Mycoplasma Lactate Dehydrogenases

To define further the biochemical properties of the fermentative mycoplasmas, representatives from each of the four fermentative subgroups were selected for examination. These fermentative mycoplasmas differ in genome size, DNA base composition, and sterol dependence. As already mentioned, these subgroups delineate major divisions among the mycoplasmas and can be expected to provide organisms with widely different properties. After exploratory studies of respiratory enzymes, detailed studies on lactate dehydrogenases (LDH) were undertaken. We searched for and characterized lactate dehydrogenases in six mycoplasmas and two acholeplasmas (Neimark and Lemcke, 1972). Lactate dehydrogenases were chosen for study because they are constitutively synthesized enzymes that catalyze an essential reaction in energy metabolism. In addition to providing information about the metabolism of fermentative mycoplasmas, it was felt that knowledge of these enzymes should be useful in examining evolutionary and taxonomic relations among mycoplasmas and bacteria. A wide variety of different lactate dehydrogenases has been found among the bacteria in contrast to the similarity of animal lactate dehydrogenases. Bacterial lactate dehydrogenases differ in such properties as stereospecificity, cofactor and activator requirements, reversibility, sensitivity to inhibitors, stability, and molecular weight.

Three completely different patterns of lactate dehydrogenase enzyme composition were found among the mycoplasmas selected for examination: (1) possession of a single nicotinamide adenine dinucleotide (NAD)-dependent L(+)-lactate dehydrogenase; (2) possession of a single NAD-dependent D(−)-lactate dehydrogenase; (3) possession of a fructose-1,6-diphosphate (FDP)-activated NAD-dependent L(+)-lactate dehydrogenase, as well as an NAD-independent D(−)-lactate dehydrogenase.

A result of the enzyme compositions is that these mycoplasmas have the characteristic property of producing a specific isomer of lactic acid. This property seems to be generally useful for characterizing fermentative mycoplasmas (Neimark, 1973; and in preparation).

The possession of only an NAD-dependent L(+)-lactate dehydrogenase and production of L(+)-lactic acid during growth was found in both large and small genome mycoplasmas and in organisms with widely different DNA base compositions. The L(+)-lactate dehydrogenases of the species examined could be readily distinguished by electrophoretic mobility. The similarity in electrophoretic mobility and enzymatic properties between the lactate dehydrogenases of *M. mycoides* subsp. *mycoides* and mycoplasma UM 30847 parallels the similarity in G + C content and growth properties of these organisms and supports the inclusion of mycoplasma UM 30847 in the biotype composed of *M. mycoides* subsp. *mycoides* and its relatives. It remains to be determined whether the lactate dehydrogenases of different strains of other species of mycoplasmas also have common electrophoretic mobilities. Intraspecific uniformity of electrophoretic mobility of lactate dehydrogenases does occur among lactobacilli and also *Leuconostoc* species where electrophoretic uniformity can extend to different species showing that certain lactate dehydrogenases have remained stable during evolution.

*Acholeplasma laidlawii* was the only mycoplasma in the study found to possess two distinct lactate dehydrogenases. The FDP-activated NAD-dependent LDH must be responsible for the accumulation of L(+)-lactate during growth. The relatively less active NAD-independent D(−)-lactic dehydrogenase is similar to enzymes occurring in several species of bacteria. The function of the NAD-independent D(−)-lactate dehydrogenase in *A. laidlawii* is unknown. Snoswell (1966) suggested that such enzymes probably do not function in the production of lactic acid in lactobacilli. In *Escherichia coli* an NAD-independent D(−)-lactate dehydrogenase functions in membrane transport.

### B. Fructose Diphosphate-Activated Lactate Dehydrogenase from Acholeplasmas and Streptococci

The finding that the NAD-dependent lactate dehydrogenase (LDH) from *A. laidlawii* is activated by low concentrations of fructose-1,6-diphosphate is particularly significant. The use of FDP was prompted by the discovery by Wolin (1964) that the lactate dehydrogenases from several species of streptococci are specifically activated by FDP. This highly specific activation is unusual and previously was known to occur only within the Lactobacillaceae, specifically in streptococci, where it has been found in every streptococcal group that has been examined (Brown and Wittenberger, 1972; Jonas *et al.*, 1972; Wittenberger and Angelo, 1970; Neimark, 1973), and in two *Lactobacillus* species, *L. bifidus* and *casei*. However, the lactate dehydrogenases from *L. bifidus* (de Vries and Stouthamer, 1968) and *L. casei* (de Vries *et al.*, 1970) require $Mn^{2+}$, in addition to FDP, for activation and thus differ from both the streptococcal and the *A. laidlawii* FDP-activated LDHs. Furthermore, the carbohydrate metabolism of *L. bifidus* differs considerably from that of the streptococci and *A. laidlawii* (Castrejon-Diez *et al.*, 1963) in that it lacks key enzymes of the Embden–Meyerhof and hexose monophosphate pathways and catabolizes glucose by the fructose-6-phosphate phosphoketolase route (de Vries and Stouthamer, 1968).

It was suggested that the FDP-activated LDHs of *A. laidlawii* and the lactic acid bacteria have molecular similarities, since all must have binding sites for the three ligands, pyruvate, FDP, and NADH. We concur with Wolin's suggestion (1964) that FDP acts as an allosteric activator of the LDH and that FDP activation could be a significant metabolic control mechanism. This type of lactate dehydrogenase is particularly well suited for studying evolutionary relations because it is essential for energy metabolism and has a regulatory function. The regulated functional site could confer additional constraint on amino acid substitution during evolution and result in relative conservation of primary structure (Smith, 1970).

The occurrence of such strikingly similar enzymes at such an important regulatory junction could result only from one of two possibilities: Either a true phylogenetic relationship exists between these sterol nonrequiring mycoplasmas and the streptococci or the similarity between their enzymes is an exceptionally interesting example of convergent evolution to identical molecular function and regulation.

Because of the similarity of these enzymes, a detailed comparative biochemical and immunochemical study was carried out. The properties of the partially purified FDP activated LDH from *A. laidlawii* were examined in detail (Neimark and Tung, 1973). Briefly, the results show that the FDP-activated LDH from *A. laidlawii* generally resembles the LDHs from streptococci. The enzyme is activated specifically by low concentrations of fructose-1,6-diphosphate and the kinetic response to FDP is hyperbolic. A number of other substrates and glycolytic intermediates fail to substitute for FDP. The enzyme is inhibited by inorganic phosphate, adenosine triphosphate, and high concentrations of reduced NAD (NADH). Low activity is demonstrable in the absence of FDP at pH 6.0–7.2, but FDP is absolutely required in the region of pH 8. FDP causes an upward shift in the optimum pH of the enzyme, which is near 7.2. Activation of the enzyme by FDP is markedly affected by substrate concentration; FDP lowers the apparent $K_m$ for pyruvate and NADH. The affinity of the enzyme for pyruvate is also influenced by $H^+$ concentration. The pyruvate analog $\alpha$-ketobutyrate serves as an effective substrate for the enzyme; when $\alpha$-ketobutyrate is utilized, the enzyme is still activated by FDP. Reversal of the pyruvate reduction reaction catalyzed by the enzyme can be demonstrated with the 3-acetylpyridine analog of NAD. Thus, the *A. laidlawii* LDH appears to function essentially as a pyruvate reductase. Its activity is influenced by FDP, pyruvate, $H^+$, and NADH concentration and is inhibited by ATP and inorganic phosphate. These properties suggest that this FDP-activated LDH, by controlling pyruvate metabolism, has an important role in regulating the carbohydrate metabolism of *A. laidlawii*.

All of the FDP-activated LDHs are similar in being specific for L(+)-lactate and NADH, but the streptococcal LDHs may differ from one another in specific enzymatic properties, such as concentration of FDP required for maximal activation, kinetic response to pyruvate, occurrence of activity in the absence of FDP, inhibition by inorganic phosphate or ATP, optimal pH, ability to oxidize lactate, and effect of activator, substrates, and inhibitors on heat inactivation of the enzyme.

The FDP-activated LDHs from *A. laidlawii* and the streptococci share one or more specific characters. For example, the enzymes from *A. laidlawii, Streptococcus mutans,* and *Streptococcus cremoris* are all inhibited by ATP, and in common with *Streptococcus bovis* and *Streptococcus cremoris,* the *A. laidlawii* enzyme can display some activity in the absence of FDP. The *A. laidlawii* LDH is unstable during various purification procedures and shares this property with the *S. bovis* LDH; both enzymes are also similar in catalyzing the essentially unidirectional reduction of pyruvate. At first, the *A. laidlawii* LDH was considered unique in being inhibited by inorganic phosphate, but later this property was found to be characteristic of the enzyme from *S. cremoris* (Jonas *et al.*, 1972). Although the *A. laidlawii* and *S. cremoris* LDHs share some properties, they differ in others, for example, in ability to oxidize lactate and in optimum pH.

Subsequently, FDP-activated LDHs were demonstrated and characterized in the remaining known acholeplasmas: *A. granularum, A. axanthum, A. modicum,* and *A. oculi* (Neimark, 1973). These acholeplasma FDP-activated LDHs also generally resemble the streptococcal LDHs and each shares one or more specific characters.

Thus, possession of an FDP-activated L(+)-lactate dehydrogenase and accumulation of L(+)-lactate acid join the properties of sterol-independent growth and possession of a relatively large genome as defining characteristics of this cluster of mycoplasmas.

## C. Immunologic Relatedness between Aldolases from Lactic Acid Bacteria and Acholeplasmas

On the basis of the early findings, I began the purification of a streptococcal LDH for use as an antigen so that an immunologic comparison between the acholeplasma and streptococcal FDP-activated LDHs could be carried out. If these two groups are indeed related, their FDP-activated LDHs would be expected to share a degree of similarity in primary structure which could be detected by immunologic procedures. At that time Dr. Jack London (National Dental Institute, National Institutes of Health) informed me that he had prepared an antiserum against the FDP-aldolase from *Streptococcus faecalis.* The specificity of this antiserum is certified by the fact that it reacts only with aldolases from streptococci, other lactic acid bacteria, and closely related allies. Upon immunoelectrophoresis it produces a single precipitin line either with purified aldolase or with crude cell extracts. Also, the antiserum does not react with crude cell extracts from heterofermentative lactic acid bacteria which lack aldolases or with extracts of a variety of other bacteria (London and Kline, 1973). It appeared that this antialdolase serum could fulfill the purpose of the anti-LDH serum I intended to prepare, and Dr. London kindly agreed to provide me with a sample of this antialdolase serum.

The rationale for applying immunologic techniques to measure amino acid sequence homology has been discussed by Prager and Wilson (1971) and Wilson *et al.* (1977). The interpretation of immunodiffusion precipitin patterns which occur in comparative immunologic analysis using antiserum against a pure protein have been discussed in detail by Gasser and Gasser (1971) and by London and Kline (1973). A brief description of precipitin patterns and their significance will simplify the presentation here of the data obtained with the antialdolase serum. In interpreting immunogel precipitin reactions utilizing this antiserum, it should be obvious that since the antistreptococcal aldolase serum was prepared against the *S. faecalis* aldolase, this enzyme becomes the point of reference for all comparisons. Three basic precipitin patterns can occur: (1) The reference protein and unknown protein produce a fused precipitin line; this reaction indicates that the two proteins possess the same antigenic determinants, and they are described as being immunologically identical. However, it is also possible for two or more proteins not immunologically identical to the reference antigen to produce fused precipitin lines; this occurs when a protein bears additional antigenic determinants not recognized by the antiserum. Such proteins have apparent immunologic identity. (2) A single-spur precipitin band is produced when one of the two antigens being compared possesses more antigenic determinant groups in common with the reference protein than its neighbor antigen. This reaction of partial identity usually occurs when two or more proteins have evolved sequentially in a unidirectional fashion. (3) The double-spur reaction occurs when two proteins share a number of determinant groups with the reference protein but very few or none with each other. This pattern of nonidentity results when two proteins have evolved in a random or divergent fashion from the reference protein.

Additional information and corroboration of immunodiffusion results can be obtained by microcomplement fixation measurements which estimate the degree of amino acid sequence resemblance between two proteins (Prager and Wilson, 1971). This method is very sensitive to small differences in protein amino acid sequence, and even single amino acid substitutions can be detected (Wilson *et al.*, 1977). The method has been calibrated to relate immunologic distance, measured by microcomplement fixation, and percentage of sites at which amino acid substitutions have been made and is generally useful for comparing proteins that differ in amino acid sequence over the range of 0–30% (Wilson *et al.*, 1977).

Cell extracts from all five *Acholeplasma* species produced strong immunodiffusion precipitin bands with the *Streptococcus* antialdolase serum. These immunodiffusion reactions demonstrate that a high degree of immunologic homology exists between the aldolases of the acholeplasmas and *S. faecalis* and, furthermore, indicate that all the acholeplasmas possess aldolases that are immunologically related to one another (Neimark, 1974a; in preparation). No immunoprecipitin reactions were obtained with extracts from any of several sterol-requiring mycoplasmas.

Fused bands of identity were obtained between *A. laidlawii* and *A. granularum;* this reaction provides independent confirmation of the close relationship between these two acholeplasmas that was previously shown by DNA–DNA hybridization (Pollock and Bonner, 1969; Neimark, 1969, 1970).

*Acholeplasma laidlawii* and *A. granularum* produced single-spur reactions over the other three acholeplasmas. On the basis of multiple cross-reactions, the sequential order of decreasing antigenic similarity to the *S. faecalis* aldolase reference appears to be: *A. laidlawii* = *A. granularum* > *A. modicum* > *A. oculi* > *A. axanthum*. This order is provisional because double-spur cross-reactions were observed between *A. modicum* (PG 49) and *A. axanthum* (S-410). This latter reaction indicates that antigenic divergence has occurred in the aldolases of these acholeplasmas (Neimark, 1974a).

Cell extracts from acholeplasmas and streptococci representing different Lancefield groups and *Pneumococcus* type 1 were compared to establish the location of the acholeplasma aldolases within the evolutionary sequence of the lactic acid bacteria aldolases. The resulting immunodiffusion reactions showed that single-spur precipitates are usually produced when each of the acholeplasma extracts are compared to extracts of streptococci. For example, all of the acholeplasma extracts produce single spurs over *S. pneumoniae* extracts, while *S. cremoris* (Lancefield group N) produces spurs against the aldolases of most of the acholeplasmas, but it is itself spurred over by the aldolase of *A. laidlawii*. These reactions, supported by others not described here, indicate that the acholeplasma aldolases have an evolutionary location between *S. faecalis* and pneumococcus and that they probably are located near the group N streptococcal aldolases (Neimark, 1974a).

These results have been essentially confirmed by quantitative microcomplement fixation measurements carried out by London on enzyme extracts supplied by the author (J. London, personal communication).

## D. Acholeplasmas Are Phylogenetically Related to the Lactobacillaceae

The inescapable conclusion that must be drawn from these enzyme and protein homology studies is that the acholeplasmas are related by evolution to the Lactobacillaceae, most probably to the streptococci.

Could the close similarity of the FDP-activated LDHs and the immunologic homology in aldolases demonstrated between the acholeplasmas and the streptococci have arisen by convergent evolution? Although in principal similarity could arise through convergent processes (and has in the case of many physiologic and morphologic features), no example of analogous enzyme proteins with extensive homologous sequences has yet been demonstrated (Smith, 1970; Wu *et al.*, 1974; Dickerson *et al.*, 1976). In any case, it seems inconceivable that the high degree of homology observed between the aldolases (representing at least 60% amino acid sequence homology) could have arisen by convergent evolution from unrelated genes. Similarly, this degree of homology could not result from unrelated enzymes sharing common immunologic determinants that might have developed through convergent evolution at the enzyme active site. These results stand by themselves, but they have been confirmed by reactions with a second pure protein reference antiserum directed against glyceraldehyde-3-phosphate dehydrogenase (J. London, personal communication). The conclusion that the acholeplasmas have a close evolutionary relationship to the lactic acid bacteria thus seems unassailable.

At this point it is interesting to observe that information on mycoplasma lipid composition are in accord with these results. Shaw and Baddiley (1968) showed that grouping bacteria on the basis of the structures of their glycolipids results in a taxonomic scheme basically very similar to that obtained by traditional methods, and Shaw (1974) pointed out that the lipid composition of those mycoplasmas that have been examined is akin to that of gram-positive bacteria. Of particular interest here is the finding that lipids from *A. laidlawii* closely resemble or are identical to those of streptococci (Smith *et al.*, 1973; this volume, Chapter 9).

### E. Acholeplasmas Evolved from Streptococci

This demonstration of a close phylogenetic relationship between acholeplasmas and the streptococci allows us to raise the fundamental question: What is the nature of the evolutionary relationship that exists between the acholeplasmas and lactic acid bacteria? There are only two possibilities: Either (1) the wall-less acholeplasmas with their relatively small genomes are the evolutionary ancestors (or surviving descendents of the ancestors) of the lactic acid bacteria, or (2) the acholeplasmas have descended from streptococci.

The immunologic evidence that demonstrates the phylogenetic relationship between these two groups indicates also that the acholeplasma and streptococcal aldolases share a common sequential evolutionary path. This evidence can be used to determine the direction of this sequential evolution from the position of the acholeplasma aldolases relative to the reference aldolase on the phylogenetic map constructed by London and Kline (1973) and London and Chace (1976). This map shows seven major lines of evolution corresponding to seven genera of the Lactobacillaceae; the lines diverge or radiate from a single point where they are related through some common ancestor.

This evolutionary map was produced from immunologic studies on the properties of one set of enzymes, the FDP aldolases, and reflects the evolutionary history of these enzymes. Other functionally similar enzymes would be expected to exhibit a similar rate of evolution (Wilson *et al.*, 1977), and indeed the ordering found in the genus *Lactobacillus* utilizing the FDP aldolase antiserum is roughly similar to that described earlier by Gasser and Gasser (1971) using antisera against lactate dehydrogenases.

All the available evidence indicates that the acholeplasmas are distal on the map to the *S. faecalis* reference, which in turn is distal to the common junction point. Consequently, the acholeplasmas appear to have diverged from the streptococci and appear not to be ancestral to the lactic acid bacteria.

Although the general linear location of the acholeplasmas on the map appears to be correct, the precise fixation of their map position in space will require a second pure protein reference antiserum.

## IV. EVOLUTIONARY IMPLICATIONS

How did the acholeplasmas arise? The demonstration that acholeplasmas are derived from streptococci immediately brings to mind the possibility that streptococcal L forms were involved in the development of acholeplasmas, but the acholeplasmas must have evolved through a far more complex process than that involved in the formation of the simple L forms we know.

Aside from the absence of a cell wall, it is evident that acholeplasmas are substantially different from the streptococci. Genome size values available for a few streptococci range from about $1.2 \times 10^9$ to $1.47 \times 10^9$ daltons (Bak *et al.*, 1970), and acholeplasma genomes are approximately 15–30% smaller than these streptococci. Interestingly, one stable L form of *Streptococcus faecalis* is known to have lost some 4–6% of DNA sequences during the transition from its parent (Hoyer and King, 1969). This loss could have occurred as a single massive deletion or through a series of deletions. Possibly, too, the average DNA base composition of acholeplasmas may have shifted down somewhat in G + C content. Also, the transfer RNAs (tRNA) of acholeplasmas (Feldmann and Falter, 1971) as well as mycoplasmas (Johnson and Horowitz, 1971; Kimball *et al.*, 1974) are known to contain fewer modified nucleotides than *E. coli* tRNA, but in many other respects the tRNAs of acholeplasmas and mycoplasmas are very similar to those of *E. coli* (Chapter 6).

Another important class of evolutionarily conserved molecules are ribosomal ribonucleic acids (Pace, 1973). Recently, Reff *et al.* (1977) examined migration rates of ribosomal RNAs (rRNA) from bacteria (including *S. faecalis* and an L form), acholeplasmas, and mycoplasmas in nondenaturing and denaturing polyacrylamide gels.

In nondenaturing gels all bacterial RNAs examined, except *Bacillus subtilis,* migrated identically. Acholeplasma 23 S RNA migrated slightly slower, while *B. subtilis* 23 S RNA migrated faster than the other bacterial 23 S RNAs. Acholeplasma and bacterial 16 S RNA had identical migration patterns while both mycoplasma RNAs migrated in a pattern distinct from both acholeplasmas and bacteria.

Under denaturing conditions the 23 S RNAs from all organisms examined comigrated, but the 16 S RNAs of acholeplasmas and mycoplasmas comigrated at a slightly faster rate than bacterial 16 S RNA. This small but regular difference in 16 S RNAs was taken to be due to a decrease in molecular weight estimated at 12,000 daltons (about 2% of the total molecular weight of 16 S RNA). This shared difference in 16 S RNA seems to link the acholeplasmas and mycoplasmas together. The authors suggest that the difference in 16 S RNA of mycoplasmas and acholeplasmas compared to bacteria seems to indicate that a significant evolutionary gap exists between these organisms and bacteria. I would suggest further, based on the evidence that acholeplasmas evolved from streptococci, that the difference in 16 S RNA arose in acholeplasmas during the course of evolution from streptococci. It will be very interesting to know if this difference in 16 S RNA is a universal property of the mycoplasmas, but in any case this property seems to join with differences in tRNA, G + C content, and genome size as clues in the puzzle of how acholeplasmas evolved.

The demonstration that acholeplasmas evolved from streptococci raises questions concerning the status of the remaining Mycoplasmatales (Neimark, 1974b). It also poses a problem in the classification of acholeplasmas, but this question is beyond the scope of this chapter.

What is the relationship between acholeplasmas and mycoplasmas? Conceivably, acholeplasmas could be related phylogenetically to certain lactate accumulating fermentative mycoplasmas in the same way that streptococci are related to other members of the Lactobacillaceae, but there is little evidence to support this notion as yet.

The remainder of the mycoplasmas would seem not to have any close relationship to acholeplasmas. Aside from physiologic differences, there is the gulf between these organisms caused by the gap in their genome sizes. In spite of this gap in genome size, the mycoplasmas and acholeplasmas seem to be connected to one another by shared properties in their RNAs. Could these differences in RNA be a commonly acquired property resulting from separate evolutionary histories? In considering this question it may be useful to mention the findings of Shine and Dalgarno (1975). They found that all bacterial 16 S RNAs examined can be divided into two groups based on homology. This division correlated with a separation of bacterial ribosomes into two catagories based on ability to translate different messenger RNA preparations. Further, this division seems to correspond partially, although not completely, with the gram-staining properties of the bacteria.

Fortunately, powerful new biochemical tools for examining phylogenetic relationships have become available. Various workers, including Woese and colleagues (Fox *et al.*, 1977) and Hori (1975), have described the value of using 5 S and 16 S RNAs for tracing both distant and close phylogenetic relationships. The procedure of Woese and coworkers involves digesting RNA with an endonuclease and determining the sequence of the resulting oligonucleotides. The method has been used to characterize over 40 bacterial species, including a few mycoplasmas and acholeplasmas, and should be useful for examining relations among the mycoplasmas and other prokaryotes. Indeed, very recent studies on mycoplasma 16 S rRNAs (Maniloff *et al.*, 1978) indicate that mycoplasmas appear to be related to a specific clostridial subline and this may be the same "subline that earlier branched to produce a common ancestor of *Bacillus* and *Lactobacillus*." The results of this study and one on tRNA (Walker, 1978) are consistent with my analysis that mycoplasmas are not descendants of primitive prokaryotes ancestral to bacteria (Neimark, 1974b). Schwartz and Dayhoff (1978) have combined recently available sequence information from bacterial proteins and 5 S ribosomal RNAs to construct a composite evolutionary tree for bacteria. The combination of information from several types of sequences provides the broad perspective required for fully examining relations among the mycoplasmas and other prokaryotes. A new method for estimating DNA sequence divergence (Upholt, 1977) from restriction nuclease digests of DNA may also prove to be of aid.

Thus we have every hope that we are on the verge of solving the puzzle of the relationship of mycoplasmas to bacteria. The solution to this problem will have a profound impact on our perception of the nature of all bacteria, for we probably will uncover a broad capacity in bacteria to undergo enormous alterations.

## REFERENCES

Allen, T. C. (1971). *J. Gen. Microbiol.* **69,** 285–286.

Askaa, G., Christiansen, C., and Ernø, H. (1973). *J. Gen. Microbiol.* **75,** 283–286.

Bak, A. L., and Black, F. T. (1968). *Nature (London)* **219,** 1044–1045.
Bak, A. L., Black, F. T., Christiansen, C., and Freundt, E. A. (1969). *Nature (London)* **224,** 1209–1210.
Bak, L., Christiansen, C., and Stenderup, A. (1970). *J. Gen. Microbiol.* **64,** 377–380.
Bode, H. R., and Morowitz, H. J. (1967). *J. Mol. Biol.* **23,** 191–199.
Bredt, W. (1973). *Ann. N.Y. Acad. Sci.* **225,** 246–250.
Brennan, P. C., Fritz, T. E., and Flynn, R. J. (1969). *J. Bacteriol.* **97,** 337–349.
Brown, A. T., and Wittenberger, C. L. (1972). *J. Bacteriol.* **110,** 604–615.
Castrejon-Diez, J., Fisher, T. N., and Fisher, E., Jr. (1963). *J. Bacteriol.* **86,** 627–636.
Chelton, E. T. J., Jones, A. S., and Walker, R. T. (1968). *J. Gen. Microbiol.* **50,** 305–312.
de Vries, W., and Stouthamer, A. H. (1968). *J. Bacteriol.* **96,** 472–478.
de Vries, W., Kapteijn, W. M. C., Vander Beek, E. A., and Stouthamer, A. H. (1970). *J. Gen. Microbiol.* **63,** 333–345.
Dickerson, R. E., Timkovich, R., and Almassy, R. J. (1976). *J. Mol. Biol.* **100,** 473–491.
Dienes, L. (1963). *Recent Prog. Microbiol.* **8,** 511–517.
Dienes, L., and Weinberger, H. J. (1951). *Bacteriol. Rev.* **15,** 245–288.
Dowell, V. R., Jr., Loper, J. C., and Hill, E. O. (1964). *J. Bacteriol.* **88,** 1805–1807.
Edward, D. G. ff. (1954). *J. Gen. Microbiol.* **10,** 27–64.
Edward, D. G. ff. (1960). *Ann. N.Y. Acad. Sci.* **79,** 308–311.
Feldmann, H., and Falter, H. (1971). *Eur. J. Biochem.* **18,** 573–581.
Fox, G. E., Pechman, K. R., and Woese, C. R. (1977). *Int. J. Syst. Bacteriol.* **27,** 44–57.
Freundt, E. A. (1957). *In* "Bergey's Manual of Determinative Bacteriology" (R. S. Breed, E. G. D. Murray, and N. R. Smith, eds.), 7th ed., pp. 914–927. Williams & Wilkins, Baltimore, Maryland.
Gasser, F., and Gasser, C. (1971). *J. Bacteriol.* **106,** 113–125.
Gourlay, R. N., and Leach, R. H. (1970). *J. Med. Microbiol.* **3,** 111–123.
Haller, G. J., and Lynn, R. J. (1969). *Bacteriol. Proc.* p. 33.
Hori, H. (1975). *J. Mol. Evol.* **7,** 75–86.
Hoyer, B. H., and King, J. R. (1969). *J. Bacteriol.* **97,** 1516–1517.
Johnson, J. D., and Horowitz, J. (1971). *Biochim. Biophys. Acta* **247,** 262–279.
Jonas, H. A., Anders, R. F., and Jago, G. R. (1972). *J. Bacteriol.* **111,** 397–403.
Jones, A. S., and Walker, R. T. (1963). *Nature (London)* **198,** 588–589.
Jones, A. S., Tittenser, J. R., and Walker, R. T. (1965). *J. Gen. Microbiol.* **40,** 405–411.
Kelton, W. H., and Mandel, M. (1969). *J. Gen. Microbiol.* **56,** 131–135.
Kenny, G. E. (1969). *J. Bacteriol.* **98,** 1044–1055.
Kimball, M. E., Szeto, K. S., and Söll, D. (1974). *Nucleic Acids Res.* **1,** 1721–1732.
Kingsbury, D. T. (1969). *J. Bacteriol.* **98,** 1400–1401.
Klieneberger-Nobel, E. (1960). *Ann. N.Y. Acad. Sci.* **79,** 483–484.
London, J., and Chace, N. M. (1976). *Arch. Microbiol.* **110,** 121–128.
London, J., and Kline, K. (1973). *Bacteriol. Rev.* **37,** 453–478.
Lynn, R. J., and Haller, G. J. (1968). *Antonie van Leeuwenhoek* **34,** 249–256.
McGee, Z. A., Rogul, M., Falkow, S., and Wittler, R. G. (1965). *Proc. Natl. Acad. Sci. U.S.A.* **54,** 457–461.
McGee, Z. A., Rogul, M., and Wittler, R. G. (1967). *Ann. N.Y. Acad. Sci.* **143,** 21–30.
Maniloff, J., Magrum, L., Zablen, L. B., Woese, C. R. (1978). *Zentralbl. Bakteriol., Parasitenkd., Infektionskr. Hyg., Abt. 1: Orig., Reihe A* **241,** 171–172. (Abstr. trans.)
Morowitz, H. J., Bode, H. R., and Kirk, R. G. (1967). *Ann. N.Y. Acad. Sci.* **143,** 110–114.
Neimark, H. (1967). *Ann. N.Y. Acad. Sci.* **143,** 31–37.
Neimark, H. (1969). *Bacteriol. Proc.* p. 32.
Neimark, H. (1970). *J. Gen. Microbiol.* **63,** 249–263.

Neimark, H. (1973). *Ann. N.Y. Acad. Sci.* **225,** 14–21.
Neimark, H. (1974a). *Abstr. Annu. Meet., Am. Soc. Microbiol.* p. 108.
Neimark, H. (1974b). *Colloq. Inst. Natl. Sante Rech. Med.* **33,** 71–78.
Neimark, H., and Lemcke, R. M. (1972). *J. Bacteriol.* **111,** 633–640.
Neimark, H. C., and Pene, J. J. (1965a). *Proc. Soc. Exp. Biol. Med.* **118,** 517–519.
Neimark, H. C., and Pene, J. J. (1965b). *Bacteriol. Proc.* p. 58.
Neimark, H. C., and Pickett, M. J. (1960). *Ann. N.Y. Acad. Sci.* **79,** 531–536.
Neimark, H., and Tung, M. C. (1973). *J. Bacteriol.* **114,** 1025–1033.
Pace, N. R. (1973). *Bacteriol. Rev.* **37,** 562–603.
Panos, C. (1965). *J. Gen. Microbiol.* **39,** 131–138.
Peterson, A. M., and Pollock, M. E. (1969). *J. Bacteriol.* **99,** 639–644.
Pollock, M. E., and Bonnor, S. V. (1969). *Bacteriol. Proc.* p. 32.
Prager, E. M., and Wilson, A. C. (1971). *J. Biol. Chem.* **246,** 7010–7017.
Reff, M. E., Stanbridge, E. J., and Schneider, E. L. (1977). *Int. J. Syst. Biol.* **27,** 185–193.
Reich, P. R., Somerson, N. L., Hybner, C. J., Chanock, R. M., and Weissman, S. M. (1966). *J. Bacteriol.* **92,** 302–310.
Rodwell, A. W. (1960). *Ann. N.Y. Acad. Sci.* **79,** 499–507.
Rodwell, A. W., and Rodwell, E. S. (1954). *Aust. J. Biol. Sci.* **7,** 18–30.
Rogul, M., McGee, Z. A., Wittler, R. G., and Falkow, S. (1965). *J. Bacteriol.* **90,** 1200–1204.
Ryan, J. L., and Morowitz, H. J. (1969). *Proc. Natl. Acad. Sci.* U.S.A. **63,** 1282–1289.
Sarov, I., and Becker, Y. (1969). *J. Mol. Biol.* **42,** 581–589.
Schwartz, R. M., and Dayhoff, M. O. (1978). *Science* **199,** 395–403.
Shaw, N. (1974). *Adv. Appl. Microbiol.* **17,** 63–108.
Shaw, N., and Baddiley, J. (1968). *Nature (London)* **217,** 142–144.
Shine, J., and Dalgarno, L. (1975). *Eur. J. Biochem.* **57,** 221–230.
Smith, E. L. (1970). *In* "The Enzymes" (P. D. Boyer, ed.), 3rd ed., Vol. 1, pp. 267–339. Academic Press, New York.
Smith, P. F., Van Demark, J., and Fabricant, J. (1963). *J. Bacteriol.* **86,** 893–897.
Smith, P. F., Langworthy, T. A., and Mayberry, W. R. (1973). *Ann. N.Y. Acad. Sci.* **225,** 22–27.
Snoswell, A. N. (1966). *In* "Methods in Enzymology" (W. A. Wood, ed.), Vol. 9, pp. 321–327. Academic Press, New York.
Somerson, N. L., Reich, P. R., Walls, B. E., Chanock, R. M., and Weissman, S. M. (1966). *J. Bacteriol.* **92,** 311–317.
Somerson, N. L., Reich, P. R., Chanock, R. M., and Weissman, S. M. (1967). *Ann. N.Y. Acad. Sci.* **143,** 9–20.
Sparrow, A. H., and Nauman, A. F. (1976). *Science* **192,** 523–530.
Sueoka, N. (1961). *J. Mol. Biol.* **3,** 31–40.
Teplitz, M. (1977). *Nucleic Acids Res.* **4,** 1505–1512.
Tourtellote, M. E., and Jacobs, R. E. (1960). *Ann. N.Y. Acad. Sci.* **79,** 521–530.
Upholt, W. B. (1977). *Nucleic Acids Res.* **4,** 1257–1265.
Van Demark, P. J. (1967). *Ann. N.Y. Acad. Sci.* **143,** 77–84.
Walker, R. T. (1978). *Zentralbl. Bakteriol., Parasitenkd., Infektionskr. Hyg., Abt. 1: Orig., Reihe A* **241,** 170–171.
Wallace, D. C., and Morowitz, H. J. (1973). *Chromosoma* **40,** 121–126.
Williams, C. O., Wittler, R. G., and Burris, C. (1969). *J. Bacteriol.* **99,** 341–343.
Wilson, A. C., Carlson, S. S., and White, T. J. (1977). *Annu. Rev. Biochem.* **46,** 573–639.
Wittenberger, C. L., and Angelo, N. (1970). *J. Bacteriol.* **101,** 717–724.
Wolin, M. J. (1964). *Science* **146,** 775–777.
Wu, T. T., Fitch, W. M., and Margoliash, E. (1974). *Annu. Rev. Biochem.* **43,** 539–566.

# 3 / MORPHOLOGY AND ULTRASTRUCTURE OF THE MYCOPLASMATALES

*Edwin S. Boatman*



*Never, in the field of human endeavor have so few species caused so much trouble for so many.*

(With apologies to Sir Winston Churchill)

## I. INTRODUCTION

Owing to their small size and to a marked tendency toward pleomorphism due to the absence of a cell wall, the mycoplasmas have presented an intriguing challenge to microscopists for over 60 years. Even with the advent of the electron microscope and other related instruments, employing an electron source for forming a high-resolution image, the challenge is still very much in evidence. The problem resides not only in morphological changes wrought by methods of preparation and examination, but also in the interpretation of the relationship of structure and function by the observer. In recent years, a variety of techniques have been utilized to



THE MYCOPLASMAS, VOL. 1

ISBN 0-12-078401-7

examine these unusual prokaryotes and the purpose of this chapter is to highlight these findings and to speculate on problems and avenues of approach for the future.

As might be expected, a number of species have been more rigorously examined than others, particularly the animal species; nevertheless, there is information to permit some discussion of various genera in terms of their morphological and ultrastructural characteristics.

By "morphology" is meant the three-dimensional shape of the cells, and "ultrastructure" as a generalization refers to the external or internal structure of cells frozen or chemically fixed at a point in time and observed by microscopes capable of a resolution between 10 nm and about 0.2 nm. Fixation, resolution, and observation (i.e., selection of fields) comprise the foundation upon which statements about morphology and ultrastructure are made.

In discussing the overall structure of the mycoplasmas one has the choice of (1) detailing the morphology and ultrastructure as it stands at present, or (2) discussing the difficulties inherent in trying to determine their structure! For the most part, the text of this chapter will consist of a mixture of the two.

## II. MORPHOLOGY AS DETERMINED BY LIGHT MICROSCOPY

With organisms as sensitive to pressure or osmotic effects as the mycoplasmas it is questionable whether these organisms can be observed free from the effects of specimen preparation. In spite of this, few would disagree about the need for observing these organisms in the living state, either in culture as a prelude to observation by electron microscopy or as single cells for following the pattern of replication.

Optical systems proven to be most useful are phase-contrast and dark-field illumination. Because the average dimensions of single cells are close to the limit of resolution of the light microscope (0.2 $\mu$m) adherence to the best principles of microscope practice is essential. Dark-field illumination in conjunction with ultraviolet light in the visible range ensures the best possible resolution as shown for *Mycoplasma pneumoniae* in Fig. 1.

Bredt and his associates (1969, 1970, 1973) have done outstanding work with the use of cinematography and phase-contrast systems to determine the growth and multiplication of mycoplasmas in coverslip chambers and to detect motile forms (see Chapter 5). They observed multilobular and filamentous forms in *Mycoplasma orale* and *Mycoplasma hominis* but not in *M. buccale, Mycoplasma salivarium,* and *Mycoplasma*

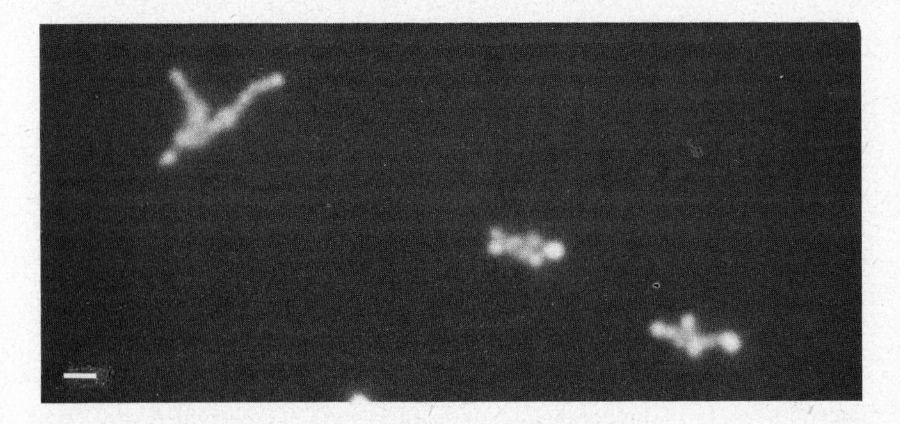

FIGURE 1. Light micrograph of multilobed organisms from a 6-day fluid culture of *M. pneumoniae* (unfixed). Dark field, uv illumination. ×4400. Marker = 1.0 μm. (Photograph by E. S. Boatman.)

*fermentans* grown and observed under identical conditions. In 10 strains of *M. hominis* three basic morphologic forms were seen: (1) coccoidal cells 0.3–0.8 μm in diameter, (2) diploforms, and (3) filamentous forms with filaments 0.3–0.4 μm thick and up to more than 40 μm long. One strain formed filaments up to 98 μm in length which fragmented into coccoid cells after about 100 min. Cinematography of single cells showed cell division occurring by binary fission, by fragmentation of filaments and rings, and by a budding process. Most surprisingly, there was at times a reversal of movement to a more condensed form and then back again to an elongated form. Hubbard and Kite (1971) observed the growth of *M. pneumoniae* and found numerous filamentous forms within 2 days and Bredt *et al.* (1975) by light microscopy found that the typical shape of *M. pneumoniae* cells growing on glass could be used as a key for the rapid detection and identification of this organism in certain clinical material. In this respect, *M. pneumoniae* is generally considered to be one of the few morphologically predictable mycoplasmas along with *Mycoplasma gallisepticum* and the type species *Mycoplasma mycoides* subsp. *mycoides*, although the pleomorphism of the latter has been questioned by Furness and DeMaggio (1973) and Furness *et al.* (1975), who concluded that filaments are a characteristic but not essential development in the growth cycle of this organism.

Interestingly, Rodwell (1965) found that the morphology of *M. mycoides* subsp. *mycoides* was substantially altered when the cells were transferred from hypertonic to hypotonic solutions and that the change in shape was reversible from filamentous to round forms.

In a study on the morphology of uncentrifuged broth cultures of *Ureaplasma urealyticum* by phase-contrast microscopy, Razin *et al.* (1977) found that filtration of the growth medium before inoculation was essential in order to remove particulate matter which could resemble *Ureaplasma* organisms. The majority of organisms were in the form of single cells, 0.3–0.8 μm diameter, in pairs or short chains; filaments or organisms attached in long chains were not found. It is the opinion of the authors that filaments are formed only under conditions appropriate for rapid growth, i.e., when cytoplasmic division lags behind genome replication. It has been suggested by Robertson *et al.* (1975a) that the more aberrant forms of mycoplasmas arise in growth media which are "not favorable for growth." A distinction must always be made between aberrant forms (primarily bizarre shapes) and filaments of a reasonable length, shape, and size.

The morphology of 10 strains of *Ureaplasma* studied by dark-field, phase-contrast, and Nomarski interference optics (Whitescarver and Furness, 1975) showed unfixed organisms to be spherical of 0.25–1.0 μm in diameter and most often with a single bud but sometimes with two.

The gross morphology of the 46 species of mycoplasmas as defined in "Bergey's Manual of Determinative Bacteriology" (Freundt, 1974) and rearranged in respect to filament formation is listed in Table I.

It should be appreciated that listing the morphology of the mycoplasmas in this manner does not imply that the morphology as stated is representative of a particular organism under all conditions of culture. However, it does serve to indicate three possible groupings: (1) 7 species which may form long filaments during growth, (2) 17 species from various genera which are coccoidal to short filamentous, and (3) 11 species with variable shapes ranging from spherical, through coccobacillary, to short filamentous. The *Ureaplasmas* (T mycoplasmas) have been included based on the findings of Whitescarver and Furness (1975) and Razin *et al.* (1977).

The term "filament," although appropriate for forms greater than perhaps 5 μm in length, loses some force when applied to "filaments" less than 2 μm long, i.e., four to six times the average length of a single mycoplasma.

It should be kept in mind that the morphology of the organisms observed in "fixed" preparations may differ from that seen in the living state. This will depend on how the organisms are treated. For example, incorrect fixation will cause *S. citri* to lose its helical shape almost completely (Cole *et al.*, 1973).

TABLE I. **Members of the Class *Mollicutes* Listed according to Gross Morphology**[a]

| Organisms | Morphology |
|---|---|
| *M. mycoides* subsp. *mycoides* | Long repeatedly branching filaments[b] about 50 μm in length |
| *M. mycoides* subsp. *capri* | Filaments of moderate length; 15 μm |
| *M. neurolyticum* | Short to long filaments |
| *M. agalactiae* | Short to moderately long filaments |
| *M. arthritidis* | Short (2–5 μm) to moderately long filaments 10–30 μm |
| *M. orale* | Moderately long filaments 8–10 μm |
| *M. fermentans* | Short (1–5 μm) to moderately long 10–40 μm |
| *A. laidlawii* | Short (2–5 μm) to longer branched filaments |
| *M. pulmonis* | Short filaments—gliding motility |
| *M. pneumoniae* | Short filaments—gliding motility |
| *M. felis* | Coccobacillary to short filamentous or lobulated cells |
| *M. canis* | Short filaments, occasional branching short |
| *M. edwardii* | Short filaments |
| *M. flocculare* | Pleomorphic, branched filaments |
| *M. maculosum* | Short, filaments (2–5 μm) with occasional branching |
| *M. spumans* | Short filaments (2–5 μm) |
| *M. hyosynoviae* | Coccoidal to filamentous |
| *M. hominis* | Short (2–5 μm) and sometimes long filaments |
| *M. bovigenitalium* | Short filaments (2–5 μm) |
| *M. primatum* | Coccoidal to short filaments |
| *A. granularum* | Short filaments |
| *A. axanthum* | Coccoidal, coccobacillary and some short filaments |
| *T. acidophilum* | Spherical to filamentous |
| *S. citri* | Spherical to ovoid to helical and nonhelical filaments |
| *M. hyorhinis* | Very short filaments with occasional branching |
| *M. hyopneumoniae* | Coccoid to short filaments |
| *M. suipneumoniae* | Coccoid to short filaments |
| *M. gallinarum* | Filaments very short, almost bacillary |
| *M. salivarum* | Short filaments (0.6–1.0 μm), almost bacillary |
| *M. gallisepticum* | Ovoid with one or two terminal ends—gliding motility |
| *M. conjunctivae* | Spherical, ring-shaped, and coccobacillary forms |
| *M. cynos* | Coccoid to coccobacillary |
| *M. alkalescens* | Spherical, ring-shaped, and coccobacillary |
| *A. modicum* | Spherical, ring-shaped, and coccobacillary |
| *U. urealyticum* | Spherical to budding forms |
| Species poorly defined or not described | *M. bovirhinis*; *M. dispar*; *M. ovipneumoniae*; *M. anatis*; *M. synoviae*; *M. feliminutum*; *M. iners*; *M. meleagridis*[c]; *M. gateae*; *M. arginini*; *A. oculi* |

[a] Adapted from Freundt (1974).

[b] As defined in *Bergey's Manual*: Filament: any long threadlike form which may or may not be segmented.

[c] The morphology and ultrastructure of *M. meleagridis* was described by Green and Hanson (1973).

It can be summarized succinctly that what is observed by light microscopy in unfixed preparations of mycoplasmas depends upon four factors: (1) the phase of growth of the culture, (2) the viability of the population, (3) the method of slide preparations, and (4) the efficiency of the optical equipment.

Finally, of great importance is the experience of the observer. Problems of interpretation will undoubtably remain because of the morphological hetereogeneity of the mycoplasmas as a whole and of the inflexible limit of resolution imposed by the light microscope itself.

## III. MORPHOLOGY AND ULTRASTRUCTURE AS DETERMINED BY ELECTRON MICROSCOPY

Since mycoplasmas have to be "fixed" prior to even the most cursory examination by the electron microscope, three factors that have a major influence on the morphology and ultimately what is seen under the microscope are (1) the phase of growth and viability of the organisms at the time of harvest; (2) whether or not the organisms are fixed prior to collection, particularly if centrifugation is a prior step; and (3) the concentrations of the buffer-fixative employed and the osmolality of the final solution. These topics have been discussed in some detail in recent years by Lemcke (1972), Boatman (1973b), Furness *et al.* (1975), and Robertson *et al.* (1975a,b). As mentioned earlier, Rodwell (1965) found the morphology of *M. mycoides* to markedly change when the cells were transferred from hypertonic to hypotonic solutions and that the change in shape was reversible. Spears and Provost (1967) suspended cells of *Acholeplasma laidlawii* and *M. hominis* in solutions of different osmolalities and found that whereas *M. hominis* formed large spheres at low osmolalities and fine filaments at high, *A. laidlawii* was less profoundly affected, altering only the size of the cells. The difference between two solutions need not be large to effect a change in cell shape. When *A. laidlawii* and *M. orale* were fixed at 270 m0sm/kg $H_2O$ and observed in thin section, the organisms appeared coccoid or coccobacillary. If however, the tonicity was increased to 360 m0sm or above, cell profiles showed invagination, vacuolation, and flattening (Lemcke, 1972).

If possible, it is advisable to "pre-fix" *in situ* by the addition of fixative directly to the culture, to give a final concentration of between 0.1 and 0.25% (Robertson *et al.*, 1975b). In instances where organisms are "fixed" by the addition of a buffered fixative, the osmolality of this solution should be kept in the region of 290 to 350 m0sm. As shown in Table II, depending upon the concentration of both buffer and fixative,

TABLE II. **Osmolalities of Some Common Buffer-Fixative Combinations**

| Fixative conc. (%) | Fixative | Buffer | pH | Osmolality (mOsm/kg $H_2O$) |
|---|---|---|---|---|
| 6.25 | Glutaraldehyde | 0.1 *M* Na cacodylate | 7.4 | 1013 |
| 5.00 | Glutaraldehyde | 0.1 *M* Na cacodylate | 7.3 | 806 |
| 3.00 | Glutaraldehyde | 0.1 *M* Na cacodylate | 7.3 | 547 |
| 2.00 | Glutaraldehyde | 0.1 *M* Na cacodylate | 7.4 | 415 |
| 1.00 | Glutaraldehyde | 0.1 *M* Na cacodylate | 7.4 | 300 |
| 2.00 | Glutaraldehyde | 0.2 *M* Na cacodylate | 7.4 | 545 |
| 2.00 | Glutaraldehyde | 0.11 *M S*-collidine | 7.4 | 340 |
| 1.50 | Glutaraldehyde | 0.11 *M S*-collidine | 7.4 | 290 |
| 1.00 | Glutaraldehyde | 0.01 *M* phosphate | 7.5 | 300 |
| 1.00 | Osmium tetroxide | 0.1 *M* Na cacodylate | 7.3 | 190 |
| 1.00 | Osmium tetroxide | Veronal acetate (RK) | 6.1 | 335 |
| 4.00 | Formaldehyde | 0.1 *M* phosphate | 7.4 | 1705 |

the osmolalities may range from under 200 to over 1000 m0sm. The latter is markedly hypertonic and usually has disastrous effects on the morphology of the organisms, often producing dumbell-shaped organisms with suggestions of connecting tubular structures.

Although cell structures may be "fixed" by the cross-linking of proteins, etc., they are not necessarily rigid nor are their enzymatic processes always destroyed (Hyatt, 1970). Thus, following "fixation" with glutaraldehyde, mycoplasmas may still be prone to suffer distortion due to osmotic pressure and to drying, and postfixation with osmium tetroxide may be advisable.

The two-dimensional morphology of "fixed" whole mycoplasmas may be determined by transmission electron microscopy without further preparation, although drying fixed cells by the "critical point" procedure (Cohen, 1974) instead of air drying is recommended. Alternatively, positive staining with uranyl acetate or negative staining with solutions of phosphotungstic acid may be employed. The dangers inherent in the use of the latter stain to determine the morphology of unfixed organisms have been discussed by Wolanski and Maramorosch (1970). Most of the problem relates to the distortion of the cell by surface-tension forces during drying of the grid. With care, however, negative staining has proved most useful not only to define the two-dimensional shape of cells but also to reveal some internal structures such as the striated structures found in some strains of *M. mycoides* (as shown in Fig. 2) (Peterson *et al.*, 1973) and aspects of the morphology and ultrastructure of *Spiroplasma citri* (as shown in Fig. 3) (Cole *et al.* 1973). The work of Cole *et al.* (1973) indicates

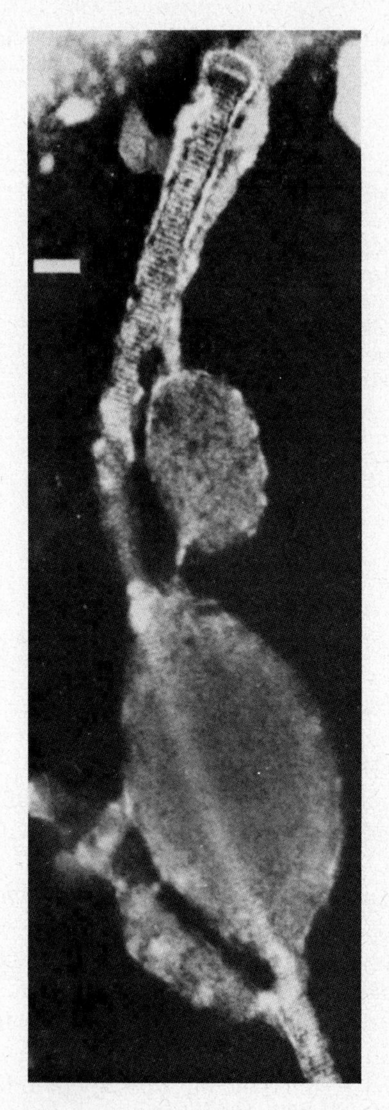

FIGURE 2. Electron micrograph of rho form of *M. mycoides* strain Y2, negatively stained, showing striated inclusion and terminal structure. ×60,000. Marker = 0.1 μm. (From Rodwell *et al.*, 1972).

---

FIGURE 3. Two-day culture of *S. citri* negatively stained with 6% ammonium molybdate. The organisms are markedly helical in shape with occasional terminal swellings. Markers = 1.0 μm (From Cole *et al.*, 1973).

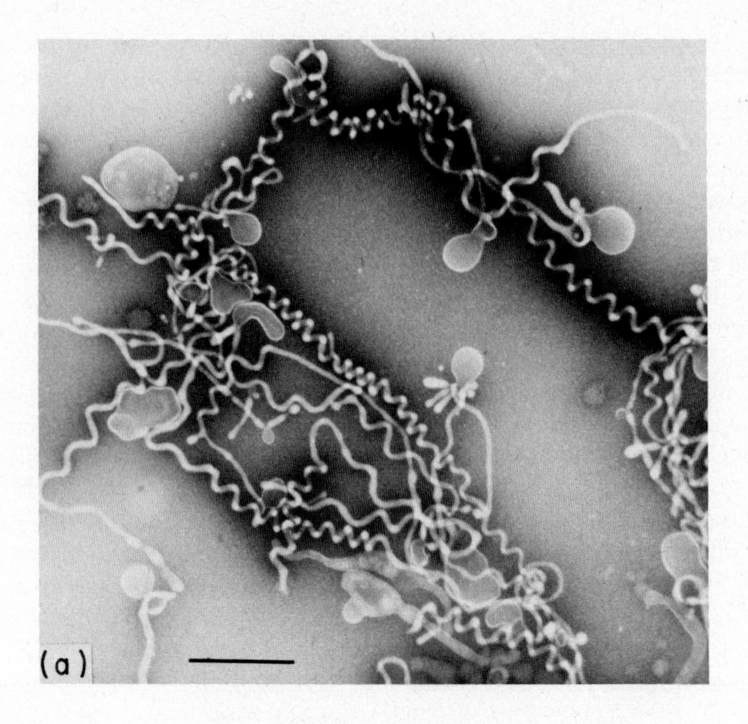
(a)

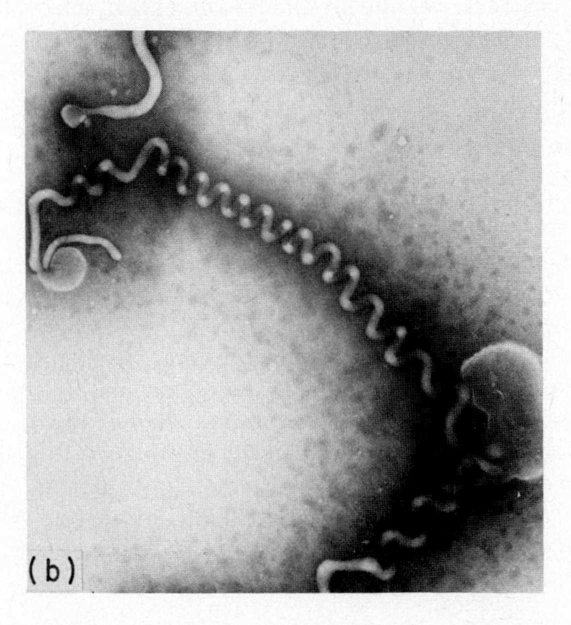
(b)

that whereas 2.5% glutaraldehyde in cacodylate buffer fails to preserve the helical shape of the organism, 10% formalin or 3.35% glutaraldehyde (final concentration) incorporated into the medium does so. Furthermore, the use of 2% sodium phosphotungstate at 54 m0sm/kg $H_2O$ obscured the helical form of the filaments; 6% ammonium molybdate at 851 m0sm enhanced the helical shape. Here, it would appear that hypertonic solutions are necessary to preserve the shape of the helical form. Thus, no single technique solves all morphological problems and a comparative study using a variety of techniques is without question necessary. Although unfavorable media and unsuitable tonicity as a cause of morphological variation can hardly be disputed, what constitutes a greater problem, however, is the manner by which synchronous growth becomes asynchronous and how soon cells of a particular culture reach a state of decline. Perhaps the problem could be alleviated by the use of a chemostat as suggested by G. K. Masover (personal communication, 1977).

In respect to shape and form, very little is known about the processes that determine shape, generate shape, and maintain shape (Henning, 1975). The surface ($\mu m^2$) to volume ($\mu m^3$) ratio (S/V) of a *Mycoplasma* cell 0.4 $\mu$m in diameter is about 15:1; if disk-shaped 0.6 $\mu$m in diameter × 0.4 $\mu$m wide, about 12; and if filamentous 0.4 $\mu$m in diameter × 20 $\mu$m long, about 10. With the filamentous forms, if the diameter remains constant, then irrespective of length, the S/V ratio is also constant.

The potentiality for reversibility in cell shape, as seen in the cinematographic studies of *M. hominis* by Bredt *et al.* (1973), makes the questions concerning processes involved in shape and form even more interesting.

### A. Surface Topography of Mycoplasmas by Scanning Electron Microscopy

Of the more recent techniques used to observe the morphology of mycoplasmas, scanning electron microscopy (SEM) is the most promising. The technique is unrivaled for the demonstration of topographical form in three-dimensions employing large populations of cells. The resolution of the standard SEM is about 7 nm and magnifications up to 35,000× are useful. Organisms may be observed in pellet form from liquid cultures (Fig. 4); as surface growths on coverslips, as shown in Fig. 5 (Biberfeld and Biberfeld, 1970; Muse *et al.*, 1976); in the form of colonies on the surface of nutrient agar (Fig. 6); or attached to cell surfaces in tissue culture (Fig. 7). Drying of cells by the "critical point" procedure to avoid phase changes in liquid–air transition is mandatory (Cohen, 1977), and to reduce charging, coating the surface of the organisms with evaporated metal is nearly always necessary.

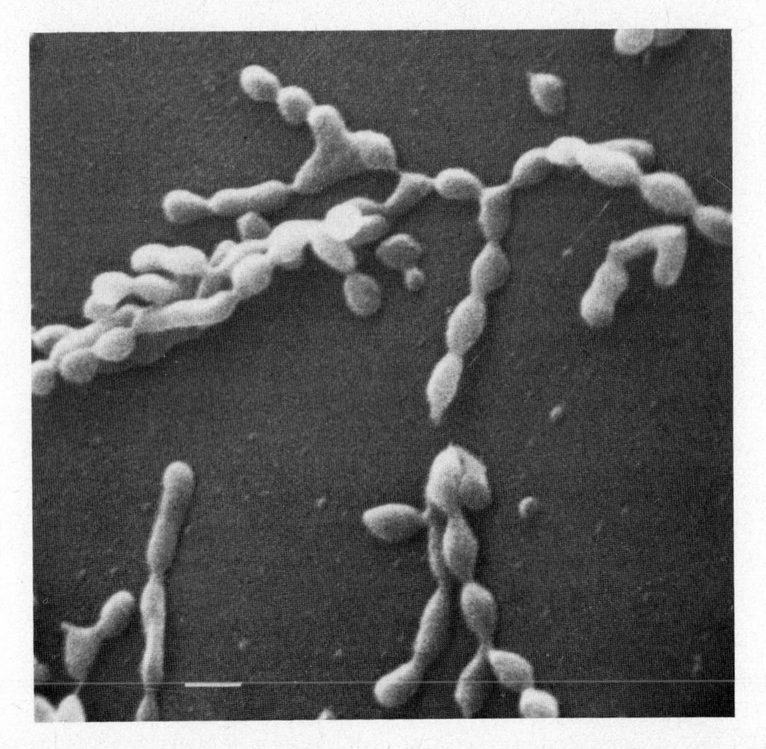

FIGURE 4. Scanning electron micrograph of organisms from a fluid culture (24 hr) of bovine *Mycoplasma* (PG50) showing mainly coccoidal cells in short branching chains. Fixed and "critical point" dried. ×14,780. Marker = 0.5 μm. (Photograph by E. S. Boatman.)

Changes in the dimensions of specimens during preparation for SEM were tabulated by Boyde *et al.* (1977); volume changes in terms of shrinkage varied from 29 to 60%. For certain types of tissues, freeze-drying (net shrinkage, 4.5% linear) was considered superior for retaining original specimen dimensions.

Predictably, perhaps, one of the first mycoplasmas to be observed by SEM was *Mycoplasma pneumoniae*. Kammer and associates (1970) used the instrument to provide observations on the cycle of morphological changes occurring in fluid cultures of *M. pneumoniae* during growth. The predominant morphology during 0.3–2.0 days growth was spherical, which after 2–6 days' additional growth became branched and filamentous. At a later stage, (6–10 days) morphology reverted to yield a predominance of "asymmetric round forms." Boatman and Kenny (1971) studied the spherule form of *M. pneumoniae* in fluid culture and, after 4 days' growth, found the spherules to be composed of ovoid forms, lobulated

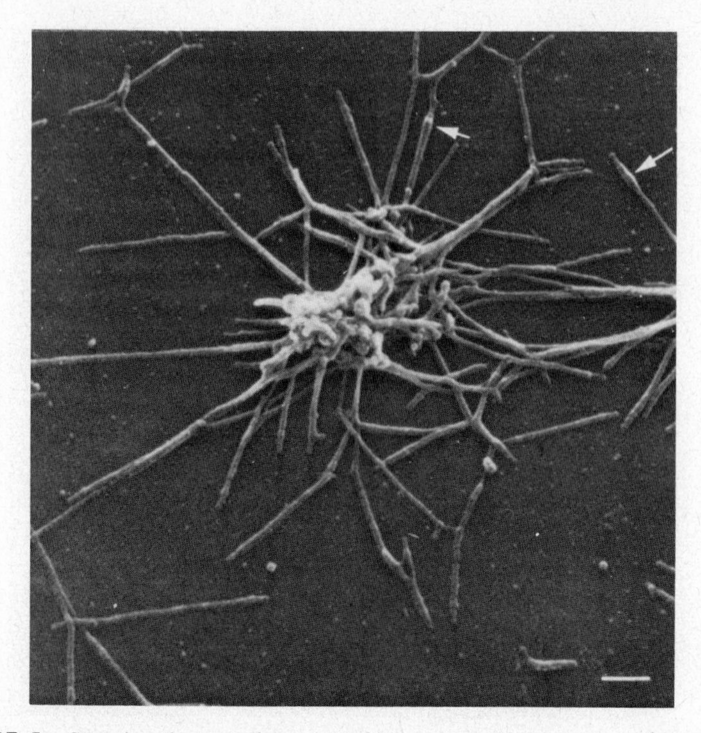

FIGURE 5. Scanning electron micrograph of *M. pneumoniae* grown on a glass coverslip for 48 hr. Organisms are slender and filamentous and many show a bulbous swelling adjacent to a tapered tip (arrow). Marker = 1 μm. (From Muse *et al.*, 1976).

forms, and some distinctly star-shaped forms. Studies by Biberfeld and Biberfeld (1970) of filtered broth cultures of *M. pneumoniae* grown on coverslips showed after 2 hr of incubation mainly spherical forms which, after 1–2 days of additional growth became star shaped and filamentous and spread over the surface, forming an irregular meshwork of filaments. For *M. pneumoniae,* the morphological findings indicate a progression of changes from round forms to filamentous forms and back to round forms depending upon the stage of growth.

Klainer and Pollack (1973) studied the surface topography of *M. pneumoniae, M. hyorhinis, M. gallisepticum. A. laidlawii, A. granularum,* and two strains of *Ureaplasma.* With all mycoplasmas examined, there was observed a diversity of size and shape and most were filamentous at some stage. *Acholeplasma laidlawii* organisms were mostly spherical of about 1.0 μm diameter with occasional large (2.5 μm diam.) forms seen. *Mycoplasma gallisepticum* was typically "bottle-shaped" with a terminal bleb. One *Ureaplasma* strain contained unusual rod-shaped helical forms of 2.0

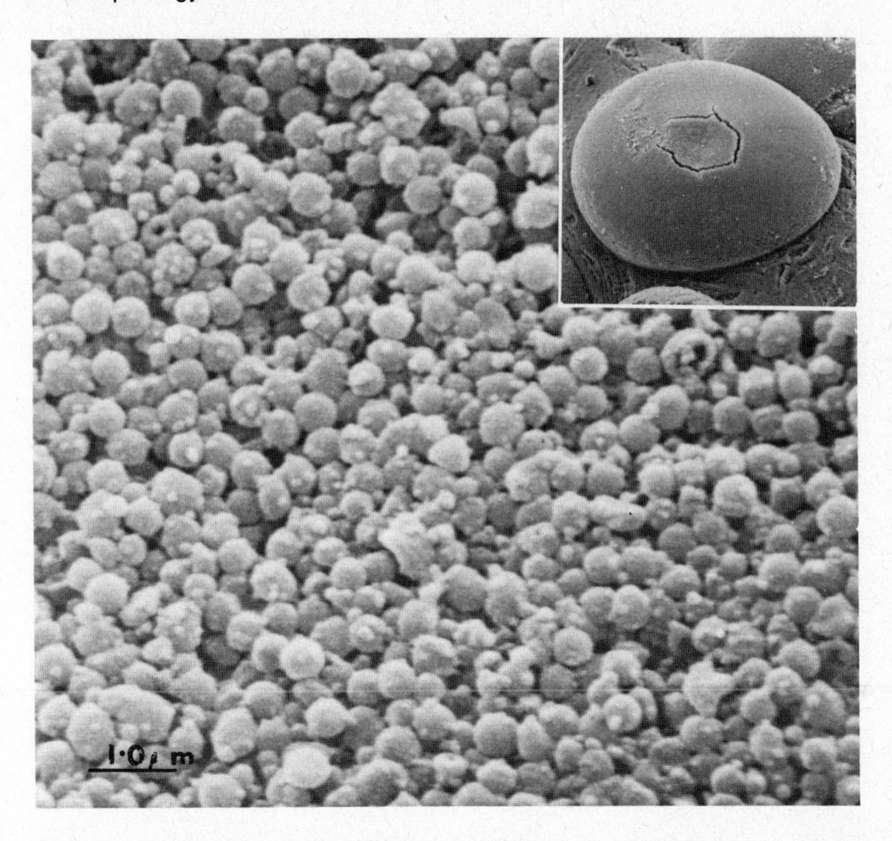


FIGURE 6. Scanning electron micrograph of the surface of a colony of *A. laidlawii* grown on soft agar for 4 days, fixed, and "critical point" dried. Organisms photographed at the colony surface are mainly spherical and fairly uniform in size: ×12,000. Inset shows whole colony ×150. (Unpublished photograph by D. Luchtel.)

× 0.25 μm in size not seen in other cultures. In some cultures, very small bodies (0.1 μm diam.) were found, the origin of which was not definitely known. The occurrence of such small bodies has been known and commented upon for as long as mycoplasmas have been observed. No evidence exists that suggests these bodies are viable (Anderson *et al.*, 1965; Razin 1967; Robertson *et al.*, 1975a). Bredt (1970) was of the opinion that the small structures associated with *M. pneumoniae* were products of degeneration; it is also possible that small contaminating particulates from the components of the growth medium may assume a spherical form upon drying and become readily visible by SEM.

Examination by SEM of tracheal epithelial cells infected with *M. pneumoniae* (Muse *et al.*, 1976) showed the organisms orientated some-

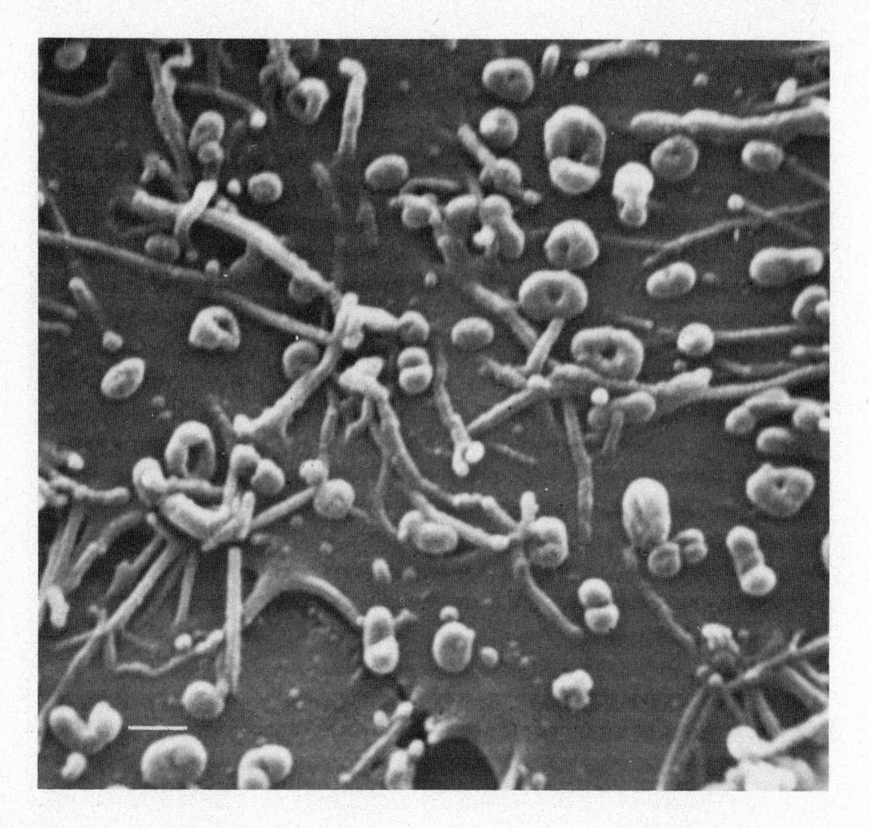

FIGURE 7. Scanning electron micrograph of an area of a HeLa cell infected with *M. hyorhinis* showing organisms of different sizes scattered over the cell surface. Fixed and "critical point" dried. ×8090. Marker = 1.0 μm (Photograph by E. S. Boatman.)

what perpendicular to the cell surface and attached by the tapered end. This terminal structure is considered to be the primary attachment point of this organism not only to eukaryotic cells (Collier and Clyde, 1971; Collier, 1972) but to the surface of glass and is, apparently, the directional indicator during motility of this organism (Bredt, 1968, 1973). This terminal structure will be discussed more fully in Section III,C.

Boatman *et al.* (1976) studied by light microscopy, TEM, and SEM the interaction of a bovine strain (PG50) of mycoplasma with HeLa cells. By SEM, the morphology of the organism was distinctly different from that of *M. pneumoniae* possessing rather wide, short lobes instead of thin filaments. Additional work (E. S. Boatman and G. E. Kenny, unpublished) with *M. hyorhinis, M. hominis, M. orale,* and *M. arginini* showed their surface topography to be similar; mainly round to oval forms and some with a central depression (e.g., Fig. 7).

Of the 15 different species of mycoplasmas observed by Ho and Ouinn (1977) in association with cell cultures, either by natural infection or artificially introduced, nearly all species appeared as single, round to oval cells 0.15–0.7 $\mu$m in diameter with a dimpled surface. In some cell lines *A. laidlawii* was seen to form spheres, rings, rods, filaments, and beadlike chains. SEM has been used to observe *M. pulmonis* attached to and ingested by mouse peritoneal macrophages. The organisms at the cell surface appeared as round, biconcave forms 1.0 $\mu$m in diameter clustered mainly in the central region of the cell (Jones *et al.*, 1977). Evidence of ingestion was indicated by the formation of "bumps" below the cell surface.

Although the shape of mycoplasmas can be readily determined by the standard SEM, actual surface structure is not easily revealed due to both limited resolution and useful magnification. Recently, a high-resolution (0.3 nm) SEM employing a "field-emission source" capable of resolving single particles of influenza virus on cell surfaces (Amako, 1975) has become available. It is likely in the near future that mycoplasmas will prove to be prime candidates for observation by this instrument.

In summary, it may be stated that at present, the one *Mycoplasma (M. pneumoniae)* that has been examined in a very thorough manner by SEM has given results that have been particularly rewarding both with the organism in culture and in association with cells and tissues. It is notable that a good many mycoplasmas "growing" on the surfaces of cells in tissue culture tend to be round or oval in form. Whether this indicates a characteristic of the organism or a characteristic of cell–organism association remains to be verified.

If a scanning electron microscope is not available, the three-dimensional shape of mycoplasmas attached to agar or glass surfaces may be examined by transmission electron microscopy by use of pseudoreplicas (Peterson *et al.*, 1973; Furness *et al.*, 1975) or by carbon replicas (Kim, 1977). These techniques, although somewhat tedious to carry out, give results somewhat comparable to those acquired by SEM. A comparison is shown in Fig. 8 (Kim, 1977, unpublished) of filamentous forms from a 4-day culture of *M. pneumoniae*. Alternatively, the growth cycle and morphological heterogeneity of mycoplasmas may be studied using samples of organisms dried by the "critical point" (SEM) procedure and observed by TEM (Green and Hanson, 1973, *M. meleagridis*; Kim, 1977, *M. pneumoniae*). An example, in respect to *M. pneumoniae*, is shown in Fig. 9.

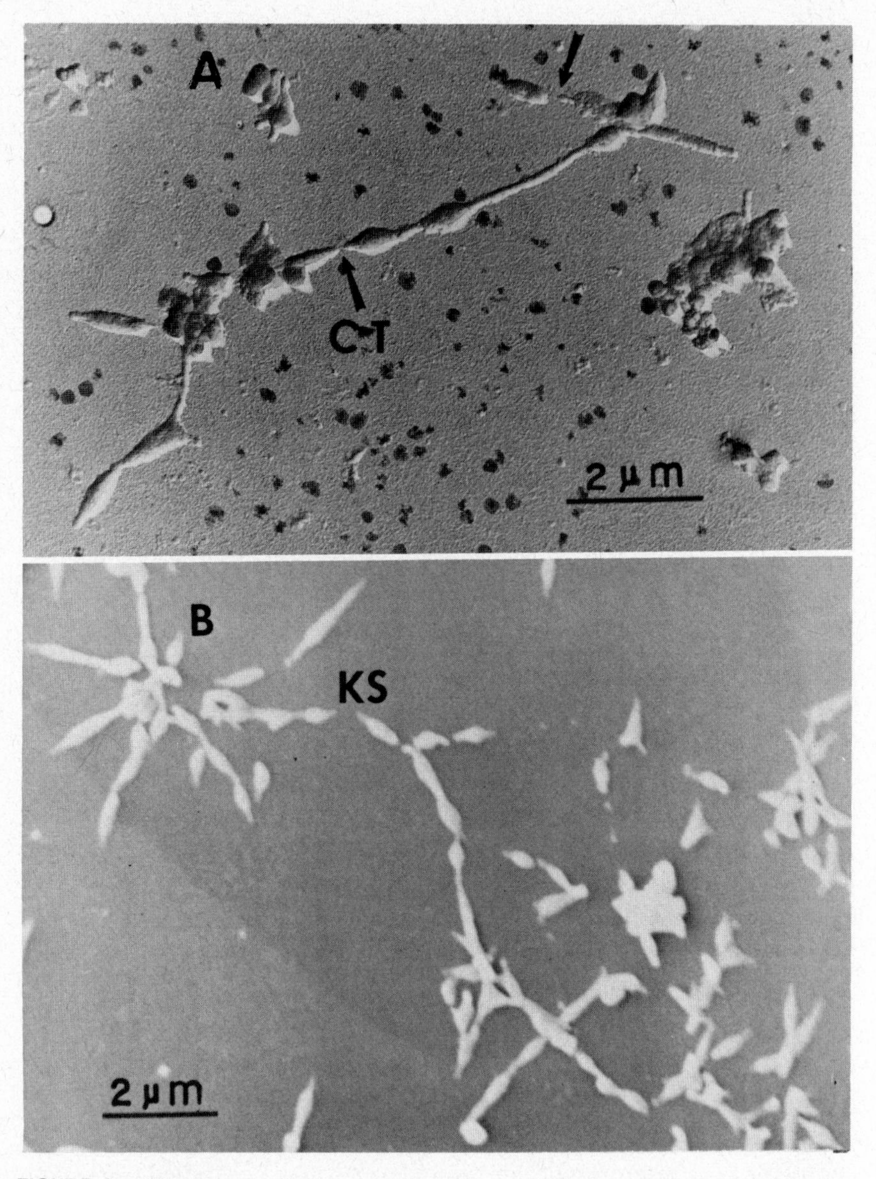


FIGURE 8. Filamentous cells from a 4-day culture of *M. pneumoniae* revealed by the carbon replica technique (A) and by scanning electron microscopy (B). Some cells appear connected by a threadlike structure (CT) and some cells show terminal knoblike structures (KS). ×7500. (Unpublished photograph by C. Kim.)

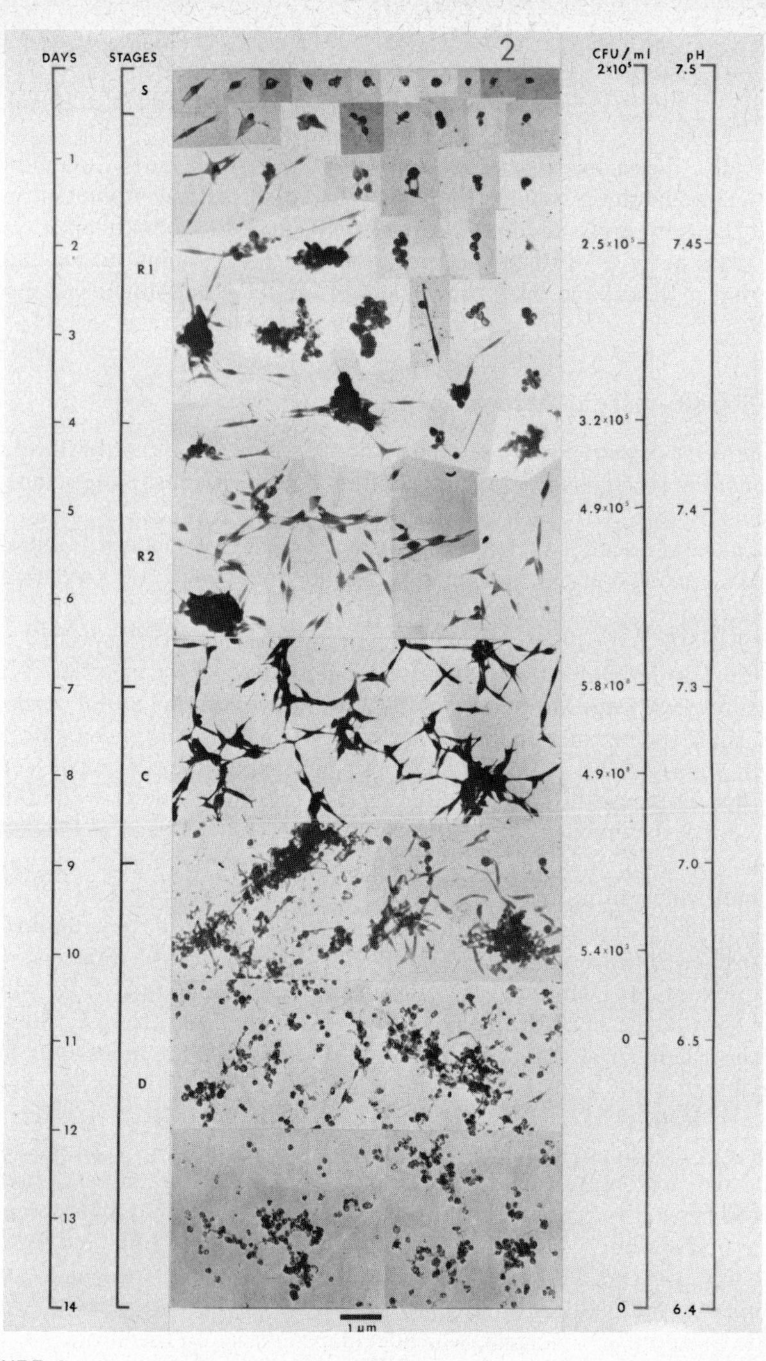


FIGURE 9. Sequence of growth of *M. pneumoniae* during one growth cycle. Morphology derived from transmission electron microscopy of "critical point" dried cells. Sequence over a period of 14 days could be divided into four stages: single cells (S), replication ($R_1$ and $R_2$, 1–7 days), confluence (C, 7–9 days), and degeneration (D, 9–14 days). Viality (CFU/ml) and pH decreased with age of the culture. (Unpublished photograph by C. Kim.)

Finally, the shape of mycoplasmas can be derived by the even more tedious technique of reconstruction by use of projected profiles of individual cells in serial section (Boatman and Kenny, 1970; Boatman, 1973a). The main purpose of this exercise, however, was to point out the errors inherent in describing three-dimensional form from two-dimensional sections.

## B. Ultrastructure of Mycoplasmas

There is, perhaps, more agreement concerning the ultrastructural aspects of the mycoplasmas than most other features of these rather unusual organisms.

In ultrathin sections, the mycoplasmas are seen to lack a traditional form of prokaryotic cell wall and are confined by a "unit" type membrane which may or may not show asymmetry.

The width of the limiting membrane of mycoplasmas varies from 7.5 to 10.0 nm (Domermuth *et al.*, 1964) and what is seen and the accuracy of the measurements depend upon many factors. However, on careful observation, some species of mycoplasmas appear to possess limiting membranes which are asymmetric; i.e., one "black line" appears less electron-dense or different in width than the other. In an extensive study on the ultrastructure of the mycoplasmas Domermuth *et al.* (1964) looked at 15 species grown on solid medium. In six species, the cell membrane appeared symmetrical, and in the remaining nine species, asymmetrical.

Razin (1973) has suggested that the protein or other material bound to the surface of the lipid bilayer would be responsible for the differences in the thickness of membranes. Membranes from *A. laidlawii,* from which over 95% of the lipid was removed by treatment with acetone, showed typical trilaminar structure. Removal of over 80% of the membrane proteins by pronase digestion resulted in a decrease in both thickness and contrast (Morowitz and Terry, 1969). By use of a polycationic ferritin label on glutaraldehyde-fixed cell membranes of *M. mycoides* subsp. *capri* and other mycoplasmas, Schiefer *et al.* (1976) found the anionic binding sites to be evenly but asymmetrically distributed, i.e., on the outer membrane surface only. If the label was applied before fixation, the ferritin molecules were located in clusters over the membrane surface, suggesting mobility of the binding sites.

The presence of amorphous or floccular material on the surface of the outer limiting membrane of mycoplasmas has been noted by a number of investigators. Nine of the fifteen species of mycoplasmas examined by Domermuth *et al.* (1964) were believed to have an additional surface layer. In some species, under certain conditions of growth, this extramembranous layer is well developed and can be further enhanced by the use of an electron-dense osmium–ruthenium red complex which has an affinity for mucopolysaccharides (Luft, 1971). This technique has been used with *M. meleagridis* (Green and Hanson, 1973), with bovine mycoplasmas, with *M. dispar* (Howard and Gourlay, 1974), and with *U. urealyticum* (Robertson and Smook, 1976). The width of the layer in *U. urealyticum* (Fig. 10) was 20–30 nm, or about twice that observed for the bovine species by Boatman *et al.* (1976) (Fig. 11). The presence of a microcapsule not seen by ordinary methods may account for a gap of 11 nm separating the cell membranes of organism and cell, as shown in Fig. 12 for *M. arginini* and seen so often in sectioned material from infected cell lines.

The observation of cell membranes in ultrathin sectioned material affords a knowledge of profiles of membranes cut in cross-section or obliquely and, from the former, permits a measurement of membrane dimensions. Rarely is the membrane face seen in sections. To visualize the

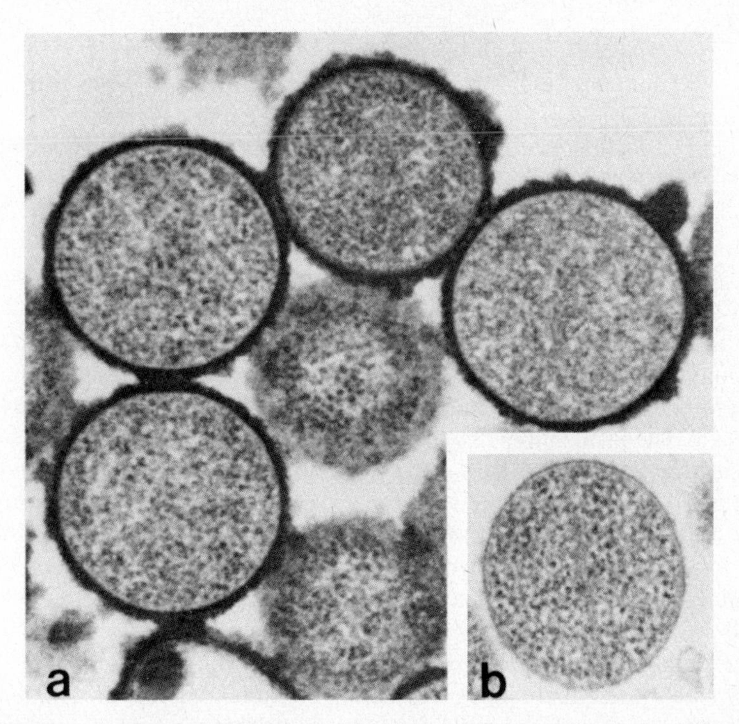


FIGURE 10. Electron micrographs of sectioned cells of *U. urealyticum*. (a) Cells stained with ruthenium red–osmium. Note well-defined layer of extracellular material. (b) Unstained control. (From Robertson and Smook, 1976.)

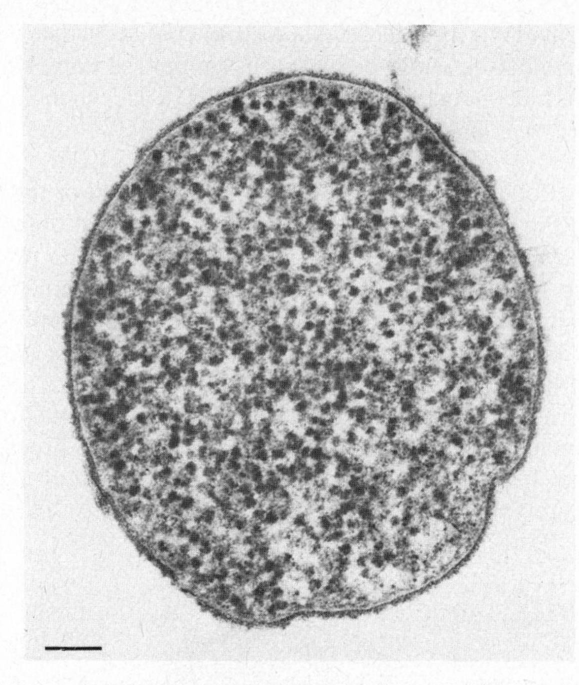

FIGURE 11. Electron micrograph of a sectioned bovine *Mycoplasma* (PG50) stained with ruthenium red–osmium. A dense 11-nm surface layer is seen external to the cell membrane. Ribosomes and less dense nuclear material present a compact cytoplasm. ×84,700. Marker = 0.1 μm. (Photograph by E. S. Boatman.)

inner and outer surfaces of membranes, either native or reconstituted, the very useful techniques of freeze-fracture and freeze-etching can be used in conjunction with transmission electron microscopy. These techniques obviate the need for chemical fixation, dehydration, and staining for contrast. The procedures call for rapid freezing of the material; fracturing the surface; etching the surface, i.e., sublimation of surface ice under a vacuum; and the production of a carbon replica. The fracture process splits the membrane down the center of the hydrophobic portion of the lipid bilayer, exposing the two inner faces of the membrane, one of which is convex and the other concave. Etching of the fractured surfaces reveals details of the surface components of the membrane. A carbon cast is made of the etched surface and examined by TEM.

By use of these techniques it has been possible to detect arrays of small globular particles of about 6.6–15.0 nm in diameter on both convex and concave surfaces of fractured native membranes from a number of mycoplasmas species. Particles have been found on membranes from *Urea-*

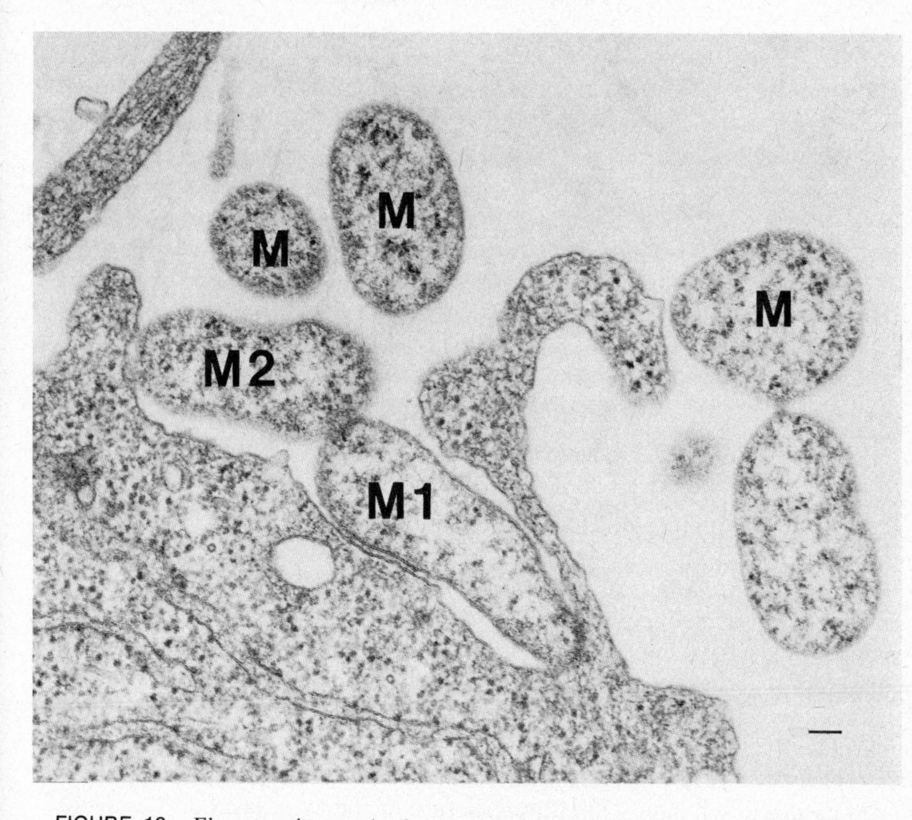


FIGURE 12. Electron micrograph of a portion of an infected HeLa cell showing profiles (M) of sectioned organisms (*M. arginini*) attached to or close to the cell. Organism (M1) shows its cell membrane adjacent to the cell membrane of the HeLa cell but separated by a gap of 11 nm. The cytoplasm of organism (M2) closely resembles that of the HeLa cell. ×46,600. Marker = 0.1 μm. (Photograph by E. S. Boatman.)

*plasma* sp. (Whitescarver and Furness, 1975) and on the convex and concave membrane surfaces of *M. gallisepticum,* including regions containing terminal blebs as shown in Fig. 13 (Bernstein-Ziv, 1969; Maniloff and Morowitz, 1972). Of particular interest was the observed lack of particles on the fracture faces of reconstituted membranes of *A. laidlawii*, suggesting a major alteration in membrane architecture (Tillack *et al.*, 1970). Green and Hanson (1973) looked at freeze-etched preparations of *M. meleagridis* and found more particles per area on the convex surface than on the concave. The surface density of particles on membranes of *A. laidlawii* could be doubled if stearate as a growth supplement was replaced by oleate (Tourtellotte *et al.*, 1970). Although the composition of the globular particles seen on fractured membranes is not known with

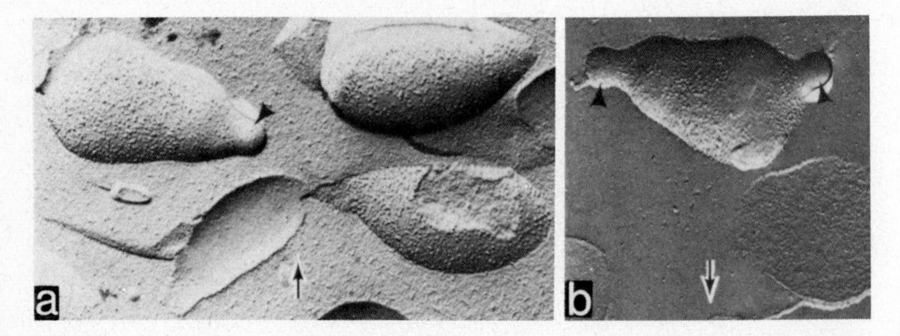


FIGURE 13. Electron micrographs of a freeze-etched preparation of *M. gallisepticum*. The single cells (a) and the dividing two-bleb cell (b) have been fractured along a plane of an internal membrane showing particles on its convex surface. Arrow heads indicate bleb regions. Long arrows indicate direction of shadowing. (Courtesy A. Ghosh.)

certainty, the findings of Tourtellotte and Zupnik (1973) suggest that they are protein and may also be involved in substrate transport. The freeze-fracture technique has also been useful for observing the effects of osmium fixation on membranes (James and Branton, 1971) a procedure which tends to reduce the number of membrane fracture faces. It is apparent that these techniques will become of major importance when the difficulties in determining the chemical composition of the membrane particles have been resolved.

Apart from a few species with specialized internal structures—which will be discussed later—the ultrastructure of the cytoplasm of the mycoplasmas as a whole is fairly uniform. A survey of profiles of sectioned organisms from various genera indicated cytoplasmic uniformity but some variability in texture, such as (1) a loose cytoplasm with well-defined ribosomes and fibrillar nuclear material; (2) a compact, moderately electron-dense cytoplasm with partially obscured nuclear material; (3) a very compact cytoplasm with nuclear material condensed into a small netlike area; and (4) a very dispersed cytoplasm with virtually no nuclear material apparent. Forms (1) and (2) would be regarded as defining optimal ultrastructure. The absence of cytoplasmic streaming in prokaryotic cells and the small size of mycoplasmas would lend support to the idea of compactness of the cytoplasm as a necessity. Form (3) is nearly always due to inappropriate tonicity of the fixative solution and form (4) is generally indicative on nonviable cells (Maniloff, 1970; Boatman and Kenny, 1970). Forms approximating those above may also be found in a single culture as shown in Fig. 14 (Anderson and Barile, 1965).

The ultrastructure of sectioned organisms from the various genera, disregarding shape, tend to show a high level of homogeneity. The large

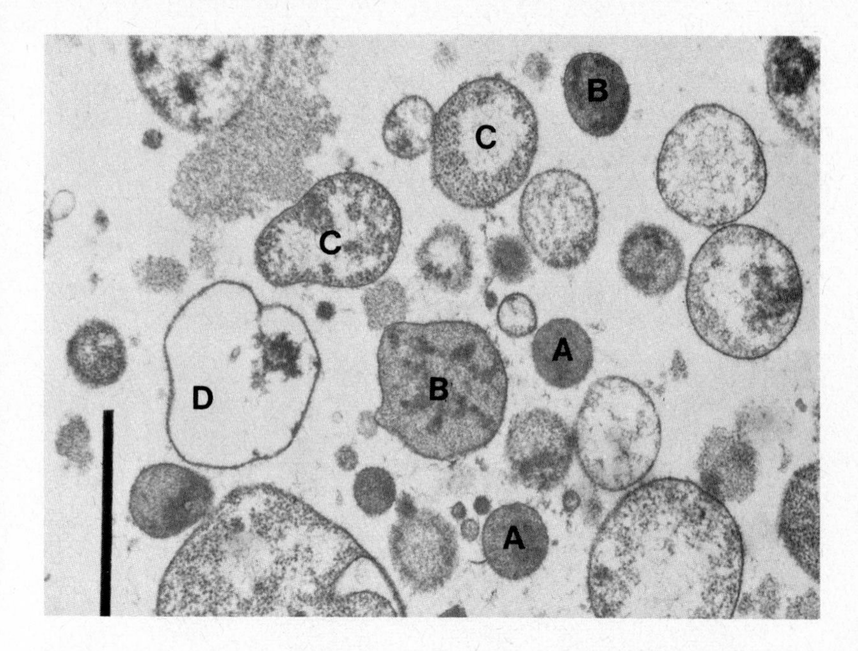


FIGURE 14. Electron micrograph of sectioned cells of *M. hominis*. Cell profiles show varying amounts of ribosomes and nuclear material ranging from high density, forms A and B; to medium, form C; to lysed cells, form D. ×53,000. (From Anderson and Barile, 1965).

genus *Mycoplasma* has been well documented and examples have been presented in Fig. 11, 12, and 14. A capsulated form of *Ureaplasma* with ultrastructure typical of the mycoplasmas is shown in Fig. 10.

A member of the genus *Acholeplasma* which has been particularly well studied, especially in terms of the structure and chemical composition of its cell membrane, is *A. laidlawii* (Razin, 1972). The ultrastructure of sectioned cells of *A. laidlawii B.* (Fig. 15) has been studied by Maniloff (1970), *A. axanthum* by Tully and Razin (1970), and *A. laidlawii A.* by Virkola (1972). All of these organisms are bounded by a triple-layered cell membrane and show, to varying degrees, ribosomes and fibrillar DNA. Filamentous forms were present in small numbers.

The ultrastructure of *Spiroplasma citri* is shown in Fig. 16. Because of its helical shape profiles of sectioned organisms show variable curvatures; its fine structure, however, is similar to mycoplasmas of other genera, but care must be taken to prevent lysis and to control cell size and internal densities (Cole *et al.*, 1973). Well-preserved helices averaged 120 nm in width and had a 7.2 nm triple-layered cell membrane and ribosomes of 17 nm in diameter; no evidence was found for cell wall partitioning and, at

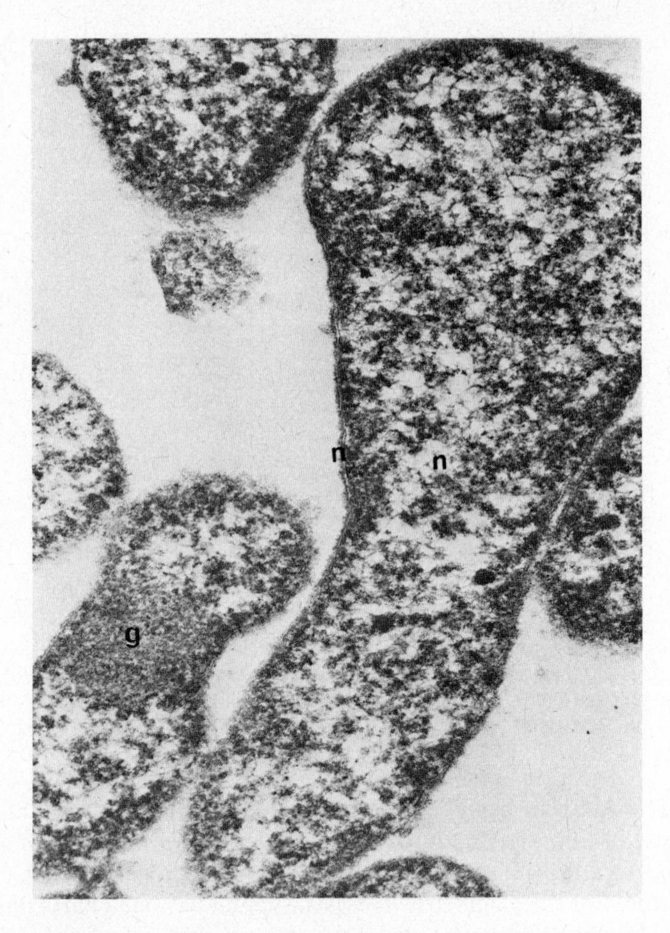


FIGURE 15. Electron micrograph of sectioned cells of *A. laidlawii*. Cells are bounded by a trilayered membrane (m) and the cytoplasm consists of ribosomes, a granular region (g), and nuclear material (n). (Courtesy of J. Maniloff.)

present, the process of cell division in this organism is unknown. Other features of the organism include (1) motility, which may be a rapid rotary motion or a slow undulation; (2) the presence of short projections on the outer surface of the cells, similar to those noted by Chu and Horne (1967) in negatively stained cells of *M. gallisepticum*; and (3) disrupted filaments in which are seen collections of subunits aligned to give a striated appearance similar to the structures observed by Rodwell *et al.* (1975) in certain strains of *M. mycoides* subsp. *mycoides*. Finally, the organisms from the original culture and from a limited number of passages thereafter were found to be infected by a bacterial type of tailed bacteriophage. Since

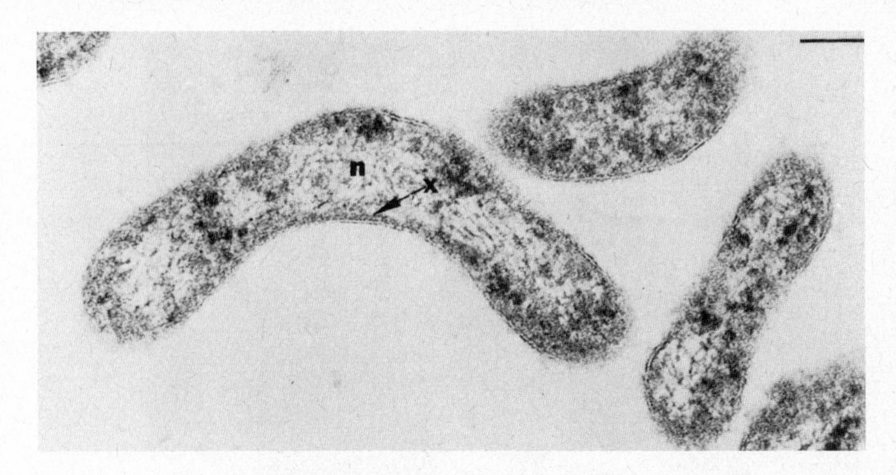


FIGURE 16. Sectioned cells of *S. citri* showing a curved portion of a filament with fibrillar nuclear material (n) and ribosomes. In one area (×) an inner layer abuts the membrane. ×12,000. Marker = 0.1 μm. (From Cole *et al.*, 1973.)

mycoplasmas do not have a bacterial type of cell wall, infection by a tailed bacteriophage is both unique and previously unknown in these organisms. The lack of a rigid cell wall poses a number of questions, such as how the helical shape of the cell is maintained, and what is the mechanism of cell motility. Razin *et al.* (1973) made a careful study of the chemical composition of membranes and cells of *S. citri* and observed cell membranes after freeze-fracture. Particles were present on both the concave and convex fracture faces of the cell membrane with more particles on the latter. Chemically and structurally, no clear evidence was obtained for the presence of cell wall material, although hexosamine components were found in isolated membrane preparations and could conceivably form a part of the external layer of projections seen in negatively stained preparations.

An organism isolated originally by Hungate (1966) as an obligate anaerobic rumen microorganism was later characterized by Robinson *et al.* (1975) as a mycoplasma and placed in the genus *Anaeroplasma*. The species *A. abactoclasticum* has been studied by electron microscopy and, in negatively stained preparations, has shown some pleomorphism, including filamentous and budding forms. Ultrathin sections have shown cells bounded by a triple-layered membrane 7.5–10 nm wide and containing cytoplasmic constituents characteristic of the mycoplasmas (Fig. 17). As yet, no unique morphologic features have been found for these organisms.

Unusual organisms are not uncommon in the mycoplasmas and a case

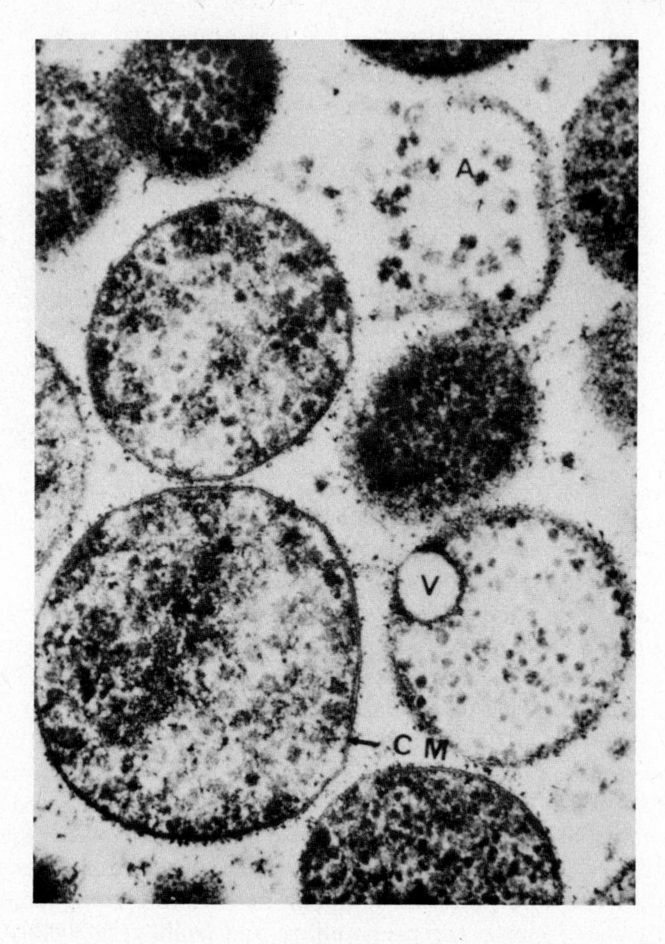


FIGURE 17. Sectioned cells of *A. abactoclasticum* showing a characteristic trilayered cell membrane (CM) and cytoplasm of ribosomes and nuclear material. A lysed cell (A) and a vacuolated cell (V) are also seen. (From Robinson *et al.*, 1975.)

in point are those in the genus *Thermoplasma*. The species *T. acidophilum* was isolated from a coal refuse pile by Darland *et al.* (1970). The organism has a temperature optimum of 59°C and an optimal pH for growth of about 2. By phase microscopy the organisms appear as spheres of 0.3–2.0 $\mu$m diameter; they appear to reproduce by budding and filamentous forms are seen, particularly in young cultures. In thin sections (Fig. 18) the organisms are devoid of a cell wall, possess a trilayered cell membrane 10–12 nm wide, and have the characteristic cytoplasmic components—i.e., dispersed nuclear material and ribosomes. Freeze-etch preparations

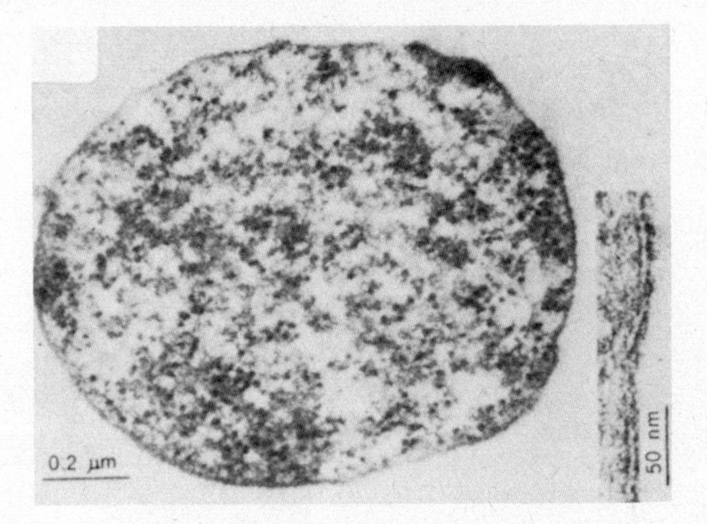


FIGURE 18. Ultrastructure of a cell of *T. acidophilum* in section. Cytoplasm contains ribosomes and nuclear material. At the right is an enlargement of the cell membrane which is 10–12 nm in width. (From Darland *et al.*, 1970.)

show particles 20 nm in diameter on the surface of the membrane. At present, their morphology has not been found to be unique.

Whatever the case for membrane particles, the dimensions and nature of cytoplasmic particles have been well documented (Maniloff and Morowitz, 1972). During normal growth of the cell, ribosomes are similar in structure to the bacterial species with a sedimentation coefficient of 70 S and are distributed fairly evenly throughout the cell cytoplasm. Under conditions where protein synthesis is stopped (Barker and Swales, 1972) or where cells of *M. gallisepticum* in particular are centrifuged prior to fixation, condensations of ribosomes are formed which align in helical or tetrahedral arrays (Allen *et al.*, 1970; Bernstein-Ziv, 1969, 1971). An example of ribosomal arrangements in *M. gallisepticum* is shown in Fig. 19 from Maniloff *et al.* (1965). Domermuth *et al.* (1964) noted a "corncob" pattern of ribosome aggregation in the cytoplasm of *M. gallisepticum* derived from colonies grown on agar. All of the other 14 species grown on the same medium did not show this pattern. During cell division in strains of ureaplasmas, Whitescarver and Furness (1975) found parallel arrangements of three to four chains of six to eight paired ribosomes per chain—large enough to be observed in the cytoplasm by phase-contrast microscopy. Somewhat similar arrays were noted by Black *et al.* (1972).

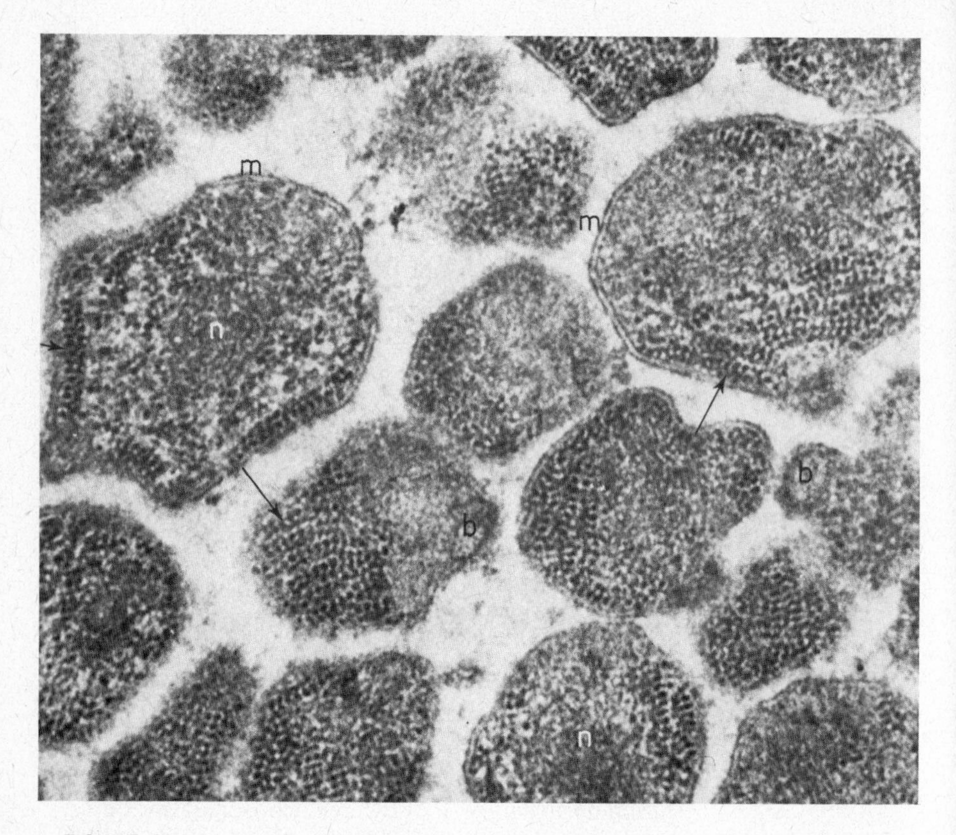


FIGURE 19. Sectioned cells of *M. gallisepticum* showing trilayered cell membranes (m), nuclear material (n), bleb regions (b), and tightly packed cylindrical arrangements of ribosomes (arrows). ×68,000. (From Maniloff *et al.*, 1965.)

Optical diffraction and rotational analyses of electron micrographs of *M. gallisepticum* (Maniloff, 1971) have shown the helix to have 10 ribosomes per three turns per repeat. Stabilization of the helix requires, apparently, interaction between 50 S subunits. It is, nevertheless, strange that of all the species so far examined none, except *M. gallisepticum,* has shown these particular helical condensations.

Because various aspects of the DNA of mycoplasmas will be covered fairly extensively in Chapter 6, the present discussion relating to structural aspects will be limited.

In 1959, Kleinschmidt and Zahn devised a protein-film technique that allowed molecules of DNA to spread on water and to be picked up and retained on the surface of an electron microscope grid without serious aggregation. Later, by use of autoradiography (Cairns, 1963), it was possible to discern the double-stranded features of bacterial DNA during replication.

The size and structure of the genome of *M. hominis* (H39) (reclassified as *M. arthritidis* H39) (Bode and Morowitz, 1967) was the first *Mycoplasma* species to be so determined. By electron microscopy of cells lysed with detergent before formation of the protein film, it was demonstrated that nonreplicating DNA molecules derived from cells in the early stationary phase had an average length of 262 $\mu$m ($5.1 \times 10^8$ daltons)[2]; the molecule was unbranched, circular, and double stranded. The overall length is quite considerable in respect to the average mycoplasma cell of about 0.4 $\mu$m diameter.

Since this initial work it has been possible to estimate the size, base composition, and other characteristics of the genome of a number of species of mycoplasmas (Maniloff and Morowitz, 1972; Razin, 1973; Morowitz and Wallace, 1973).

For the cell to accommodate this nuclear material it is clear that the genome must be packaged in the cell in an intricate yet obviously functional manner. Following the procedure of Kellenberger *et al.* (1958) it was possible to visualize the DNA in ultrathin sections of bacteria as resembling a skein of wool. This analogy, although simple, is pertinent because if the correct piece of wool is pulled, the skein can be unwound to its full length without knotting. The prokaryotic cell has to accomplish a similar but exceedingly more complex maneuver on replication and segregation of its genome during cell division and, possibly, under circumstances where growth is filamentous. In order for the cell to segregate replicating DNA during cell division and to ensure that each daughter cell receives a complete genome, it has been proposed that in prokaryotic cells the "chromosome" is attached to the cell membrane (Jacob *et al.*, 1963). In respect to the mycoplasmas it has been shown with *M. gallisepticum* (Maniloff and Quinlan, 1974) that the membrane–bleb–infrableb region of this organism is the site of the DNA replication complex. The evidence is derived from biochemical assay of gradient fractions and electron microscopy of epon embedded and sectioned gradient banded material. Two major bands were found; one contained membrane vesicles and the other membrane-bound bleb structures. Most of the ATPase activity and nascent DNA was found in the latter band. On the basis of this and other evidence, it was thought that these bleb complexes might be a "prokaryotic analogue of eukaryotic centrioles." Whether this possibility also exists for *M. pneumoniae* with its polar structure and whether some sort of membrane–DNA binding site is common to all species of

[2] The average bacterial genome would be about 1200 $\mu$m in length and about $2.6 \times 10^9$ daltons.

mycoplasmas remains to be seen. Separation of daughter cells following DNA segregation occurs by constriction of the cell membrane without the formation of a cell "cross wall" septum.

### C. Specialized Structures in the Mycoplasmas

Of the few species of mycoplasmas that have been found to possess intracytoplasmic structures, all but one have been found to be motile; thus the assumption has been that these structures are related to the function of motility in these species. However, this belief is complicated by the observation that certain of these species attach avidly to animal cells, and especially to ciliated epithelial cells of the airways of the lungs, and that this structure may be primarily a structure for attachment.

The interaction of *M. pneumoniae* with human and animal respiratory epithelium in organ cultures has been studied with notable success by Collier and Clyde (1971), Collier (1972), Collier and Baseman (1973), Wilson and Collier (1976), and Powell *et al.* (1976). A consistent feature of all these studies was that the organisms were found to be attached to respiratory epithelium by a structure located at one end of the cell. This differentiated portion of the organism consists of an extension of the unit membrane containing an electron-dense core surrounded by a lucent space, as shown in Fig. 20. Recent work by Wilson and Collier (1976) has shown the structured terminal bleb of *M. pneumoniae* to contain basic protein but not nucleic acid as judged by use of indium trichloride staining. Tannic acid incorporated into the fixative further suggested the presence of a periodicity to the central core, indicating a specific substructure. Staining both with the tannic acid complex and with ruthenium-red–osmium has revealed evidence of a mucoprotein extracellular layer which was particularly heavy in the region of the specialized tip. In respect to the facility for attachment of *M. pneumoniae* to tracheal rings in organ culture, it is notable that the ability to attach is markedly reduced using killed organisms, on incubation at 4°C, and following the use of neuraminidase or sodium periodate (Powell *et al.*, 1976), suggesting the necessity for a biochemical factor other than purely a structural site for facilitating attachment of these organisms. The fact that an avirulent strain of *M. pneumoniae* also had terminal structures and yet was not found to attach or cytadsorb (Collier, 1972) poses the question of a requirement for some sort of biochemical mediator. Quite recent work by Hu *et al.* (1977) implicates a membrane protein on the surface of *M. pneumoniae* as the mediator in such attachment.

The specialized structure in *M. gallisepticum* is the terminal bleb–infrableb region. The bleb itself is hemispherical, about 80 × 125 nm, the

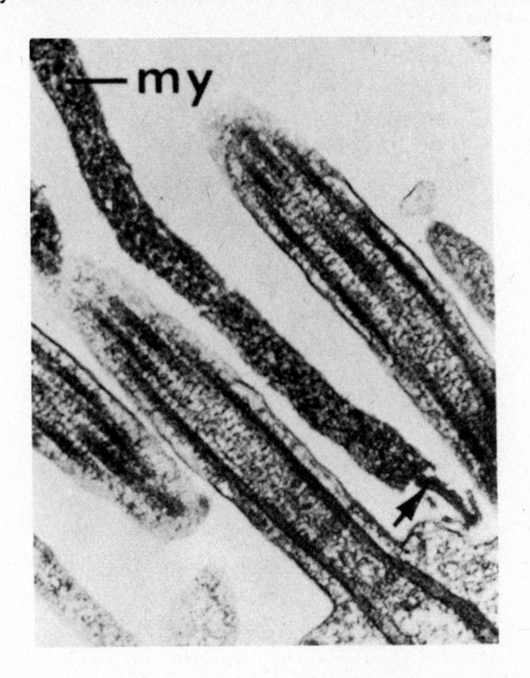


FIGURE 20. Electron micrograph of a portion of a sectioned ciliated epithelial cell from a human fetal trachea infected in organ culture with virulent *M. pneumoniae*. A filamentous organism (my) is seen aligned with its terminal structure in close association with the cell membrane. A short dense filament (arrow) is seen in the center of the terminal structure. ×45,000. (From Collier, 1972.)

base of which is attached to the infrableb region by a flat, circular plate. The infrableb region is about 200 nm in diameter. In the appropriate plane of sections (Fig. 21) cells are seen either with a single bleb or, in predivision cells, with two blebs. In association with mammalian cells *M. gallisepticum* is found to attach to the eukaryotic cell membrane by the blebbed end of the organism (Zucker-Franklin, 1966a,b). An additional similarity that *M. gallisepticum* has to *M. pneumoniae* is that of motility, for which the underlying mechanism is also not known. Apart from being sites of enzymatic activity (Munkres and Wachtel, 1967) cell fractions containing the bleb complex were also shown to contain cellular DNA which, incidentally, could not be detected by the indium staining method (Maniloff and Quinlan 1974) and, as mentioned earlier, the bleb region is considered to contain the membrane site and growing point of the "chromosome." In all, this is an impressive array of functions. In respect to *M. pulmonis*, however, less evidence is available for a specialized terminal structure, even though it has been shown to be motile (Nelson and Lyons,

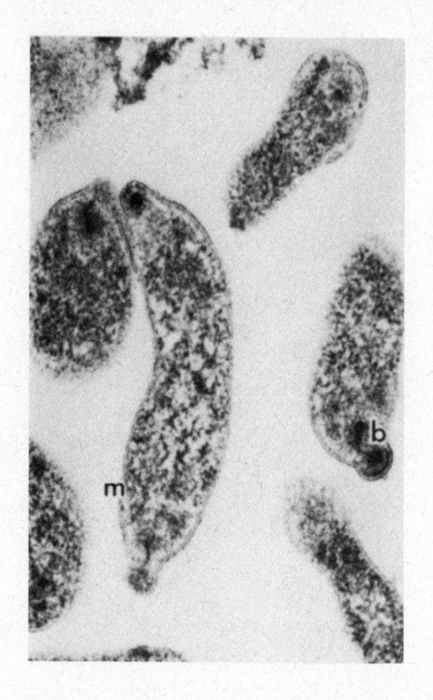


FIGURE 21. Section of log-phase cells of *M. gallisepticum*. Several one-bleb daughter cells and a two-bleb predivision cell are shown. The trilayered cell membrane (m), ribosomes, nuclear material, and bleb and infrableb region (b) are evident. (Courtesy of A. Ghosh.)

1965) and to attach to various mammalian cells (Organick *et al.*, 1966). The profiles of *M. pulmonis* in sections tend to be round, often with a suggestion of a terminal structure or, filamentous (Nelson and Lyons, 1965; Tanaka *et al.*, 1965). In a study on the motility of this organism Bredt and Radestock (1977) observed two morphologically distinct forms. One was a round cell with a protruding flexible stalk thickened at its end, and the other form was an elongated cell with a tapered leading end. Both forms showed gliding motility. Although it appears that some form of cellular extension is necessary for motility in *M. pulmonis* a requirement for this structure as a mediator of attachment to cells and tissues has yet to be shown.

On occasions, and mostly following negative staining of whole cells or cell fractions, some strains of *M. pulmonis* (Hummler *et al.*, 1965), *M. gallisepticum* (Chu and Horne, 1967), and *S. citri* (Cole *et al.*, 1973) have been shown to have short uniform projections (spikes) extending from the cell surface similar to those seen on the outer coat of myxoviruses by Chu and Horne (1967). The function of this surface layer is unclear as is the point concerning whether or not the layer is an artifact of the techniques used prior to observation.

Another intriguing specialized structure, termed the rho-form, is found in certain caprine and bovine strains of mycoplasmas and is first described by Rodwell *et al.* (1972). The rho-form of *Mycoplasma* is a rodlike organism (assumed to be relatively rigid) in which an intracytoplasmic striated fiber ($\rho$ fiber) extends axially throughout the cell and terminates at one end of the cell, or sometimes both ends, in a knoblike structure as shown in Fig. 22. The cell, which is often 15 $\mu$m or more in length and is multinucleated, is characterized by discoidal swellings at regular intervals. The fiber appears to be composed of fibrils which are aligned in such a manner as to give a cross-banded pattern (Fig. 23). The dimensions of the fiber range from 40 to 120 nm in diameter and the periodicity of the banding from 12 to 14.5 nm. Two requirements are necessary for the selection and expression of the rho character, i.e., a medium with a high tonicity and a nonlimiting energy source (Peterson *et al.*, 1973). Under hypotonic conditions the $\rho$ fiber cannot be demonstrated, nor can it be found after osmotic lysis. Under other conditions, fibers can be made to disassemble into subunits and to reassemble into aggregates having the same ultrastructure as native $\rho$ fibers (Rodwell *et al.*, 1973, 1975). The function of this structure and why only certain strains of mycoplasmas are seen to have them is not known. Additional details of the rho-form of mycoplasmas are to be found in Chapter 4.

## D. High-Voltage Electron Microscopy

Until recently, the virtue of the high-voltage electron microscope (1–3 meV) was its use in the field of materials science, where its high accelerating voltage enabled greater penetration of electrons into relatively thick materials as never before. With accumulating knowledge as to the merits of these instruments, they are now being put to use to observe biological materials (Hama and Porter, 1969). The instruments are costly and demanding in space and auxiliary equipment.

The main advantages of the instrument, due to its exceptionally high accelerating voltage, are that (1) thick specimens of about 1.0 $\mu$m can be observed; (2) resolution is improved for sections of a given thickness; (3) damage to the specimen caused by the electron beam is generally less than that by the traditional 50–100 kV instruments; (4) stereo pairs of images with a large depth of focus may be obtained; and (5) under certain conditions, living cells in a wet atmosphere and confined in a special chamber may be observed (Parsons *et al.*, 1974). Some disadvantages are that contrast is reduced and focusing may be a problem.

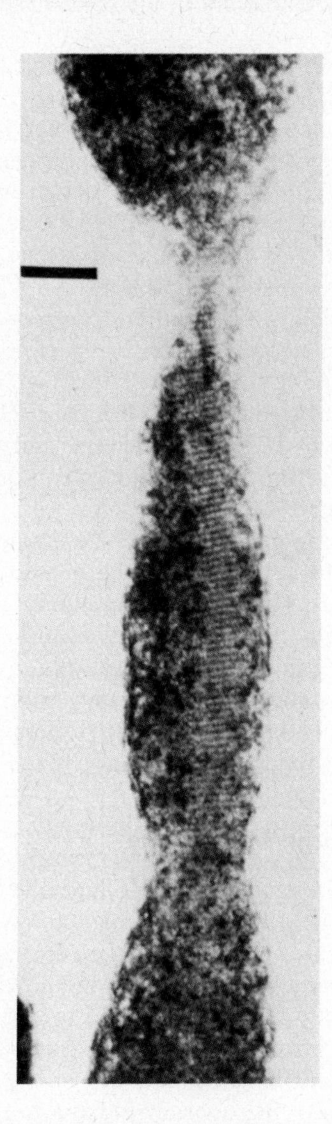

FIGURE 22. Section of rho form of *M. mycoides* strain Y2, showing longitudinally orientated intracytoplasmic striated structure. ×105,000. Marker = 0.1 μm. (From Rodwell *et al.*, 1972.)

Wolosewick and Porter (1976) obtained stereo images at 1 meV of fixed whole cells of a human diploid cell line (W1-38) and were able to discern nuclei, mitochondria, microtubules, ribosomes, and other components.

The "state of the art" in terms of the observation of living cells is in its infancy and the present images may be no better than those obtained by light microscopy; the possibilities for marked improvements, however, are good.

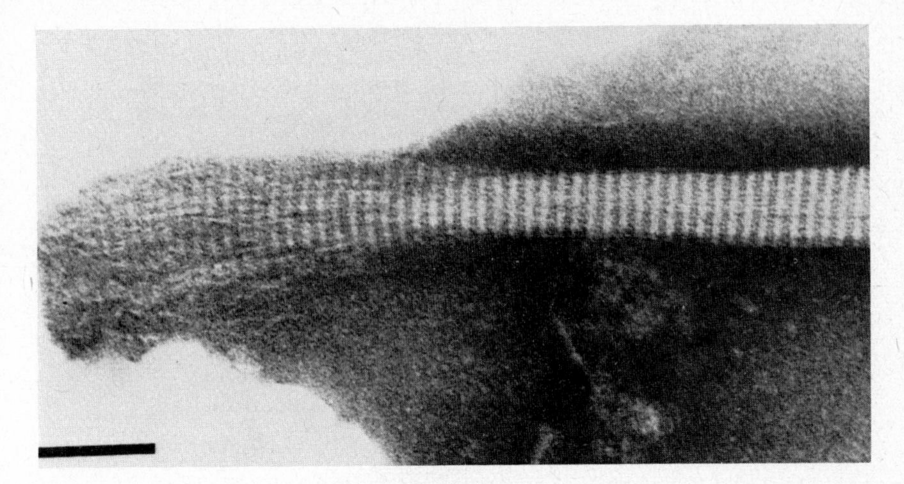

FIGURE 23. Negatively stained preparation of a fixed rho fiber from *M. mycoides*. The free end of the cylindrical fiber has frayed out to reveal its fibrillar nature. Note also the regular periodicity of cross-striations. Marker = 100 nm. (From Rodwell *et al.*, 1972.)

As far as the present writer is aware, the only study that has employed high-voltage electron microscopy for the study of mycoplasmas is that of Green and Hanson (1973), with *M. meleagridis*. It is an excellent study, in that a variety of techniques have been utilized to probe into the organisms morphology and ultrastructure. A triple fixation was used with a reported osmolality of 2010 mOsm. In spite of the hypertonicity of the fixative and the condensed cytoplasm resulting from this, the overall fine structure is good. Cultures of organisms were processed by the "critical point" drying procedure to demonstrate three-dimensional morphology; others were frozen, fractured and freeze-etched, or embedded in resin for thick and thin sectioning. Osmium–ruthenium red and also potassium tellurite was used to demonstrate extramembranous (capsular) material.

Apart from the thick sectioned material, all other preparations were observed by a 50–100 kV electron microscope. Thick sections (0.5 μm) observed by a 1 meV electron microscope revealed chains of spherical and rod-shaped cells and chains of "streptococcal-like" cells (Fig. 24). Cell diameters ranged from 0.25 to 0.75 μm; in some cases the whole thickness of a cell could be observed. Areas of low electron density were also seen, suggesting nuclear material. It remains to be seen what details may be derived from fixed whole cells observed at 1 meV.

High-voltage electron microscopy should prove rewarding in the examination of long filaments of mycoplasmas for the location of DNA and in the study of the terminal processes of the motile species. It would be particularly interesting to see how well one could locate mycoplasmas either at the surface or intracellularly on eukaryotic cells by stereoimag-

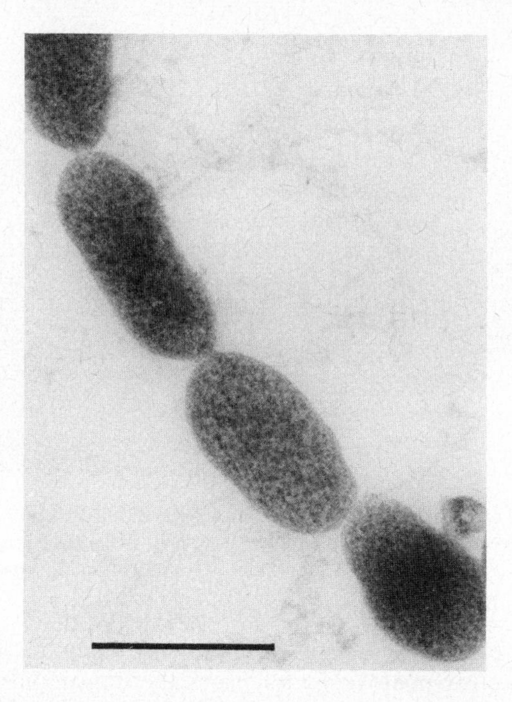

FIGURE 24. Thick section (0.5 μm) of cells of *M. meleagridis* examined at 1 meV. A chain of rod-shaped cells with a compact cytoplasm containing nuclear material and ribosomes is seen. Marker = 1.0 μm. (From Green and Hanson, 1973.)

ing. Many promising results are to be expected in this area of microscopy and the mycoplasmas as objects for scrutiny should not be left out. The three-dimensional form of the aster yellows agent (Hirumi and Maramorosch, 1973) in thick sections of the sieve tube elements might be worth examination.

## IV. FUTURE PROSPECTS

It is clearly essential to continue with studies similar to the elegant work involving the microcinematography of mycoplasmas during cycles of growth. With more species under examination, the promise of clarifying the puzzling back and forth gymnastics of certain species during cell division appears good.

In recent years, the care with which mycoplasmas have been processed for morphological and ultrastructural studies, bearing in mind the substantial susceptibility of these organisms to environmental factors, has noticeably improved. Particularly noteworthy is the desire of investigators to utilize more of the numerous and different techniques available for such studies.

The use of scanning electron microscopy to study virtually undisturbed growth on the surfaces of many substrates, especially growth of colonies, is expected to be intensified. With the gradual improvement of the resolution of this instrument more of the surface fine structure of the mycoplasmas should be revealed, including the form of the extramembranous layer found on some organisms and even, perhaps, the location of mycoplasma viruses at the cell surface.

The use of freeze-etching techniques has proven to be most useful and it is probably the only way to study membranes *en face*.

High-voltage electron microscopy, because of the cost and space requirements, is available to only a few investigators; nevertheless, some unique findings are likely to be forthcoming if only in respect to eukaryotic cells and select bacteria. Whether mycoplasmas will tolerate the intense electron beam and X rays produced by these instruments during observation in a "wet" chamber is a moot point.

When more of the mycoplasma infecting plants are isolated, they also will have to undergo the rigors of morphological and ultrastructural examination. For steady progress to be maintained, it behooves the mycoplasma morphologists to grasp new techniques and apply them to the study of these intriguing organisms whenever the opportunities arise.

## ACKNOWLEDGMENTS

The author wishes to thank Dr. D. Luchtel of the University of Washington for permission to use Fig. 6, and Dr. Kim of Ohio State University for Figs. 8 and 9. Thanks are also due to Dr. J. G. Holt, Editor and Secretary of Bergey's Manual Trust and the publishers of "Bergey's Manual of Determinative Bacteriology, 8th ed., 1974, Academic Press Inc., for permission to abstract portions of the text relating to the morphology of mycoplasmas. The work in this chapter relating to the author was carried out under NIAID grant #A109586.

## REFERENCES

Allen, T. C., Stevens, J. O., Florance, E. R., and Hampton, R. O. (1970). *J. Ultrastruct. Res.* **33,** 318–331.

Amako, K. (1975). *Jpn. J. Microbiol.* **19,** 452–455.

Anderson, D. L., Pollock, E. M., and Brower, L. F. (1965). *J. Bacteriol.* **90,** 1768–1777.

Anderson, D. R., and Barile, M. F. (1965). *J. Bacteriol.* **90,** 180–192.

Barker, D. C., and Swales, L. S. (1972). *Cell Differ.* **1,** 307–315.

Bernstein-Ziv, R. (1969). *Can. J. Microbiol.* **110,** 1154–1162.

Bernstein-Ziv, R. (1971). *Can. J. Microbiol.* **17,** 1203–1205.

Biberfeld, G., and Biberfeld, P. (1970). *J. Bacteriol.* **102,** 855–861.

Black, F. T., Birch-Andersen, A., and Freundt, E. A. (1972). *J. Bacteriol.* **111,** 254–259.
Boatman, E. S. (1973a). *J. Infect. Dis.* **127,** Suppl., 12–14.
Boatman, E. S. (1973b). *Ann. N.Y. Acad. Sci.* **225,** 172–180.
Boatman, E. S., and Kenny, G. (1970). *J. Bacteriol.* **101,** 262–277.
Boatman, E. S., and Kenny, G. (1971). *J. Bacteriol.* **106,** 1005–1015.
Boatman, E. S., Cartwright, F., and Kenny, G. (1976). *Cell Tissue Res.* **170,** 1–16.
Bode, H. R., and Morowitz, H. J. (1967). *J. Mol. Biol.* **23,** 191–199.
Boyde, A., Bailey, E., Jones, S. J., and Tamarin, A. (1977). *In* "Scanning Electron Microscopy: Proc. 10th Ann. S.E.M. Symposium," Vol. I, pp. 507–518. IIT Research Institute, Chicago.
Bredt, W. (1968). *Proc. Soc. Exp. Biol. Med.* **128,** 338–340.
Bredt, W. (1969). *Experientia* **25,** 1118–1119.
Bredt, W. (1970). *Z. Med. Mikrobiol. Immunol.* **155,** 248–274.
Bredt, W. (1973). *Ann. N.Y. Acad. Sci.* **225,** 246–250.
Bredt, W., and Radestock, U. (1977). *J. Bacteriol.* **130,** 937–938.
Bredt, W., Heunert, H. H., Hofling, K. H., and Milthaler, B. (1973). *J. Bacteriol.* **113,** 1223–1227.
Bredt, W., Lam, W., and Berger, J. (1975). *J. Clin. Microbiol.* **2,** 541–545.
Cairns, J. (1963). *J. Mol. Biol.* **6,** 208–213.
Chu, H. P., and Horne, R. W. (1967). *Ann. N.Y. Acad. Sci.* **143,** 190–203.
Cohen, A. L. (1974). *In* "Principles and Techniques of Scanning Electron Microscopy" (M. A. Hayatt, ed.), Vol. 1, pp. 44–112. Van Nostrand-Reinhold, Princeton, New Jersey.
Cohen, A. L. (1977). *In* "Scanning Electron Microscopy: Proc. 10th Ann. S.E.M. Symposium," Vol. I, pp. 525–586. IIT Research Institute, Chicago.
Cole, R. M., Tully, J. G., Popkin, T. J., and Bové, J. M. (1973). *J. Bacteriol.* **115,** 367–386.
Collier, A. M. (1972). *Pathogenic Mycoplasmas, Ciba Found. Symp., 1972* pp. 307–327.
Collier, A. M., and Baseman, J. B. (1973). *Ann. N.Y. Acad. Sci.* **225,** 227–289.
Collier, A. M., and Clyde, W. A. (1971). *Infect. Immun.* **3,** 694–701.
Darland, G., Brock, T. D., Samsonoff, W., and Conti, S. F. (1970). *Science* **170,** 1416–1418.
Domermuth, C. H., Nielsen, M. H., Freundt, E. A., and Birch-Andersen, A. (1964). *J. Bacteriol.* **88,** 727–744.
Freundt, E. A. (1974). *In* "Bergey's Manual of Determinative Bacteriology" (R. E. Buchanan and N. E. Gibbons, eds.), 8th ed., Part 19, pp. 929–955. Williams & Wilkins, Baltimore, Maryland.
Furness, G., and DeMaggio, M. (1973). *J. Infect. Dis.* **127,** 563–566.
Furness, G., DeMaggio, M., and Whitescarver, J. (1975). *Tex. Rep. Biol. Med.* **33,** 415–422.
Green, F., III, and Hanson, R. P. (1973). *J. Bacteriol.* **116,** 1011–1018.
Hama, K., and Porter, K. R. (1969). *J. Microsc. (Paris)* **8,** 149–158.
Henning, U. (1975). *Annu. Rev. Microbiol.* **29,** 45–60.
Hirumi, H., and Maramorosch, K. (1973). *Ann. N.Y. Acad. Sci.* **225,** 201–222.
Ho, T. Y., and Ouinn, P. A. (1977). *In* "Scanning Electron Microscopy: Proc. 10th Ann. S.E.M. Symposium," Vol II, pp. 291–300. IIT Research Institute, Chicago.
Howard, C. J., and Gourlay, R. N. (1974). *J. Gen. Microbiol.* **83,** 393–398.
Hu, P. C., Collier, A. M., and Baseman, J. B. (1977). *J. Exp. Med.* **145,** 1328–1343.
Hubbard, J. C., and Kite, J. H. (1971). *Appl. Microbiol.* **22,** 120–130.
Hummeler, K., Tomassini, N., and Hayflick, L. (1965). *J. Bacteriol.* **90,** 517–523.
Hungate, R. E. (1966). "The Rumen and its Microbes." Academic Press, New York.
Hyatt, M. A., ed. (1970). "Principles and Techniques of Electron Microscopy," Vol 1, p. 412. Van Nostrand-Reinhold, Princeton, New Jersey.

Jacob, F., Brenner, S., and Cuzin, F. (1963). *Cold Spring Harbor Symp. Quant. Biol.* **28,** 329–348.
James, R., and Branton, D. (1971). *Biochim. Biophys. Acta* **233,** 504–512.
Jones, T. C., Minick, R. C., and Yang, L. (1977). *Am. J. Pathol.* **87,** 347–358.
Kammer, G. M., Pollack, J. D., and Klainer, A. S. (1970). *J. Bacteriol.* **104,** 499–502.
Kellenberger, E., Ryter, A., and Séchaud J. (1958). *J. Biophys. Biochem. Cytol.* **4,** 671–678.
Kim, C. K. (1977). Ph.D. Dissertation, Ohio State University, Columbus (unpublished).
Klainer, G. M., and Pollack, J. D. (1973). *Ann. N.Y. Acad. Sci.* **225,** 236–245.
Kleinschmidt, A., and Zahn, R. (1959). *Z. Naturforsch., Teil B* **16,** 770–779.
Lemcke, R. (1972). *J. Bacteriol.* **110,** 1154–1162.
Luft, J. (1971). *Anat. Rec.* **171,** 347–368.
Maniloff, J. (1970). *J. Bacteriol.* **102,** 561–572.
Maniloff, J. (1971). *Proc. Nat Acad. Sci. U.S.A.* **68,** 43–47.
Maniloff, J., and Morowitz, H. H. (1972). *Bacteriol. Rev.* **36,** 263–290.
Maniloff, J., and Ouinlan, D. C. (1974). *J. Bacteriol.* **120,** 495–501.
Maniloff, J., Morowitz, H. J., and Barrnett, R. J. (1965). *J. Bacteriol.* **90,** 193–204.
Morowitz, H. J., and Terry, T. M. (1969). *Biochim. Biophys. Acta* **183,** 276–294.
Morowitz, H. J., and Wallace, D. C. (1973). *Ann. N.Y. Acad. Sci.* **225,** 62–73.
Munkres, M., and Wachtel, A. (1967). *J. Bacteriol.* **93,** 1096–1103.
Muse, K. E., Powell, D. A., and Collier, A. M. (1976). *Infect. Immun.* **13,** 229–237.
Nelson, J. B., and Lyons, M. J. (1965). *J. Bacteriol.* **90,** 1750–1763.
Organick, A. B., Siegesmund, K. A., and Lutsky, I. I. (1966). *J. Bacteriol.* **92,** 1164–1176.
Parsons, D. F., Uydess, I., and Matricardi, V. R. (1974). *J. Microsc. (Paris)* **100,** 153–167.
Peterson, J. E., Rodwell, A. W., and Rodwell, E. S. (1973). *J. Bacteriol.* **115,** 411–425.
Powell, D. A., Hu, P. C., Wilson, M., Collier, A. M., and Baseman, J. B. (1976). *Infect. Immun.* **13,** 959–966.
Razin, S. (1967). *Ann. N.Y. Acad. Sci.*. **143,** 115–129.
Razin, S. (1972). *Biochim. Biophys. Acta* **265,** 241–296.
Razin, S. (1973). *Adv. Microb. Physiol.* **10,** 28–66.
Razin, S., Hasin, M., Ne'eman, Z., and Rottem, S. (1973). *J. Bacteriol.* **116,** 1421–1435.
Razin, S., Masover, G. K., Palant, M., and Hayflick, L. (1977). *J. Bacteriol.* **130,** 464–471.
Robertson, J., and Smook, E. (1976). *J. Bacteriol.* **128,** 658–660.
Robertson, J., Gomersall, M., and Gill, P. (1975a). *J. Bacteriol.* **124,** 1007–1018.
Robertson, J., Gomersall, M., and Gill, P. (1975b). *J. Bacteriol.* **124,** 1019–1022.
Robinson, M. I., Allison, M. J., and Hartman, P. A. (1975). *Int. J. Syst. Bacteriol.* **25,** 173–181.
Rodwell, A. W. (1965). *J. Gen. Microbiol.* **40,** 227–234.
Rodwell, A. W., Peterson, J. E., and Rodwell, E. S. (1972). *Pathogenic Mycoplasmas, Ciba Found. Symp., 1972* pp. 123–144.
Rodwell, A. W., Peterson, J. E., and Rodwell, E. S. (1973). *Ann. N.Y. Acad. Sci.* **225,** 190–200.
Rodwell, A. W., Peterson, J. E., and Rodwell, E. S. (1975). *J. Bacteriol.* **122,** 1216–1229.
Schiefer, H., Krauss, H., Brunner, H., and Gerhardt, U. (1976). *J. Bacteriol.* **127,** 461–468.
Spears, D. M., and Provost, P. J. (1967). *Can. J. Microbiol.* **13,** 213–225.
Tanaka, H., Hall, W. T., Sheffield, J. B., and Moore, D. H. (1965). *J. Bacteriol.* **90,** 1735–1749.
Tillack, T. W., Carter, R., and Razin, S. (1970). *Biochim. Biophys. Acta* **219,** 123–130.
Tourtellotte, M. E., and Zupnik, J. S. (1973). *Science* **179,** 84–86.
Tourtellotte, M. E., Branton, D., and Keith, A. (1970). *Proc. Natl. Acad. Sci. U.S.A.* **66,** 909–916.

Tully, J. G., and Razin, S. (1970). *J. Bacteriol.* **103,** 751–754.
Whitescarver, J., and Furness, G. (1975). *J. Med. Microbiol.* **8,** 349–355.
Wilson, H. M., and Collier, A. M. (1976). *J. Bacteriol.* **125,** 332–339.
Wolanski, B., and Maramorosch, K. (1970). *Virology* **42,** 319–327.
Wolosewick, J. J., and Porter, K. R. (1976). *Am. J. Anat.* **147,** 303–324.
Virkola, P. (1972). *Acta Pathol. Microbiol. Scand.* **80,** 388–396.
Zucker-Franklin, D., Davidson, M., and Thomas, L. (1966a). *J. Exp. Med.* **124,** 521–532.
Zucker-Franklin, D., Davidson, M., and Thomas, L. (1966b). *J. Exp. Med.* **124,** 533–542.

# 4 / NUTRITION, GROWTH, AND REPRODUCTION

*A. W. Rodwell, and Alana Mitchell*



## I. NUTRITION

## A. General Considerations

### 1. Problems in Defining Nutritional Requirements of Mycoplasmas

Knowledge of the nutritional requirements of a microorganism provides extensive information on its biosynthetic capabilities. It also provides a valuable experimental tool by allowing controlled manipulation of the nutrients supplied for growth of the organism and hence an estimation of the effects of these alterations on cellular metabolism. Without such information, assessment of the mechanisms for control of macromolecular syntheses and regulation of metabolic pathways is not possible. Despite the obvious benefits in defining the nutritional requirements of mycoplasmas, our knowledge in this field is still very limited. Completely defined media have been described for only two species: *Mycoplasma mycoides* and *Acholeplasma laidlawii* (Section I, B). Some of the difficulties in defining the nutritional requirements have been discussed previously (Rodwell, 1974).

**a. Complex requirements for nutrients and vitamins.** In keeping with the small size of their genome, it is to be expected that most mycoplasmas will have limited biosynthetic capabilities and, accordingly, will require a wide array of precursor molecules for macromolecular syntheses. Vitamins, nucleic acid precursors, amino acids, and lipids may have to be supplied, often in a complex yet membrane-permeable form. For example, neither pantothenate nor pantotheine can serve as a precursor of coenzyme A in *M. mycoides* (Rodwell, 1969b), whereas *A. laidlawii* (Rottem *et al.*, 1973a) and *Mycoplasma* strain J (Lund and Shorb, 1966) can use pantotheine. *Mycoplasma synoviae* requires the complete nicotinamide adenine dinucleotide (NAD) molecule or at least a precursor more complex than nicotinamide (Chalquest and Fabricant, 1960), although strains which can synthesize NAD from nicotinamide can be selected (DaMassa and Adler, 1975).

Uptake of some amino acids may be restricted due to unfavorable competition for a common transport system, or, following uptake, a rapidly catabolized amino acid may become limiting as an essential nutrient. These difficulties are not peculiar to mycoplasmas but also arise in relation to the nutrition of other bacteria. [See Guirard and Snell (1962) for a discussion of the amino acid requirements of microorganisms].

**b. Provision of lipids and lipid precursors.** A problem even more complicated than the supply of amino acids is that of providing lipids in an assimilable but nontoxic form. Apart from acholeplasmas and thermoplasmas, all of the Mollicutes require exogenous sterol. Many also require long-chain fatty acids but present exposed plasma membranes which are susceptible to damage by surface active agents. Sterols can be supplied together with a carrier protein, such as the defatted serum fraction C (Rodwell, 1969a) or lipoprotein fraction (Smith and Boughton, 1960). For the growth of yeast, Thompson *et al.* (1973) supplied ergosterol complexed to yeast phosphomannan. Phosphomannan was used successfully for the growth of mycoplasmas (A. W. Rodwell, unpublished; Archer, 1975), although growth rate and cell yield were lower than with fraction C. Sterols may also be dispersed in nonionic surfactants, such as Tweens, or in mixed micelles of water-dispersible fatty acid esters, such as the diacetoxysuccinoyl esters of monoglycerides, which can also serve as less toxic sources of fatty acids (Lund and Shorb, 1966).

Long-chain unsaturated fatty acids cause lysis of mycoplasmas at extremely low concentrations, so if required, they must be supplied in a nontoxic form. If they are supplied as water-dispersible esters they must be released by mycoplasma lipase (Razin and Rottem, 1963; Rottem and Razin, 1964) at just the right rate to avoid the accumulation of free fatty acids to toxic levels. Unesterified fatty acids can be complexed with serum albumin. Where strict control over the fatty acid composition of the membrane lipids is desired, the endogenous fatty acids bound to albumin, some of which are probably present in esterified form, must first be removed. The efficient removal of both free and esterified fatty acids, without loss of the fatty acid-binding properties of the protein, can be achieved by mild alkaline methanolysis (Rodwell and Peterson, 1971). While it is probably the most convenient available means for supplying lipids in a nontoxic form, the use of carrier proteins has the disadvantage of obscuring the requirements for amino acids of those mycoplasmas which can obtain amino acids from the degradation of protein.

None of the above solutions to the problem of lipid provision is entirely satisfactory. The elaboration of synthetic, water-soluble, nondegradable polymers with fatty acid and sterol-binding properties similar to those of serum albumin and serum lipoproteins would fulfill an urgent need. Crosslinked polyethyleneimines, in which fatty acid side chains were introduced in amide linkage, had high fatty acid-binding properties (Klotz and Sloniewsky, 1968). One such polymer, PEI-6, containing approximately 10% lauroyl side chains, failed to replace serum albumin in medium C2 (Rodwell, 1969a) for growth of *M. mycoides* strain Y (A. W. Rodwell, unpublished). Possibly, this polymer binds fatty acids too strongly, so rendering them unavailable for incorporation, or the polymer may have a deleterious effect on the organism due to its strong basic charge. In either case further investigation of similar polymers seems justified.

## 2. Energy Sources, Tonicity, and Accumulation of Metabolic Products

**a. Energy sources.** Mycoplasmas generally possess an inefficient energy-yielding mechanism and so must consume large amounts of substrate to supply sufficient energy for macromolecular syntheses. On the basis of acid production from carbohydrate, the genus *Mycoplasma* can be divided into fermentative or nonfermentative species. The energy-yielding metabolism of both groups has been reviewed (Rodwell, 1969c). Briefly, those species which ferment carbohydrate usually produce acid from glucose, fructose, mannose, maltose, starch, and glycogen. Anaerobically, lactate is the end product of glycolysis and aerobically, glucose is oxidized to acetate and carbon dioxide. Glycolysis alone would yield 2 moles of ATP per mole of glucose metabolized, and the demonstration of phosphate acetyl transferase (EC 2.3.1.8.) and acetate kinase (EC 2.7.2.1.) activities in both fermentative and nonfermentative species of mycoplasma (Kahane *et al.* 1978) suggests that under aerobic conditions these species might derive a further 2 moles of ATP from pyruvate oxidation *via* acetyl CoA.

The nature of the energy sources of the nonfermentative species is less clear. Schimke *et al.* (1966) found that growing cells of *Mycoplasma arthritidis* (formerly reported as *Mycoplasma hominis* 07)' a representative nonfermenting mycoplasma, can derive sufficient energy for synthesis of macromolecules from the metabolism of arginine via the arginine dihydrolase pathway. Most nonfermentative species tested have been found to possess arginine deiminase (Tully and Razin, 1977), the enzyme catalyzing the first reaction in this pathway; and of those species examined, all have possessed the second enzyme, ornithine transcarbamylase (Barile *et al.*, 1966). However, the presence of the complete dihydrolase pathway in strains shown to possess only arginine deiminase can only be assumed. A recent examination of the effect of arginine supplementation on the yield of cellular protein and specific activity of arginine deiminase in several mycoplasma species, and particularly *M. hominis* (Fenske and Kenny, 1976), has provided evidence that the arginine dihydrolase pathway is not the major energy-yielding system in arginine-requiring, nonglycolytic strains, and that arginine is used only as an alternative source of energy. *Mycoplasma hominis* may derive additional ATP from the pathway proposed by Kahane *et al.* (1978) *via* phosphate acetyl transferase and acetate kinase and, as discussed in Section I, C, 2, *M. arthritidis* is able to use a number of other substrates as energy sources. These observations imply that if arginine functions as a source of energy and as an essential nutrient, it may tend to be rapidly depleted from the medium through energy-providing metabolism. Thus, the maintenance of an adequate supply of arginine is important for continued growth of those organisms possessing the arginine dihydrolase pathway.

The energy source for *Ureaplasma* is not known. All known species of *Acholeplasma, Spiroplasma, Thermoplasma,* and *Anaeroplasma* catabolize glucose. *Spiroplasma citri* also catabolizes arginine (Townsend, 1976).

**b. Tonicity and pH of growth media.** Strongly growing fermentative mycoplasmas produce large amounts of acid during sugar fermentation. Growth is best at pH values in the range of 7–8 and a decrease in pH to less than 6.5 causes cessation of growth followed by rapid death of cells. Nonfermenting, arginine-utilizing mycoplasmas growing in arginine-supplemented media produce ammonia, as do ureaplasmas, by hydrolysis of urea (Section I, C, 3), so the problem is then to prevent an excessive increase in pH.

The growth medium must therefore be well buffered. Apart from pH, a second consideration in the choice of a suitable buffer is the tonicity of the medium. The maximum osmotic pressure tolerated by most mycoplasmas lies in the range 12–14 atm (Leach, 1962), and so imposes a limitation on the concentration of buffer salts present. In addition, cellular metabolism, such as the consumption of carbohydrate during fermentation or the release of ammonia from arginine or urea, causes an increase in osmolarity during growth to further limit the choice of an appropriate buffer concentration.

Sodium phosphate is a suitable buffer for growth of *M. mycoides* (Rodwell, 1967), but is inhibitory for *A. laidlawii* (Razin and Cohen, 1963; Tourtellotte *et al.*, 1964). Since some samples of phosphate have been found to contain toxic metal ions (A. W. Rodwell, unpublished), good quality salts should be used. As shown in Fig. 1, growth of *A. laidlawii* in an undefined medium is markedly improved by the addition of sodium *N'*-2-hydroxyethylpiperazine-*N'*-2-ethanesulfonate (HEPES). This buffer at 0.075 *M* was also satisfactory for growth of *M. mycoides* and allowed a decrease in phosphate concentration of the medium from 140 m*M* to 0.5 m*M*.

**c. Accumulation of toxic metabolic products.** Gentle aeration causes an increase in growth rate for some mycoplasmas and under these conditions hydrogen peroxide is produced either from glycerol oxidation by the flavoprotein glycerol-3-phosphate oxidase in *M. mycoides* (Rodwell and Rodwell, 1954; Rodwell, 1967) and *Mycoplasma pneumoniae* (Low *et al.*,

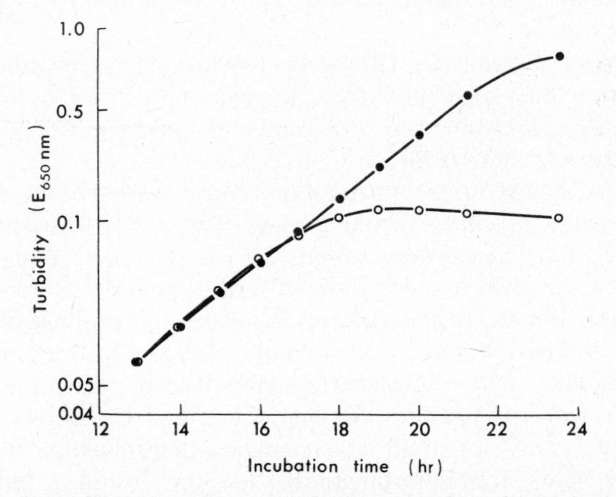


FIGURE 1. The effect of sodium *N'*-2-hydroxyethylpiperazine-*N'*-2-ethanesulfonate (HEPES) buffer on the growth of *A. laidlawii* (oral strain) in modified Edward medium containing 2% v/v PPLO serum fraction (Razin, 1963). An absorbance of 0.5 corresponds to 0.2 mg cell protein/ml. ●, 0.075 *M* sodium HEPES; ○, 0.075 *M* sodium chloride.

1968), or from flavin-terminated oxidation of NADH in other fermentative species. Hydrogen peroxide may accumulate to toxic levels since the addition of catalase to *M. mycoides* growing in an undefined medium increased the growth rate and prolonged viability (Buttery, 1967).

In addition to causing increase in the pH of cultures, the accumulation of ammonia from metabolism of arginine or urea may be toxic to cells. From work on ureaplasmas, attempts to resolve contributions from pH changes and ammonia toxicity to growth inhibition have led to conflicting reports (Section I, C, 3).

### B. Nutritional Requirements of *M. mycoides* and *A. laidlawii*

In a completely defined medium for *M. mycoides* subsp. *mycoides* (strain Y), fatty acids were supplied as a mixture of synthetic diacetoxysuccinoyl esters of monoolein and monopalmitin (Rodwell, 1969b). Since this medium supported a growth rate and yield similar to those obtainable in the best nondefined media, it enabled analysis of the nutritional requirements of strain Y. Strain V5 did not grow in the fully defined medium. Growth of other mycoplasmas in this medium was not tested, but several strains of *M. mycoides* subsp. *mycoides* and *capri* and bovine arthritis strains grew well in a partly defined medium (designated medium C3; Rodwell, 1969b), in which fatty acids were complexed with defatted serum albumin. Their requirements are probably similar to those of strain Y.

The use of medium C2 (Rodwell, 1969a) enabled modification and control of the fatty acid and sterol composition of strain Y (Rodwell and Peterson, 1971; Rodwell *et al.*, 1972) and of the PG3 strain of *M. mycoides* subsp. *capri* (Archer, 1975).

Extensive studies on the nutrition of *A. laidlawii* led to the description of a fully defined medium for the growth of strain B (Tourtellotte *et al.*, 1964). Free fatty acids were supplied in the medium, along with two peptides of unknown function derived from ribonuclease. It was shown subsequently that the peptides were serving the dual purpose of providing a readily assimilable source of glutamic acid and also of binding and detoxifying fatty acids (Tourtellotte, 1969). These studies together with the earlier work of Razin (1962) and Razin and Cohen (1963) on the nutrition of *A. laidlawii* strain A have enabled determination of the essential requirements. It is unfortunate that the nutrition of *A. laidlawii* has been the subject of little further investigation, since it seems likely that with the knowledge now available it might not be very difficult to devise a fully defined medium supporting abundant growth. Such a medium would greatly facilitate studies on the macromolecular syntheses and regulatory mechanisms of this organism.

It is instructive to compare, as far as they are known, the nutritional requirements of the two species, *M, mycoides* and *A. laidlawii* (Table I).

### 1. Inorganic Requirements

The inorganic requirements have not been studied in detail. As expected, a growth requirement was found for $K^+$, $Mg^+$, and $PO_4^{2-}$. At concentrations greater than 0.04–0.06 *M*, KCl was reported to inhibit growth of *A. laidlawii* (Razin and Cohen, 1963) and *M. mycoides* (Leach, 1962; Rodwell, 1967). However, Leach (1962) and Cho and Morowitz (1969) reported that at a concentration of 0.14 *M* or greater, KCl did not inhibit growth of *Mycoplasma gallisepticum* or *A. laidlawii. Mycoplasma mycoides* could be adapted to grow with 0.14 *M* KCl (A. W. Rodwell, unpublished).

### 2. Amino Acids

The amino acid requirements of both species are complex. *Mycoplasma mycoides* strain Y required all amino acids apart from cystine, aspartic acid, and glutamic acid, although cysteine, asparagine, and glutamine were essential. Alanine was only essential in the absence of pyridoxal. An observed nutritional antagonism between alanine and glycine could be overcome either by increasing the ratio of glycine to alanine or by supplying alanine as a suitable peptide, such as L-$Ala_3$ or L-$Ala_4$. Alanylalanine inhibited growth (Rodwell, 1969b).

*Acholeplasma laidlawii* strain B is slightly less exacting in its requirements for amino acids. When provided with alanine, it did not require tyrosine or phenylalanine, suggesting the presence of an operative shikimic acid pathway for the synthesis of aromatic amino acids (Tourtellotte *et al.*, 1964). Supporting evidence for the presence of this pathway was the incorporation of $^{14}C$-labeled shikimic acid (M. E. Tourtellotte, personal communication). Strain B readily incorporated glutamine and asparagine and, although glutamic acid could not be used, aspartic acid supported growth in the absence of asparagine. Cysteine and cystine were interchangeable medium constituents (Tourtellotte *et al.*, 1964). Strain A possesses a glutamate dehydrogenase resembling the mammalian enzyme in that it has dual coenzyme specificity for NAD and NADP (Yarrison *et al.*, 1972). This is a significant finding: In other microorganisms glutamate dehydrogenase is a key enzyme in that it links carbohydrate and nitrogen metabolism and is commonly a control point. It is not known whether the glutamate dehydrogenase in *A. laidlawii* is an allosteric enzyme: Its activity was unaffected by AMP, ADP, or ATP, in which respect the enzyme

TABLE I. **Nutritional Requirements of *M. mycoides* and *A. laidlawii***

| | *M. mycoides* | *A. laidlawii* |
|---|---|---|
| Inorganic | | |
| Cations | $K^+$, $Mg^{2+}$ | $K^+$, $Mg^{2+}$ |
| Anions | $PO_4^{3-}$ | $PO_4^{3-}$ |
| Amino acids | Ala, Arg, Asn, Cys, Gln, Gly, His, Ile, Leu, Lys, Met, Phe, Pro, Ser, Thr, Trp, Tyr, Val | Ala, Arg, Asn, cystine, Gln, Gly, His, Ile, Leu, Lys, Met, Pro, Ser, Thr, Trp, Val |
| Saccharides | Glucose | Glucose |
| Lipids and precursors | | |
| Fatty acids | Essential | Unsaturated fatty acid essential |
| Acetate | Not incorporated | Incorporated into long-chain saturated fatty acids[a] and carotenoids[b] |
| Sterol | Essential | Not required |
| Glycerol | Essential for synthesis of glycerides | Not required |
| Nucleic acid precursors | Guanine, uracil, thymine | Adenosine, guanosine, cytidine |
| Vitamins and coenzymes | Coenzyme A, riboflavin, nicotinic acid, thiamine, (pyridoxamine)[c], α-lipoic acid | Pantetheine, riboflavin, nicotinic acid, thiamine, pyridoxal, biotin, folinic acid |
| Polyamines | Spermine or spermidine stimulate growth | Not known to be required |

[a] Acetate is normally formed endogenously from glucose. Exogenous odd-numbered and branched-chain fatty acids of up to five carbon atoms can serve as primers for the synthesis of odd-numbered and branched-chain fatty acids, respectively, by strain B (Saito *et al.*, 1977).

[b] Smith and Rothblat (1962).

[c] Pyridoxamine was required only in the absence of alanine or alanyl peptides.

resembled those from other microorganisms. Its presence in *A. laidlawii* points to the probable ability of this mycoplasma to synthesize or degrade glutamate and other amino acids, but so little is known about the intermediary metabolism of *A. laidlawii* that it is difficult to assess the importance of glutamate dehydrogenase in amino acid synthesis or degradation. How for example, are keto acids other than pyruvate generated or metabolized? There is no evidence for a functional TCA cycle in *A. laidlawii* (Tourtellotte and Jacobs, 1960).

### 3. Carbohydrates

Glucose, or some other fermentable sugar, is essential for both species. *Mycoplasma mycoides* incorporated glucose carbon into its extracellular galactan, and into glycolipids and pentoses (Plackett, 1967). Slater and Folsome (1971), in the only demonstration of control of enzyme synthesis in a mycoplasma, showed that transfer of *A. laidlawii* strain A from glucose to maltose resulted in the induction of $\alpha$-glucosidase.

### 4. Lipids and Precursors

**a. Fatty acids.** *Mycoplasma mycoides* is unable to synthesize or alter the chain length of either saturated or unsaturated fatty acids. This complete lack of metabolic activity toward fatty acids greatly facilitated interpretation of experiments examining the fatty acid requirements of *M. mycoides* (Rodwell and Peterson, 1971) since only the fatty acids supplied were incorporated into membranes. Elaidate, *trans*-vaccenate, *cis*-12-octadecenoate, and the branched-chain saturated fatty acids, isopalmitate, isostearate, and anteisoheptadecanoate, each supported optimal growth. With most *cis*-monoenoic acids there was little or no growth unless a straight-chain saturated fatty acid was also supplied. Both the chain length and the position of the double bond of the monoenoic acid markedly affected the chain length of the saturated acids optimal for growth. In some cases, shifting of cells to growth on different fatty acids caused the onset of a lag period. The biochemical basis for this lag is not known, although several possible reasons for its occurrence have been discussed (Rodwell and Peterson, 1971). The above results are consistent with the idea that any single fatty acid or combination of fatty acids supports growth provided that, at the growth temperature, the fluidity of the membrane lipids lies within certain limits.

Tourtellotte *et al.* (1964) reported that *A. laidlawii* strain B would not grow in a defined medium completely devoid of exogenous long-chain fatty acids. However, strain B (and *Acholeplasma* sp. strain KHS) was found to be capable of sustained growth in a defatted medium containing only small amounts of exogenous fatty acids (Henrickson and Panos,

1969). The amounts of unsaturated fatty acids in the lipids of cells grown in this medium did not exceed about 5% of the total and could probably be accounted for by the incorporation of exogenous unsaturated fatty acids present in the defatted medium. Laurate, myristate, and palmitate together accounted for about 90% of the total, myristate predominating. The addition of exogenous monoenoic fatty acids failed to alter the growth rate, the yield, the fatty acid composition, or the morphology of strain B. However, the concentrations added (0.5–1.0 $\mu$g/ml) may have been too low to produce a significant effect on any of these parameters. McElhaney and Tourtellotte (1969) found that palmitate was the predominant fatty acid in the lipids of strain B grown in a defatted, undefined medium. Large amounts of exogenous unsaturated and branched-chain fatty acids were incorporated when provided, and the resultant changes in the fatty acid composition of the cellular lipids were accompanied by marked morphologic changes (Chapter 10).

*Acholeplasma laidlawii* strain A differed from strain B in that the addition of an octadecenoic acid was essential for growth in the defatted medium of Henrickson and Panos (Panos and Rottem, 1970). An octadecenoic acid was replaceable by a cyclopropane-ring fatty acid (Panos and Leon, 1974) and, by analogy with the results obtained with *M. mycoides* strain Y, it should also be replaceable by a suitable branched-chain saturated fatty acid.

This apparent difference in the fatty acid growth requirements of strains A and B cannot be accounted for by any known differences in their ability to synthesize fatty acids. Both strains incorporated $^{14}$C-labeled acetate into straight-chain saturated fatty acids (Pollack and Tourtellotte, 1967; Rottem and Razin, 1967). Recent studies (Saito *et al.*, 1977) have shown that strain B is capable of *de novo* biosynthesis of saturated fatty acids from acetate, a normal product of glucose metabolism. A variety of exogenous precursors of up to five carbon atoms, including iso- and anteiso monomethyl branched-chain fatty acids, could serve as primers. When strain B was supplied with a branched-chain primer, branched-chain fatty acids comprised up to 80% of the fatty acids in the glycerolipids. It is likely that strain A also is capable of *de novo* synthesis of saturated fatty acids. A soluble system for the synthesis of saturated long-chain fatty acids from acetate required malonyl-coenzyme A (Rottem and Panos, 1970). In intact cells, malonyl-coenzyme A is no doubt a normal intermediate in saturated fatty acid biosynthesis. It would be of interest to determine whether exogenous branched-chain precursors (isobutyrate, isovalerate, and anteisovalerate) could serve as primers for the synthesis of branched-chain fatty acids by strain A and, if so, whether they would replace the growth requirement for an exogenous octadecenoic acid.

Both strains can elongate shorter chain monoenoic acids. Strain A elongated dodecenoic and tetradecenoic acids, but only to the hexadecenoic acid (Panos and Rottem 1970), whereas strain B (and *Acholeplasma* sp. strain KHS) elongated shorter chain precursors to octadecenoic acid (Romijn *et al.*, 1972; Panos and Henrickson, 1969). The inability of strain A to elongate monoenoic acid precursors beyond hexadecenoic acid is unlikely to account for the reported difference in the fatty acid growth requirements of strains A and B since the amounts of such precursors present in the defatted media would be extremely small.

Strain B can, to some extent, compensate for an imbalance in the supply of exogenous fatty acids by regulating the amounts and chain lengths of the saturated fatty acids incorporated into the glycerolipids (Silvius *et al.*, 1977). It does this by two mechanisms: by excreting biosynthesized fatty acids into the growth medium as free fatty acids and by altering the total amount or the average chain length of the saturated fatty acids synthesized. Thus it appears that strain B compensates for a relative deficiency of exogenous unsaturated fatty acids or their equivalent by incorporating larger amounts of shorter chain saturated fatty acids, chiefly myristic acid, into the glycerolipids. The extent to which strain A can similarly compensate is not known. The explanation for its greater dependence on an exogenous supply of unsaturated fatty acids may lie in its more limited ability to compensate for a deficiency of unsaturated fatty acids. An alternative and perhaps more likely explanation for the apparent difference in the fatty acid growth requirements of the two strains may result from a potential for greater growth of strain A (oral strain) in the undefined defatted medium of Panos and Rottem (1970) as compared with that of strain B. It is possible that residual unsaturated fatty acids were incorporated from the medium to allow limited growth of strain A before eventual lysis due to a fatty acid imbalance in the membrane lipids. The growth of strain B, on the other hand, may have been limited by a nutrient other than an unsaturated fatty acid. Thus, there may be no difference in the fatty acid growth requirements of strains A and B.

**b. Sterols.** A growth requirement for a suitable sterol is a major criterion distinguishing the Mycoplasmataceae from the Acholeplasmataceae. Sterols are incorporated into the plasma membrane of mycoplasmas without change in structure (Smith, 1962; Rodwell, 1963; Argaman and Razin, 1965). However, according to Lynn and Smith (1960), some mycoplasmas are capable of esterifying cholesterol with short-chain fatty acids, such as butyrate, and *Mycoplasma gallinarum* strain J is capable of glycosylating cholesterol (Smith, 1971). Thus the sterol composition of mycoplasma membranes can be modified by altering the sterol supplied for growth. Cholesterol could be replaced for growth of mycoplasmas by 3βcholestanol, 7-cholesten-3β-ol, β-sitosterol, stigmasterol, and ergosterol (Rodwell, 1963; Smith, 1964; Archer, 1975). The essential requirements are that the sterol molecule should have a planar nucleus, an unesterified 3β-hydroxyl group, and a side chain at C17 (Smith, 1964). These are precisely the requirements necessary to exert an ordering effect on the acyl chains of fatty acids in lipid monolayers (Demel *et al.*, 1972). Some other sterols which do not possess these essential features, such as 3α-cholesterol, 3α-cholestanol, coprostan-3β-ol, coprostan-3α-ol, 4-cholesten-3-one, and 5-cholesten-3-one, not only could not replace cholesterol but competitively inhibited growth in the presence of cholesterol (Smith, 1964; Rodwell, 1963). It is not clear why sterols are essential for growth of mycoplasmas, but not for acholeplasmas and other bacteria. Sterol normally accounts for about 20% by weight of the total membrane lipids of a mycoplasma. *Mycoplasma mycoides* strains Y and PG3 were adapted to grow with very low concentrations of cholesterol (Rodwell *et al.*, 1972; Rottem *et al.*, 1973b), but even when these organisms were supplied with a mixture of fatty acids which might be expected to confer the right degree of fluidity on their membranes, it was not possible to further train them to dispense fully with sterol. Cells grown at low cholesterol concentration had only about 0.4% cholesterol by weight of total membrane lipid, which presumably accounts for their greater fragility relative to cells grown on normal levels of cholesterol.

**c. Glycerol.** Glycerol is an essential nutrient for *M. mycoides* subsp. *mycoides* and *capri*. This species, lacking NAD-linked glycerol-3-phosphate dehydrogenase, has no reversible pathway for the reduction of triosephosphate and therefore requires exogenous glycerol for the synthesis of glycerides. Glycerol is phosphorylated to glycerol 3-phosphate which may function as the precursor of glycerides or may undergo aerobic, flavoprotein-mediated oxidation to triosephosphate, with the concomitant formation of hydrogen peroxide (Rodwell, 1967). Consequently, under aerobic growth conditions, larger amounts of glycerol carbon enter the triosephosphate pool, and more glycerol must be supplied to ensure that it does not become growth-limiting. The flavoprotein pathway for glycerol oxidation is extremely active and even in unstirred cultures only about 10% of the glycerol consumed was incorporated (Plackett, 1961). Of this, about 90% was present in the glycerides and the remainder, greatly diluted by carbon from glucose, was present in the pentoses of nucleic acids.

## 5. Nucleic Acid Precursors

***a. Mycoplasma mycoides.*** The minimal requirements of *M. mycoides* for precursors of nucleic acids are met by the bases guanine, thymine, and uracil. Investigation of the pathways for ribonucleotide biosynthesis in this organism (Mitchell and Finch, 1977; Mitchell *et al*., 1978) has indicated that it does not possess pathways for the synthesis of nucleotides *de novo* but is capable of many nucleotide interconversions, as shown in Fig. 2.

Uracil alone provided all pyrimidine ribonucleotides. Cytosine was not incorporated in the presence of uracil, whereas cytidine competed effectively with uracil to provide most of the cytidine nucleotide and also an appreciable proportion of uridine nucleotide. Uridine was rapidly degraded to uracil. All purine nucleotides could be derived from guanine. Adenine completely excluded, and hypoxanthine partially excluded guanine from incorporation into adenine nucleotides, but neither adenine nor hypoxanthine could act as precursor of guanine nucleotides. Exogenous guanosine, inosine, and adenosine underwent rapid phosphorolysis to the corresponding bases (Fig. 2). The growth requirement for thymine suggests that *M. mycoides* is incapable of methylating deoxyuridine monophosphate but does possess thymidine phosphorylase and thymidine kinase. Pentoses are not required, but little is known of their synthesis in mycoplasmas.

**b. *Acholeplasma laidlawii.*** The requirements for precursors of nucleic acids and the pathways of nucleotide interconversion in *A. laidlawii* are less well understood. Apparent discrepancies regarding growth requirements may reflect strain differences rather than inconsistencies. Thus the four nucleosides adenosine, guanosine, cytidine, and thymidine could

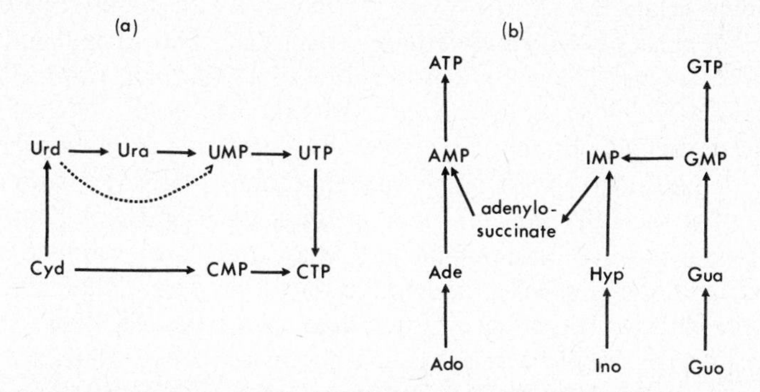


FIGURE 2. Proposed pathways for nucleotide biosynthesis by *M. mycoides* subsp. *mycoides*. (a) Pyrimidine nucleotides; (b) purine nucleotides. Symbols: probable major pathways, ⟶; possible minor pathway ---→; Cyd, cytidine; Urd, uridine; Ura, uracil; UMP, uridine 5′-monophosphate; UTP, uridine 5′-triphosphate; CTP, cytidine 5′-triphosphate; CMP, cytidine 5′-monophosphate; Ado, adenosine; Ade, adenine; AMP, adenosine 5′-monophosphate; ATP, adenosine 5′-triphosphate; Ino, inosine; Hyp, hypoxanthine; IMP, inosine 5′-monophosphate; Guo, guanosine; Gua, guanine; GMP, guanosine 5′-monophosphate; GTP, guanosine 5′-triphosphate. (From Mitchell and Finch, 1977.)

supply the nucleic acid precursors for strain A (Razin, 1962) and strain B (Tourtellotte *et al.*, 1964). In the presence of folinic acid, strain A did not require thymidine (Razin, 1962). Smith and Hanawalt (1968) found strain B to have an absolute requirement for uridine and thymidine, while deoxyguanosine and guanosine stimulated growth. This requirement for thymidine, which could not be met by thymine, might reflect a deficiency of folinic acid. Liška and Smith (1974) found that yet another strain (LA-1) is able to grow on guanine, thymine, and either uracil or cytosine.

### 6. Vitamins

*Mycoplasma mycoides* was unable to use pantotheine for the synthesis of coenzyme A. Growth response to pyridoxamine was only observed in the absence of alanine or alanyl peptides, suggesting that pyridoxal phosphate is required for the transamination of pyruvate. A deficiency of $\alpha$-lipoic acid had little effect on growth, but both $\alpha$-lipoic acid and coenzyme A were essential for the functioning of the pyruvate oxidase system (Rodwell, 1969b). The requirements for nucleic acid precursors and amino acids were determined both in the presence and in the absence of folinic acid and biotin, but neither vitamin elicited a growth response under either condition. Possibly, *M. mycoides* is unable to carry out any of the carboxylation reactions requiring biotin as cofactor. The lack of need for a derivative of folic acid is more difficult to understand. The methyl groups in the modified nucleosides of *M. mycoides* subsp. *capri* (Walker, 1971) and *M. hominis* (Johnson and Horowitz, 1971) would be donated by methionine (Hall, 1971). However, tetrahydrofolate should be required for the synthesis of *N*-formylmethionyl-$tRNA_f^{Met}$. This formylated amino acyl tRNA species has been demonstrated in *A. laidlawii* B, *M. gallisepticum,* and *Mycoplasma capricolum* (Kid) (Hayashi *et al.*, 1969). Presumably, these species obtained folic acid or a derivative from the undefined growth media for the synthesis of formylmethionyl tRNA. It would seem unlikely that the fully defined medium for growth of *M. mycoides* contained folic acid as a contaminant, or that this organism is able to synthesize folic acid. One possible explanation for this apparent anomaly is that, in the presence of thymine or thymidine, some mycoplasmas may be capable of bypassing the requirement for a derivative of folic acid by initiating protein synthesis with methionyl–$tRNA_f^{Met}$ rather than its formylated product. Indirect support for this proposal comes from reports that in cell-free protein-synthesizing systems, the incorporation of amino acids can occur independently of formyl tetrahydrofolate or formylmethionyl–$tRNA_f^{Met}$ provided the $Mg^{2+}$ concentration is of the order of 10 m*M* (Eisenstadt and Lengyel, 1966; Kolakofsky and Nakamoto, 1966).

Biotin was present in the defined medium of *A. laidlawii,* but the effect of its omission was not reported (Tourtellotte *et al.*, 1964). A requirement for folinic acid, which could not be met by folic acid, was found only in the absence of thymidine (Razin, 1962). Thus, in the fully defined medium of Tourtellotte *et al.* (1964), which contained thymidine but not folinic acid, the above considerations pertaining to initiation of protein synthesis in *M. mycoides* are also applicable to *A. laidlawii.* The coenzymes needed to allow pyruvate oxidation by *A. laidlawii* have not been investigated. Acetate is a known product of glucose catabolism (Castrejon-Diez *et al.*, 1963; Neimark and Pickett, 1960; Tourtellotte and Jacobs, 1960) so this strain may resemble *M. mycoides* in that it requires α-lipoic acid for pyruvate oxidation but not for growth. A deficiency of α-lipoic acid might explain the requirement for exogenous saturated fatty acids reported for growth of strain B in fully defined medium by Tourtellotte *et al.* (1964), since acetate would not have been formed endogenously and the medium contained neither acetate nor α-lipoic acid. If this were the case, carotenogenesis would also have been suppressed, causing the cells to be colorless, but this observation was not mentioned.

### 7. Polyamines

Spermine and spermidine, while not essential, markedly stimulated growth of *M. mycoides* strain Y (Rodwell, 1967). There are no other reports of a polyamine requirement for growth of a mycoplasma, but since polyamines are found in a wide variety of bacteria (Tabor and Tabor, 1972) and have been shown to stimulate the growth of several species of lactobacilli (Guirard and Snell, 1964), it would be surprising if such a requirement did not exist.

## C. Nutrition of Other Mycoplasmas

This section, referring to cells grown in partly defined media, summarizes scattered pieces of information from studies which include incorporation of radioactive precursors, growth responses to various nutrients and *in vitro* enzyme assays on cell-free extracts.

### 1. *Mycoplasma gallinarum* Strain J

Lund and Shorb (1966) described a partly defined medium in which fatty acids were provided by diacetoxysuccinoyl esters of tallow monoglycerides. Their medium, containing a tryptic digest of casein, supported good growth and, when modified by replacement of the digest with the amino acid mixture described by Razin and Cohen (1963), continued to allow near optimal growth of strain J. Maltose was preferred to glucose as carbon and energy source. Adenosine, guanosine, cytidine, and thymidine were provided for the synthesis of nucleic acids, but these may not have been the minimal requirements. This organism resembled *A. laidlawii* in being able to use pantotheine interchangeably with coenzyme A. Biotin, and, to a lesser extent, vitamin $B_{12}$ (Lund and Shorb, 1966) and $\alpha$-lipoic acid (Lund and Shorb, 1967) stimulated growth in this medium. Several other vitamins, for which growth responses were not reported, were also included in the medium. With this work as a basis, strain J would seem to be a favorable organism (available to workers in the United States) to select for further nutritional studies, in which case tris (hydroxymethyl) aminomethane–HCl buffer and NaCl might with advantage be replaced by the sodium salt of a nontoxic buffer, such as HEPES.

### 2. *Mycoplasma arthritidis*

In an early attempt to devise a partly defined medium for the growth of *M. arthritidis* strain 07 (Smith, 1955) lipids were provided by a serum lipoprotein fraction. The medium was supplemented with nine amino acids, among which glutamate and glutamine were interchangeable. Adenosine triphosphate, guanine, and hypoxanthine were provided in addition to deoxyribo- and ribonucleic acids, so the purine and pyrimidine requirements could not be defined. *Mycoplasma arthritidis* strain 07 required several vitamins, including biotin, folic acid, thiamine, and pyridoxine, and was able to use pantothenate as a precursor of coenzyme A. There was no requirement for added carbohydrate, although ribose was present in the medium as a constituent of nucleic acids at a level of about 4 mg/liter. Despite this description of a largely defined medium, most biochemical investigations of *M. arthritidis* have been performed on broth-grown cells.

Aspects of the intermediary metabolism and respiratory pathways of strains 07 and Campo have been studied in some detail. Since *M. arthritidis* is able to use a wide variety of substrates as energy sources, this species appears to be far more versatile in its oxidative metabolism than other species so far investigated. Thus, strain 07 was capable of oxidizing short-chain fatty acids (Lynn, 1960) and VanDemark and Smith (1965) have presented evidence for a cyclic mechanism for butyrate oxidation. Strain 07 possessed a complex respiratory chain (VanDemark and Smith, 1964a) and probably also a functional TCA cycle (VanDemark and Smith, 1964b). *Mycoplasma arthritidis* strain 07 has been observed to carry out oxidative phosphorylation (VanDemark, 1969), but whether this process can yield significant amounts of energy remains in doubt. Studies of Holländer *et al.* (1977) have indicated that the quinone content of this mycoplasma might be too low to maintain phosphorylative electron transport in the membrane. Certainly, if oxidative phosphorylation were available to it, *M. arthritidis* strain 07 could derive considerably more energy from fatty acid oxidation than from substrate level phosphorylations. From evidence on the oxidation of various intermediates of the glycolytic and pentose phosphate pathways it appears that *M. arthritidis* strain 07 and other nonfermentative mycoplasmas possess only those portions of the pathways required for synthetic purposes (Gewirtz and VanDemark, 1966).

### 3. *Ureaplasma*

The only nutritional requirement defined for growth of ureaplasmas is urea (Ford and MacDonald, 1967) and even this requirement has been questioned (Masover *et al.*, 1974). The function of urea in the growth of ureaplasmas remains a perplexing problem. No significant incorporation of carbon from $^{14}C$-labeled urea could be detected, while approximately 95% of the label was recovered as $^{14}CO_2$ (Ford *et al.*, 1970). Neither $CO_2$ nor ammonia, the products of urea hydrolysis, possessed growth-promoting properties (Masover *et al.*, 1977a). Further doubt concerning the need for urea for the growth of ureaplasmas arises from the observations that, in medium containing only trace amounts of contaminating urea, putrescine and allantoin could substitute for additional urea to allow somewhat slower growth (Masover *et al.*, 1974) and that exhaustion of urea from the medium appeared not to limit growth (Masover *et al.*, 1977a). In contrast to these results, Kenny and Cartwright (1977), using 2-(*N*-morpholino) ethanesulfonic acid (MES) as buffer, found that the amount of growth was directly proportional to urea concentration over the range 0.03–30 m*M*. At optimal concentrations of MES and urea the yield of cellular protein was approximately 0.6 mg/liter, corresponding to an extremely low molar growth yield of 20 mg cellular protein per mole urea hydrolyzed. A similarly low molar growth yield occurred in cultures buffered with $CO_2$ (Masover *et al.*, 1977b). It is not easy to determine the growth yield in such sparsely growing cultures. Masover *et al.* (1976) noted that the membrane fraction prepared by lysis of cells by osmotic shock or by digitonin contained from four to ten times as much protein as the cytoplasmic fraction and suggested as an explanation that the cell pellet was heavily contaminated with proteins from the growth medium. An alternative possibility, which does not appear to have been excluded, is that the cell pellet was contaminated with membranous debris from cells which had undergone lysis during growth of the culture. In the latter case the growth yield would have been underestimated.

The source of energy for growth of ureaplasmas is not known, although the possibility that urea may serve this function has been considered. In

*Streptococcus allantoicus* urea is utilized to give carbamyl phosphate and NAD (Valentine and Wolfe, 1961). However, as Kenny and Cartwright (1977) pointed out, unless the molar growth yield is seriously underestimated the hydrolysis of urea to yield energy is performed by an extraordinarily inefficient process. Some properties of the urease activity of *Ureaplasma urealyticum* have been examined both in whole cultures and cell-free preparations (Masover *et al.*, 1976). In either case the enzyme had a pH optimum of between 5 and 6 and the rate of hydrolysis was maximal at a urea concentration of 5.6 m*M*. It was not inhibited by ammonium ions and differed from jack bean urease in being less susceptible to inhibition by acetohydroxamic acid. Delisle (1977) subjected cell-free extracts of *U. urealyticum* to polyacrylamide gel electrophoresis and, on developing the gels with a catalytic urease stain (Fishbein, 1969), demonstrated urease isozymes in several strains examined. No special cofactors were required for activity. Thus, available evidence renders unlikely the existence of phosphate-dependent degradation of urea by ureaplasmas. An alternative hypothesis proposed by Masover *et al.* (1977b) is that ammonia produced in the cytoplasm by the action of urease on urea may provide a mechanism for proton elimination or acid balance and may also allow establishment of an ion gradient coupled to an energy-producing ion transport system (Harold, 1972).

The finding that a strain of ureaplasma incorporated radioactivity from $^{14}C$-acetate into both saturated and unsaturated fatty acids (Romano *et al.*, 1976) suggests that ureaplasmas may have greater biosynthetic capabilities than many other mycoplasmas.

### 4. *Thermoplasma*

All known thermoplasmas were isolated from coal refuse heaps where the temperature is approximately 60°C and the pH ranges from 1 to 2. The extreme harshness of their environment suggests that the thermoplasmas have very different requirements from all other mycoplasmas. Indeed, Smith *et al.* (1975) succeded in growing *Thermoplasma acidophilum* in an inorganic salt medium at pH 2 supplemented with 1% glucose and 0.1% yeast extract. A peptide fraction containing one or more peptides of eight to ten amino acids was isolated from the yeast extract. At a concentration of 0.04% this fraction abolished the requirement for the remainder of the yeast extract. The function of the oligopeptides is not known, although the observation that they bound cations avidly and were required in relatively large amounts led Smith *et al.* (1975) to suggest that their function may be related to the high $H^+$ concentration of the growth environment.

These results clearly indicate that *T. acidophilum* has considerably greater biosynthetic capability than other mycoplasmas.

## II. GROWTH AND REPRODUCTION

### A. Effects of Nutrition on Morphology

The morphology of mycoplasmas is extremely variable. While a particular species may show a tendency to favor growth in one of the various morphologic forms, the composition of the culture medium markedly influences which form predominates. For example, Freundt (1958) stressed the importance of cholesterol in the formation of filaments. Rodwell and Abbot (1961) extended this observation to include the lipid precursors, glycerol and fatty acids, as requirements for the formation of filaments by *M. mycoides* strain V5. Thus, exponential-phase cultures of *M. mycoides* subsp. *mycoides* (small colony type[1]) consist characteristically of long and often branching filaments, many of which may be wholly or partly composed of coccoid units (Fig. 3). Cultures of *M. mycoides* strain Y grown with a mixture of oleate and palmitate contained numerous elongated forms with discoidal swellings (Fig. 4a), whereas growth on a mixture of the saturated fatty acids laurate and behenate caused the cells to appear large and roughly spherical (Fig. 4b)

*Mycoplasma gallisepticum* is an example of a mycoplasma which rarely exhibits filamentous growth, being characteristically pear-shaped and possessing a polar bleb structure in most liquid media (Maniloff *et al.* 1965). Yet supplementation of undefined media with unsaturated fatty acids induced filament formation in *M. gallisepticum,* as well as in other species, including *A. laidlawii* (Razin and Cosenza, 1966; Razin *et al.*, 1966, 1967). When *A. laidlawii* was grown with an excess of saturated fatty acids, the cells were predominantly coccoid. All of the above examples demonstrate the importance of the lipid composition of the membrane in determining the morphology of mycoplasmas. Freundt's suggestion (1958, 1960) that filament formation is a feature shared by all myco-

[1] *Mycoplasma mycoides* subsp. *mycoides* may be divided into two types on the basis of growth rate, colony size, and some other characteristics (see Cottew, Chapter 3, Vol. 2). Strain V5, isolated from bovine pleuropneumonia, belongs to the slower growing small colony type. The faster growing large colony type to which strain Y belongs resembles *M. mycoides* subsp. *capri* in these characteristics as well as its potentiality for growth in the $\rho$ form.

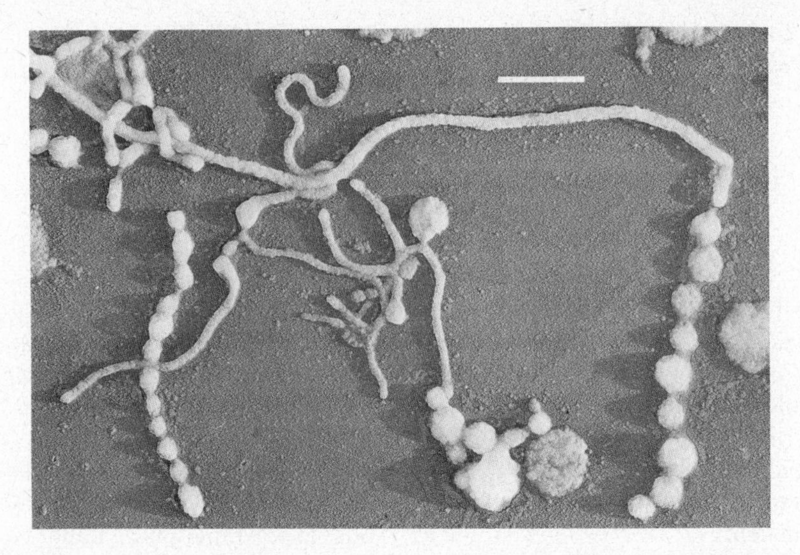

FIGURE 3. Typical morphology of *M. mycoides* subsp. *mycoides* (strain V5) showing segmentation of filaments. This is a reversible process and may occur many times before the cell ultimately divides by pinching off coccoid mononucleate cells. Metal-shadowed electron micrograph. Bar represents 1 μm.

plasmas should perhaps be modified: filamentous growth can occur in most, if not all, mycoplasmas under the appropriate conditions.

## B. Control of Macromolecular Syntheses

The effects of glycerol and thymine deprivation on macromolecular syntheses and growth of *M. mycoides* strain Y have been investigated (Rodwell *et al.*, 1972). Previously, examination of the effects of glycerol deprivation on *M. mycoides* strain V5 (Rodwell and Abbot, 1961) showed that the cells became swollen and eventually underwent lysis unless RNA or protein synthesis was inhibited by the respective omission of uracil or addition of chloramphenicol (CAP) concomitantly with glycerol deprivation (Fig. 5). These findings favored the suggestion that "glycerolless death" and lysis might arise from unbalanced growth in which cytoplasmic synthesis continued in the absence of membrane lipid synthesis.

Rodwell *et al.* (1972) measured lipid, DNA, and protein synthesis by the incorporation of $^{14}C$-palmitate, $^{3}H$-thymidine, and $^{3}H$-glycine, respectively. As shown in Fig. 6, when cells of strain Y were incubated in medium without glycerol, lipid synthesis ceased immediately. Increase in cell mass (as measured by turbidity) and DNA synthesis continued at about the same rate as in the control culture for about 6 hr, or more than

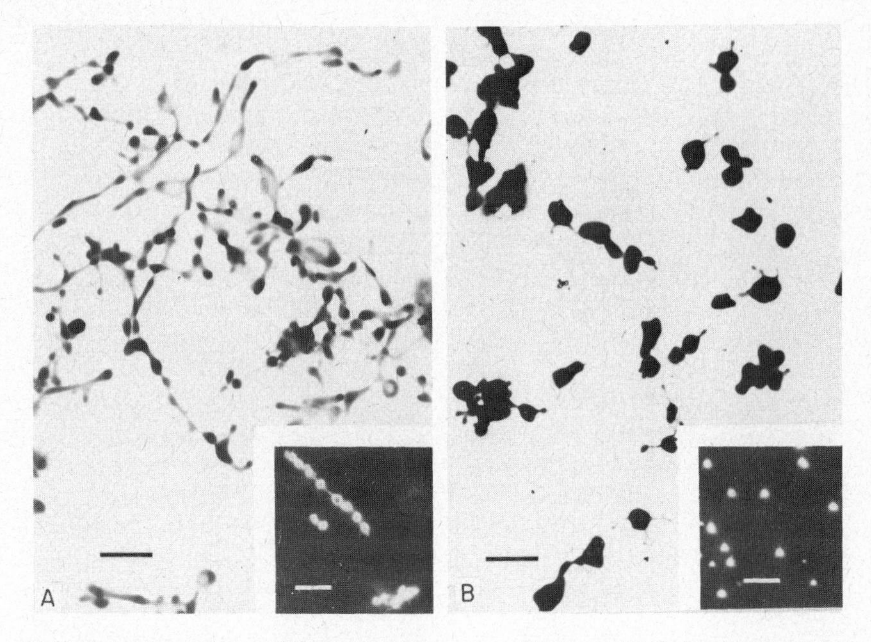


FIGURE 4. Morphology of *M. mycoides* subsp. *mycoides* (strain Y) during the exponential growth phase in medium with different fatty acids. (A) Organisms grown with oleate and palmitate; (B) organisms grown with laurate and behenate. Electron micrographs: bar represents 1 $\mu$m. Darkground photographs (inset): bar represents 5 $\mu$m. Many of the organisms are $\rho$ forms. (From Rodwell and Peterson, 1971.)

two doublings. The number of colony-forming units (CFU) increased almost threefold and then slowly declined. The cells became markedly distended after glycerol deprivation but, unlike the results obtained with the V5 strain (Rodwell and Abbot, 1961), there was little lysis. It is likely that the growth conditions for the experiments with strain V5 resulted in an unfavorable ratio of unsaturated to saturated fatty acids such that a glycerol deficiency was superimposed on a fatty acid imbalance.

When DNA synthesis was prevented by thymine deprivation, lipid synthesis continued at an arithmetic rate for more than 12 hr; cell mass increased more slowly and the numbers of CFU remained constant to indicate that there was no thymineless death (TLD). In a separate experiment the incorporation of $^{3}$H-glycine into protein continued during incubation without thymine at a somewhat higher rate than that of $^{14}$C-palmitate. The continued synthesis of protein and lipid which occurred in the absence of DNA synthesis was accompanied by morphologic changes which were quite different from those resulting from glycerol deprivation. Small spherical forms of varying size, thin threadlike filaments, and other

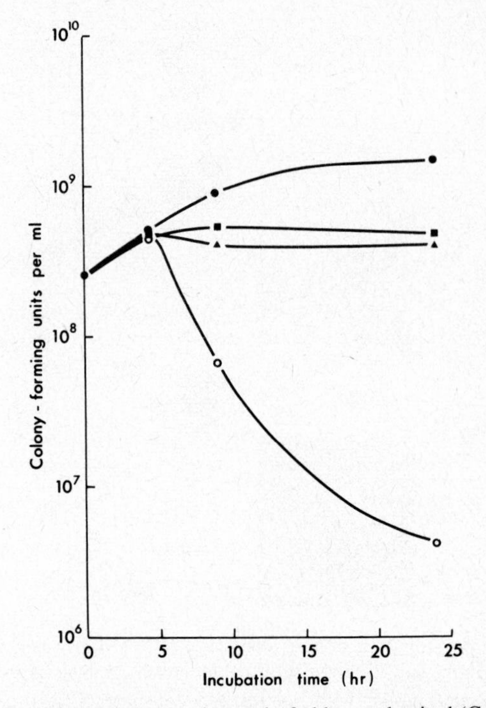


FIGURE 5. Effect of glycerol deprivation and of chloramphenicol (CAP) on the numbers of colony-forming units during incubation of *M. mycoides* subsp. *mycoides* (strain V5) in various media. ●, Complete, partly defined medium; ■, complete medium with CAP added; ○, medium lacking glycerol; ▲, medium lacking glycerol but with CAP added. (Adapted from Rodwell and Abbot, 1961.)

bizarre forms as well as some typical $\rho$ forms were all present (Fig. 7). The small cells were anucleate (Fig. 8). In section they were seen to contain ribosomelike particles and to be bounded by a typical unit membrane. They may be regarded as the mycoplasmal equivalent of minicells.

Rodwell *et al.* (1972) concluded that lipid synthesis is not necessary for DNA replication, and that DNA synthesis is not required for continued protein and lipid synthesis, nor for cell division: That is, the syntheses of lipids, proteins, and DNA are not necessarily coordinated. This conclusion is not meant to imply that no regulatory mechanisms exist in mycoplasmas for controlling the rates of macromolecular syntheses. Indeed, the finding that strain Y did not undergo "glycerolless death" and lysis is indirect evidence that some regulatory process functioned to inhibit cytoplasmic synthesis. The effects of amino acid starvation have not been examined, so we have no knowledge of whether mycoplasmas possess

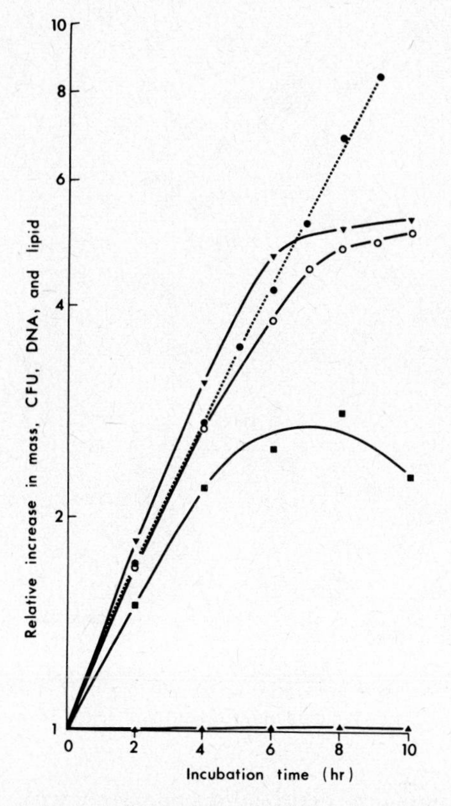


FIGURE 6. Effect of glycerol deprivation on relative increases in mass, numbers of colony-forming units (CFU), DNA, and lipid during incubation of *M. mycoides* subsp. *mycoides* (strain Y). ▲, Lipid; ■, numbers of CFU; ○, mass; ▼, DNA; --●--, mass, control culture in complete medium. (Adapted from Rodwell *et al.*, 1972.)

stringent control mechanisms for regulating macromolecular syntheses as does *Escherichia coli* (Cashel, 1975).

Smith and Hanawalt (1968) studied the effects of thymidine deprivation in *A. laidlawii* strain B. In this organism rapid TLD took place at an exponential rate after a very variable lag period ranging from 2 to 15 hr. During TLD, which only occurred in the presence of the energy source (glucose), the cell mass increased about 10-fold. Inhibition of protein synthesis by the addition of CAP, or the removal of the essential amino acid tryptophan, allowed continued DNA synthesis to an extent which could be accounted for by the completion of rounds of DNA replication already begun. In the absence of protein synthesis, TLD occurred and proceeded at a somewhat slower rate, with a survival level after 60 hr of

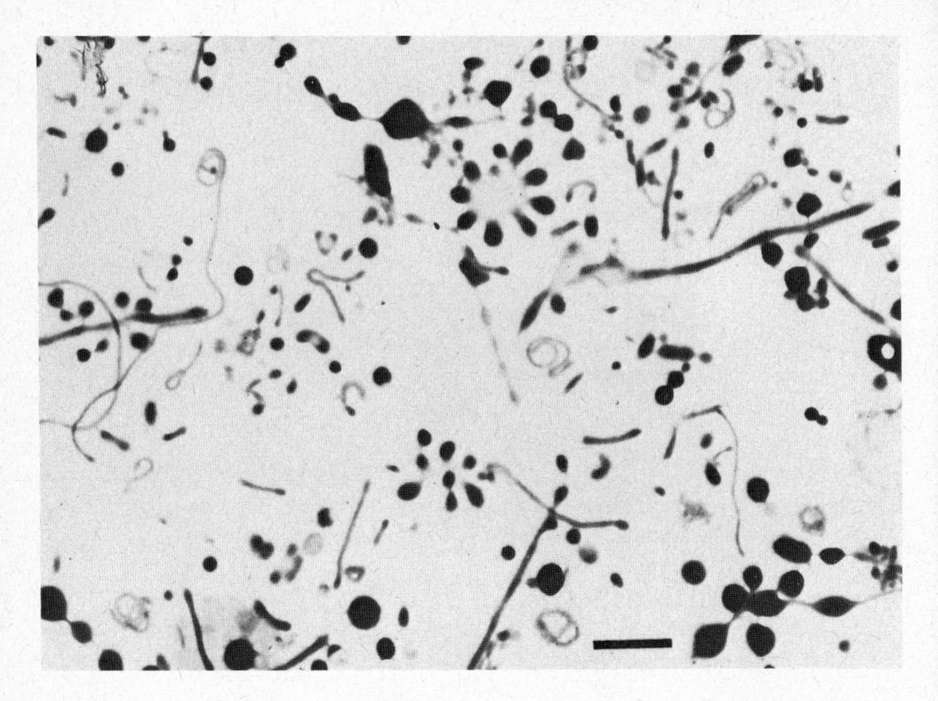

FIGURE 7. Effect of thymine deprivation on morphology of *M. mycoides* subsp. *mycoides* (strain Y). Cells were incubated in medium without thymine for 24 hr. Bar represents 1 μm. (From Rodwell *et al.*, 1972.)

less than 0.5%. Thymidine deprivation concurrent with inhibition of RNA synthesis by the omission of all pyrimidine nucleosides allowed greater than 10% survival after 60 hr. Thus, continued RNA synthesis may be essential for the lethal effectiveness of TLD. In many respects TLD in *A. laidlawii* resembles this phenomenon in thy$^-$ auxotrophs of *E. coli* (Barner and Cohen, 1954).

## C. Reproduction

### 1. Replication of the Genome

Present knowledge concerning replication of the genome in mycoplasmas comes almost entirely from studies on *M. gallisepticum*. The genome is replicated essentially as in other prokaryotes. Quinlan and Maniloff (1972) presented evidence that in *M. gallisepticum* the DNA growing point is membrane-bound. Cells were pulse-labeled with $^3$H-thymidine and lysed by freezing and thawing. After brief sonication to shear DNA, a fraction (P2) was isolated which was enriched with the

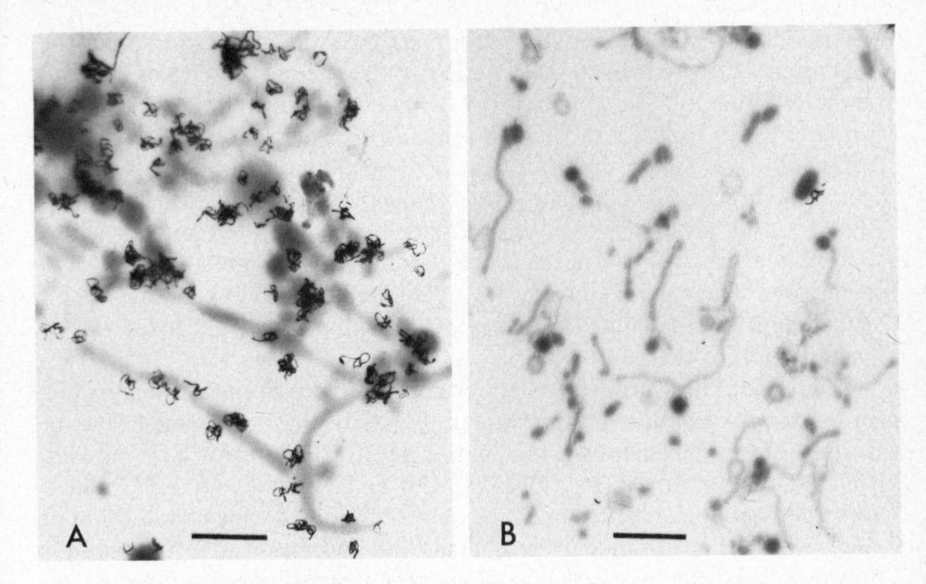


FIGURE 8. Absence of DNA from "minicells" of *M. mycoides* subsp. *mycoides* (strain Y). Cells labeled by growth with $^3$H-thymidine were incubated in medium without thymine or thymidine for 24 hr. The smaller cells were then separated from the larger $\rho$-form cells by density gradient centrifugation. (A) Autoradiograph of cells in the large cell fraction; (B) autoradiograph of cells in the small cell fraction exposed for the same period. Bar represents 1 $\mu$m.

pulse-labeled DNA. In addition, fraction P2 contained most of the membrane lipid and apparently intact bleb structures. If, before lysis, the pulse-labeled cells were incubated for 7–8 min in unlabeled medium, about half of the label was "chased" from the P2 fraction into a supernatant fraction (S) which contained the bulk of the DNA. Upon further incubation, about 80% of the label was recoverable from the S fraction. In examining the time course of DNA synthesis during the replication cycle of synchronously dividing cells, Quinlan and Maniloff (1973) found that after incubation of pulse-labeled cells in unlabeled medium for approximately one doubling time (100–115 min), almost 75% of the label from a pulse of $^3$H-thymidine was again found in the P2 fraction. These results were interpreted to indicate that the pulse-labeled region of the chromosome had returned to the DNA synthesis site for a second round of replication. In the synchronized culture there was a period which seemed to correspond to the phase of daughter cell separation, when DNA was not synthesized. A similar period of decreased DNA synthesis during the *E. coli* cell cycle has been observed (Helmstetter, 1967).

In later studies, a partial purification of the replication complex of the

P2 fraction was achieved to further support the suggestion that the membrane–bleb–infrableb region is in fact the site of the DNA replication complex (Maniloff and Quinlan, 1974). Whether or not the bleb structure plays any part in DNA replication remains uncertain. The bleb structure (Maniloff *et al.*, 1965) does not appear to be unique to *M. gallisepticum,* since an unrelated mycoplasma (*Mycoplasma alvi*) isolated from the intestinal tract of cattle displayed a similar cell shape and terminal structure (Gourlay *et al.*, 1977). The bleb may function as an organelle of attachment, concerned with motility (see Chapter 5). Possibly both the cell attachment site and the DNA replication complex are localized in the same region of the cell.

We do not know how the daughter chromosomes of mycoplasmas, nor of other prokaryotes, are segregated. Possibly the membrane plays an active role in chromosome segregation by a primitive form of mitosis. Related to this possibility, Bredt (1970) and Bredt *et al.* (1973) have observed active constrictions in the cell membrane during cell division in *M. hominis.* The absence of both a cell wall and mesosomes may render mycoplasmas simpler experimental material than other bacteria for investigating the mechanism of daughter chromosome segregation in prokaryotes.

### 2. Cell Division

**a. Modes of reproduction.** The process of cell division of mycoplasmas has long been controversial. Early ideas based on microscopic observations included descriptions of complex life cycles in which, for example, small elementary particles formed within large bodies. The large bodies were supposed later to disintegrate, liberating the elementary particles and so beginning a new cycle (Kleineberger-Nobel, 1962). These ideas have now been abandoned; the so-called large bodies and the small elementary particles (in the 100–250 nm size range) were nonviable involution forms. Reproduction by binary fission (Morowitz and Maniloff, 1966; Maniloff and Morowitz, 1967; Furness *et al.*, 1968), by fragmentation of filaments (Freundt, 1969), and by budding (Liebermeister, 1960) all had their protagonists and were often described as fundamentally different processes. Indeed, a classification of mycoplasmas was proposed in which those reproducing by binary fission were placed in a separate genus, *Schizoplasma* (Furness *et al.*, 1968; Furness, 1970). As Razin (1973) has pointed out, there are really no grounds for controversy: The process of cell multiplication of mycoplasmas does not differ in any fundamental way from that of other prokaryotes.

Binary fission has been described as the characteristic mode of reproduction in *M. gallisepticum* (Morowitz and Maniloff, 1966), *M. pneumo-*

*niae* (Bredt, 1968; Furness *et al.*, 1968), *M. hominis* (Robertson *et al.*, 1975), and other species. While this may be true for the first two species, in which a specialized terminal structure appears to play a central role in cell division (see Section II. C. 2. *b*), it is doubtful if it is the sole mode in any species. For binary fission to occur, division of the cell must be fully synchronized with replication of the genome, and this is by no means always the case. Frequently, cell division lags behind genome replication to give rise to multinucleate filaments. This event is in keeping with the findings that the syntheses of lipids, proteins, and DNA are not necessarily coordinated. As mentioned (Section II, A), the tendency toward formation of branching filaments during growth is strongly influenced by nutritional factors, particularly the lipid composition of the medium. Once formed, the multinucleate filaments segment to the characteristic chains of beads which subsequently fragment into a number of single, mononucleate daughter cells (Turner, 1935; Tang *et al.*, 1935; Freundt, 1958; Razin and Cosenza, 1966). A third means of cell replication is the process of budding, in which a bi- or multinucleate cell protrudes one or more budlike processes which later become "pinched off" to form daughter cells.

Is reproduction by fragmentation of filaments and by budding to be regarded as abnormal? Robertson *et al.* (1975) dismissed reproduction in *M. hominis* by these modes as being abnormal and occurring only under adverse growth conditions. The growth of exponential-phase cells of *M. hominis* has been observed continuously in the beautiful microcinematographic studies of Bredt (1970) and Bredt *et al.* (1973). All three modes of reproduction were shown to occur side by side in the same culture. Bredt's observations led him to conclude that separation of the daughter cells "seemed to be caused by active constrictions of the cell membrane or the cell plasma" (Bredt, 1970). Changes in cell form took place continuously during growth. For example, a ring form changed to a three- or four-lobed form and back to the ring form numerous times while the cell increased in size until finally the change to the lobed form was followed by separation of the daughter cells. The configurational changes were quite rapid; the time taken for a ring form to change to a lobed form was usually less than 7.5 sec and a long filament could become beaded along its entire length within 2.5 min.

The propensity of *M. mycoides* subsp. *mycoides* to change its shape quickly was first observed by Ørskov (1927). Reversible changes in cell form associated with reproduction by fragmentation of filaments and by budding were later described by Turner (1935) and Tang *et al.* (1936) in dark-field studies on living cultures of the same organism, and their descriptions agree well with Bredt's observations on *M. hominis*. The final "pinching off" of the coccoidal elements from the beaded filaments occurred, according to Turner, at the rate of one per half minute. Turner used his observations to postulate complex life cycles, which may help to explain why important observations on shape changes were overlooked by later investigators.

**b. Reproduction in mycoplasmas with terminal structures.** Mycoplasmas such as *M. gallisepticum, M. pneumoniae,* and $\rho$ forms have a definite polarity. A detailed description of reproduction by binary fission in *M. gallisepticum* was reconstructed by arranging negatively stained electron micrographs of cells at various stages of division in order to give a continuous picture of the life cycle (Morowitz and Maniloff, 1966; Maniloff and Morowitz, 1967). Cell division was preceeded by the appearance of a second bleb, usually but not always, at the opposite pole (Figs. 9 and 10), and the cell then divided by constriction about the middle. Irrespective of whether the bleb is directly involved in cell replication, the formation of the second bleb certainly appears to be well-synchronized with this process. Because these cultures were filtered through a Millipore filter (pore size 0.45 $\mu$m) "to disperse aggregates" before fixation and examination, it should be noted that binary fission may not have been the sole mode of reproduction. Larger multinucleate cells, had they been present, would have been removed from the population during filtering.

*Mycoplasma pneumoniae,* when growing in liquid medium on a solid surface (Biberfeld and Biberfeld, 1970) or when attached to host cells in organ culture (Collier, 1972), is an elongated organism with a terminal tip structure and an intracytoplasmic electron-dense core. Biberfeld and Biberfeld observed, by scanning electron microscopy, cells which appeared to bifurcate at one end into two terminal tip structures. By transmission electron microscopy, they also observed cells with two parallel electron-dense cores. Together these observations led to the suggestion that *M. pneumoniae* might start dividing at the tip and then split longitudinally (Biberfeld, 1972). From studies by light microscopy on living cells growing on a glass surface, Bredt (1968) concluded that reproduction was by binary fission. In further studies by light and electron microscopy Bredt and Bierther (1974) suggested that every cell growing under these conditions contains a specialized terminal tip structure which is intimately concerned with both reproduction and motility. Replication of the tip structure precedes cell division and may occur very early in the cell cycle. Some cells were bifurcated at the tip as described by Biberfeld and Biberfeld (1970), while in other dividing cells the tip was located at the opposite end of the cell. The tip structures appeared to draw apart during cell division.

Thus, the process of division in the two motile species (*M. gallisep-*

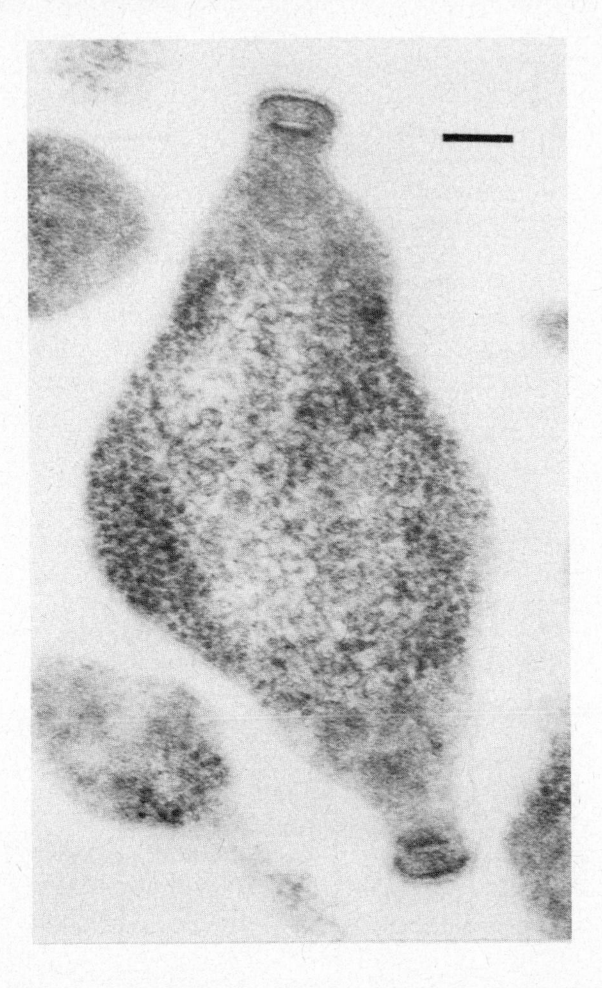

FIGURE 9. Section of a dividing cell of *M. gallisepticum* in which the two bleb structures are at opposite poles. Bar represents 100 nm. (From Morowitz and Maniloff, 1966.)

*ticum* and *M. pneumoniae*) appears to be very similar. Both have a terminal structure which appears to be intimately concerned with reproduction and motility. The infrableb region of *M. gallisepticum* and the electron-dense core of *M. pneumoniae* (which should not be thought of as a rigid organized structure that splits longitudinally during cell division) may also be related in function but little structural organization has been discerned in either. The description of the life cycle of *M. gallisepticum* (Morowitz and Maniloff, 1966) was obtained with cells dividing in suspension, whereas that of *M. pneumoniae* (Bredt and Bierther, 1974) was with

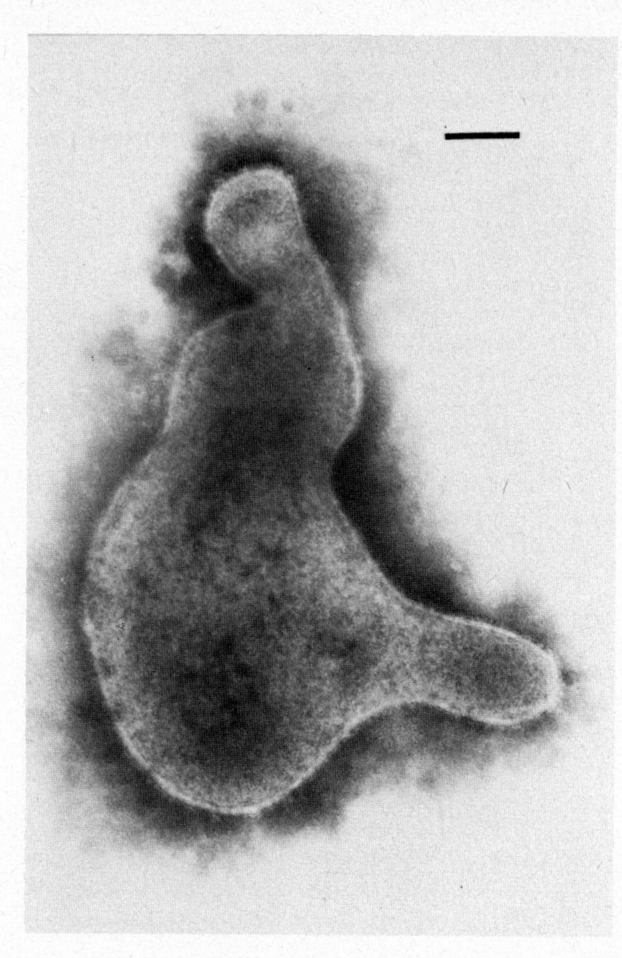

FIGURE 10. A dividing cell of *M. gallisepticum* in which the two bleb structures are not diametrically opposite. Negatively stained after fixation. Bar represents 100 nm. (Courtesy of J. E. Peterson.)

cells attached to a substratum. Cell attachment might to some extent modify the shape of the cells and the appearance of the terminal structures and underlying regions as seen by electron microscopy.

There is one important difference between the cell cycle of *M. gallisepticum* as reconstructed from electron micrographs by Maniloff and Morowitz (1967) and the division process postulated for *M. pneumoniae*. Maniloff and Morowitz (1967) suggested that an early event in cell division of *M. gallisepticum* was the formation *de novo* of a second bleb structure at the opposite end of the cell, whereas the evidence in *M.*

*pneumoniae* suggests that the tip structure replicates and the two then draw apart. We have reviewed the evidence that in *M. gallisepticum* the chromosome origin is membrane-attached and that the DNA replication complex is located in the bleb–infrableb region. Thus, if the cell cycle begins with the replication of the chromosome and bleb structure, the original and newly formed blebs with their attached chromosomes should initially be close together as in the bifurcated cells of *M. pneumoniae*. When cells in suspension are dividing, as in the study of Morowitz and Maniloff (1966), some internal mechanism, which does not involve cell adhesion to a substratum, must move the newly formed bleb–daughter chromosome complex to the opposite pole. Dividing cells of *M. gallisepticum* in which the two blebs do not lie opposite are sometimes seen (Bernstein-Ziv, 1969; and Fig. 10), but their comparative rarity suggests that the postulated movement of the bleb–daughter chromosome complex to the opposite pole is relatively rapid. Ghosh, Maniloff and Gerling (1978) observed that cytochalasin B inhibited cell division in *M. gallisepticum,* and their preliminary studies indicating a similar effect of cytochalasin B on *M. pneumoniae* support a previous report by Neimark (1977) that *M. pneumoniae* synthesizes an actinlike protein. *Acholeplasma laidlawii,* which lacks a terminal structure, was not inhibited. These observations led the authors to postulate that the movement of the newly replicated bleb–daughter chromosome structure is mediated by an actomyosinlike system in *M. gallisepticum* (and presumably also in *M. pneumoniae*).

The third known motile species of *Mycoplasma, Mycoplasma pulmonis,* also has a specialized structure which is the leading part during movement (Bredt, 1973). Whether motile cells of this species divide by binary fission or whether the specialized structure plays a similar role in the process is as yet unknown.

$\rho$ forms have been found in all recently isolated strains of *M. mycoides* subsp. *mycoides* (large colony type), *M. mycoides* subsp. *capri,* and *M. capricolum* when grown under appropriate conditions. They are characterized by an intracytoplasmic, banded fiber which extends throughout the length of the cell. The fiber has a substructure composed of fibrils of about 3 nm diameter, in parallel alignment and terminating at one end of the cell in a structure (TS) which is closely associated with the plasma membrane (Peterson *et al.*, 1973; Rodwell *et al.*, 1972, 1973, 1975). Division of the fiber lengthwise during division of the cell is highly unlikely, since it was never seen to branch. An occasional cell was seen with a TS at both ends, but the incidence was too rare to suggest that the TS is replicated before division of the cell. The elongated $\rho$-form cells were multinucleate, as shown by autoradiography of cells grown in the presence of $^3$H-thymidine (Fig. 11). The distribution of silver grains on the autoradiographs corresponded either to the discoidal swellings or to the budlike outgrowths, the threadlike connecting portions of the cell and the TS being unlabeled. Rho-form cultures always contained a proportion of smaller, more or less spherical cells which lacked fiber and TS, suggesting a scheme of reproduction whereby budding from the multinucleate $\rho$-form cells occurs. The mononucleate daughter cells, initially devoid of $\rho$ structures, then elongate and form $\rho$ fibers and TS. Figure 12 is a section of what appears to be a newly developed $\rho$-form cell in the process of elongating.

Helical *Spiroplasma* cells also display polarity, having a pointed and a blunt end (see, for example, Williamson and Whitcomb, 1974, Fig. 2). It is difficult to envisage how these cells could reproduce except by a scheme like that which we postulate for $\rho$ forms, in which daughter cells, initially without polarity, subsequently differentiate and become helical. The reproduction of *Spiroplasma* is discussed by Bové and Saillard in Volume 3, Chapter 4.

**c. Conclusions and outlook.** Just as some species of *Mycoplasma* have a tendency to favor growth in a particular morphologic form, so also do they have a tendency to favor a particular mode of reproduction. Both cell form and mode of reproduction are strongly influenced by nutritional factors. As in other bacteria, the separation of the daughter cells may often lag behind replication and segregation of the genome. Reproduction by fragmentation of long branching filaments occurs only in rich media containing an abundant supply of all lipid precursors.

Although replication of the genome and separation of the daughter cells are not closely coordinated events in most species, cell division in *M. gallisepticum* by binary fission does appear to be closely integrated with replication of the genome. The machinery for synchronizing these events seems to be associated with the specialized terminal structure.

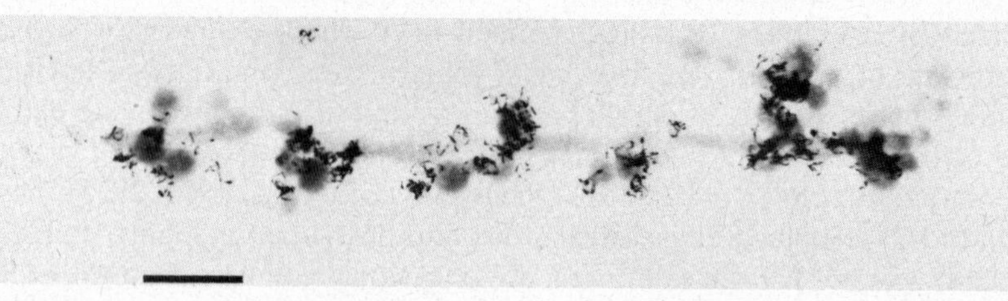

FIGURE 11. Autoradiograph of $\rho$-form cell of *M. mycoides* subsp. *mycoides* grown in the presence of $^3$H-thymidine. Labeling was restricted to the swollen regions of the cell and to the budlike outgrowths. Bar represents 1 $\mu$m. (From Peterson *et al.*, 1973.)

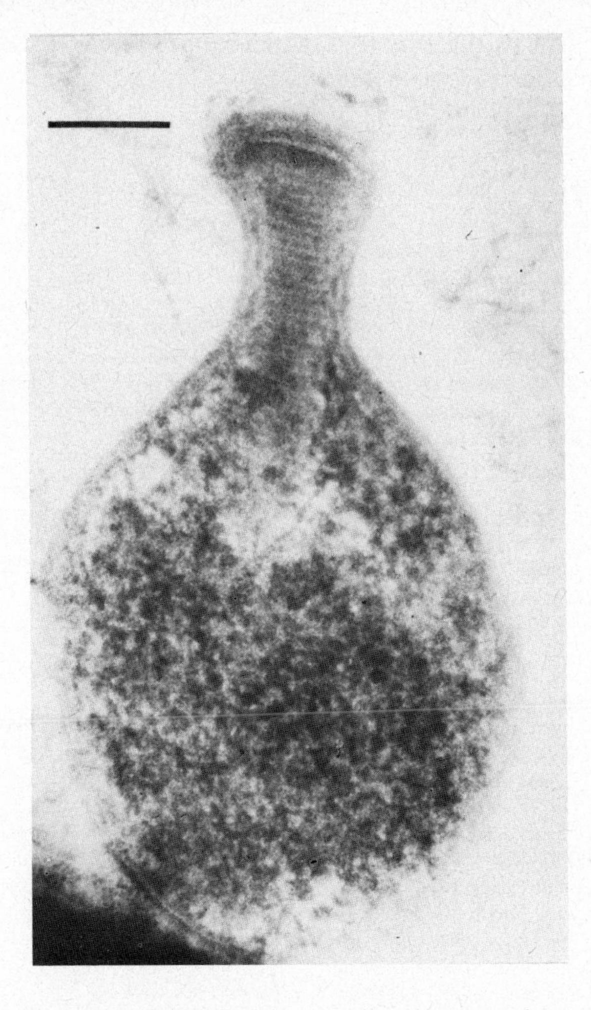

FIGURE 12. Section of ρ-form cell of *M. mycoides* subsp. *mycoides* cut longitudinally through the terminal structure and portion of the ρ fiber. Bar represents 100 nm. (From Peterson *et al.*, 1973.)

The mechanism for the rapid and reversible changes in cell form which occur during growth and reproduction of mycoplasmas and for the gliding motility in some species will be one of the most exciting areas for future investigation. Both phenomena argue the presence of some form of contractile protein system. If such a system is present in mycoplasmas, it seems likely that it will be present also in other bacteria and that it will be found to play a fundamental role in other cell processes. Recent evidence

(Neimark, 1977) for the synthesis of an actinlike protein by *M. pneumoniae* is thus of considerable interest.

## REFERENCES


Archer, D. B. (1975). *J. Gen. Microbiol.* **88,** 329–338.

Argaman, M., and Razin, S. (1965). *J. Gen. Microbiol.* **38,** 153–168.

Barile, M. F., Schimke, R. T., and Riggs, D. B. (1966). *J. Bacteriol.* **91,** 189–192.

Barner, H. D., and Cohen, S. S. (1954). *J. Bacteriol.* **68,** 80–88.

Bernstein-Ziv, R. (1969). *Can. J. Microbiol.* **15,** 1125–1128.

Biberfeld, G. (1972). *Pathog. Mycoplasmas Ciba Found. Symp. 1972.* Discussion, p. 322.

Biberfeld, G., and Biberfeld, P. (1970). *J. Bacteriol.* **102,** 855–861.

Bredt, W. (1968). *Pathol. Microbiol.* **32,** 321–326.

Bredt, W. (1970). *Z. Med. Mikrobiol. Immunol.* **155,** 248–274.

Bredt, W. (1973). *Ann. N.Y. Acad. Sci.* **225,** 246–250.

Bredt, W., and Bierther, M. F. W. (1974). *Zentralbl. Bakteriol., Parisitenkd., Infektionskr. Hyg., Abt. 1: Orig., Reihe A* **229,** 249–255.

Bredt, W., Heunert, H. H., Höfling, K. H., and Milthaler, B. (1973). *J. Bacteriol.* **113,** 1223–1227.

Buttery, S. H. (1967). *Bull. Epizoot. Dis. Afr.* **15,** 227–230.

Cashel, M. (1975). *Annu. Rev. Microbiol.* **29,** 301–318.

Castrejon-Diez, J., Fisher, T. N., and Fisher, E. (1963). *J. Bacteriol.* **86,** 627–636.

Chalquest, R. R., and Fabricant, J. (1960). *Avian Dis.* **4,** 515–539.

Cho, H. W., and Morowitz, H. J. (1969). *Biochim. Biophys. Acta* **183,** 295–303.

Collier, A. M. (1972). *Pathog. Mycoplasmas, Ciba Found. Symp. 1972.* pp. 307–327.

DaMassa, A. J., and Adler, H. E. (1975). *Avian Dis.* **19,** 544–555.

Delisle, G. J. (1977). *J. Bacteriol.* **130,** 1390–1392.

Demel, R. A., Bruckdorfer, K. R., and Van Deenen, L. L. M. (1972). *Biochim. Biophys. Acta* **255,** 311–320.

Eisenstadt, J., and Lengyel, P. (1966). *Science* **154,** 524–527.

Fenske, J. D., and Kenny, G. E. (1976). *J. Bacteriol.* **126,** 501–510.

Fishbein, W. N. (1969). *Proc. Int. Symp. Chromatog. Electrophoresis, 5th, 1968,* pp. 238–242.

Ford, D. K., and MacDonald, J. (1967). *J. Bacteriol.* **93,** 1509–1512.

Ford, D. K., McCandlish, K. L., and Gronlund, A. F. (1970). *J. Bacteriol.* **102,** 605–606.

Freundt, E. A. (1958). "The Mycoplasmataceae (the Pleuropneumonia Group of Organisms)." Munksgaard, Copenhagen.

Freundt, E. A. (1960). *Ann. N.Y. Acad. Sci.* **79,** 312–325.

Freundt, E. A. (1969). *In* "The Mycoplasmatales and the L-Phase of Bacteria" (L. Hayflick, ed.), pp. 281–315. Appleton, New York.

Furness, G. (1970). *J. Infect. Dis.* **122,** 146–157.

Furness, G., Pipes, F. J., and McMurtrey, M. J. (1968). *J. Infect. Dis.* **118,** 7–13.

Gewirtz, M., and VanDemark, P. J. (1966). *Bacteriol. Proc.* p. 77, Abstr. P45.

Ghosh, A., Maniloff, J., and Gerling, D. (1978). *Cell* **13,** 57–64.

Gourlay, R. N., Wyld, S. G., and Leach, R. H. (1977). *Int. J. Syst. Bacteriol.* **27,** 86–96.

Guirard, B. M., and Snell, E. E. (1962). *Bacteria,* **4,** 67–71.

Guirard, B. M., and Snell, E. E. (1964). *J. Bacteriol.* **88,** 72–80.

Hall, R. H. (1971). *In* "The Modified Nucleosides in Nucleic Acids," p. 296. Columbia Univ. Press, New York.
Harold, F. M. (1972). *Bacteriol. Rev.* **36,** 172–230.
Hayashi, H., Fisher, H., and Söll, D. (1969). *Biochemistry* **8,** 3680–3686.
Helmstetter, C. E. (1967). *J. Mol. Biol.* **24,** 417–427.
Henrickson, C. V., and Panos, C. (1969). *Biochemistry* **8,** 646–651.
Holländer, R., Wolf, G., and Mannheim, W. (1977). *Antonie Van Leeuwenhoek; J. Microbiol. Serol.* **43,** 177–185.
Johnson, J. D., and Horowitz, J. (1971). *Biochim. Biophys. Acta* **247,** 262–279.
Kahane, I., Razin, S., and Muhlrad, A. (1978). *FEMS Microbiol Lett.* **3,** 143–145.
Kenny, G. E., and Cartwright, F. D. (1977). *J. Bacteriol.* **132,** 144–150.
Kleineberger-Nobel, E. (1962). "Pleuropneumonialike Organisms (PPLO), Mycoplasmataceae." Academic Press, New York.
Klotz, I. M., and Sloniewsky, A. R. (1968). *Biochem. Biophys. Res. Commun.* **31,** 421–426.
Kolakofsky, D., and Nakamoto, T. (1966). *Proc. Natl. Acad. Sci. U.S.A.* **56,** 1786–1793.
Leach, R. H. (1962). *J. Gen. Microbiol.* **27,** 345–354.
Liebermeister, K. (1960). *Ann. N.Y. Acad. Sci.* **79,** 326–343.
Liška, B., and Smith, P. F. (1974). *Folia Microbiol. (Prague)* **19,** 107–117.
Low, I. E., Eaton, M. D., and Proctor, P. (1968). *J. Bacteriol.* **95,** 1425–1430.
Lund, P. G., and Shorb, M. S. (1966). *Proc. Soc. Exp. Biol. Med.* **121,** 1070–1075.
Lund, P. G., and Shorb, M. S. (1967). *J. Bacteriol.* **94,** 279–280.
Lynn, R. J. (1960). *Ann. N.Y. Acad. Sci.* **79,** 538–542.
Lynn, R. J., and Smith, P. F. (1960). *Ann. N.Y. Acad. Sci.* **79,** 493–498.
McElhaney, R. N., and Tourtellotte, M. E. (1969). *Science* **164,** 433–434.
Maniloff, J., and Morowitz, H. J. (1967). *Ann. N.Y. Acad. Sci.* **143,** 59–65.
Maniloff, J., and Quinlan, D. C. (1974). *J. Bacteriol.* **120,** 495–501.
Maniloff, J., Morowitz, H. J., and Barrnett, R. J. (1965). *J. Bacteriol.* **90,** 193–204.
Masover, G. K., Benson, J. R., and Hayflick, L. (1974). *J. Bacteriol.* **117,** 765–774.
Masover, G. K., Sawyer, J. E., and Hayflick, L. (1976). *J. Bacteriol.* **125,** 581–587.
Masover, G. K., Razin, S., and Hayflick, L. (1977a). *J. Bacteriol.* **130,** 292–296.
Masover, G. K., Razin, S., and Hayflick, L. (1977b). *J. Bacteriol.* **130,** 297–302.
Mitchell, A., and Finch, L. R. (1977). *J. Bacteriol.* **130,** 1047–1054.
Mitchell, A., Sin, I. L., and Finch, L. R. (1978). *J. Bacteriol.* **134,** 706–712.
Morowitz, H. J., and Maniloff, J. (1966). *J. Bacteriol.* **91,** 1638–1648.
Neimark, H. (1977). *Proc. Natl. Acad. Sci. U.S.A.* **74,** 4041–4045.
Neimark, H., and Pickett, M. J. (1960). *Ann. N.Y. Acad. Sci.* **79,** 531–537.
Ørskov, J. (1927). *Ann. Inst. Pasteur, Paris* **41,** 473–482.
Panos, C., and Henrickson, C. V. (1969). *Biochemistry* **8,** 652–658.
Panos, C., and Leon, O. (1974). *J. Gen. Microbiol.* **80,** 93–100.
Panos, C., and Rottem, S. (1970). *Biochemistry* **9,** 407–412.
Peterson, J. E., Rodwell, A. W., and Rodwell, E. S. (1973). *J. Bacteriol.* **115,** 411–425.
Plackett, P. (1961). *Nature (London)* **189,** 125–126.
Plackett, P. (1967). *Ann. N.Y. Acad. Sci.* **143,** 158–164.
Pollack, J. D., and Tourtellotte, M. E. (1967). *J. Bacteriol.* **93,** 636–641.
Quinlan, D. C., and Maniloff, J. (1972). *J. Bacteriol.* **112,** 1375–1379.
Quinlan, D. C., and Maniloff, J. (1973). *J. Bacteriol.* **115,** 117–120.
Razin, S. (1962). *J. Gen. Microbiol.* **28,** 243–250.
Razin, S. (1963). *J. Gen. Microbiol.* **33,** 471–475.
Razin, S. (1973). *Adv. Microb. Physiol.* **10,** 1–80.
Razin, S., and Cohen, A. (1963). *J. Gen. Microbiol.* **30,** 141–154.

Razin, S., and Cosenza, B. J. (1966). *J. Bacteriol.* **91,** 858–869.
Razin, S., and Rottem, S. (1963). *J. Gen. Microbiol.* **33,** 459–470.
Razin, S., Cosenza, B. J., and Tourtellotte, M. E. (1966). *J. Gen. Microbiol.* **42,** 139–145.
Razin, S., Cosenza, B. J., and Tourtellotte, M. E. (1967). *Ann. N.Y. Acad. Sci.* **143,** 66–72.
Robertson, J., Gomersall, M., and Gill, P. (1975). *J. Bacteriol.* **124,** 1007–1018.
Rodwell, A. W. (1963). *J. Gen. Microbiol.* **32,** 91–101.
Rodwell, A. W. (1967). *Ann. N.Y. Acad. Sci.* **143,** 88–109.
Rodwell, A. W. (1969a). *J. Gen. Microbiol.* **58,** 29–37.
Rodwell, A. W. (1969b). *J. Gen. Microbiol.* **58,** 39–47.
Rodwell, A. W. (1969c). *In* "The Mycoplasmatales and the L-Phase of Bacteria" (L. Hayflick, ed.), pp. 413–449. Appleton, New York.
Rodwell, A. W. (1974). *Colloq. Inst. Natl. Sante Rech. Med.* **33,** 79–85.
Rodwell, A. W., and Abbot, A. (1961). *J. Gen. Microbiol.* **25,** 201–214.
Rodwell, A. W., and Peterson, J. E. (1971). *J. Gen. Microbiol.* **68,** 173–186.
Rodwell, A. W., and Rodwell, E. S. (1954). *Aust. J. Biol. Sci.* **7,** 18–30.
Rodwell, A. W., Peterson, J. E., and Rodwell, E. S. (1972). *Pathog. Mycoplasmas, Ciba Found. Symp. 1972* pp. 123–139.
Rodwell, A. W., Peterson, J. E., and Rodwell, E. S. (1973). *Ann. N.Y. Acad. Sci.* **225,** 190–200.
Rodwell, A. W., Peterson, J. E., and Rodwell, E. S. (1975). *J. Bacteriol.* **122,** 1216–1229.
Romano, N., Rottem, S., and Razin, S. (1976). *J. Bacteriol.* **128,** 170–173.
Romijn, J. C., Van Golde, L. M. G., McElhaney, R. N., and Van Deenen, L. L. M. (1972). *Biochim. Biophys. Acta* **280,** 22–32.
Rottem, S., and Panos, C. (1970). *Biochemistry* **9,** 57–63.
Rottem, S., and Razin, S. (1964). *J. Gen. Microbiol.* **37,** 123–134.
Rottem, S., and Razin, S. (1967). *J. Gen. Microbiol.* **48,** 53–63.
Rottem, S., Muhsam-Peled, O., and Razin, S. (1973a). *J. Bacteriol.* **113,** 586–591.
Rottem, S., Yashouv, J., Ne'eman, Z., and Razin, S. (1973b). *Biochim. Biophys. Acta* **323,** 495–508.
Saito, Y., Silvius, J. R., and McElhaney, R. N. (1977). *J. Bacteriol.* **132,** 497–504.
Schimke, R. T., Berlin, C. M., Sweeney, E. W., and Carroll, W. R. (1966). *J. Biol. Chem.* **241,** 2228–2236.
Silvius, J. R., Saito, Y., and McElhaney, R. N. (1977). *Arch. Biochem. Biophys.* **182,** 455–464.
Slater, M. L., and Folsome, C. E. (1971). *Nature (London), New Biol.* **229,** 117–118.
Smith, D. W., and Hanawalt, P. C. (1968). *J. Bacteriol.* **96,** 2066–2076.
Smith, P. F. (1955). *Proc. Soc. Exp. Biol. Med.* **88,** 628–631.
Smith, P. F. (1962). *J. Bacteriol.* **84,** 534–538.
Smith, P. F. (1964). *J. Lipid Res.* **5,** 121–125.
Smith, P. F. (1971). *J. Bacteriol.* **108,** 986–991.
Smith, P. F., and Boughton, J. E. (1960). *J. Bacteriol.* **80,** 851–860.
Smith, P. F., and Rothblat, G. H. (1962). *J. Bacteriol.* **83,** 500–506.
Smith, P. F., Langworthy, T. A., and Smith, M. R. (1975). *J. Bacteriol.* **124,** 884–892.
Tabor, H. and Tabor, C. W. (1972). *Adv. Enzymol. Relat. Areas Mol. Biol.* **36,** 203–268.
Tang, F. F., Wei, H., McWhirter, D. L., and Edgar, J. (1935). *J. Pathol. Bacteriol.* **40,** 391–406.
Tang, F. F., Wei, H., and Edgar, J. (1936). *J. Pathol. Bacteriol.* **42,** 45–51.
Thompson, E. D., Knights, B. A., and Parks, L. W. (1973). *Biochim. Biophys. Acta* **304,** 132–141.

Tourtellotte, M. E. (1969). *In* "The Mycoplasmatales and the L-Phase of Bacteria" (L. Hayflick, ed.), pp. 451–468. Appleton, New York.
Tourtellotte, M. E., and Jacobs, R. E. (1960). *Ann. N.Y. Acad. Sci.* **79,** 521–530.
Tourtellotte, M. E., Morowitz, H. J., and Kasimer, P. (1964). *J. Bacteriol.* **88,** 11–15.
Townsend, R. (1976). *J. Gen. Microbiol.* **94,** 417–420.
Tully, J. G., and Razin, S. (1977). *In* "CRC Handbook of Microbiology" (A. J. Laskin and H. Lechevalier, eds.), 2nd ed., Vol. 1, pp. 405–459. CRC Press, Cleveland, Ohio.
Turner, A. W. (1935). *J. Pathol. Bacteriol.* **41,** 1–32.
Valentine, R. C., and Wolfe, R. S. (1961). *Nature (London)* **191,** 925–926.
VanDemark, P. J. (1969). *In* "The Mycoplasmatales and the L-Phase of Bacteria" (L. Hayflick, ed.), pp. 491–501. Appleton, New York.
VanDemark, P. J., and Smith, P. F. (1964a). *J. Bacteriol.* **88,** 122–129.
VanDemark, P. J., and Smith, P. F. (1964b). *J. Bacteriol.* **88,** 1602–1607.
VanDemark, P. J., and Smith, P. F. (1965). *J. Bacteriol.* **89,** 373–377.
Walker, R. T. (1971). *J. Bacteriol.* **107,** 618–622.
Williamson, D. L., and Whitcomb, R. F. (1974). *Colloq. Inst. Natl. Sante Rech. Med.* **33,** 283–290.
Yarrison, G., Young, D. W., and Choules, G. L. (1972). *J. Bacteriol.* **110,** 494–503.

# 5 / MOTILITY

*Wolfgang Bredt*



## I. INTRODUCTION

For a long time, organisms of the mycoplasma group were considered as primitive prokaryotes possessing no special organelles and carrying out only the most primitive functions of life. However, as early as 1946 Andrewes and Welch described gliding movements of pleuropneumonia-like organisms (PPLO) isolated from mice. All data obtained on animal mycoplasmas after this first report suggest that the movements of these species are surface-dependent and that there is no flagellum or other external organelle related to the gliding.

With the recognition of new and rather unique microorganisms as Mollicutes the spectrum of possible types of locomotion widened considerably. The movements of *Thermoplasma acidophilum* are probably caused by a flagellum (Freundt, 1972) and the recent discovery of spiro-



THE MYCOPLASMAS, VOL. 1

ISBN 0-12-078401-7

plasmas added another type of rapid rotatory motion and flexional movements (Cole *et al.*, 1973a; Davis and Worley, 1973). This diversity characterizes the Mollicutes as a highly differentiated group of prokaryotes and permits interesting speculations about the relationship to similar types of locomotion in bacteria.

## II. GLIDING MOTILITY OF *Mycoplasma*

### A. Description and Morphology of Motile Species

#### 1. Discovery of Motility

In 1946 Andrewes and Welch recognized that the movements of their PPLO strains were surface-dependent and were no longer evident when the organisms were detached from the slides. The report described exactly the typical gliding movements of what is now named *Mycoplasma pulmonis*. This property of mycoplasmas went unnoticed until 1960, when Nelson reported the same type of gliding movements for *M. pulmonis* strains freshly isolated from mice which showed signs of infectious catarrh. He, too, could observe the motility only up to the sixth *in vitro* passage. Descriptions of movements and of cell morphology resembled those given by Andrewes and Welch (1946). These observations were confirmed in a paper by Nelson and Lyons (1965), who found five of the six strains examined to be motile. The authors emphasized motility as "commonly disregarded". In fact, reports on gliding motility were not cited in reviews on mycoplasmas for a long time.

Improved techniques for continuous observation of living mycoplasma cells (Bredt, 1968a) provided more convincing results. Gliding movements of *Mycoplasma pneumoniae* were first seen rather accidentally on a glass surface in a liquid medium (Bredt, 1968b). The motility of *M. pneumoniae* strain FH was confirmed and documented by use of a more viscous medium (Bredt, 1968c). Improved techniques of cultivation and observation permitted recording by cinematography (Bredt *et al.*, 1970a). First measurements were obtained from this film (Bredt *et al.*, 1970b). Subsequently the motility of freshly isolated strains of *M. pulmonis* could be confirmed and similar but very slow movements of *Mycoplasma gallisepticum* were reported (Bredt, 1972). Other human and animal mycoplasma species (including *Mycoplasma suipneumoniae*) have been found nonmotile under similar conditions, and the microscopic examination of these species did not indicate any morphologic details possibly related to motility (W. Bredt, unpublished observations).

## 2. Microscopic Morphology of Moving Cells

Gliding mycoplasmas are different from nonmotile species in cell shape and ultrastructure. The most striking aspect is a pronounced head structure (leading part, terminal or specialized structure) which shows peculiar ultrastructural details (Bredt, 1973, 1974).

**a. *Mycoplasma pulmonis.*** Motile cells of this species can be observed in two different basic forms. One is a coccoid or globular cell with a rodlike projection or stalk (Andrewes and Welch, 1946; Nelson and Lyons, 1965), which is often slightly thickened at the front end (Bredt and Radestock, 1977). Its length varies and often exceeds the diameter of the cell (Fig. 1). Occasionally several globules connected by thin filaments follow the leading stalk-carrying cell (Andrewes and Welch, 1946; Bredt and Radestock, 1977). The other typical form is a more filamentous structure of varying length with tapered ends (Fig. 1). The stalk and filamentous forms are highly flexible.

**b. *Mycoplasma gallisepticum.*** The morphology of this organism was examined mainly by electron microscopy (Maniloff and Morowitz, 1972). The cells are pear-shaped with a distinct "bleb" structure on their forward ends. This appearance was confirmed by microscopic examination of living cells (Fig. 2) in coverslip chambers (Erdmann, 1976; W. Bredt, unpublished observations).

**c. *Mycoplasma pneumoniae.*** The typical moving cell of this species appears elongated, consisting of a frontal projection (tip structure), a thicker "body" part, and a longer taillike rear end (Bredt, 1973; Radestock and Bredt, 1977) (Fig. 3). Contact of moving and resting cells resulted often in temporary attachment, followed by stretching, especially of the tail parts. When the cells were moving, the extended tails sticking to other cells detached and then contracted to normal length. Cells mov-

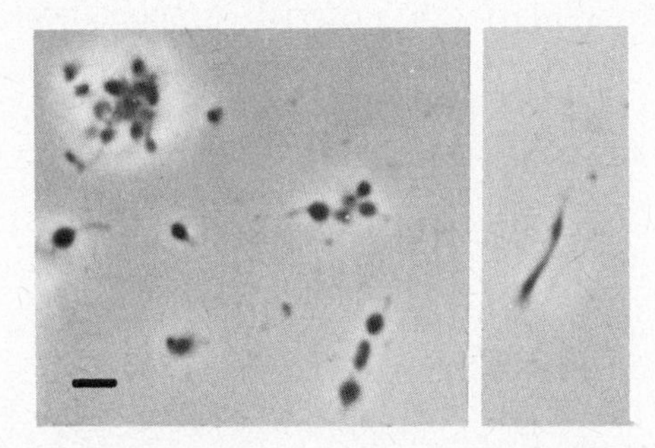

FIGURE 1. Motile forms of *M. pulmonis*. Left, coccoid; right, filamentous.

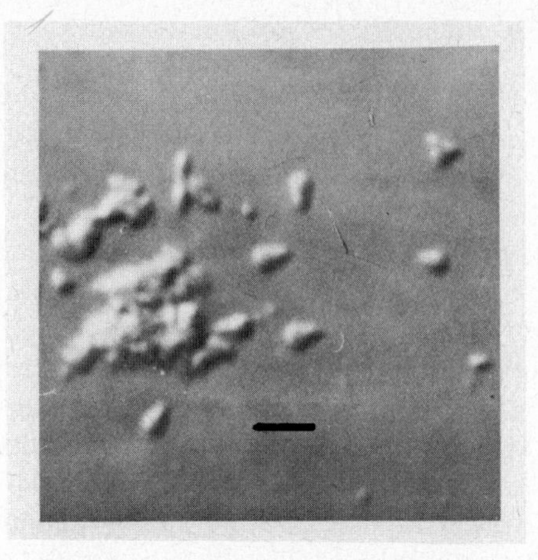

FIGURE 2. Living *M. gallisepticum* cells.

ing during the process of multiplication may have two tip structures and body parts, being connected by the tails.

### 3. Ultrastructural Details Possibly Related to Motility

**a. Specialized structures.** The terminal structures of the motile species are different in electron-microscopic appearance. Many details are known for *M. gallisepticum*. The bleb of this species is hemispherical, about 800 Å × 1250 Å. It is attached to the so-called infrableb region by a flat circular plate (Maniloff, 1972). However, nothing is known about the relationship of the different structures to motility.

A related structure in *M. pneumoniae* was first described in 1970 by Biberfeld and Biberfeld. Their report and further investigations (Collier, 1972) showed the tip structure to consist of a central dark rod surrounded by adjacent electron-translucent material. A nonpathogenic mutant was

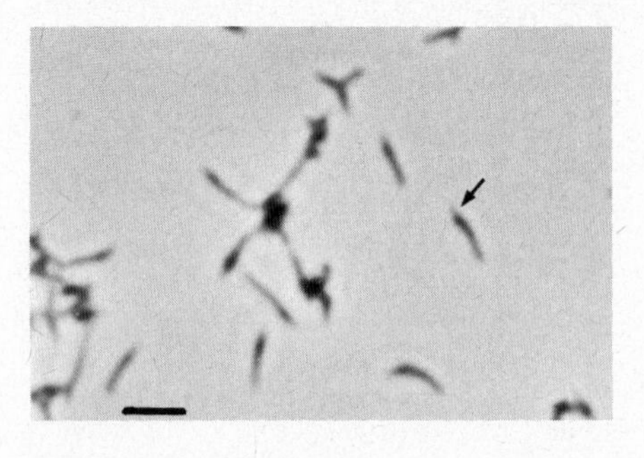

FIGURE 3. Motile cells of *M. pneumoniae*.

found to possess a similar tip structure (Collier, 1972), although this mutant proved to be nonmotile.

Nothing is known about the ultrastructural details of the stalk of *M. pulmonis*. Existing electron microscopic photographs do not reveal details of this structure.

Recently a new mycoplasma species from the bovine intestinal and urogenital tract, *Mycoplasma alvi*, has been described. Electron microscopic pictures show structural details which resemble the tip structure of *M. pneumoniae* (Gourlay *et al.*, 1977). If examined by phase contrast, some of the living cells carried a rodlike structure resembling the stalk of *M. pulmonis*. However, the stalk did not attach firmly to the glass and there were no signs of motility (W. Bredt, unpublished observations). The strain examined (kindly provided by R. N. Gourlay) had been subcultured *in vitro* several times, however, leaving open the question of motility of freshly isolated strains.

The structure of the membrane covering the tip areas of motile mycoplasmas seems not to be different from other parts of the cell envelope. However, the surface of this area may have some special properties related to attachment (see Section C.2.a).

**b. Cytochemistry.** Few data are available concerning the cytochemistry of tip structures. ATPase was found by electron microscopy (Munkres and Wachtel, 1967) and by chemical examination of isolated blebs of *M. gallisepticum* (Maniloff and Quinlan, 1974). For the other mycoplasmas no comparable data are available.

**c. Contractile structures.** Similarly scarce are data on ultrastructural details related to contractility. A recent study of Wilson and Collier (1976) on *M. pneumoniae* shows a possible periodicity in the central rod of the tip structure, suggesting the existence of substructures. However, the nature of this periodicity remains open and at present it cannot be related to contractile substances.

### 4. Motility and Contractility

Living cells of *M. hominis* are able to change their cell shape rapidly and reversibly (Bredt, 1970; Bredt *et al.*, 1971, 1973), suggesting the presence of a contractile substance forming a sort of cytoskeleton. Similar processes have been observed as well with *Mycoplasma orale* (W. Bredt, unpublished observations) as with *M. mycoides* (Tang *et al.*, 1936) and can be assumed by morphologic analogy for many other species. Contractile processes are apparently needed for the shaping of daughter cells and for final separation. Moreover, the presence of such a cytoskeleton would adequately explain the ability of mycoplasmas to maintain their variable but distinct cell shape. Motile mycoplasmas also maintain their very characteristic shape and therefore should possess contractile substance in or on their membrane. However, one may speculate whether the gliding apparatus consists mainly of a local concentration of such "normal" contractile material or differs from it in structure and organization. No data are available to support either suggestion.

## B. Physiology of Movements

The movements of animal mycoplasmas can only be observed on surfaces. They can easily be examined qualitatively in simple coverslip chambers (Bredt, 1968a). The standardization desirable for quantitative studies, however, is difficult to maintain (Radestock and Bredt, 1977) and comparisons can only be made within the respective experiment.

### 1. Description of Movements

**a. *Mycoplasma pneumoniae.*** The cells of *M. pneumoniae* always move in the direction of their tip structure. If the tip has changed direction, the remaining part of the cell will follow (Bredt, 1973). The direction is never reversed. The movements mostly occur in relatively narrow circles, but straight motion can also be seen. The periods of movement are interrupted by resting periods at irregular intervals. The average speed (including resting periods) was found to be about 0.3 μm/sec. When the resting periods were omitted from the calculations the average speed was about 0.4 μm/sec (corrected average speed). Maximum speed for short periods (2-sec intervals) was occasionally 1.5–2.0 μm/sec but was more often found around 1 μm/sec. Resting periods of cells under normal experimental conditions varied widely between 0 and 80% (Radestock and Bredt, 1977). Nothing is known about the relationship between movements and the cell replication cycle.

Motility of *M. pneumoniae* seems to be a stable property. All quantitative studies were done with the FH strain, which had been subcultured for several hundred passages on artificial media. However, movements and the typical morphology were also observed with freshly isolated strains and with temperature-sensitive mutants (W. Bredt, unpublished observation).

**b. *Mycoplasma pulmonis.*** The movements of this species resemble those of *M. pneumoniae* in speed and general properties. The coccoid or globular forms always move in the direction of the stalk. The filamentous forms occasionally reverse their direction (Bredt and Radestock, 1977), but not as often as described by Nelson and Lyons (1965). The "spinning" of stalk-carrying coccoid cells, as described by the earlier authors, could not be observed in the more recent study. The reason for this difference is unknown. The general pattern of the movements is irregular, but the cells seem to move more often in straight paths than does *M. pneumoniae*. Curves are always initiated by the stalk or the tapered end of the filament; the other parts of the cell follow the new direction specified by the leading part.

The speed of *M. pulmonis* is comparable with that of *M. pneumoniae:* The corrected average speed was slightly higher (0.4–0.7 μm/sec) and the maximum speed was found to be about 1 μm/sec. There was no difference in speed between rod-carrying spheres and longer filaments. The percentage of pauses was low (0–25%), and the cells were in general more active than those of *M. pneumoniae*. The cells of *M. pulmonis* remained motile only up to the sixth (Andrewes and Welch, 1946), ninth, or fifteenth (Nelson and Lyons, 1965) passage. Andrewes and Welch were able to revive the movements for some time by mouse passage, but after more subcultures even this method could not prevent the loss of motility.

**c. *Mycoplasma gallisepticum.*** The cells of this species move only in the direction of their bleb. Their movements are very slow, rather wriggling, and interrupted by numerous periods without motion (40–80%). The average speed (without pauses) varies around 2–3 μm/min (0.03–0.05 μm/sec) and the maximum speed observed for 15 sec was about 9 μm/min (0.15 μm/sec) (Erdmann, 1976).

The percentage of moving and resting cells in a given population and at a certain age of the culture is not known for either species. Qualitative examination of cultures sometimes suggests an increase of motionless periods during cell multiplication, but this still has to be confirmed.

### 2. Influence of Exogenous Factors

Motile mycoplasmas are highly fastidious and sensitive organisms, so that environmental conditions can only be varied within certain limits. Moreover, it is hardly possible to differentiate between specific effects on the gliding apparatus and less specific reactions affecting the viability of the cell.

**a. Physical factors.** Gliding of *M. pulmonis* has so far only been observed in normal liquid medium. The cells did not detach from the glass during their movements. In contrast, moving *M. pneumoniae* cells in liquid media frequently lose contact with the glass surface and disappear in the medium (Radestock and Bredt, 1977). An increase of viscosity by addition of gelatin kept the cells better attached to the glass; however, in higher concentrations (up to 5%) it reduced the corrected average speed significantly and increased the percentage of resting periods.

Variations in temperature seem to influence the movements of all three species. *Mycoplasma pneumoniae* moved slower at 32°C compared with

37°C; however, an increase up to 39°C did not influence the velocity. At both temperatures the cells showed more resting periods than the control cells at 37°C. The only data available for *M. pulmonis* show a correlation between lower temperature (32°C) and reduced speed. However, the cells were gliding with considerable speed even at room temperature (Bredt and Radestock, 1977). Some experiments with *M. gallisepticum* suggested a similar influence of temperature on cell velocity, but the number of cells observed did not permit statistical analysis (Erdmann, 1976).

*Mycoplasma pneumoniae* responded to lower pH (6.5 vs 7.2) with a slightly increased average speed and reduced resting periods. Increase to pH 7.9 had the opposite effect (Radestock and Bredt, 1977).

**b. Nutritional factors.** Variations in nutritional factors within tolerable limits (reduction of serum content to 5%, addition of glucose up to 1%) had no significant effect on *M. pneumoniae*. The reported cessation of movements in medium with a reduced concentration of yeast extract (Radestock and Bredt, 1977) was not reproducible in following experiments. Removal of calcium ions from the medium by ethylenglycoltetraacetic acid (EGTA) or addition of calcium was without effect. Addition of ethylenediaminetetraacetic acid (EDTA) damaged the cells, apparently by removing magnesium ions from the cell membrane. Other chemicals (cholesterol, ATP) or the use of conditioned medium (preincubated with *M. pneumoniae*) had no visible effect (Radestock and Bredt, 1977). No movements could be seen after the cells were transferred to buffer. The movements of *M. gallisepticum* are apparently not affected by the addition of 0.1% glucose to the medium (Erdmann, 1976).

**c. Inhibitory substances.** The average speed of moving *M. pneumoniae* cells is significantly reduced by iodoacetate at inhibitory or subinhibitory concentrations (Radestock and Bredt, 1977). Furthermore, the percentage of resting periods was increased. The latter effect was also seen with *p*-chloromercuribenzoate (PCMB). Ouabain, sodium cyanide, sodium fluoride, and dinitrophenol did not affect movement (Bredt, 1973; Radestock and Bredt, 1977).

Of the inhibitors of nucleic acid and protein synthesis tested, only mitomycin C reduced the speed significantly at inhibitory levels. Actinomycin D and chloramphenicol were not effective at nonkilling concentrations. Motile cells were found even when inhibitory concentrations of chloramphenicol had been present for 24 hr, suggesting a slow turnover of proteins involved in motility.

Finally, "specific" inhibitors of contractile elements (cytochalasin B, colchicine) were without effect on *M. pneumoniae* (Bredt, 1973; Radestock and Bredt, 1977). The movements of *M. gallisepticum* were not affected by high concentrations of cytochalasin B either (W. Bredt, unpublished results), but quantitative data are not available. This is in agreement with negative effects on gliding bacteria (Henrichsen, 1972).

**d. Antisera.** In contrast to the metabolic inhibitors, which act mainly on enzyme systems throughout the cell, antibodies primarily affect the cell surface, impairing membrane functions necessary for motility, e.g., transport mechanisms, membrane fluidity, or attachment. Gliding of *M. pneumoniae* was effectively inhibited by antiserum directed against this species. The effect was reversible. The respective structural elements seem to be present even in nonmotile mutants, because antiserum against such a mutant also stopped the movements (Radestock and Bredt, 1977).

## C. Gliding and Adherence

### 1. Adherence of Motile Species

Attachment to a given surface is a prerequisite of gliding motion, although the ability to adhere is not restricted to motile mycoplasma species. Adherence to glass and plastic has been observed for *M. pulmonis*, *M. pneumoniae*, and *M. gallisepticum*. Furthermore, *M. pneumoniae* and *M. gallisepticum* attach to a variety of cells, including erythrocytes (Gorski and Bredt, 1977; Sobeslavsky *et al.*, 1968), HeLa cells (Manchee and Taylor-Robinson, 1969a), spermatozoa (Taylor-Robinson and Manchee, 1967), and tracheal organ cultures (Collier and Clyde, 1971). Jones and Hirsch (1971) reported attachment of *M. pulmonis* to macrophages and L cells. However, motility has so far only been observed on glass.

### 2. Binding Sites and Receptors

**a. Nature and localization of the mycoplasma binding sites.** The binding of motile mycoplasmas to the various substrates is possibly mediated by a protein or proteinlike substance on the mycoplasma surface. This has been established for *M. pneumoniae* (Hu *et al.*, 1977; Gorski and Bredt, 1977) and for *M. gallisepticum* (Gesner and Thomas, 1965; Banai *et al.*, 1978). Attachment of *M. pneumoniae* to either sheep erythrocytes or protein-covered polystyrene beads was similarly affected by the various treatments, indicating an identical or closely related binding site (Gorski and Bredt, 1977). Adherence of *M. pulmonis* to macrophages was not prevented by protease treatment of the mycoplasma cells (Jones *et al.*, 1972). However, attachment to macrophages is possibly different from adherence to other cell types, and its role in motility is so far unknown.

The important role of the tip, bleb, or stalk for the direction of movements (Bredt, 1973) suggests that the binding sites are either restricted to this area or at least are found there in higher concentration. The uniform arrangement of *M. pneumoniae* cells attached to tracheal cells (Collier, 1972) or inert material (Biberfeld, 1972) also indicates a special role of the tip in the binding process. Besides this more functional evidence only a few additional data are available. Pronounced "shrinking figures" seen at the tip of *M. pneumoniae* after fixation suggest a stronger attachment of this area to the glass (Bredt and Bierther, 1974). Wilson and Collier (1976) noted a material with stronger affinity for ruthenium red outside the tips of *M. pneumoniae* and found more intensive staining of these structures after fixation with tannic acid. Although the substances stained by these methods may not constitute the binding sites themselves, they could represent material attached to these sites, thus demonstrating indirectly their high concentration on the tip structure.

**b. Receptors of host cells.** The receptors of the respective host cells tested for attachment are neuraminidase-sensitive and seem to contain sialic acid as a major component (Gorski and Bredt, 1977; Manchee and Taylor-Robinson, 1969a,b). However, blocking of the *M. gallisepticum*–erythrocyte attachment is only effective if the sialic acid is still part of the glycoprotein molecule. Treatment of human erythrocytes with neuraminidase did not prevent attachment to *M. pneumoniae,* suggesting the presence of different adherence mechanisms (J. Feldner and W. Bredt, unpublished observations).

**c. Nature of binding forces.** The nature of the binding forces related to gliding motion is largely unknown. Besides the assumed specific interactions between binding site and receptor there must be rather unspecific, perhaps electrostatic, binding to such inert materials as glass and plastic.

## D. Mechanism of Gliding

The mechanism of gliding motion of prokaryotes is still unknown (Burchard *et al.*, 1977). No morphologic changes indicating contractile processes can be seen in either gliding mycoplasmas or gliding bacteria. If these changes do occur but are suppressed in myxobacteria by the cell wall, they should at least be visible in the wall-less mycoplasmas. Their absence suggests that they either do not occur, they are too frequent, or their extent is below the level of visibility.

The fibers recently found in myxobacteria (Burchard *et al.*, 1977) are of considerable length, indicating a possible involvement of the whole cell in the movements. In contrast, the mycoplasmas have developed specialized organelles—the different terminal structures which lead the gliding motion of these cells. However, the only morphologic evidence for contractile material so far is the rather preliminary suggestion of periodicity in the central rod of the *M. pneumoniae* tip structure (Wilson and Collier, 1976). Similarly scarce is information on the chemical nature of the gliding apparatus. Generally the motility mechanism possibly consists of long-lasting structural proteins as suggested by the persistence of movements under the influence of chloramphenicol (Radestock and Bredt, 1977).

The report of Neimark (1977) on actinlike material in *M. pneumoniae* does support the assumed presence of contractile material; however, it does not contain information on the localization of this substance. Furthermore, the presence of the typical tip structure even in a nonmotile strain (Collier, 1972) and the similarity of motile and nonmotile strains in gel electrophoresis analysis of cell protein suggest that this terminal structure is only one of the structures involved in motility.

The mechanism of the other prerequisite for gliding, attachment to surfaces, is also largely unknown. If contractility is the gliding force, then there should be, in certain membrane areas, an alternate attachment to and detachment from the foreign surface. Neither this nor the other possibility, namely streaming of the membrane, can be substantiated by any data. Chemotactic properties were not examined in mycoplasmas because a screening test has not yet been developed.

## E. Gliding Motion and Pathogenicity

Mycoplasmas are host-dependent parasites and the ability to move actively probably gives them some advantage over nonmotile competitors in their struggle to survive in the host. All three species known at present to be motile are pathogens of the respiratory tract, suggesting motility as a pathogenicity factor for this specific area. The concomitant loss of motility and of pathogenicity in *M. pulmonis* after a few passages *in vitro* and the avirulence of the nonmotile mutant of *M. pneumoniae* strongly indicate such a correlation. However, not all mycoplasmas pathogenic for the respiratory tract are motile (e.g., *M. suipneumoniae*). Moreover, the motile laboratory strain FH of *M. pneumoniae* is not considered to be pathogenic. So motility is certainly just one of the many factors which may contribute to the virulence of pathogenic mycoplasma species. Motility may help the organisms to penetrate the mucus layer of tracheal and bronchial epithelium and to reach the cell surface for final attachment. It could help them to actively invade intercellular spaces and fissures or membrane crypts, e.g., in macrophages, in which they can survive the attack of such host defense factors as complement (P. Erb and W. Bredt, unpublished observations). Finally, it may enable the parasites to leave areas in which the environment has become unfavorable. Considering the possible applications of locomotion in the host–parasite relationship it seems likely that *in vivo* movements occur only temporarily, ceasing as soon as the cell has attached firmly to host tissue. However, it should be kept in mind that all studies on gliding motion of mycoplasmas so far have only been done on glass or plastic. Although all available data suggest similar movements *in vivo*, we still lack experimental confirmation.

## III. MOTILITY OF *Spiroplasma*

### A. Morphology and Physiology of Movements

The discovery of wall-less helical-shaped microorganisms in diseased plants (Saglio *et al.*, 1973) not only added a quite unique agent to the Mollicutes but also introduced a totally different type of motility into this class. Two types of movements were observed. The first was rapid rotatory motion which could reverse, leading to minimal back and forth progress. The second was "a slow undulation and bending of filaments sometimes with a slight rotatory component, but not leading to change of position of the filament" (Cole *et al.*, 1973a). Elevated viscosity or contact with a solid surface resulted in striking translational locomotion (Davis, 1978). No flagella, axial filaments, or other organelles were seen under the electron microscope. Thus, in contrast to animal mycoplasmas, the spiroplasmas are able to move freely in liquid medium or in agar, as shown by the formation of satellite colonies (Townsend *et al.*, 1977). For the agent of corn stunt disease Davis and Worley (1973) described a similar whirling or spinning around the long axis and flexional motions but no translational motion. The contractile movements stopped after the organisms had been exposed to glutaraldehyde or mild heat (40° or 56°C). Straightening of a folded cell occurred within 1/9 sec. Attachment of cells to glass occurred but no gliding was seen.

Townsend *et al.* (1977) described a nonhelical strain of *Spiroplasma citri* which was also nonmotile. This correlation suggests simultaneous involvement of a contractile substance in maintenance of both shape and motility. Evidence for such contractile material has been presented by Williamson (1974) in the sex-ratio spiroplasma and by Cole *et al.* (1973a,b) in *S. citri*. Furthermore, Neimark (1976) reported the presence of actinlike material in *S. citri*. The nonmotile mutant of *S. citri* described by Townsend *et al.* (1977) lacked a membrane protein present in motile strains. It is too early, however, to definitely link these findings to the mechanism of motility.

### B. Motility and Pathogenicity in Spiroplasmas

The ability of spiroplasmas to move freely in liquid or more viscous environments suggests an advantage of motile cells in infection. It can be imagined that the particular movements of spiroplasmas enable these parasites to colonize the plant more rapidly than nonmotile organisms could. A similar ecologic advantage in viscous environments is discussed for the coiling movements of spirochetes (Greenberg and Canale-Parola, 1977). However, the proven infectivity and pathogenicity of the nonmotile strain (Townsend *et al.*, 1977) does not permit such a simple explanation for the existence and role of motility in spiroplasmas.

## IV. CONCLUSIONS

Motility and contractility are properties not necessarily expected in such small microorganisms as mycoplasmas. Moreover, the organisms have developed sophisticated organelles for that purpose. The existence of such specialized organelles and the diversity of locomotion phenomena within the Mollicutes suggest a long phylogenetic evolution of this class comparable to that of walled bacteria. The ability to move and, in gliding bacteria, to attach firmly to surfaces mainly by the special organelle may have been developed under selective pressure of environmental factors. We can only speculate on the nature of these forces. Did these mycoplasmas "flee" competition of other microorganisms by becoming motile? Or did they invade areas poorly colonized and accessible only to motile microorganisms? Certainly the parasitic nature of mycoplasmas would evolutionarily reward any property opening up previously uninhabited tissues or even permitting the infection of new host species.

It is interesting to note that motility has developed somewhat differently in bacteria and in mycoplasmas. Whereas spirochetes are found mainly in animal hosts, helical spiroplasmas usually infect plants and insects. On the other hand, gliding, mostly observed in free-living bacteria, has been used by the mycoplasmas to facilitate the colonization of animal hosts. At present, however, we can only see the results and are not able to explain or to understand the causative interactions.

Motility of mycoplasmas, either gliding or rotatory, is an interesting phenomenon for basic studies on locomotion mechanisms of small-sized cells. Furthermore, it can be used to examine the role of such properties in host–parasite interactions. The field of research is relatively new and further experiments will, hopefully, expand our knowledge.

## REFERENCES

Andrewes, C. H., and Welch, F. V. (1946). *J. Pathol. Bacteriol.* **58,** 578–580.
Banai, M., Kahane, I., Razin, S., and Bredt, W. (1978). *Infect. Immun.* **21,** 365–372.
Biberfeld, G. (1972). *Pathog. Mycoplasmas, Ciba Found. Symp., 1972,* pp. 322–323.
Biberfeld, G., and Biberfeld, P. (1970). *J. Bacteriol.* **102,** 855–861.
Bredt, W. (1968a). *Proc. Soc. Exp. Biol. Med.* **128,** 338–340.
Bredt, W. (1968b). *Zentralbl. Bakteriol., Parasitenkd., Infektionskr. Hyg., Abt. 1: Orig., Reihe A* **208,** 549–562.
Bredt, W. (1968c). *Pathol. Microbiol.* **32,** 321–326.
Bredt, W. (1970). *Z. Med. Mikrobiol. Immunol.* **155,** 248–274.
Bredt, W. (1972). *Med. Microbiol. Immunol.* **157,** 169.
Bredt, W. (1973). *Ann. N.Y. Acad. Sci.* **225,** 246–250.
Bredt, W. (1974). *In* "Mycoplasmas of Man, Animals, Plants and Insects" (J. M. Bové and J. F. Duplan, eds.), pp. 47–52. INSERM, Paris.
Bredt, W., and Bierther, F. W. (1974). *Zentralbl. Bakteriol., Parasitenkd., Infektionskr. Hyg., Abt. 1: Orig., Reihe A* **229,** 249–255.
Bredt, W., and Radestock, U. (1977). *J. Bacteriol.* **130,** 937–938.
Bredt, W., Höfling, K. H., and Heunert, H. H. (1970a). Encyclopaedia cinematographica Film E 1633, Inst. Wiss. Film, Göttingen.
Bredt, W., Höfling, K. H., Heunert, H. H., and Milthaler, B. (1970b). *Z Med. Mikrobiol. Immunol.* **156,** 39–43.
Bredt, W., Höfling, K. H., and Heunert, H. H. (1971). Encyclopaedia cinematographica Film E 1813, Inst. Wiss. Film, Göttingen.
Bredt, W., Heunert, H. H., Höfling, K. H., and Milthaler, B. (1973). *J. Bacteriol.* **113,** 1223–1227.
Burchard, A. C., Burchard, R. P., and Kloetzel, J. A. (1977). *J. Bacteriol.* **132,** 666–672.
Cole, R. M., Tully, J. G., Popkin, T. H., and Bové, J. M. (1973a). *J. Bacteriol.* **115,** 367–386.
Cole, R. M., Tully, J. G., Popkin, T. H., and Bové, J. M. (1973b). *Ann. N.Y. Acad. Sci.* **225,** 471–493.
Collier, A. M. (1972). *Pathog. Mycoplasmas, Ciba Found. Symp. 1972,* pp. 307–320.
Collier, A. M., and Clyde, W. A., Jr. (1971). *Infect. Immun.* **3,** 694–701.
Davis, R. E. (1978). *Proc. Meeting Int. Council Lethal Yellowing, 3rd.* Univ. Florida, Publ. 2, 19.
Davis, R. E., and Worley, J. F. (1973). *Phytopathology,* **63,** 403–408.
Erdmann, T. (1976). M.D. Thesis, Univ. Mainz.
Freundt, E. A. (1972). *Pathog. Mycoplasmas, Ciba Found. Symp. 1972*, p. 10.
Gesner, B., and Thomas, L. (1965). *Science* **151,** 590–591.
Gorski, F., and Bredt, W. (1977). *FEMS Microbiol. Lett.* **1,** 265–267.
Gourlay, R. N., Wyld, S. G., and Leach, R. H. (1977). *Int. J. Syst. Bacteriol.* **27,** 86–96.
Greenberg, E. P., and Canale-Parola, E. (1977). *J. Bacteriol.* **131,** 960–969.
Henrichsen, J. (1972). *Acta Pathol. Microbiol. Scand., Sect. B* **80,** 623.
Hu, P. C., Collier, A. M., and Baseman, J. B. (1977). *J. Exp. Med.* **145,** 1328–1343.
Jones, T. C., and Hirsch, J. G. (1971). *J. Exp. Med.* **133,** 231–259.
Jones, T. C., Yeh, A., and Hirsch, J. (1972). *Proc. Soc. Exp. Biol. Med.* **139,** 464–470.
Manchee, R. J., and Taylor-Robinson, D. (1969a). *J. Bacteriol.* **98,** 914–919.
Manchee, R. J., and Taylor-Robinson, D. (1969b). *Br. J. Exp. Pathol.* **50,** 66–75.
Maniloff, J. (1972). *Pathog. Mycoplasmas, Ciba Found. Symp., 1972,* pp. 67–87.
Maniloff, J., and Morowitz, H. J. (1972). *Bacteriol. Rev.* **36,** 263–290.
Maniloff, J., and Quinlan, D. C. (1974). *J. Bacteriol.* **120,** 495–501.

Munkres, M., and Wachtel, A. (1967). *J. Bacteriol.* **93,** 1096–1103.
Neimark, H. C. (1976). *Am. Soc. Microb. Abstr. Annu. Meet.*, p. 61.
Neimark, H. C. (1977). *Proc. Natl. Acad. Sci. U.S.A.* **74,** 4041–4045.
Nelson, J. B. (1960). *Ann. N.Y. Acad. Sci.* **79,** 450–457.
Nelson, J. B., and Lyons, M. J. (1965). *J. Bacteriol.* **90,** 1750–1763.
Radestock, U., and Bredt, W. (1977). *J. Bacteriol.* **129,** 1495–1501.
Saglio, P., L'Hospital, M., La Fleche, D., Dupont, G., Bové, J. M., Tully, J. G., and Freundt, E. A. (1973). *Int. J. Syst. Bacteriol.* **23,** 191–204.
Sobeslavsky, O., Prescott, B., and Chanock, R. M. (1968). *J. Bacteriol.* **96,** 695–705.
Tang, F. F., Wei, H., and Edgar, J. (1936). *J. Pathol. Bacteriol.* **42,** 45–51.
Taylor-Robinson, D. and Manchee, R. J. (1967). *Nature (London)* **215,** 484–487.
Townsend, R., Markham, P. G., Plaskitt, A. K., and Daniels, M. J. (1977). *J. Gen. Microbiol.* **100,** 15–21.
Williamson, D. L. (1974). *J. Bacteriol.* **117,** 904–906.
Wilson, M. H., and Collier, A. M. (1976). *J. Bacteriol.* **125,** 332–339.

# 6 / THE MOLECULAR BIOLOGY OF MYCOPLASMAS

*Eric J. Stanbridge and Mitchell E. Reff*



## I. INTRODUCTION

Interest in the molecular biology of mycoplasmas has been somewhat fickle in nature. From the discovery of mycoplasmas in 1898 until almost 20 years ago, scientific interest in these microorganisms was primarily the domain of veterinarians and the enlightened few who staked their careers



THE MYCOPLASMAS, VOL. 1

ISBN 0-12-078401-7

on the pursuit of characterizing these intriguing microorganisms. Interest in mycoplasmas burgeoned in the early 1960s with the isolation of the first human pathogen, *Mycoplasma pneumoniae* (Chanock *et al.*, 1962), and the recognition that they are the smallest known free-living organisms (Morowitz, 1966; Pirie, 1973). The latter property prompted a series of investigations into the nature of their molecular properties in the hope that they might offer unique insights into the origins of life. As evidence accumulated that the physiology of mycoplasmas is similar to that of other prokaryotes, interest in these microorganisms on the part of molecular biologists has waned.

In this review we present recent information concerning the molecular biology of mycoplasmas and have taken the liberty of emphasizing those areas where mycoplasmas seem to have relatively unique properties. In doing so, we hope to stimulate a renaissance of scientific inquiry into the fundamental properties of these interesting organisms.

## II. THE MYCOPLASMA CHROMOSOME

Using a modification of the protein film technique of Kleinschmidt and Zahn (1959), Bode and Morowitz (1967) demonstrated that the nonreplicating chromosome of *Mycoplasma arthritidis* is a single, unbranched, circular, double-stranded deoxyribonucleic acid (DNA) molecule. The size of the DNA was calculated to be in the region of $5.1 \times 10^8$ daltons, a value obtained by calibration of the linear array of DNA threads. This value confirmed an earlier observation (Riggs, 1966) that the mycoplasma genome was the smallest known in a prokaryotic organism. Forks and double forks were also seen and interpreted as evidence that the mode of replication of the mycoplasma genome is essentially the same as that observed for bacteria (Cairns, 1963).

This initial report was soon followed by others confirming the structural and molecular weight similarities of the chromosomes of a number of mycoplasma species. Extensive studies by Bak *et al.* (1969) revealed that the genome size of acholeplasmas is approximately double that of mycoplasmas, being in the order of $1 \times 10^9$ daltons. A list of the known genome sizes for members and affiliates of the class Mollicutes is given in Table I. It should be noted for comparison that the lowest reported value for bacterial DNA is $8 \times 10^8$ daltons, and the values range from this low to genome sizes of several billion daltons (Morowitz, 1966).

Wallace and Morowitz (1973) have commented on the striking fact that there are no intermediates between mycoplasmal and acholeplasmal genome sizes. Based upon this observation, they have suggested that it is

TABLE I. **Genome Sizes of Members of the Class Mollicutes**

| Genus | Genome size (daltons) | Range of guanine + cytosine content (% G + C) |
|---|---|---|
| *Mycoplasma* | $5 \times 10^8$ | 23–41 |
| *Ureaplasma* | $5 \times 10^8$ | 27–30 |
| *Acholeplasma* | $1 \times 10^9$ | 30–33 |
| *Anaeroplasma* | Not known | 30–33 and 40[a] |
| *Spiroplasma* | $1 \times 10^9$ | 25 |
| *Thermoplasma* | $8.4–9.5 \times 10^8$ | 46 |

[a] Although the G + C content of the sterol-requiring *Anaeroplasma abactoclasticum* has been reported as 30% (Robinson *et al.*, 1975), other candidates for the genus *Anaeroplasma* which are sterol-nonrequiring have a G + C content of approximately 40% (Robinson *et al.*, 1975). However, it is unlikely that these latter strains will be included in the genus *Anaeroplasma* (International Committee on Systematic Bacteriology, Subcommittee on the Taxonomy of Mycoplasmatales, 1977).

unlikely that mycoplasmas are merely degenerate bacteria since, if this were the case, one would expect to see values in between the two. This question of the phylogenetic diversity between mycoplasmas and other prokaryotes is covered in considerable detail in Chapter 2 by Dr. Neimark.

Based upon the size of their genomes, the informational contents of mycoplasmal and acholeplasmal DNAs have been calculated as approximately 650 and 1300 cistrons, respectively (Morowitz, 1969), assuming that the entire genome encodes for gene products. This relatively low number of gene functions correlates well with the observed fastidious nature of these microorganisms.

## A. DNA Packaging

The way in which a cell manages to package its relatively enormous length of DNA in a form accessible to transcription and replication has fascinated investigators for many years. The chromosome of *Escherichia coli* can be isolated under conditions of gentle lysis as a highly organized, compact structure (Stonington and Pettijohn, 1971; Worcel and Burgi, 1972). The isolated folded chromosomes have dimensions similar to *in vivo* nucleoids, indicating that they may be structurally intact (Hecht *et al.*, 1975). Both RNA and protein appear to play roles in maintaining the structure of the folded chromosomes (Stonington and Pettijohn, 1971; Worcel and Burgi, 1972; Drlica and Worcel, 1975). Modifications of either DNA–RNA or DNA–protein interactions cause changes in the sedimentation value of the folded chromosomes, indicating an altered conformation.

Folded chromosomes have also been isolated from mycoplasma species (Teplitz, 1977). However, preliminary results indicate that, unlike *E. coli*, conditions expected to dissociate protein and RNA from DNA do not change the sedimentation properties of the folded chromosomes. The significance of these differences with respect to DNA packaging is as yet unclear.

A finding of possible interest is that *Thermoplasma acidophilum* (a thermophilic, acidophilic mycoplasma) DNA is closely associated with a histone-like protein (Searcy, 1975). This organism has a guanine plus cytosine (G + C) content of 46% (Christiansen *et al.*, 1975; Searcy and Doyle, 1975), in contrast to other thermophiles whose G + C contents are usually much higher (Williams, 1975). It is possible that this histone-like protein may protect *T. acidophilum* from thermal denaturation. The effect this basic protein has on the DNA packaging of *T. acidophilum* is unknown.

## B. DNA Base Composition

Another interesting property of the mycoplasma genome is the extremely low guanine plus cytosine (G + C) content seen in certain species (Jones and Walker, 1963; Neimark, 1970). Although G + C contents range from 23 to 41%, the organisms can be subgrouped into sets which indicate some homogeneity within a given group. For example, many members of the genus *Mycoplasma* have G + C contents ranging from 25 to 29% (Neimark, 1970). The eight human ureaplasma serotypes examined have very close values of 27.7–28.5% (Bak and Black, 1968; Black *et al.*, 1972). Acholeplasmas have G + C contents somewhat higher than most mycoplasmas, ranging from 30 to 33% (Neimark, 1970; Askaa *et al.*, 1973). A few mycoplasma species have G + C values which overlap with acholeplasmas. *Mycoplasma pneumoniae* is unique among mycoplasmas in having a G + C content of 39–41%. Finally, the prototype organisms *Spiroplasma citri, Anaeroplasma abactoclasticum,* and *Thermoplasma acidophilum* have G + C values of 25.9, 30, and 46%, respectively (Saglio *et al.*, 1974; Robinson *et al.*, 1975; Christiansen *et al.*, 1975).

The extremely low G + C content, characteristic of many mycoplasma species, is of interest both from an evolutionary and a genetic code restraint standpoint (Subak-Sharpe *et al.*, 1974; Russell *et al.*, 1973; Woese, 1967; Woese and Bleyman, 1972). The theoretical maximum and minimum limits of G + C contents of coding DNA are 71 and 26%, respectively (Elton, 1973). This range is strikingly close to that actually found in natural organisms, with mycoplasmas representing the lowest limit.

It is hard to imagine what evolutionary advantage there would be in having a low genomic G + C content. Singer and Ames (1970) identified a correlation between G + C content and the degree of exposure to ultraviolet irradiation in the environmental habitats of bacteria. Organisms with a low G + C would obviously be at higher risk due to a greater likelihood of thymidine dimerization. It is perhaps not surprising that mycoplasmas have entered into a protected host–parasite relationship with animals and plants (Stanbridge, 1976) which would shield them from radiation damage.

Woese (1967) has suggested that organisms with low G + C values are forced to take on very specific nearest neighbor patterns in their coding DNA. Subak-Sharpe and colleagues (1974; Russell *et al.*, 1973) in an extensive analysis of the doublet frequencies of the DNAs of bacteria and other organisms, confirmed this prediction. Figure 1 illustrates the doublet frequencies or general design patterns of groups with very low G + C. Here we see that evolutionarily unrelated organisms, such as *Clostridium* and *Mycoplasma,* have very similar patterns. The same is true for the protozoan *Tetrahymena,* which has a doublet pattern with the same characteristics (Swartz *et al.*, 1962). This underscores the fact that although general design similarities between organisms reflect their evolutionary relationship, this is not so for organisms of extreme G + C values. In this case, because of the constraints on their genetic codes, molecular or evolutionary relatedness should not be assumed from doublet pattern similarities (Subak-Sharpe *et al.*, 1974). This is true for organisms possessing both low and high G + C contents.

An interesting feature of the nearest neighbor frequencies in Fig. 1 is the very low value for the doublet CG. Assuming a random distribution of the doublet in the DNA sequence, this would indicate that codons containing CG are rarely used in mycoplasmas with 25% G + C contents. Thus, arginine, in particular, would be coded for almost exclusively by its AGA codon rather than its CGX ones (Woese, 1967).

The degeneracy of the genetic code is another way in which an organism may overcome a deficiency of G + C. The degeneracy involves only the third nucleotide of any codon. Thus, it is possible that codons with G or C in the third position are present at a much lower frequency. In 1961, before the genetic code had been determined, Sueoka (1961) suggested that the amino acid composition of the total protein of an organism correlated with the G + C content of its DNA. Walker and colleagues (Chelton *et al.,* 1968) examined the total protein of *Mycoplasma mycoides* subsp. *capri* and showed that the amino acid composition

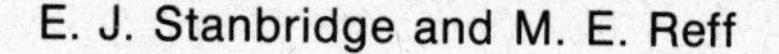

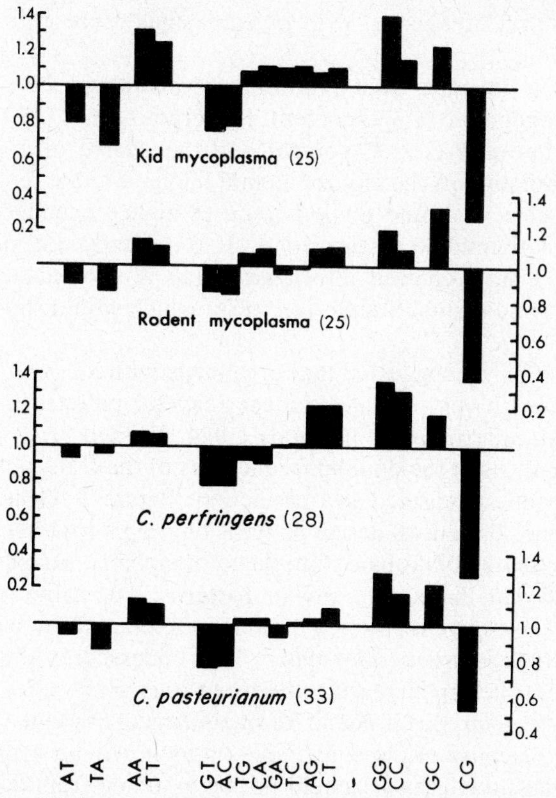


FIGURE 1. General design patterns of bacterial DNAs of the low G + C group. Values on the ordinate are the ratios of the observed doublet frequencies to the expected random frequencies. The values in parentheses are the G + C contents determined from doublet frequency analysis. (From Russell *et al.*, 1973; with permission of the authors and publishers.)

agreed with the prediction for an organism with a DNA base composition of 25% G + C. They further noted that to obtain the amino acid composition, it would be necessary for the code degeneracy to discriminate against G–C base pairs and to use A–T base pairs wherever possible.

Another property of mycoplasmas which would support this contention is the reduced number of isoaccepting transfer ribonucleic acid (tRNA) species. One report has claimed that there is only enough DNA complementary to tRNA to code for 44 different tRNA molecules (Ryan and Morowitz, 1969). Since it is becoming evident that wobble in *E. coli* is more extensive (Mitra *et al.*, 1977; Holmes *et al.*, 1977, 1978; Goldman *et al.*, 1978) than previously hypothesized (Crick, 1966) it will be interesting to examine the tRNAs and codon assignments of mycoplasmas in more detail.

### C. DNA Replication

The early electronmicroscopic studies of Bode and Morowitz (1967) indicated that the mode of replication of mycoplasmas was essentially similar to that of bacteria. Smith (1969) substituted 5-bromodeoxyuridine (BrdUrd) for thymidine in the growth medium of *Acholeplasma laidlawii* and demonstrated that DNA replication in this organism is semiconservative and proceeds unidirectionally along the chromosome. Comparison of the buoyant density distribution of intact chromosomes in cesium chloride gradients with that of mildly sheared chromosomes indicated that replication proceeds from only a few growing points. The conclusion derived from these experiments, and from related ones demonstrating thymineless death in *A. laidlawii* (Smith and Hanawalt, 1968), was that DNA replication in acholeplasmas, at least, is very similar to that observed in bacteria. Furthermore, it was concluded that the DNA of *A. laidlawii* is replicated throughout the generation time of the cell and that the replication rate is considerably slower (about 7%) than that found in *E. coli*.

Smith and Hanawalt (1968) lysed *A. laidlawii* cells directly on a sucrose gradient and obtained preliminary evidence that, as in bacteria, newly synthesized DNA, representing the growing point of the chromosome, was membrane associated. This observation was confirmed and extended by Quinlan and Maniloff (1972, 1973), using *Mycoplasma gallisepticum*. This organism has a unique and readily identifiable bleb–infrableb region at one end of the cell (Maniloff *et al.*, 1965). Using pulse-chase techniques and gentle lysis procedures, Quinlan and Maniloff (1972) demonstrated that newly synthesized, or nascent, DNA was associated with these membrane bleb–infrableb structures. Interestingly, the bleb–infrableb region appears at opposite poles of each cell prior to division. The unique nature of the bleb–infrableb region, and the relative ease with which it can be purified, should make this organism an attractive model in which to study DNA replication complexes.

In summary, it would seem that DNA replication in mycoplasmas proceeds in a manner very similar to that observed in bacteria. Since the bleb–infrableb is a structure unique to *M. gallisepticum* it would be of interest to examine the possible association of nascent DNA with membrane in other mycoplasma species.

### D. DNA Polymerase

To date, there has been only one report on the isolation and partial characterization of DNA-dependent DNA polymerase activity from mycoplasmas. Mills *et al.* (1976, 1977) identified a single DNA polymerase species from crude extracts of two mycoplasmas, *Mycoplasma hyorhinis* and *Mycoplasma orale*. The purification procedure consisted of separation of enzyme activity on the following columns: Sephadex G-25, DEAE–cellulose, phosphocellulose, and DNA–cellulose. With this relatively rapid and simple purification protocol highly purified preparations of DNA polymerase activity were obtained. The polymerases from the two mycoplasma species demonstrated some minor differences in their column chromatographic behavior but the most purified fractions of each were essentially identical in their physical and catalytic properties. Thus, this enzyme may be prototypic of mycoplasma DNA polymerases. The polymerase has a sedimentation coefficient of 5.6 S in high salt and an apparent molecular weight of 130,000, a size similar to that reported for *E. coli* DNA polymerases I, II, and III (Gefter, 1974; Kornberg and Kornberg, 1974). The specific activity was found to be >50,000 units per milligram of protein and compares favorably with the best values recorded for *E. coli* DNA polymerase I.

A comparison of the properties of the mycoplasma polymerase with *E. coli* polymerases is given in Table II. The strong preference of the mycoplasmal DNA polymerase for activated DNA primer templates suggests that these enzymes, like bacterial polymerases II and III, carry out a repair type of gap-filling synthesis.

TABLE II. **Comparison of the Properties of the DNA Polymerases of *E. coli* and *Mycoplasma***

| | | *E. coli*[b] | | |
|---|---|---|---|---|
| | *Mycoplasma*[a] | Pol I | Pol II | Pol III |
| Size (daltons) | 130,000 | 109,000 | 120,000 | 160,000 |
| Functions | | | | |
| Polymerization 5′ → 3′ | + | + | + | + |
| Exonuclease 5′ → 3′ | − | + | − | − |
| Exonculease 3′ → 5′ | − | + | + | + |
| Template primer | | | | |
| Intact duplex DNA | − | − | − | − |
| Activated DNA | + | + | + | + |
| (*dA* − *dT*)*n* | + | + | − | − |
| General | | | | |
| KCl optimum (m*M*) | 80–100 | 100 | 50 | 20 |
| *K*m for triphosphate | Low | Low | Low | High |
| Inhibition by SH blocking agents | Partial | − | + | + |

[a] Information derived from Mills *et al*. (1977).
[b] Information derived from Kornberg (1974).

The most surprising finding was the apparent absence of associated exonuclease activities, particularly the 3′ → 5′ exonuclease function. All other eubacterial DNA polymerases studied, with the possible exception of *Bacillus subtilis* DNA polymerase II (Low *et al.*, 1976), have been shown to contain this activity. By contrast, this exonuclease function has not been found associated with highly purified eukaryotic DNA polymerases (Bollum, 1975), with the exception of a high molecular weight polymerase in yeast (Chang, 1977).

To date, no evidence has been obtained for the existence of more than a single mycoplasmal DNA polymerase. This is not a surprising result given the small size and limited complexity of the mycoplasma genome. Consideration, however, should be made of the fact that it was only after the isolation and characterization of *pol A* mutants of *E. coli* that polymerases II and III were recognized (DeLucia and Cairns, 1969; Kornberg and Gefter, 1970, 1971). It would be premature to conclude that additional DNA polymerases may not be present in mycoplasmas. However, if these organisms truly possess a single DNA polymerase, they would be very favorable models for the elucidation of DNA replication and repair.

## E. DNA Repair

The majority of investigations into DNA repair mechanisms in mycoplasmas have utilized *A. laidlawii*. Folsome (1968) reported that *A laidlawii* possessed a photoreactivating capacity to repair ultraviolet (uv) damage but was unable to find evidence for dark repair mechanisms. Smith and Hanawalt (1969) assayed for repair replication in this organism following uv irradiation. They showed, as with bacterial systems, that uv irradiation causes a drastic reduction in DNA synthesis and also leads to DNA degradation, which is related to a uv dose-dependent lysis of the organisms. Using a radiation dose calculated to give about 70% survival (85 ergs/mm$^2$) and labeling the cells postirradiation with BrdUrd and $^{32}PO_4$ for varying lengths of time, followed by isolation of DNA in cesium chloride gradients, the authors were able to assay repair ahead of, and behind, the replication growing point. Their results were compatible with nonconservative repair-type replication following irradiation. Replication consisted of small single-strand regions in the parental DNA strand, and the newly synthesized DNA was capable of subsequent semiconservative DNA replication. The authors calculated the extent of excision repair to be 150–600 nucleotides around each radiation-induced pyrimidine dimer.

Smith and Hanawalt drew attention to the fact that the extent of DNA degradation following uv irradiation is much greater in *A. laidlawii* than in wildtype *E. coli*. It is more comparable to that observed in a "reckless" *rec*$^-$ strain of *E. coli* JC1569 (Clark *et al.*, 1966). Rec$^-$ strains of *E. coli* can still repair uv-damaged DNA. This observation is compatible with the unsuccessful attempts by Folsome (1968) to establish transformation in *A. laidlawii,* leading him to conclude that this strain lacked the necessary enzymes for recombination.

When cultures were exposed to varying doses of uv, followed by visible light illumination of 3 hr, it was found that photoreactivation of the organisms reduced the amount of dark repair. This phenomenon has also been observed in *E. coli* (Pettijohn and Hanawalt, 1964).

The possession of dark and light repair mechanisms were confirmed by Das *et al.* (1972), who also showed that sensitivity to uv irradiation and the degree of photoreversal varied with the cells' growth phase, maximum sensitivity occurring during middle and late logarithmic phases. Folsome's failure to detect dark repair was ascribed to the gradual loss in viability of mycoplasmas held in buffer, either in the light or in the dark.

Although it has been suggested that uv survival curves indicate that certain mycoplasma cells contain multiple genomes (Das *et al.*, 1972), caution should be exercised. Furness and colleagues (1968a,b; Furness, 1968) have demonstrated that there are differential responses of single cells and aggregates of mycoplasmas to uv irradiation. Multiorder survival curves are seen when aggregates are irradiated, whereas exponential death is seen when single cells are exposed to uv or X rays.

DNA repair mechanisms have only very recently been studied in *Mycoplasma* species. Ghosh *et al.* (1977) have reported on their somewhat surprising finding that *M. gallisepticum* possesses neither dark repair nor photoreactivation mechanisms for the repair of uv-induced DNA damage.

## F. Other Radiobiologic Properties of Mycoplasmas

As mentioned earlier, mycoplasmas are very sensitive to irradiation, single cells being killed exponentially by uv irradiation and X rays. Survival rates can be influenced by medium components (Chelack *et al.*, 1974), sulfhydryl groups (Petkau and Chelack, 1974a,b), pH, temperature (Chelack *et al.*, 1974), oxygen content (Drasil *et al.*, 1972), and the phase of the life cycle (Das *et al.*, 1972; Hutkova *et al.*, 1975).

The $D_0$ values (defined as the dose necessary to reduce survival to $1/e$ in the exponential region) for *A. laidlawii* vary from strain to strain. Reported values have ranged from 5 to 23.5 krads. These differences in $D_0$ are probably due, in part, to variation in the experimental conditions outlined above.

Das *et al.* (1977) have recently reported that *A. laidlawii* has mechanisms for both host cell and uv reactivation of uv-irradiated mycoplasmaviruses. Host cell reactivation was inhibited by acriflavine treatment, a property also noted in bacteria (Feiner and Hill, 1963). Acriflavine is considered to be an inhibitor of excision repair. The finding that the survival of uv-irradiated double-stranded DNA mycoplasmavirus MVL2 was decreased by acriflavine treatment of the host *A. laidlawii* cells, whereas the survival of uv-irradiated single-stranded DNA mycoplasmavirus MVL51 was unaffected is consistent with this presumed action.

## III. EXTRACHROMOSOMAL DNA

The possible existence of viruses in mycoplasmas was suggested as early as 1960 by Edwards and Fogh who noted electron-dense "virus-like" particles in electron micrographs of mycoplasma cells. However, it was not until 1970 that the first mycoplasmavirus was isolated and characterized (Gourlay, 1970). Since then several different mycoplasma viruses have been isolated. These viruses are described in considerable detail in a separate chapter and will not be discussed further here.

Extrachromosomal DNA not associated with virus particles has also been reported. Morowitz (1969) described the presence of small circles in electron microscopic preparations of *M. arthritidis* DNA. Satellite bands of DNA were also observed in cesium chloride (Haller and Lynn, 1968) and sucrose gradients (Dugle and Dugle, 1971; Zouzias *et al.*, 1973). In all these cases no virus particles were detected. These "plasmids" have been detected in *A. laidlawii, M. hominis,* and *M. arthritidis*. Their size has been calculated as approximately $20 \times 10^6$ daltons and, based on the amount of labeled thymidine in the satellite DNA versus chromosomal DNA, it was calculated that there was an average of 50–100 plasmids per cell (Maniloff *et al.*, 1977).

There has been no attempt thus far to correlate antibiotic resistance with the presence of these plasmids. It would be interesting to see if the more promiscuous bacterial plasmids (Novick *et al.*, 1976) are able to enter and survive in mycoplasmas, thereby facilitating marker transfer.

## IV. MYCOPLASMA GENETICS

At first glance the small size of the mycoplasma genome would appear to render it an appealing model for genetic studies. Unfortunately, this is not the case. Gene mapping requires mutants and, in bacteria, has benefited from a vast array of auxotrophic markers. Because of the fastidious nature of mycoplasmas, very few such markers are available. In fact, defined media are available only for *A. laidlawii* and *M. mycoides* subsp. *mycoides* (Rodwell, 1969; Tourtellotte *et al.*, 1964). Also, it would seem that *A. laidlawii,* at least, behaves as a rec$^-$ organism, precluding stable integration of transferred genetic material.

Several genetic markers have been reported, including antibiotic resistance (Stanbridge, 1971; Stanbridge and Doersen, 1978), temperature-sensitive growth mutants of *Mycoplasma pneumoniae* (Steinberg *et al.*, 1969), and host range mutants for mycoplasmaviruses (Nowak *et al.*, 1976). It would seem imperative that, in order to even attempt genetic studies in mycoplasmas, rec$^+$ strains should be sought, plus vehicles for DNA transfer. Of the three methods generally used in genetic studies, namely conjugation, transformation, and transduction, the latter would seem to be the most promising.

It remains to be seen whether the recent advances in gene cloning via plasmid vehicles will prove useful in mycoplasma genetic studies.

## V. NUCLEIC ACID HYBRIDIZATION STUDIES

The phylogenetic status of mycoplasmas, and their relationship to L forms of bacteria in particular, have intrigued investigators for years. The conventional Adansonian system, which is based upon a series of morphologic and biochemical properties, has not been very helpful with respect to mycoplasma taxonomy, owing to the rather limited number of biochemical characteristics. This problem has been circumvented by the development of nucleic acid hybridization techniques which have been used successfully in determining the genetic relatedness of bacterial species (McCarthy and Bolton, 1963; Gerloff *et al.*, 1966). These techniques are based upon the principle that the sequence of nucleotide bases in an organism's nucleic acids serves as a blueprint for the identity of the respective organism. Therefore, the proportion of nucleotide base sequences held in common by two organisms indicates the extent of their genetic relatedness.

Several investigators (Reich *et al.*, 1966a,b; Somerson *et al.*, 1966, 1967; McGee *et al.*, 1965, 1967; Walker, 1967) have applied nucleic acid hybridization techniques, both DNA–DNA and DNA–RNA, in attempts to clarify the genetic relatedness of mycoplasmas to other bacteria. A summary of these efforts is outlined in Table III.

Clearly, there is no apparent homology between any of the mycoplasma species and bacterial species examined. As expected, L forms exhibited a high degree of nucleic acid homology with their parent bacteria. Additionally, as we see in Table IV, there is limited relatedness between myco-

TABLE III. **Genetic Relatedness between Mycoplasmas and Bacteria as Measured by Nucleic Acid Homology**[a]

| Source of "donor" DNA/RNA | Source of "recipient" DNA | % Homology |
|---|---|---|
| *Streptococcus* MG | *Streptococcus* MG | 100 |
| *Streptococcus* MG | *Mycoplasma pneumoniae* | 0 |
| *Mycoplasma arthritidis* | *Mycoplasma arthritidis* | 100 |
| *Mycoplasma arthritidis* | Diptheroid Campo | 0.6 |
| *Mycoplasma arthritidis* | L form Diptheroid Campo | 0 |
| *Haemophilus gallinarum* | *Haemophilus gallinarum* | 100 |
| *Haemophilus gallinarum* | *Mycoplasma gallinarum* | 0 |
| *Haemophilus gallinarum* | *Mycoplasma gallisepticum* | 1.1 |
| *Proteus mirabilis* | *Proteus mirabilis* | 100 |
| *Proteus mirabilis* L form | *Proteus mirabilis* | 101 |
| *Mycoplasma hominis* | *Escherichia coli* | 0 |

[a] Data obtained from McGee *et al*. (1965, 1967), Rogul *et al*. (1965), and Somerson *et al*. (1967).

plasma species. These results confirm the heterogeneous nature of mycoplasmas, as evidenced by their range of G + C contents, and add support to the contention that mycoplasmas are phylogenetically distinct from bacteria. However, it should be noted that measurements of genetic relatedness using nucleic acid hybridization of total DNA will distinguish only between relatively closely related organisms.

## VI. TRANSFER RNA

Mycoplasma transfer RNA (tRNA) molecules have been examined in some considerable detail. An early indication of the simplicity of mycoplasmas was the report by Ryan and Morowitz (1969) that *Mycoplasma* sp. (Kid), subsequently identified as *M. capricolum* (Tully *et al*., 1974), contains only enough DNA complementary to tRNA to code for 44 different tRNA molecules. By contrast, *E. coli* has enough DNA to code for 60–80 tRNA genes (Smith, 1972). The partial purification of mycoplasma rRNA and tRNA cistrons was accomplished by a differential melting technique. *Mycoplasma capricolum* has a G + C content of 25%, whereas the G + C value of rDNA and tDNA is approximately 46%. Thus, a second hyperchromic rise, corresponding to 1.4% of the total DNA, occurred during the melting of purified *M. capricolum* DNA at 88°C, while the bulk of the DNA melted at 79.5°C. Ryan and Morowitz sonicated total DNA to free the high G + C regions from the large amounts of contaminating low G + C DNA and separated them by hydroxyapatite chromatography. This method has been modified some-

TABLE IV. **Genetic Relatedness between Mycoplasmas as Measured by Nucleic Acid Homology**[a]

| Source of "donor" DNA/RNA | Source of "recipient" DNA | % Homology |
|---|---|---|
| *M. hominis* | *M. hominis* | 39–100[b] |
| *M. hominis* | *M. arthritidis* | 3 |
| *M. hominis* | *M. salivarium* | 3 |
| *M. hominis* | *M. orale* | 3 |
| *M. hominis* | *M. fermentans* | 2 |
| *M. hominis* | *M. pneumoniae* | 0 |
| *M. hominis* | *A. laidlawii* | 0 |
| *M. hyorhinis* | *M. hyorhinis* | 100 |
| *M. hyorhinis* | *M. pulmonis* | 0 |
| *M. hyorhinis* | *M. hominis* | 1 |
| *M. hyorhinis* | *A. laidlawii* | 0 |
| *M. hyorhinis* | *M. gallisepticum* | 0 |
| *M. salivarium* | *M. salivarium* | 100 |
| *M. salivarium* | *M. orale* | 0 |
| *M. pneumoniae* | *M. pneumoniae* | 100 |
| *M. pneumoniae* | *M. orale* | 0 |
| *M. pneumoniae* | *M. hominis* | 0 |

[a] Data obtained from Reich *et al.* (1966b) and Somerson *et al.* (1966).
[b] Genetic heterogeneity between different isolates of *M. hominis* (Somerson *et al.*, 1966).

what to provide milligram quantities of tDNA (Feldmann, 1973), which should prove useful for further studies.

The finding that a mycoplasma apparently contains only enough DNA to code for 44 tRNA molecules would lead one to expect to find fewer isoaccepting tRNAs in this class of organisms. Preliminary reports are in agreement with this prediction. Only one $tRNA^{Phe}$ was detected in *M. capricolum* (Kimball and Söll, 1974) and only one species of $tRNA^{Met}_{f}$, $tRNA^{Met}_{m}$, $tRNA^{Gly}$, $tRNA^{Lys}$, and $tRNA^{Val}$ in *M. mycoides* subsp. *capri* (Walker, 1976). In *A. laidlawii,* an organism with twice as much DNA as *Mycoplasma* spp., one or more tRNA species were found (Feldmann and Falter, 1971) for the following amino acids (with the number of isoaccepting species in parentheses): alanine (1), arginine (2), cysteine (1), glycine (1), isoleucine (2), leucine (3), methionine (2), phenylalanine (1), serine (3), tyrosine (1), and valine (3). However, it would seem prudent to reexamine the distribution of mycoplasma tRNA genes with the more sophisticated techniques now available, rather than to rely upon the single report by Ryan and Morowitz.

The physical properties of mycoplasma tRNAs are very similar to those of *E. coli* (Hayashi *et al.*, 1969). Thermal denaturation curves of unfractionated *M. capricolum* and *E. coli* tRNAs were found to be similar, and the extent of hyperchromicity for the tRNAs from both organisms indicated a similar degree of secondary structure. Sedimentation properties were also similar. Charged seryl-tRNAs from *A. laidlawii, M. gallisepticum, M. capricolum,* and *E. coli* cosedimented in a deuterium oxide–water gradient.

Functional similarities between the tRNAs of mycoplasmas and other prokaryotes are also reflected in their cross-reactivities in heterologous charging systems. Hayashi *et al.* (1969) showed that *E. coli* aminoacyl–tRNA synthetases could replace those of mycoplasmas in the aminoacylation of mycoplasma tRNA with valine, serine, and methionine, and *vice versa.* Feldmann and Falter (1971) studied aminoacylation and transformylation cross-reactions between mycoplasma, *E. coli,* and yeast. In addition to confirming the observations of Hayashi *et al.*, they showed that yeast synthetase aminoacylated $tRNA^{Phe}$ (mycoplasma) completely, whereas $tRNA^{Phe}$ (yeast) was not aminoacylated by mycoplasma synthetase under standard conditions. This lack of reciprocity has also been noted in *E. coli*–yeast cross-charging experiments (Doctor and Mudd, 1963; Thiebe and Zachau, 1970). The transformylase reactions also showed an interesting restriction. *E. coli* was able to transfer formate to Met-tRNA (*E. coli*), Met-$tRNA_f^{Met}$ (mycoplasma) and Met-$tRNA_{ia}^{Met}$ (yeast), whereas *A. laidlawii* was unable to formylate any of the methionine-specific tRNAs of yeast.

The most striking feature of mycoplasma tRNAs is the reduced amount of modified nucleosides. At one time it was thought that modified nucleosides could not be detected in mycoplasmas (Hall *et al.,* 1967). However, it has now been shown conclusively that they are present, albeit in reduced amounts (Ryan and Morowitz, 1969; Feldmann, 1973; Johnson *et al.*, 1970; Kimball *et al.*, 1974; Walker and RajBhandary, 1975, 1978). Table V lists the modified nucleosides which have been found in mycoplasma tRNAs.

Modified nucleosides have been found as true constituents of tRNAs of all prokaryotes examined. However, their role in tRNA-related functions is poorly understood. Several investigators have studied mycoplasma tRNAs that lack one or more of these minor nucleosides in an effort to clarify their function. Johnson and colleagues (1970) isolated $tRNA^{Ile}$ from *M. capricolum* which was deficient in ribothymidine. Despite this deficiency $tRNA^{Ile}$ (mycoplasma) and $tRNA^{Ile}$ (*E. coli*) had similar thermal denaturation and sedimentation profiles. When mycoplasma $tRNA^{Ile}$ was methylated with heterologous *E. coli* tRNA methylase, aminoacylation of the newly methylated $tRNA^{Ile}$ proceeded to the same extent as the

TABLE V. **Modified Nucleosides Present in Mycoplasma Transfer Ribonucleic Acids**

| Modified nucleoside | Nucleosides per tRNA molecule | | | | | |
|---|---|---|---|---|---|---|
| | *A. laidlawii*[a] total tRNA | *M. hominis*[b] total tRNA | *M. capricolum*[c] $tRNA^{Phe}$ | *M. mycoides*[d] $tRNA^{Met}$ | *E. coli*[e] $tRNA^{Phe}$ | *E. coli*[e] $tRNA_f^{Met}$ |
| 1-Methyladenosine } | | 0.22 | — | — | — | — |
| $N^6$-Methyladenosine } | 0.4 | 0.35 | — | — | — | — |
| $N^6$-Isopentenyladenosine | <0.3 | — | — | — | 1.0 | — |
| 6-6-*N*-Dimethyladenosine | — | 0.04 | — | — | — | — |
| 1-Methylguanosine } | 0.8 | — | 1.0 | — | — | — |
| $N^2$-Methylguanosine } | | 0.02 | — | — | — | — |
| $N^7$-Methylguanosine | 1.0 | 0.64 | 1.0 | — | 1.0 | 1.0 |
| 2′-*O*-Methylguanosine | 0.4 | — | — | — | — | — |
| Inosine | — | 0.12 | — | — | — | — |
| 5-Methylcytidine | 0.6 | 0.02 | — | — | — | — |
| Pseudouridine | 2.4 | 1.02 | 2.0 | 1.0 | 3.0 | 1.0 |
| Ribothymidine | 0.9 | 0.01 | — | — | 1.0 | 1.0 |
| Dihydrouridine | 0.8 | 0.94 | 1.0 | 1.0 | 2.0 | 1.0 |
| 4-Thiouridine | 0.2 | 0.32 | — | 1.0 | 1.0 | 1.0 |
| 2′-*O*-Methyluridine | <0.2 | — | — | — | — | — |
| 2′-*O*-Methylcytidine | — | — | — | — | — | 1.0 |

[a] From Feldmann and Falter (1971).
[b] From Johnson and Horowitz (1971).
[c] From Kimball *et al*. (1974).
[d] From Walker and RajBhandary (1978).
[e] From Barrel and Clark (1974).

ribothymidine-deficient $tRNA^{Ile}$. Thus, at least in this case it would seem that ribothymidine is not required for the recognition of tRNA by aminoacyl tRNA synthetases.

Ribothymidine is present in bacterial tRNAs in the common tetranucleotide sequence GpTpψpC (Zachau, 1969). It has been postulated that this common sequence is responsible for the specific interaction of tRNA with ribosomes (Ofengand and Henes, 1969). Johnson *et al.* (1970), however, found that mycoplasma ribothymidine-deficient $tRNA^{Ile}$ and remethylated $tRNA^{Ile}$ performed equally well in protein synthesis and that neither form of tRNA was particularly error prone.

Another hypermodified nucleoside, $N^6$-(*N*-threonyl carbonyl) adenosine, appears to be lacking in bulk tRNA from *M. capricolum* (Miller and Schweizer, 1972). Most tRNAs which recognize codons starting with A contain this modified nucleoside, which is located next to the 3′ end of the anticodon. Thus, based upon the analysis of bulk mycoplasma tRNA, this particular nucleoside is presumably not absolutely necessary for codon recognition. However, the confirmation of this supposition must await the sequencing of the appropriate tRNA molecules.

To date, two mycoplasma tRNAs have been fully sequenced: $tRNA^{Phe}$ (Kimball *et al.*, 1974) and $tRNA_f^{Met}$ (Walker and RajBhandary, 1978). Kimball *et al.* have found that mycoplasma $tRNA^{Phe}$ isolated from *M. capricolum* lacks isopentenyladenosine, a hypermodified nucleoside which has been found in the tRNAs of almost all organisms investigated (Hall, 1970). Its occurrence is restricted to those tRNAs that correspond to codons beginning with U, which includes $tRNA^{Phe}$. Phenylalanine tRNAs from other organisms contain isopentenyladenosine or a highly modified fluorescent nucleoside (Y or derivatives) (Kimball *et al.*, 1974). Sequence analysis has shown that one or other of these hypermodified nucleosides is usually found adjacent to the 3′-terminal necleotide of the anticodon (Söll, 1971; Nishimura, 1972). Certain experiments (Hirsch and Zachau, 1970; Furuichi *et al.*, 1970) have indicated that isopentenyladenosine at this site is involved in the interaction of tRNA with messenger RNA on the ribosome. In mycoplasma $tRNA^{Phe}$ the nucleoside next to the anticodon was found to be 1-methylguanosine. Despite the fact that this single methylated guanosine substituted for the hypermodified methylthioisopentenyladenosine there was no detectable difference in the relative efficiency of the tRNA in *in vitro* protein synthesis, and mycoplasma $tRNA^{Phe}$ was fully equivalent to *E. coli* $tRNA^{Phe}$ even in direct incorporation competition experiments. Kimball *et al.* (1974) have suggested that 1-methylguanosine is a precursor of the Y base and that *M. capricolum* may not possess the necessary enzymes to modify this base further.

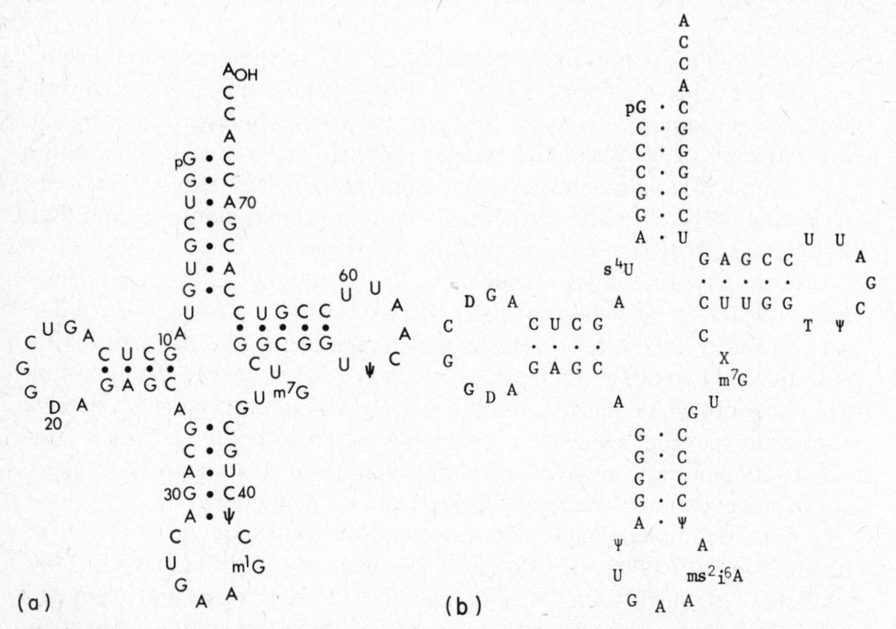


FIGURE 2. Comparison of the cloverleaf structures (a) *M. capricolum* tRNA$^{Phe}$ and (b) *E. coli* tRNA$^{Phe}$. (Reproduced from Kimball *et al.*, 1974; and Barrell and Clark, 1974; with permission of the authors and publishers.)

In addition to isopentenyladenosine the nucleotide sequence of *M. capricolum* tRNA$^{Phe}$ (see Fig. 2) is characterized by the lack of ribothymidine and the presence of only five modified nucleosides, compared to the ten found in *E. coli*. Apart from the low content of modified nucleosides the primary nucleotide sequence is similar to tRNA$^{Phe}$ of *E. coli* (Barrell and Sanger, 1969), yeast (RajBhandary *et al.*, 1967), and higher eukaryotes (Dudock *et al.*, 1969; Keith *et al.*, 1973).

Walker and RajBhandary (1978) have recently published the nucleotide sequence of *M. mycoides* subsp. *capri* formylmethionine tRNA. Once again, although the sequence is typical of a prokaryotic organism (see Fig. 3) it contains only a few modified nucleosides. The only nucleoside modified in mycoplasma tRNA$_f^{Met}$ is uridine; modifications include 4-thiouridine, pseudouridine, and dihydrouridine, but ribothymidine is not present.

Whereas the initiator tRNAs of other prokaryotes contain the common sequence G–T–Ψ–C in loop IV, *M. mycoides* subsp. *capri* contains the sequence G–U–Ψ–C, a property which, again, may be due to the lack of

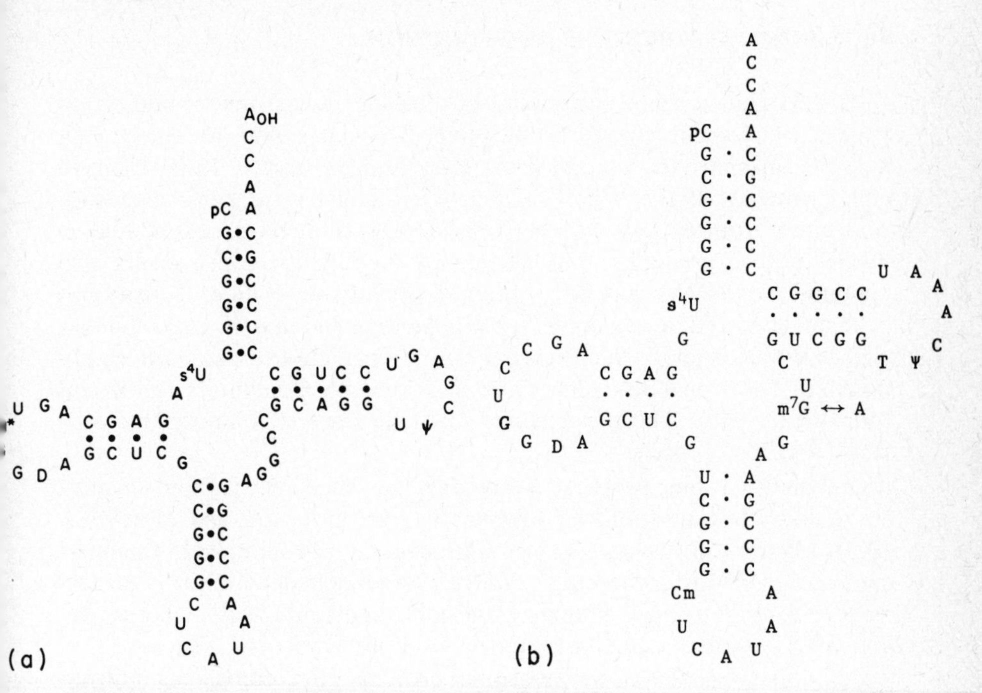


FIGURE 3. Comparison of the cloverleaf structures of (a) *M. mycoides* subsp. *capri* tRNA$_f^{Met}$ and (b) *E. coli* tRNA$_f^{Met}$. (Reproduced from Walker and RajBhandary, 1978; and Barrell and Clark, 1974; with permission of the authors and publishers.)

appropriate methylase activity. This tRNA also contains only eight bases in loop I, whereas other prokaryotic initiator tRNAs contain nine bases in this loop. Despite these differences, the conservation of the nucleotide sequence of initiator tRNAs during evolution is reflected in the sequence of mycoplasma initiator tRNA and the lack of a Watson–Crick base pair between the first nucleotide of the 5′ end and the fifth nucleotide of the 3′ end.

Perhaps the most interesting feature of mycoplasma tRNAs in general is the consistent lack of modified nucleosides. It would appear from these data that hypermodification is unnecessary for tRNAs to participate effectively in protein synthesis. Mycoplasmas provide potential model systems for tRNA studies which are of current high interest among molecular biologists—for example, the role of tRNAs in autoregulation (Calhoun and Hatfield, 1975) and the effect of hypermodification on this role, plus the exciting new developments in the reevaluation of the wobble effect (Mitra *et al.*, 1977; Holmes *et al.*, 1977).

## VII. RIBOSOME STRUCTURE AND FUNCTION

At first glance there appears to be nothing remarkable about mycoplasma ribosomes. The 70 S ribosome dissociates into 30 S and 50 S particles in appropriate conditions (Kirk and Morowitz, 1969; Johnson and Horowitz, 1971). When mixing experiments were performed with ribosomes from *E. coli* and *M. hominis* the subunits cosedimented in linear sucrose gradients. The ribosomes of both these organisms also contained 61% RNA and 39% protein. Some degree of heterogeneity has been encountered; for example, it has been reported that *M. gallisepticum* possesses ribosomes which have a value (corrected to standard conditions) of 74 S and dissociate into 56 S and 36 S subunits (Kirk and Morowitz, 1969). These particular ribosomes were composed of 55% RNA and 45% protein.

One distinguishing property of mycoplasma ribosomes is their sensitivity to low concentrations of divalent cations (Johnson and Horowitz, 1971). Mycoplasma ribosomes were dissociated into subunits in the presence of 5 m*M* $Mg^{2+}$, whereas *E. coli* ribosomes remained intact. Furthermore, *E. coli* ribosomal subunits were stable at 0.1 m*M* $Mg^{2+}$, whereas *M. hominis* ribosomal subunits lost their structural integrity.

Preliminary studies of the ribosomal proteins of mycoplasmas indicate they are very similar in composition to bacterial ribosomal proteins. Polyacrylamide gel patterns indicated that the proteins isolated from *M. hominis* ribosomes were as complex as those found in *E. coli*. Eighteen bands were resolved from the 30 S subunits and 21 from the 50 S subunit. Similar results were obtained when the ribosomal proteins from *M. gallisepticum* were examined.

Typical monosomes and polyribosomes can be visualized and isolated from mycoplasma cells. Helical ribosomal arrays have been noted in *M. gallisepticum* cells (Maniloff *et al.*, 1965; Maniloff, 1971) following centrifugation. However, this helical array does not constitute the active polyribosome and may well be an artifact of preparation. Similar helical arrays have been reported in *E. coli* (Nauman *et al.*, 1971) under conditions of stress.

Early reports (Ryan and Morowitz, 1969; Kirk and Morowitz, 1969; Johnson and Horowitz, 1971) indicated that mycoplasma ribosomes contained the usual 23 S, 16 S, and 5 S species of ribosomal RNA (rRNA) found in prokaryotes. The base composition of the rRNAs, and hence ribosomal DNA (rDNA), is of interest because, like bacterial rRNA, the G + C content is relatively constant irrespective of the G + C contents of the total DNA (Table VI). The clustering of values around 50% G + C among diverse bacterial species is also indicative of evolutionary conservation, a prediction that has been borne out by sequencing studies of bacterial rRNA species (Sogin *et al.*, 1972; Woese *et al.*, 1974, 1975). The G + C contents of mycoplasma rRNAs are lower than those found in bacteria but are also relatively constant between mycoplasma species (Table VI). *Mycoplasma hominis* 5 S rRNA is particularly striking in this regard, with a G + C content of 43% compared with 64% in *E. coli* 5 S rRNA (Ryan and Morowitz, 1969).

Ryan and Morowitz took advantage of the difference between the G + C content of rDNA versus total DNA in *Mycoplasma capricolum* in order to isolate ribosomal cistrons by sonicating the DNA and eluting DNA fragments from hydroxyapatite columns. They determined that this mycoplasma species contained only enough ribosomal DNA to code for one set of 23 S plus 16 S rRNA. *Escherichia coli* by comparison has enough ribosomal DNA to encode for five sets of 16 S and 23 S rRNA (Kohne, 1968).

Nucleic acid hybridization and sequencing studies have demonstrated that substantial sequence homologies exist among the rRNAs of diverse species of bacteria. This conservation presumably reflects the maintenance of a unique conformation of the rRNA essential for functional ribosome assembly and ribosome function. The conservation of conformation of prokaryotic rRNA is also apparent from the nondenaturing gel electrophoretic studies of Loening (1968), which showed only very minor migrational differences between 23 S and 16 S rRNAs from a variety of unrelated bacterial species and blue-green algae.

In addition to differing from bacterial rRNA in base composition,

TABLE VI. **Guanine + Cytosine Content of Ribosomal RNA and Total Genome DNA of Various Species of Mycoplasmas and Bacteria**

| Organism | % G + C content ribosomal RNA | % G + C content total DNA | Reference |
|---|---|---|---|
| *M. hominis* | 46.4 | 28 | Johnson and Horowitz (1971) |
| *M. capricolum* | 47.9 | 25 | Ryan and Morowitz (1969) |
| *M. gallisepticum* | 42.8 | 30 | Kirk and Morowitz (1969) |
| *A. laidlawii* | 48.0 | 33 | Feldman (1973) |
| *Streptococcus pneumoniae* | 50.1 | 39 | Pace (1973) |
| *Vibrio marinus* | 51.1 | 40 | Pace (1973) |
| *Proteus vulgaris* | 53.1 | 38 | Pace (1973) |
| *Bacillus subtilis* | 53.3 | 43 | Pace (1973) |
| *Pseudomonas aeruginosa* | 53.3 | 65 | Pace (1973) |
| *Escherichia coli* | 53.7 | 52 | Pace (1973) |
| *Micrococcus luteus* | 53.9 | 72 | Pace (1973) |

mycoplasmal rRNA also differs with respect to sedimentation properties in sucrose gradients (Kirk and Morowitz, 1969; Johnson and Horowitz, 1971; Reich, 1967) and electrophoretic mobility in polyacrylamide gels (Harley *et al.*, 1973; Reff *et al.*, 1977). Reff *et al.* (1977), in an extensive comparative study of the electrophoretic behavior of mycoplasmal and bacterial 23 S and 16 S rRNAs in polyacrylamide gels, detected significant differences between these two groups of organisms. In agreement with other investigators they found that rRNAs from diverse bacterial species comigrated in both nondenaturing gels and fully denaturing warm formamide gels. Also, L-form rRNA comigrated with parental bacterial rRNA (Fig. 4). Mycoplasmal rRNAs, however, migrated in nondenaturing gels more slowly than bacterial 23 S and 16 S rRNA (Fig. 5). Acholeplasmal 23 S rRNA also migrated more slowly than bacterial 23 S rRNA but the respective 16 S rRNAs comigrated. In fully denaturing formamide gels the 23 S rRNAs of all the organisms comigrated, but both mycoplasma and acholeplasma 16 S rRNA comigrated slightly ahead of the bacterial 16 S rRNA. These results suggest that mycoplasmal and bacterial rRNAs may exist in different conformational states and also that the 16 S rRNA of

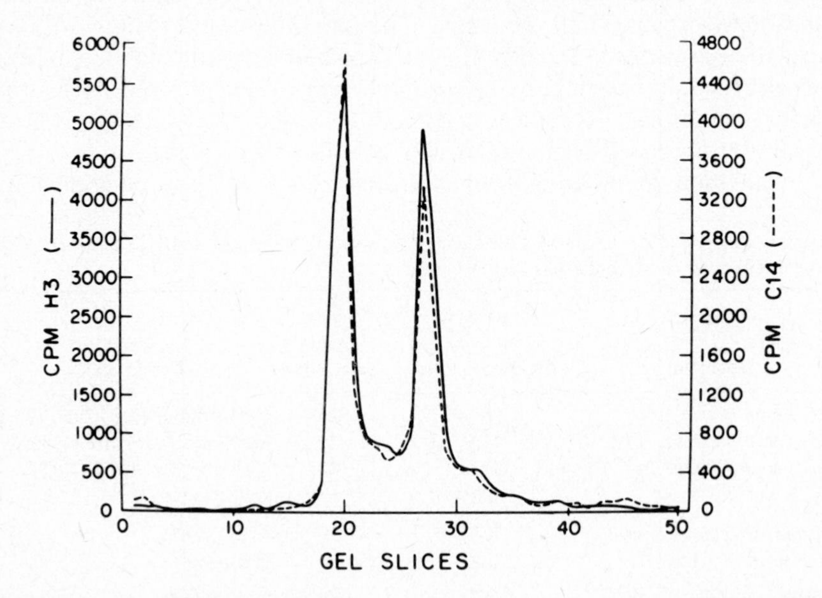


FIGURE 4. $^{3}$H-Labeled rRNA of *Streptococcus faecalis* L form subjected to coelectrophoresis with $^{14}$C-labeled rRNA of *S. faecalis* parent bacterium in a nondenaturing gel. ——, $^{3}$H-Labeled *S. faecalis* L form; ----, *S. faecalis*. (From Reff *et al.*, 1977; with permission of the authors and publishers.)

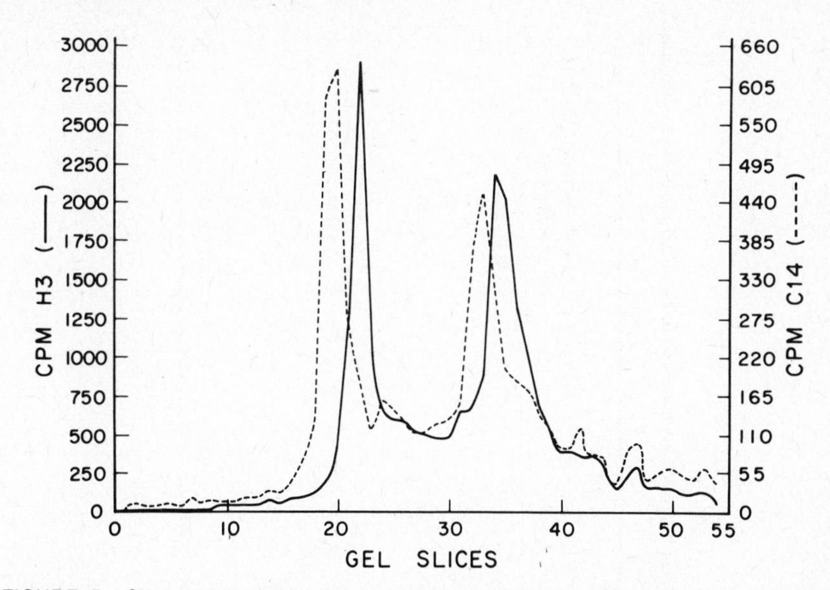


FIGURE 5. $^3$H-Labeled rRNA of *S. faecalis* subjected to coelectrophoresis with $^{14}$C-labeled rRNA of *M. hyorhinis* in a nondenaturing gel. Separation of both 23 S and 16 S peaks is apparent. ——, $^3$H-labeled *S. faecalis;* ----, $^{14}$C-labeled *M. hyorhinis*. (From Reff *et al.*, 1977; with permission of the authors and publishers.)

mycoplasmas may differ in molecular weight from that of bacteria (Table VII).

Preliminary studies with spiroplasma rRNAs indicate that these molecules migrate in a pattern identical to that of acholeplasma species. Thermoplasma rRNAs, however, migrate in a pattern totally distinct from mycoplasmas. The rRNAs of this organism migrate in a fashion similar to, if not identical with, an acidophilic bacterium, *Sulfolobus acidocaldarius* (M. Reff and E. Stanbridge, unpublished observations).

Reff *et al.* (1977) have suggested, based upon the differences discussed above, that a significant evolutionary gap exists between mycoplasmas and bacteria or bacterial L forms. Preliminary information regarding the degrees of nucleic acid homology between the ribosomal DNAs of mycoplasmas and bacteria is in agreement with this suggestion (discussed in Stanbridge, 1976). It will be interesting to see if sequencing studies of mycoplasma ribosomal RNA will support this theory. Work is beginning in this area. For example, Woese and Maniloff (discussed at the Second International Organization of Mycoplasmology Conference, Freiburg, 1978) have recently analyzed the 16 S RNA of *M. gallisepticum* by olignu-

TABLE VII. **Apparent Molecular Weights of the Ribosomal RNAs of Mycoplasmas and Bacteria Calculated from Their Migration in Nondenaturing and Denaturing Gels**[a]

| Organism | Species of rRNA (S units) | Apparent MW of rRNA species calculated from nondenaturing gels | MW of rRNA species calculated from denaturing gels |
|---|---|---|---|
| *Streptococcus faecalis* L-phase variant, *S. faecalis*, *S. pneumoniae*, *Escherichia coli* | 23 | 1,050,000 (standard) | 1,050,000 |
| | 16 | 525,000 (standard) | 525,000 |
| *Bacillus subtilis* | 23 | 983,000 | 1,050,000 |
| | 16 | 525,000 | 525,000 |
| *Mycoplasma* spp. | 23 | 1,190,000 | 1,050,000 |
| | 16 | 555,000 | 513,000 |
| *Acholeplasma* spp. | 23 | 1,130,000 | 1,050,000 |
| | 16 | 525,000 | 513,000 |

[a] From Reff *et al.* (1977), with permission of the authors and publishers.

cleotide sequencing. They found that the oligonucleotides of the 16 S RNA of this species closely resemble a *Clostridium* subgroup that is viewed as being phylogenetically somewhat distinct from other eubacteria. Also, a library of EcoRI DNA fragments from *M. hominis* has been prepared using P3 physical containment and the HV-2 certified vector λgtWES.λB (W. M. Sugino, D. T. Kingsbury, and D. C. Tiemeier, personal communication). Clones containing mycoplasma 23 S and 16 S rRNA gene sequences have been characterized. These gene sequences, with other cloned fragments, are being used to examine gene organization in *M. hominis* and other mycoplasma species.

Mycoplasma ribosomal RNA, like transfer RNA, contains less modified nucleosides than bacterial rRNA (Johnson and Horowitz, 1971). In 16 S rRNA 1.02 mol % of *M. hominis* bases are modified and 0.58 mol % in 23 S rRNA versus 1.85 mol % and 1.12 mol % in *E. coli* 16 S and 23 S rRNA, respectively.

## VIII. PROTEIN SYNTHESIS

The protein-synthesizing systems of mycoplasmas are sensitive to the usual inhibitors of prokaryotic protein synthesis. Chloramphenicol and erythromycin, which specifically inhibit protein synthesis on 70 S ribosomes, inhibit mycoplasma protein synthesis, whereas cycloheximide has no effect (Tourtellotte, 1969; Stanbridge and Doersen, 1978; Doersen and Stanbridge, 1978). The presence of $tRNA_f^{Met}$ in all mycoplasmas examined (Feldmann and Falter, 1971; Walker and RajBhandary, 1975, 1978) suggests typical prokaryotic initiation of peptide chains.

Messenger RNA is short-lived, with a calculated half-life of less than 4 min (Tourtellotte, 1969). Although mycoplasma ribosomes appear to function in protein synthesis in a manner similar to other prokaryotes, there have been no detailed studies in this area.

## IX. CONCLUDING REMARKS

It is clear from the preceding discussion that although certain aspects of the molecular biology of mycoplasmas have been well studied, there are large gaps in our knowledge of these microorganisms. Although they superficially resemble bacteria, there are enough differences in the components of their DNA, RNA, and protein synthesizing machinery to warrant further investigation. By study of these simple organisms, we may gain further insights into areas of current active interest in molecular biology; for example, DNA replication and the role of tRNA in the regulation of biologic functions.

## ACKNOWLEDGMENTS

We are indebted to numerous colleagues who furnished us with unpublished reports and helpful discussion. While this work was in progress, E. Stanbridge was the recipient of a Special Fellowship from the Leukemia Society of America and Public Health Service Research Career Development Award 1 KO4 CA 00271-01A1 from the National Cancer Institute.

## REFERENCES

Askaa, G., Christiansen, C., and Ernø, H. (1973). *J. Gen. Microbiol.* **75,** 283–286.

Bak, A. L., and Black, F. T. (1968). *Nature (London)* **224,** 1209–1210.

Bak, A. L., Black, F. T., Christiansen, C., and Freundt, E. A. (1969). *Nature (London)* **224,** 1209–1210.

Barrell, B. G., and Clark, B. F. C. (1974). "Handbook of Nucleic Acid Sequences." Joynson-Bruvvers, Ltd., Oxford.

Barrell, B. G., and Sanger, F. (1969). *FEBS Lett.* **3,** 275–278.

Black, F. T., Christiansen, C., and Askaa, G. (1972). *Int. J. Syst. Bacteriol.* **22,** 241–242.

Bode, H. R., and Morowitz, H. J. (1967). *J. Mol. Biol.* **23,** 191–199.

Bollum, F. J. (1975). *Prog. Nucleic Acid Res. Mol. Biol.* **15,** 109–144.

Cairns, J. (1963). *J. Mol. Biol.* **6,** 208–213.

Calhoun, D. H., and Hatfield, G. W. (1975). *Annu. Rev. Microbiol.* **29,** 275–299.

Chang, L. M. S. (1977). *J. Biol. Chem.* **252,** 1873–1880.

Chanock, R. M., Hayflick, L., and Barile, M. F. (1962). *Proc. Natl. Acad. Sci. U. S. A.* **48,** 41–49.

Chelack, W. S., Forsyth, M. P., and Petkau, A. (1974). *Can. J. Microbiol.* **20,** 307–320.

Chelton, E. T. J., Jones, A. S., and Walker, R. T. (1968). *J. Gen. Microbiol.* **50,** 305–312.

Christiansen, C., Freundt, E. A., and Black, F. T. (1975). *Int. J. Syst. Bacteriol.* **25,** 99–101.

Clark, A. J., Chamberlin, M., Boyce, R. P., and Howard-Flanders, P. (1966). *J. Mol. Biol.* **19,** 442–454.

Crick, F. H. C. (1966). *J. Mol. Biol.* **19,** 548–555.

Das, J., Maniloff, J., and Bhattacharjee, S. B. (1972). *Biochim. Biophys. Acta* **159,** 189–197.

Das, J., Nowak, J. A., and Maniloff, J. (1977). *J. Bacteriol.* **129,** 1424–1427.

DeLucia, P., and Cairns, J. (1969). *Nature (London)* **224,** 1164–1166.

Doctor, B. P., and Mudd, J. A. (1963). *J. Biol. Chem.* **238,** 3677–3681.

Doersen, C. J., and Stanbridge, E. J. (1978). Submitted for publication.

Drasil, V., Hutkova, J., and Liska, B. (1972). *Folia Biol. (Prague)* **18,** 424–433.

Drlica, K., and Worcel, A. (1975). *J. Mol. Biol.* **98,** 393–411.

Dudock, B. S., Katz, G., Taylor, E. K., and Holley, R. W. (1969). *Proc. Natl. Acad. Sci. U.S.A.* **62,** 941–945.

Dugle, D. L., and Dugle, J. R. (1971). *Can. J. Microbiol.* **17,** 433–434.

Edwards, G. A., and Fogh, J. (1960). *J. Bacteriol.* **79,** 267–276.

Elton, R. A. (1973). *J. Mol. Evol.* **2,** 263–276.

Feiner, R. R., and Hill, R. F. (1963). *Nature (London)* **200,** 291–293.
Feldmann, H. (1973). *Hoppe-Seyler's Z. Physiol. Chem.* **354,** 189–202.
Feldmann, H., and Falter, H. (1971). *Eur. J. Biochem.* **18,** 573–581.
Folsome, C. E. (1968). *J. Gen. Microbiol.* **59,** 43–53.
Furness, G. (1968). *J. Infect. Dis.* **118,** 436–442.
Furness, G., Pipes, F. J., and McMurtrey, M. J. (1968a). *J. Infect. Dis.* **118,** 1–6.
Furness, G., Pipes, F. J., and McMurtrey, M. J. (1968b). *J. Infect. Dis.* **118,** 7–13.
Furuichi, Y., Wataya, Y., Hayatsu, H., and Ukita, T. (1970). *Biochem. Biophys. Res. Commun.* **41,** 1185–1191.
Gefter, M. L. (1974). *Prog. Nucleic Acid Res. Mol. Biol.* **14,** 101–115.
Gerloff, R. K., Ritter, D. B., and Watson, R. O. (1966). *J. Infect. Dis.* **116,** 197–200.
Ghosh, A., Das, J., and Maniloff, J. (1977). *J. Mol. Biol.* **116,** 337–344.
Goldman, E., Holmes, W. M., and Hatfield, G. W. (1979). *J. Mol. Biol.* (in press).
Gourlay, R. N. (1970). *Nature (London)* **225,** 1165.
Hall, R. H. (1970). *Prog. Nucleic Acid Res. Mol. Biol.* **10,** 57–84.
Hall, R. H., Mittelman, A., Horoszewicz, J., and Grace, J. T., Jr. (1967). *Ann. N.Y. Acad. Sci.* **143,** 799–800.
Haller, G. J., and Lynn, R. J. (1968). *Bacteriol. Proc.* p. 68.
Harley, E. H., White, J. W., and Rees, K. R. (1973). *Biochim. Biophys. Acta* **229,** 253–263.
Hayashi, H., Fisher, H., and Söll, D. (1969). *Biochemistry* **8,** 3680–3686.
Hecht, R. M., Taggart, R. T., and Pettijohn, D. E. (1975). *Nature (London)* **253,** 60–62.
Hirsch, R., and Zachau, H. G. (1970). *Hoppe-Seyler's Z. Physiol. Chem.* **351,** 563–566.
Holmes, W. M., Goldman, E., Miner, T. A., and Hatfield, G. W. (1977). *Proc. Natl. Acad. Sci. U.S.A.* **74,** 1393–1397.
Holmes, W. M., Hatfield, G. W. and Goldman, E. M. (1978). *J. Biol. Chem.* **253,** 3482–3486.
Hutkova, J., Drăsil, V., and Liska, B. (1975). *Zentralbl. Bakteriol. Parasitenkd., Infektionskr. Hyg., Abt. 2* **130,** 424–432.
International Committee on Systematic Bacteriology, Subcommittee on the Taxonomy of Mycoplasmatales (1977). *Int. J. Syst. Bacteriol.* **27,** 392–394.
Johnson, J. D., and Horowitz, J. (1971). *Biochim. Biophys. Acta* **247,** 262–279.
Johnson, L., Hayashi, H., and Söll, D. (1970). *Biochemistry* **9,** 2823–2831.
Jones, A. S., and Walker, R. T. (1963). *Nature (London)* **198,** 588–589.
Keith, G., Picaud, R., Weissenbach, J., Ebel, J. P., Petrissant, G., and Dirheimer, G. (1973). *FEBS Lett.* **31,** 345–347.
Kimball, M. E., and Söll, D. (1974). *Nucleic Acids Res.* **1,** 1713–1720.
Kimball, M. E., Szeto, K. S., and Söll, D. (1974). *Nucleic Acids Res.* **1,** 1721–1732.
Kirk, R. G., and Morowitz, H. J. (1969). *Am. J. Vet. Sci.* **30,** 287–293.
Kleinschmidt, A., and Zahn, R. (1959). *Z. Naturforsch. Teil B* **14,** 770–779.
Kohne, D. E. (1968). *Biophys. J.* **8,** 1104–1118.
Kornberg, A. (1974). "DNA Synthesis." Freeman, San Francisco, California.
Kornberg, T., and Gefter, M. L. (1970). *Biochem. Biophys. Res. Commun.* **40,** 1348–1355.
Kornberg, T., and Gefter, M. L. (1971). *Proc. Natl. Acad. Sci. U.S.A.* **68,** 761–764.
Kornberg, T., and Kornberg, A. (1974). *In* "The Enzymes" (P. D. Boyer, ed.), 3rd ed., Vol. 10, pp. 119–144. Academic Press, New York.
Loening, U. (1968). *J. Mol. Biol.* **38,** 355–365.
Low, R. L., Rashbaum, S. A., and Cozzarelli, N. R. (1976). *J. Biol. Chem.* **251,** 1311–1325.
McCarthy, B. J., and Bolton, E. T. (1963). *Proc. Natl. Acad. Sci. U.S.A.* **50,** 156–164.
McGee, Z. A., Rogul, M., Falkow, S., and Wittler, R. G. (1965). *Proc. Natl. Acad. Sci. U.S.A.* **54,** 457–461.
McGee, Z. A., Rogul, M., and Wittler, R. G. (1967). *Ann. N.Y. Acad. Sci.* **143,** 21–30.

Maniloff, J. (1971). *Proc. Natl. Acad. Sci. U.S.A.* **68,** 43–47.
Maniloff, J., Morowitz, H. J., and Barnett, R. J. (1965). *J. Bacteriol.* **90,** 193–204.
Maniloff, J., Das, J., and Christensen, J. R. (1977). *Adv. Virus Res. 21,* 343–380.
Miller, J. P., and Schweizer, M. P. (1972). *Anal. Biochem.* **50,** 327–336.
Mills, L. B., Stanbridge, E. J., Korn, D., and Sedwick, W. D. (1976). *Fed. Proc., Fed. Am. Soc. Exp. Biol.* **35,** 1590.
Mills, L. B., Stanbridge, E. J., Sedwick, W. D., and Korn, D. (1977). *J. Bacteriol.* **132,** 641–649.
Mitra, S. K., Lustig, F., Åkesson, B., Lagerkvist, U., and Strid, L. (1977). *J. Biol. Chem.* **252,** 471–478.
Morowitz, H. J. (1966). *Prin. Biomol. Organ., Ciba Found. Symp., 1965*, pp. 446–459.
Morowitz, H. J. (1969). *In* "The Mycoplasmatales and the L-Phase of Bacteria" (L. Hayflick, ed.), pp. 405–412. Appleton, New York.
Nauman, R. K., Silverman, D. J., and Voelz, H. (1971). *J. Bacteriol.* **107,** 358–360.
Neimark, H. C. (1970). *J. Gen. Microbiol.* **63,** 249–263.
Nishimura, S. (1972). *Prog. Nucleic Acid Res. Mol. Biol.* **12,** 50–86.
Novick, R. P., Clowes, R. C., Cohen, S. N., Curtiss, R., Datta, N., and Falkow, S. (1976). *Bacteriol. Rev.* **40,** 168–189.
Nowak, J. A., Das, J., and Maniloff, J. (1976). *J. Bacteriol.* **127,** 832–836.
Ofengand, J., and Henes, C. (1969). *J. Biol. Chem.* **244,** 6241–6253.
Pace, N. R. (1973). *Bacteriol. Rev.* **37,** 562–603.
Petkau, A., and Chelack, W. S. (1974a). *Int. J. Radiat. Biol.* **25,** 321–328.
Petkau, A., and Chelack, W. S. (1974b). *Int. J. Radiat. Biol.* **26,** 421–426.
Pettijohn, D. E., and Hanawalt, P. C. (1964). *J. Mol. Biol.* **9,** 395–410.
Pirie, N. W. (1973). *Annu. Rev. Microbiol.* **27,** 119–132.
Quinlan, D. C., and Maniloff, J. (1972). *J. Bacteriol.* **112,** 1375–1379.
Quinlan, D. C., and Maniloff, J. (1973). *J. Bacteriol.* **115,** 117–120.
RajBhandary, U. L., Chang, S. H., Stuart, A., Faulkner, R. D., Hoskinson, R. M., and Khorana, H. C. (1967). *Proc. Natl. Acad. Sci. U.S.A.* **57,** 751–758.
Reff, M. E., Stanbridge, E. J., and Schneider, E. L. (1977). *Int. J. Syst. Bacteriol.* **27,** 185–193.
Reich, P. R. (1967). *Ann. N.Y. Acad. Sci.* **143,** 113.
Reich, P. R., Somerson, N. L., Rose, J. A., and Weissman, S. M. (1966a). *J. Bacteriol.* **91,** 153–160.
Reich, P. R., Somerson, N. L., Hybner, C. J., Chanock, S. M., and Weissman, S. M. (1966b). *J. Bacteriol.* **92,** 302–310.
Riggs, A. D. (1966). Ph.D. thesis, California Institute of Technology, Pasadena.
Robinson, I. M., Allison, M. J., and Hartman, P. A. (1975). *Int. J. Syst. Bacteriol.* **25,** 173–181.
Rodwell, A. W. (1969). *J. Gen. Microbiol.* **58,** 39–47.
Rogul, M., McGee, Z. A., Wittler, R. G., and Falkow, S. (1965). *J. Bacteriol.* **90,** 1200–1203.
Russell, G. F., McGeoch, D. J., Elton, R. A., and Subak-Sharpe, J. H. (1973). *J. Mol. Evol.* **2,** 277–292.
Ryan, J. L., and Morowitz, H. J. (1969). *Proc. Natl. Acad. Sci. U.S.A.* **63,** 1282–1289.
Saglio, P., Davis, R. E., Dalibart, R., Dupont, G., and Bové, J. M. (1974). *Colloq. Inst. Natl. Sante Rech. Med.* **33,** 27–34.
Searcy, D. G. (1975). *Biochim. Biophys. Acta* **395,** 535–547.
Searcy, D. G., and Doyle, E. K. (1975). *Int. J. Syst. Bacteriol.* **25,** 286–289.
Singer, C. E., and Ames, B. (1970). *Science* **170,** 822–826.
Smith, D. W. (1969). *Biochim. Biophys. Acta* **179,** 408–421.

Smith, D. W., and Hanawalt, P. C. (1968). *J. Bacteriol.* **96,** 2066–2076.

Smith, D. W., and Hanawalt, P. C. (1969). *J. Mol. Biol.* **46,** 57–72.

Smith, J. D. (1972). *Annu. Rev. Genet.* **6,** 235–256.

Sogin, S. J., Sogin, M. L., and Woese, C. R. (1972). *J. Mol. Evol.* **1,** 173–184.

Söll, D. (1971). *Science* **173,** 293–299.

Somerson, N. L., Reich, P. R., Walls, B. E., Chanock, R. M., and Weissman, S. M. (1966). *J. Bacteriol.* **92,** 311–317.

Somerson, N. L., Reich, P. R., Chanock, R. M., and Weissman, S. M. (1967). *Ann. N.Y. Acad. Sci.* **143,** 9–20.

Stanbridge, E. (1971). *Bacteriol. Rev.* **35,** 206–227.

Stanbridge, E. J. (1976). *Annu. Rev. Microbiol.* **30,** 169–187.

Stanbridge, E. J., and Doersen, C. J. (1978). *In* "Mycoplasma Infection in Cell Cultures" (G. McGarrity, ed.), pp. 119–134. Plenum, New York.

Steinberg, P., Horswood, R. L., and Chanock, R. M. (1969). *J. Infect. Dis.* **120,** 217–224.

Stonington, O., and Pettijohn, D. E. (1971). *Proc. Natl. Acad. Sci. U.S.A.* **68,** 6–9.

Subak-Sharpe, J. H., Elton, R. A., and Russell, G. J. (1974). *Symp. Soc. Gen. Microbiol.* **24,** 131–150.

Sueoka, N. (1961). *Proc. Natl. Acad. Sci. U.S.A.* **47,** 1141–1149.

Swartz, M. N., Trautner, T. A., and Kornberg, A. (1962). *J. Biol. Chem.* **237,** 1961–1967.

Teplitz, M. (1977). *Nucleic Acids Res.* **4,** 1505–1512.

Thiebe, K., and Zachau, H. G. (1970). *Biochim. Biophys. Acta* **217,** 294–304.

Tourtellotte, M. E. (1969). *In* "The Mycoplasmatales and the L-Phase of Bacteria" (L. Hayflick, ed.), pp. 451–468. Appleton, New York.

Tourtellotte, M. E., Morowitz, H. J., and Kasimir, P. (1964). *J. Bacteriol.* **88,** 11–15.

Tully, J. G., Barile, M. F., Edward, D. G.ff., Theodore, T. S., and Ernø, H. (1974). *J. Gen. Microbiol.* **85,** 102–120.

Walker, R. T. (1967). *Nature (London)* **216,** 711–712.

Walker, R. T. (1976). *In* "Synthesis, Structure and Chemistry of Transfer Ribonucleic Acids and Their Components," pp. 291–305. Dymaczewo, Poland.

Walker, R. T., and RajBhandary, U. L. (1975). *Nucleic Acids Res.* **2,** 61–78.

Walker, R. T., and RajBhandary, U. L. (1978). *Nucleic Acids Res.* **5,** 57–70.

Wallace, D. W., and Morowitz, H. J. (1973). *Chromosoma* **40,** 121–126.

Williams, R. A. D. (1975). *Sci. Prog. (Oxford)* **62,** 373–393.

Woese, C. R. (1967). *In* "The Genetic Code." Harper, New York.

Woese, C. R., and Bleyman, M. A. (1972). *J. Mol. Evol.* **1,** 223–229.

Woese, C. R., Sogin, M. L., and Sutton, L. A. (1974). *J. Mol. Evol.* **3,** 293–299.

Woese, C. R., Fox, G. E., Zablen, L., Uchida, T., Bover, L., Peckham, K., Lewis, B. L., and Stahl, D. (1975). *Nature (London)* **254,** 83–86.

Worcel, A., and Burgi, E. (1972). *J. Mol. Biol.* **71,** 127–147.

Zachau, H. G. (1969). *Angew. Chem.* **8,** 711–727.

Zouzias, D., Mazaitis, A. J., Seinberkoff, M., and Rush, M. (1973). *Biochim. Biophys. Acta* **312,** 484–491.

# 7 / RESPIRATORY PATHWAYS AND ENERGY-YIELDING MECHANISMS

*J. D. Pollack*



How do Mollicutes obtain energy? What pathways do they possess which demonstrate or suggest their ability to synthesize ATP? Can all Mollicutes manufacture enough ATP to meet their synthetic needs?

The response to these questions will include a general discussion and selected examination of some of the known respiratory components and other energetically useful pathways in the Mollicutes.



THE MYCOPLASMAS, VOL. 1

ISBN 0-12-078401-7

# I. RESPIRATORY PATHWAYS AND THE MOLLICUTES: GENERAL VIEW

## A. Flavin-Terminated Respiration (FTR) in Fermentative Mycoplasmas

Fermentative Mollicutes are described as having a flavin-terminated respiratory chain (VanDemark, 1969; Smith, 1971). This characterization is based, in part, on the presence of flavins and the absence of cytochromes and quinones.

The FTR chain is diagrammatically represented as

$$\begin{array}{c} \text{NAD} \rightarrow \text{flavoprotein} \rightarrow O_2 \\ \uparrow \\ \text{NADP} \end{array}$$

### The Presence of Flavins

The presence of flavin and flavin-related respiratory activity in the Mollicutes has been frequently demonstrated.

S. L. Smith *et al.* (1963) found by manometry that the oxidation of lactate by *Mycoplasma gallisepticum* 293 was stimulated by riboflavin-5-phosphate (FMN) or flavin adenine dinucleotide (FAD) and was inhibited by atabrine. The study also showed that the rate of lactate oxidation increased as the oxygen concentration increased, over the range 21–100% the $Q_{O_2}$ (dry weight) rose from 15.6 to 36.7. This was considered a characteristic response of an FTR system. It was also calculated that the average flavin level was $2.1 \times 10^{-10}$ mol/mg dry weight which was a level comparable to that found in some lactic acid bacteria. VanDemark and Smith (1964a) treated some extracts of *Mycoplasma hominis* 07, now called *Mycoplasma arthritidis* 07, at acid pH to obtain the NADH oxidase apoenzyme. FAD but not FMN could restore activity. Flavin adenine dinucleotide and cysteine were required to reactivate the NADH oxidase activity of *Mycoplasma mycoides* V5 kept at −15°C for 4 months (Rodwell, 1967).

Morowitz and Terry (1969) published absorption spectra of oxidized and of reduced flavins released from *Acholeplasma laidlawii* B membranes by digestion with pronase. The oxidized spectra exhibited peaks at 366 and 445 nm (Fig. 1). J. D. Pollack and A. J. Merola (unpublished data) also examined the membranes and whole cells of a number of *Acholeplasma* species: *A. laidlawii* A and B, *A. oculi* 19L, *A. axanthum* S743, *A. granularum* BTS39, *A. modicum* PG49, and *A. equifetale* N93. By differential spectroscopy of dithionite reduced minus oxidized mem-

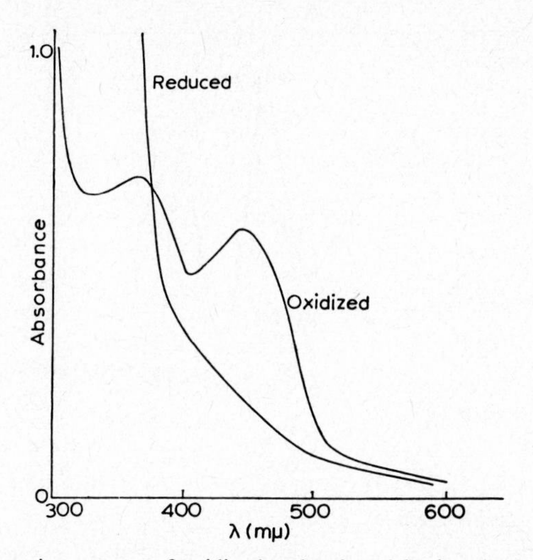


FIGURE 1. Absorption spectra of oxidized and reduced flavin released from *A. laidlawii* membranes by digestion with pronase. (From Morowitz and Terry, 1969.)

branes at liquid nitrogen temperature the characteristic flavin bleaching spectrum (Singer, 1963; Singer and Gutman, 1971) was observed for all *Acholeplasma*. No evidence for cytochromes in the membranes or whole cells of any *Acholeplasma* species was found. Almost identical findings were observed when NADH was used as the reductant, as reported in an earlier study (Pollack, 1975). In Fig. 2 is an example of these data obtained with osmotically prepared *A. laidlawii* B membranes.

Jinks and Matz (1976a) reported the presence of both FMN and FAD (2 : 1) in a NADH dehydrogenase preparation from *A. laidlawii* membranes. The dehydrogenase was insensitive to stimulation by exogenous FAD or FMN. The presence of more than one flavin suggested the presence of more than one flavoprotein.

## B. Diaphorases

Diaphorases are enzymes that couple the reduction of ''artificial electron acceptors'' as dyes, to the oxidation of NADH. Dolin (1961) has noted that many are flavoproteins. Diaphorase activities have been found in many Mollicutes as noted by VanDemark (1969) and Smith (1971). The presence of diaphorase activity has been used for the detection and identification of Mollicutes and their antibodies.

The ability of certain Mollicutes to anaerobically reduce triphenyltetrazolium chloride (TPTC) was first detected by Somerson and Morton

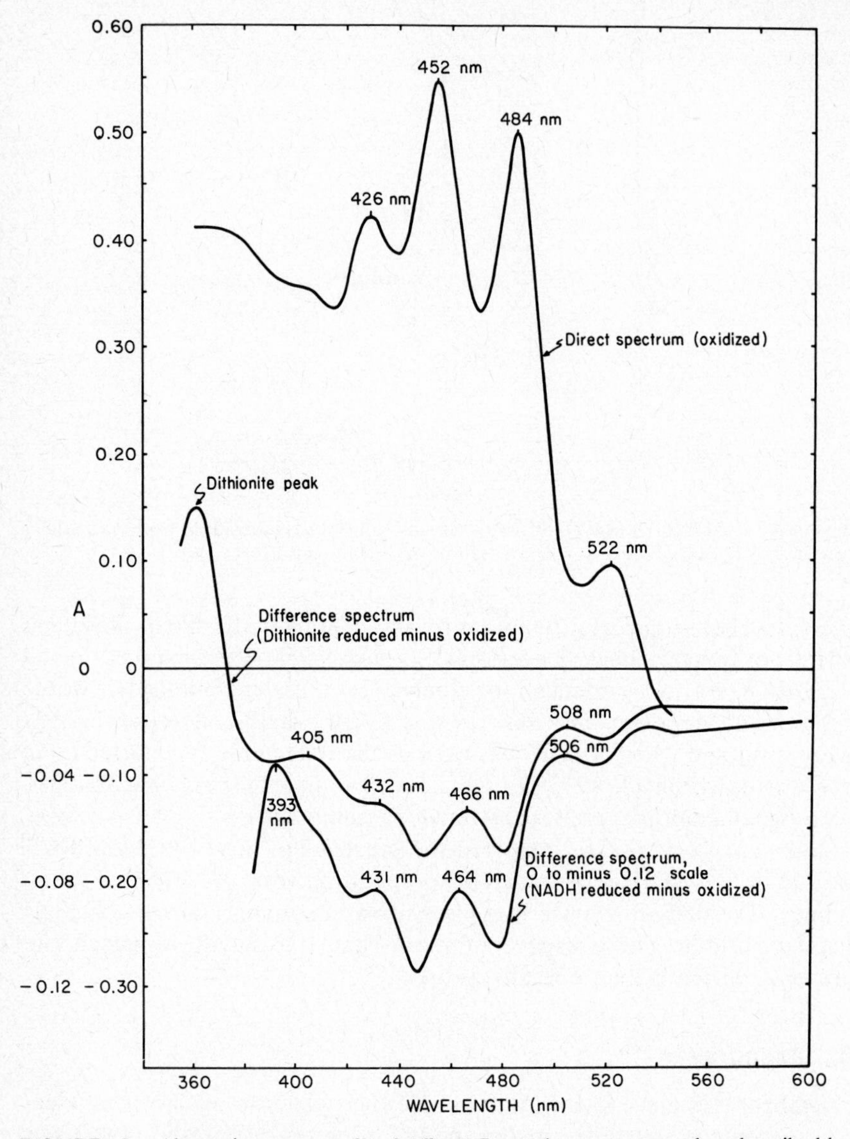


FIGURE 2. Absorption spectra of *A. laidlawii* B membranes, prepared as described by Pollack (1975). Upper curve: direct spectrum of oxidized preparation at 26°C. Lower curves: difference spectra (reduced by dithionite or NADH minus oxidized) at liquid nitrogen temperature. (Unpublished data of J. D. Pollack and A. J. Merola.)

(1953). It is thought that the TPTC accepts electrons from reduced flavoproteins.

Diaphorase activity has also been detected by the reduction of methylene blue. In this application the abilities of Mollicutes' dehydrogenases to mediate the transfer of electrons from substrate donors have been revealed (Holmes and Pirie, 1932; Pirie and Holmes, 1933; Warren, 1942; Edward, 1954; Rodwell and Rodwell, 1954a,b; Kandler and Kandler, 1955; Freundt, 1958). Dichlorophenol indophenol (DCPIP) has also been used as an "artificial electron acceptor" by some workers, most recently by Larraga and Razin (1976) and Jinks and Matz (1976a), to detect the NADH–DCPIP oxidoreductase activity in preparations of *A. laidlawii* oral strain, *M. mycoides* subsp. *capri,* and in membranes of *A. laidlawii* ATCC 14192.

The use of ferricyanide as an "artificial electron acceptor" has special significance because its reduction has been related to site I and the NADH dehydrogenase of the mitochondrial electron transport oxidative phosphorylation system (Hatefi, 1976). Ferricyanide oxidoreductase activity has been reported in *M. gallisepticum* 293 (S. S. Smith, 1963), *M. mycoides* Y (Rodwell, 1967), *Mycoplasma pneumoniae* (Low and Zimkus, 1973), and *A. laidlawii* B and *Mycoplasma capricolum* 14 (Pollack, 1975).

### C. The Involvement of Quinones

Quinones have only been reported in *Mycoplasma* and *Thermoplasma* species. If autooxidizable they may act as a direct or intermediate link between reduced pyridine nucleotides and oxygen.

The presence and role of respiratory quinones in Mollicutes was first indicated with *M. arthritidis* 07 by VanDemark and Smith (1964a). Gerwitz and Baum, cited by VanDemark (1969), reported the isolation of a napthoquinone-like material with an ultraviolet (uv) spectrum similar to $K_2$ from *Mycoplasma* strain C56R. Gale *et al.* (1964) were unable to detect quinones or naphthoquinones in *M. gallisepticum* J. Smith *et al.* (1963) detected menadione reductase activity in *M. gallisepticum* 293.

Langworthy *et al.* (1972) reported the presence of menaquinone-7 in *Thermoplasma acidophilum* 122–132. Holländer *et al.* (1977) studied the lipoquinones of *Acholeplasma, Mycoplasma, Spiroplasma,* and *Thermoplasma* strains by thin-layer chromatography and difference spectra (reduced minus oxidized). *Thermoplasma acidophilum* strains 122-1B2, 122-1B3, and 124-1 produced 400–1100 nmol menaquinone-7 per gram cell protein. *Acholeplasma axanthum* S743, *A. granularum* BTS39, *A. laid-*

*lawii* PG8, *M. arthritidis* PG27 and 07, *Mycoplasma gallinarum* PG16, *M. hominis* Overweg, *Mycoplasma neurolyticum* PG39, and *Spiroplasma citri* Morocco R8-A2 produced from 0.8 to 6.8 nmol naphthoquinones per gram cell protein. The authors questioned the presence of energetically useful respiratory chain systems in *Acholeplasma, Mycoplasma,* and *Spiroplasma* since their isolates only contained quinones in nanomolar concentrations or no quinones. In respiring bacteria that were examined the levels were in the micromolar range. The authors reported that *M. arthritidis* 07 produces a naphthoquinone-like substance with peaks at 276 and 290 nm and troughs at 272, 283, and 296 nm, reminiscent of the substance reported by VanDemark (1969). The view of Holländer *et al.* (1977) contradicts the impression that *M. arthritidis* 07 has a functional quinone locus (VanDemark and Smith, 1964a; VanDemark, 1969).

### D. The Involvement of Cytochromes

VanDemark and Smith (1964a) reported the detection by difference spectroscopy, of "*a*-" and "*b*-type" cytochromes in sonic preparations of whole cells of the nonfermentative human genital isolate *M. arthritidis* 07.

Holländer *et al.* (1977) could not demonstrate cytochromes using methods of medium sensitivity. Holländer (1978) reiterated these findings and suggested that the presence of cytochromes in *M. arthritidis* 07, reported by VanDemark and Smith (1964a), and in a bovine strain C56R reported by VanDemark (1969) seemed doubtful since cytochrome contamination by medium components could not be ruled out. Holländer *et al.* (1977) did not find cytochromes in other nonfermentative mycoplasmas: *M. arthritidis* PG27, as well as 07, *M. hominis* Overweg, and *M. gallinarum* PG16.

It was suspected that the 07 strain might be unique as neither membranes nor whole cells of either *M. arthritidis* ATCC 14124 or ATCC 14152 showed any evidence for cytochromes by difference spectroscopy (J. D. Pollack, unpublished data).

All attempts to find cytochromes in fermentative *Mycoplasma* and *Acholeplasma* species have been unsuccessful. The following fermentative mycoplasmas have been examined: *M. mycoides* V5 (Rodwell and Rodwell, 1954a), *M. gallisepticum* 293 (S. S. Smith, 1963; S. L. Smith *et al.*, 1963), *A. laidlawii* Köller (Tarshis *et al.*, 1976), and, by Holländer *et al.* (1977), *M. neurolyticum* PG39, *S. citri* Morocco R8-A2, *A. granularum* BTS39, *A. laidlawii* PG8, and *A. axanthum* S743. J. D. Pollack and A. J. Merola (unpublished data) failed to find cytochromes in membranes or whole cells of *A. laidlawii* PG8 and PG9, *A. granularum* BTS39, *A.*

*axanthum* S743, *A. oculi* 19L, *A. modicum* PG49, and *A. equifetale* N93 by differential spectroscopy at liquid nitrogen temperatures.

Cytochromes have been identified in the obligately thermophilic and acidophilic *T. acidophilum*. Belly *et al.* (1973) first demonstrated *a*- and *c*-type cytochromes and apparently an *o*-cytochrome oxidase in five strains. Ruwart and Haug (1975) indicated that malate may act as an electron donor for *Thermoplasma*. Holländer (1978) examined strains 122-1B2 and 122-1B3 and found one *b*-type and two *c*-type cytochromes, and cytochrome oxidases $d(a_2)$ and *o*. Holländer concluded that in *T. acidophilum* the presence of cytochromes, oxidases, and menaquinones in amounts comparable to those found in respiring walled bacteria indicated the presence of an intact terminal electron transport system, and that this property distinguishes *Thermoplasma* from all other Mollicutes.

It will be necessary to examine other mycoplasmas for cytochromes by techniques of maximum sensitivity and confirm the absence of the heme moiety by alternate procedures, such as the production of reduced pyridine hemochrome. It is worth considering that the cytochrome composition may vary both qualitatively and quantitatively as the nutriment varies, and the presence or absence of oxygen may directly control the production or function of cytochromes (Whittenbury, 1960; White and Sinclair, 1971; Ritchie and Seeley, 1976).

## E. Oxidative Phosphorylation

Phosphorylation in cytochrome-free bacteria occurs almost entirely at the substrate level (Dolin, 1961). If it is to occur at all in a respiratory-linked electron transport process independent of cytochromes it should, by processes understood for microbial systems, take place at the NADH–flavoprotein–nonheme iron locus (Haddock and Jones, 1977). As almost all nonthermophilic Mollicutes examined lack cytochromes and quinones, conservation of energy linked to a respiratory-like electron transport system should occur at a similar locus.

Hardly any evidence has been offered to indicate the presence of oxidative phosphorylation in cytochrome-deficient Mollicutes. Tarshis *et al.* (1976) have inferred that oxidative phosphorylation in *A. laidlawii* strain is extremely low. This opinion was substantiated by experiments, reported without details, indicating no ATP synthesis using the luciferin–luciferase assay.

Monovalent thallous ion acts as an uncoupler of energized mitochondria (Melnick *et al.*, 1976). The use of thallium(ous) acetate as an inhibitor of bacterial contamination of some Mollicutes cultures may bear on the detection of oxidative phosphorylation in mycoplasmas.

Oxidative phosphorylation has been reported in two *Mycoplasma* isolates thought to contain cytochromes. VanDemark (1967, 1969), using the hexokinase–glucose trap system of Pinchot (1953), found phosphorylation concomitant with the oxidation of NADH by crude extracts of *M. arthritidis* 07 and a bovine *Mycoplasma* strain WID. The determination of the presence and magnitude of oxidative phosphorylation is sensitive to the interfering activity not only of adenylate kinase, which was considered by VanDemark (1967), but also, as more recently recognized, of polynucleotide phosphorylase, ADP-hydrolyzing enzymes, and ATPase.

The presence of cytochromes in *Thermoplasma acidophilum* suggests that this Mollicutes may be capable of oxidative phosphorylation.

## F. The Presence of Catalase and Peroxide

The oxidation by oxygen of NADH or other reduced terminal acceptors of electrons does not have to be directly coupled to the synthesis of ATP to be useful. Oxygen may serve merely as an electron acceptor or sink, in order to permit further metabolism of intermediates or substrate. Gunsalus and Umbreit (1945) reported that *Streptococcus faecalis* 24 used oxygen as an electron acceptor in the oxidation of glycerol phosphate. The sequence became energetically useful through the coupling of glyceraldehyde to ATP synthesis. Somewhat analagously, Rodwell (1967) found with *M. mycoides* V5 that the phosphorylation of glycerol to L-glycerol-3-phosphate (GP) is followed by the oxidation of GP to glyceraldehyde-3-phosphate. In this concept other substrates normally not metabolized because of lack of hydrogen acceptors now become usable. This view emphasizes the usefulness of certain oxidations linked to oxygen without phosphorylation and may characterize the main path by which cytochrome-independent Mollicutes usefully dispose of their electrons.

The hallmark of oxygen-linked electron disposal mediated by flavoprotein catalysis is peroxide formation. Peroxide is toxic and its necessary removal is mediated by heme-containing enzymes: peroxidases or catalases. The production of catalases is susceptible to the level of hemin in the growth medium of a variety of *Haemophilus* isolates (Bieberstein and Gills, 1961). This is important as the absence of catalase and peroxide in Mollicutes may be related to the media in which they are grown. The absence of catalase could be due to hemin insufficiency, while the absence of peroxide could be due to the presence of either contaminating serum catalase or a system possessed by some clostridia where no peroxide intermediate is formed (Dolin, 1961).

### 1. Catalase

Pirie (1938) did not detect catalase in *A. laidlawii* C. Using heavy suspensions of *M. hominis* Campo cells Lecce and Morton (1953) could detect catalase activity. Rodwell and Rodwell (1954a) noted apparent catalase activity in *M. mycoides* V5. Neither Kandler and Kandler (1955) nor Freundt (1958) were able to find catalase in their experiments. Weibull and Hammarberg (1962) did not find catalase activity in six mycoplasmas. S. L. Smith *et al.* (1963) did not find catalase or peroxidase activities or hemes in *M. gallisepticum* 293. VanDemark and Smith (1964a) demonstrated cyanide-sensitive catalase activity in *M. arthritidis* 07. Cole *et al.* (1968) indicated the presence of catalase in some Mollicutes. VanDemark (1967) employed a highly sensitive spectrophotometric test for catalase activity and a benzidine test for the heme moiety and obtained positive results in both cases for *Mycoplasma iners* and seven other unidentified *Mycoplasma* isolates.

### 2. Peroxide

Somerson *et al.* (1965) first identified the hemolysin of *M. pneumoniae* as peroxide. Low and co-workers also studied the production of $H_2O_2$ in mycoplasmas (Low and Eaton, 1965; Low *et al.*, 1968; Low, 1971; Low and Zimkus, 1973). Low and Zimkus (1973) found that $H_2O_2$ production was associated with the aerobic oxidation of NADH by cell-free extracts of *M. pneumoniae* FH and Mac. $H_2O_2$ production paralleled $O_2$ consumption and was negligible unless FMN was added; FAD was only slightly inhibitory. The FMN–$NADH_2$-dependent $H_2O_2$ formation was inhibited by catalase but not affected by atabrine, rotenone, amytal, antimycin A, oligomycin, azide, or cyanide. In this work Low and Zimkus (1973) observed a second NADH–peroxide activity apparently not requiring FAD. These findings will be considered later.

Thomas and Bitensky (1966) detected peroxide in *M. gallisepticum* S6 reaction mixtures. Rodwell (1967) and Plackett and Rodwell (1970) reported the presence of peroxide in cultures of *M. mycoides* associated with the oxidation of glycerol. Cole *et al.* (1968) examined a variety of Mollicutes for peroxide. Twelve isolates were found to give a positive benzidine–blood agar reaction, which was inhibited by catalase and then could be reactivated by aminotriazole: *M. pneumoniae* EA, ATCC 15492, ATCC 15531; *M. arthritidis* L4P; *Mycoplasma pulmonis* T1; *M. neurolyticum* TA; *M. gallinarum*; *M. gallisepticum* B11; *M. mycoides* var.

*capri*; *Mycoplasma bovigenitalium* ATCC 14173; *Mycoplasma felis* B2; and *A. laidlawii* B. Cole *et al.* (1968) found that five other species were negative for peroxide by all tests. Cherry and Taylor-Robinson (1970) confirmed the ability of many of these species to produce peroxide in chicken tracheal organ cultures. Johnson and Muscoplat (1972) reported the presence of peroxide in *M. bovigenitalium, Mycoplasma bovirhinis, Mycoplasma bovimastitidis, Mycoplasma salivarium, Mycoplasma hyorhinis, M. pneumoniae, A. granularum, A. oculi,* and *A. laidlawii.*

Peroxidase activity has not been reported in mycoplasmas.

In summary, peroxide has been found more frequently than catalase in Mollicutes. Peroxide may be the terminal sink for substrate electrons and be associated with the production of oxidized NAD.

## G. Electron Donors

Electrons may be supplied to the flavin-terminated path of fermentative Mollicutes through the action of NAD-linked dehydrogenases that are associated with the oxidation of glyceraldehyde-3-phosphate (S. L. Smith *et al.*, 1963), lactate (Rodwell and Rodwell, 1954b; Tourtellotte and Jacobs, 1960; S. L. Smith *et al.*, 1963), and glutamate (Yarrison *et al.*, 1972). The NADP-linked dehydrogenases of glucose 6-phosphate (Tourtellotte and Jacobs, 1960; Castrejon-Diez *et al.*, 1963; Pollack *et al.*, 1965b), glyceraldehyde-3-phosphate (Tourtellotte and Jacobs, 1960), α-glycerophosphate (Tourtellotte and Jacobs, 1960; Rodwell, 1967), hydroxymethyl glutaryl-CoA (Smith and Henrikson, 1965), and glutamate (Yarrison *et al.*, 1972) may also be involved in the contribution of electrons to the FTR path. There are no reports of transhydrogenase activity in the Mollicutes. It should be emphasized that all fermentative mycoplasmas "assigned" to the FTR pattern do not possess all of these dehydrogenase activities. In a comparative study, NADP–glucose-6-phosphate dehydrogenase activity was only found in *A. laidlawii* A and B and not in other fermentative *Mycoplasma* species (Pollack *et al.*, 1965b).

In the nonfermentative *M. arthritidis* 07 NAD- and NADP-linked dehydrogenases for the following substrates have been reported: ethanol (Lecce and Morton, 1953), malate (VanDemark and Smith, 1965), β-hydroxybutyrate (VanDemark and Smith, 1965), and glyceraldehyde-3-phosphate, glycerol phosphate, and lactate (Gerwitz and VanDemark, 1966). The reduction of NADP with proline, malate, and isocitrate was reported (Smith, 1957; VanDemark and Smith, 1964a). A few studies have been published without noting the intervention of pyridine nucleotides in the *M. arthritidis* 07 oxidation of such substrates as lactate (Lecce and

Morton, 1953), succinate and α-ketoglutarate (VanDemark and Smith, 1964b), and butyryl-CoA (VanDemark and Smith, 1965).

## II. RESPIRATORY COMPONENTS

The presumption that mycoplasmas may be capable of oxidative phosphorylation is built on evidence compiled for *M. arthritidis* 07 as already noted. Also, some data suggest the possibility that phosphorylation associated with oxygen consumption may occur in cytochromeless Mollicutes. In this case, phosphorylation would occur at a locus analagous to Site I of bovine mitochondria, between NADH and either flavoprotein or nonheme iron (Ragan, 1976; Beinert, 1976).

The evidences presented for cytochromeless Mollicutes include the localization and nature of the NADH oxidase and "primary" flavoprotein dehydrogenase activities, which are discussed below, and the presence and role of flavin, consumption of oxygen, and the action of respiratory and phosphorylation inhibitors, which have already been cited.

### A. Localization of Respiratory Activities

The first study on the localization of respiratory activities in the Mollicutes was conducted by S. S. Smith (1963). *M. gallisepticum* 293 was shown to contain three times more NADH oxidase and 2.5 times more DCPIP activities in high-speed supernatants (125,000 *g* for 90 min) of Hughes press-fractured cells than in membrane fractions. The author stated that NADH oxidase activity (NOA) was associated with particles which only pass through 450 nm pored filters. Some of these findings were also observed in cell-free extracts (CFE) (20,000 *g* for 1 hr) fractured by sonic oscillation (S. L. Smith *et al.*, 1963). In this latter study, the oxidation of NADH by cell-free extracts was stimulated by menadione or ferricyanide.

In a search for electron transport systems in Mollicutes it was observed that NOA was localized in washed membrane fractions of *A. laidlawii* B prepared after osmotic lysis and sedimentation (37,000 *g* for 45 min). The NOA of *Mycoplasma* species was localized in supernatant fractions (144,000 *g* for 2 hr) (Pollack *et al.*, 1965b). Osmotic lysis had been shown to lead to CFE with the least membrane contamination (Pollack *et al.*, 1965a). Our use of osmotic lysis was suggested by the influential studies of Shmuel Razin which have been summarized by Razin and Rottem (1974). We have ascribed the success of some of our studies to the use of a

$\beta$-mercaptoethanol-containing buffer called B2 or $\beta$ buffer (Pollack *et al.*, 1965b). A superior formulation now in use is $\kappa$ (kappa) buffer (Pollack, 1975). The localization studies of respiratory-associated enzymes have been almost entirely concerned with NOA and these data are summarized in Table I. This table reports the ratios of NADH oxidase activity ($M/S$) in membrane ($M$) and supernatant ($S$) fractions from Mollicutes. All *Acholeplasma* species have ratios higher than 3, indicating membrane localization of this activity, and all *Mycoplasma* species and *S. citri* have ratios of 0.3 or lower, indicating supernatant localization. The localization of NOA in *Ureaplasma urealyticum* strain 960 has been studied but no conclusion was drawn because of the small amounts of material available for study (Masover *et al.*, 1977). Mudd *et al.* (1977) reported that in *S. citri* California 189 the NOA was either readily removed from the membrane or not associated with the membrane.

TABLE I. **Ratio of NADH Oxidase Activity[a] in Membrane ($M$) and Supernatant ($S$) Fractions from Mollicutes[b].**

| Organism | $M/S$ |
|---|---|
| *Acholeplasma laidlawii* B (PG9) | 763.22 |
| *A. laidlawii* B (PG9-PS) | 22.05 |
| *A. laidlawii* A (PG8) | 12.78 |
| *A. granularum* (BTS39) | 172.78 |
| *A. axanthum* (S743) | 25.83 |
| *A. modicum* (PG49) | 54.13 |
| *A. oculi* (19L) | 118.27 |
| *A. equifetale* (N93) | 9.46 |
| *Mycoplasma pneumoniae* (65-2052) | 0.02 |
| *M. pulmonis* (N3) | 0.03 |
| *M. neurolyticum* (PG28) | 0.23 |
| *M. bovigenitalium* (PG11) | 0.01 |
| *M. gallisepticum* (S6) | 0.03 |
| *M. gallisepticum* (A5969) | 0.04 |
| *M. gallisepticum* (293) (S. S. Smith, 1963) | 0.30[c] |
| *M. arthritidis* (PG27, ATCC 14152) | 0.09 |
| *M. arthritidis* (ATCC 14124) | 0.10 |
| *M. arthritidis* (H39) | 0.10 |
| *M. capricolum* (14) | 0.01 |
| *M. mycoides* (Y) (Rodwell, 1967) | 0.01[c] |
| *Spiroplasma citri* (ATCC 27556) (Kahane *et al.*, 1977) | 0.05[c] |

[a] Activity was expressed either as the decrease in $A_{340}$ nm or micromoles NADH oxidized per minute per milligram protein.

[b] Data taken from the work of Pollack (1975, 1978) and Pollack *et al.* (1965b, 1970).

[c] Calculated from cited work.

NADH oxidase activity of acholeplasmas is considered to be localized on the inner membrane surface. This view rests in part on the observations that the whole cell does not oxidize NADH and that as the membrane is dissolved by deoxycholate the specific activity increases (Pollack *et al.*, 1965b). When *A. laidlawii* whole cells are treated with pronase or antiserum to *A. laidlawii* membranes there is little or no effect on NOA of subsequently isolated membranes; however, when isolated membranes are so treated NOA decreases (Ne'eman and Razin, 1975).

NADH oxidase activity is not substantially released from membranes by low-ionic strength buffers of Tris with or without EDTA (Ne'eman *et al.*, 1971). NOA could not be detected in membranes of *M. mycoides* subsp. *mycoides* isolated by digitonin in the presence of $Mg^{2+}$ (Razin, 1975). These methods minimize the release of loosely associated membrane proteins.

Larraga and Razin (1976) solubilized membranes of *M. mycoides* subsp. *capri* and reaggregated them in the presence of the cytoplasmic fraction containing NOA and NADH–DCPIP activities. The reaggregated membranes possessed 16 and 26% of the cytoplasmic activities, respectively. In similar experiments reaggregated *A. laidlawii* membranes contained about 90% of the activities.

A more direct approach to the localization of respiratory activity in Mollicutes was reported by Vinther and Freundt (1977). Using deposits of tellurium as indicators of respiratory activity in electron and light microscopic studies, these workers concluded that the bulk of the *Acholeplasma* activity was at the periphery of the cells and appeared to be on (or near) the inner surface of the membrane. In *M. mycoides* subsp. *mycoides* electron-dense patches were not associated with the membrane but were localized in the cytoplasm. This latter observation was not observed in *M. mycoides* subsp. *capri* or in *M. bovigenitalium*. Maniloff (1972) found intracellular clusters of tellurium crystals in *M. gallisepticum* in proximity to the membrane. Green and Hanson (1973) did not observe tellurium deposits in their studies with *Mycoplasma meleagridis*. These observations support the membrane localization of respiratory-associated activities in *Acholeplasma*. In *Mycoplasma* species these activities may be near or loosely associated with the membrane or generally distributed in the cytoplasm.

NADPH oxidase activity has been localized in the supernatant fractions of *M. gallisepticum* A5969 and S6; *M. capricolum* 14; *M. neurolyticum* PG28; *M. arthritidis* ATCC 14124, ATCC 14152, and H39; and *M. bovigenitalium* PG11 (Pollack *et al.*, 1965b). The *M/S* ratios were all $<0.01$. NADPH oxidase activity was not detected in any fraction from *A. laidlawii* A or B. Parenthetically, only the two acholeplasmas demonstrated glucose-6-phosphate dehydrogenase activity, which was localized in their supernatant fractions.

NADH:ferricyanide oxidoreductase activity has been localized in the membrane of *A. laidlawii* B and in the supernatant fraction of *M. capricolum* 14 with *M/S* ratios of 8.66 and 0.01, respectively (Pollack, 1975). Ferricyanide reductase activity was also reported in supernatants of *M. gallisepticum* 293 (S. S. Smith, 1963). NADH:ferricyanide oxidoreductase activity (EC 1.6.99.1) is a relatively specific component of the mitochondrial energy-conserving Site I locus. The presence of "primary" flavoprotein dehydrogenase-like activity characteristic of Site I has suggested that an analogous oxidative phosphorylation step may also be present in these mycoplasmas.

### B. The Respiratory-Associated Electron Feeder Paths of the Mollicutes

Electrons can enter the mitochondrial system by three paths (Beinert, 1976), one leading from succinate by succinate dehydrogenase or from the fatty acyl-CoA dehydrogenase system of electron transferring flavoprotein, or from NADH by NADH dehydrogenase. These paths are similar in that they all contain centers of differing midpoint potentials containing nonheme iron and labile sulfur. These paths also contain a flavin, either FAD or FMN but not both, and all converge at the level of ubiquinone (Q). These paths constitute the substrate side of the system; the unbranched cytochrome-containing side of Q constitutes the oxygen side of the system. The substrate side contains more components than the cytochrome arm and a much greater total capacity for electron uptake. Beinert (1976) suggested that the multiplicity of iron–sulfur centers, with presumably differing midpoint potentials, might be useful in "tuning" in at the potential level required for differing donor substrate and acceptor systems.

The mitochondrial model is not the only example to consider. Dolin (1961) and Haddock and Jones (1977) described a number of microbial systems for the transfer of electrons to oxygen. In the microbial cell electrons may enter the terminal path by routes containing flavoproteins and possibly iron–sulfur centers. These branches are pictured as meeting at some common redox locus, i.e., iron–sulfur associated with NADH dehydrogenase at Q. In some cases, the paths lack Q and the flavoproteins are considered to communicate directly with electron acceptors as inorganic ions, cytochromes, or oxygen. In aerobic systems Q may transfer electrons to cytochromes or directly to oxygen. Brodie and Gutnick

(1972) and Haddock and Jones (1977) have reviewed microbial electron transport and oxidative phosphorylation.

### 1. The Succinate Pathway

The oxygen consumption of *M. arthritidis* 07 (VanDemark and Smith, 1964a) and *M. gallisepticum* (Tourtellotte and Jacobs, 1960) increases in the presence of succinate. Little else is known about the place of succinate in the metabolism of mycoplasmas.

### 2. Oxidation of Fatty Acids

Lynn (1960) reported that *M. arthritidis* 07 and other nonfermentative species were capable of oxidizing short-chain fatty acids. The 07 strain showed maximum consumption of oxygen with butyrate and caprylate. The oxidation of certain other fatty acids, $C_2$ to $C_{12}$, was accompanied by lower oxygen consumption. VanDemark and Smith (1964b, 1965) showed that cell-free extracts of butyrate-grown *M. arthritidis* 07 possessed a $\beta$-oxidative pathway and proposed a cyclic mechanism involving an acetyl-CoA synthetase, short-chain CoA transferase (in order to activate fatty acids other than acetate), an acyl-CoA dehydrogenase, an enoyl-CoA hydratase, an hydroxyacyl-CoA dehydrogenase, and a ketothiolase. The presence of acetyl-CoA synthetase in *M. arthritidis* and *M. gallinarum* (VanDemark and Smith, 1965), acetokinases (Castrejon-Diez *et al.*, 1963; Rottem and Razin, 1967), and phosphotransacetylase (Smith and Henrikson, 1965) activities have been reported. As Smith (1971) has pointed out these reactions may be involved in the metabolism of nonfermentative species, leading to the activation of fatty acids for $\beta$ oxidation or processing through the TCA cycle. In fermentative species these reactions may be associated with pyruvate oxidation or synthesis.

### 3. The "Primary" Flavoprotein Dehydrogenase

A third possible route toward the conservation of energy may involve the oxidation of reduced pyridine nucleotides. Either the membrane or the supernatant fractions of all Mollicutes can oxidize NADH. It is not known in any case what the immediate electron acceptor is. Oxygen is presumably the final acceptor.

In the general absence of evidence for cytochromes and phosphorylation it is not unreasonable to ask: How do Mollicutes obtain energy? Possibly NADH oxidation is not directly coupled to ATP synthesis at all. NADH oxidation may serve to supply NAD or, in concert with the apparently ubiquitous membrane-bound ATPase of the Mollicutes, maintain membrane pH and/or proton gradients. This apparently nonphosphorylating pathway of NADH oxidation may be similar to the soluble "bypass" enzymes which mediate the oxidation of NADPH in *Mycobacterium phlei* (Asano and Brodie, 1965; Bogin *et al.*, 1969).

Of little consolation is the observation that the mechanism of mammalian mitochondrial NADH oxidation is not completely understood (Ragan, 1976). Worse, it is uncertain that a single enzyme is responsible for the reaction; the "enzyme" may be metastable (Ragan, 1976).

The oxidation of NADH by Mollicutes has been frequently examined. Some studies have shown that NADH oxidase activity is localized in either the membrane fraction of *Acholeplasma* species or the supernatant fractions of *Mycoplasma* species and *S. citri*.

Ne'eman *et al.* (1972) found that proteins and lipids of *A. laidlawii* (oral) membranes solubilized by Triton X-100 could be separated on Sephadex G-200. The first eluted peak contained almost no lipid and all the NOA. Solubilization by deoxycholate, Lubrol W, or Brij 58 produced similar data. NOA was not affected by removal of deoxycholate. The results suggested that NOA does not require the presence of membrane lipids. In a number of microbial systems and mitochondria the NOA–dehydrogenase activities require lipid (Machinist and Singer, 1965; Dancey and Shapiro, 1977). NOA in *A. laidlawii* is not influenced by the physical state of the membrane lipids (de Kruyff *et al.*, 1973) or the presence of phosphatidylglycerol (Bevers *et al.*, 1977).

NOA either released from the membranes or found in the supernatant fractions of Mollicutes may be particulate (S. S. Smith, 1963; Pollack, 1975; Jinks and Matz, 1976b). The activity and its relationship to the small membrane-associated particles observed after freeze fracturing deserves study.

Low and Zimkus (1973) working with the FH and Mac strains of *M. pneumoniae* presented evidence for the presence of two soluble NOAs. One, as described earlier, was flavin independent, slower moving by polyacrylamide gel electrophoresis (PAGE), and precipitated by 50–70% saturated ammonium sulfate (AS). An FMN-dependent NOA was associated with the more rapidly moving PAGE band, was not completely precipitated by 70–100% AS, and was associated with $H_2O_2$ formation. Cell-free extracts (CFE) or the flavin-independent, 50–70% AS fraction responded similarly to inhibitors in spectrophotometric or manometric assay of NADH oxidation. The NOA of the 50–70% AS fraction was reported to have a $K_m$ of $2.8 \times 10^{-5}$ $M$ (NADH).

Pollack (1975) examined the NADH:ferricyanide oxidoreductase activity of osmotically prepared membrane and supernatant fractions of *A. laidlawii* B and *M. capricolum* 14. The activity was localized in the *Acholeplasma* membrane and *Mycoplasma* supernatant. The supernatant activity of *M. capricolum* 14 at 1.50 m*M* NADH exhibited a biphasic response in Lineweaver–Burk plots with two $K_m$ values for ferricyanide: $5.2 \times 10^{-4}$ *M* and $5.3 \times 10^{-5}$ *M* (Fig. 3). By difference spectroscopy, at room temperature, no respiratory pigments were detected using various inhibitors; with amytal, bleaching typical of flavin was observed. Little activity was lost by passage of the supernatant fractions through 0.1-$\mu$m pore size filters. The kinetic data indicate the presence of two enzymes with different requirements for ferricyanide or one enzyme with two ferricyanide reaction sites with different Michaelis–Menten constants.

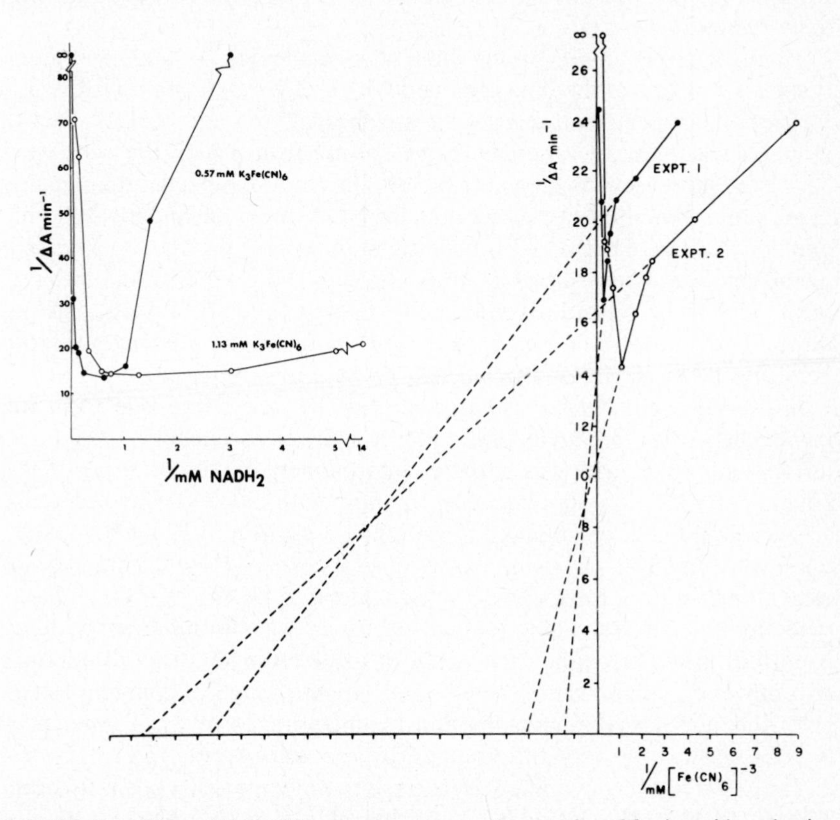


FIGURE 3. Effect of $K_3Fe(CN)_6$ concentration on the velocity of ferricyanide reduction using *M. capricolum* 14 supernatant fraction concentrate at 1.50 m*M* $NADH_2$. In experiment 1 the cuvettes contained 14 mg of protein (closed circles); in experiment 2, with another batch of concentrate, the cuvettes contained 22 mg of protein (open circles). Insert: effect of NADH concentration on the velocity of ferricyanide reduction using *M. capricolum* 14 supernatant fraction concentrate. In both experiments the cuvettes contained 14 mg of protein from the same batch of concentrate. In one experiment the $K_3Fe(CN)_6$ was 0.57 m*M* (closed circles); in the other experiment it was 1.13 m*M* (open circles). (From Pollack, 1975.)

One activity may have been "derived" from the other, preparatively induced, perhaps lacking flavin or an iron–sulfur center or both.

Larraga and Razin (1976) examined the NOA in osmotically prepared membrane fractions of *A. laidlawii* (oral) and cytoplasm of *M. mycoides* subsp. *capri*. The specific activities of both NOA and NADH–DCPIP reductase of the *Mycoplasma* cytoplasmic fraction decreased as the age of the culture from which the cells were harvested increased. In the same fraction two pH maxima for NOA were observed, pH 7.8 and 9.4, but only one for NADH–DCPIP reductase, pH 7.4. By PAGE two bands were detected which gave a strong positive NADH–nitroblue tetrazolium oxidoreductase reaction.

Many of these observations were also made for *A. laidlawii* membranes. Contrary to the data obtained with the *Mycoplasma* cytoplasmic fraction, the specific activities of the membrane NOA and NADH–DCPIP reductase increased as the age of the culture from which the cells were harvested increased. Both whole membranes or solubilized–reaggregated membranes exhibited two pH optima for NOA at pH 8.6 and 9.6, but only one for NADH–DCPIP reductase at about pH 7.4. The *A. laidlawii* membrane was also resolvable, after Triton X-100 solubilization, into two positive NADH–nitroblue tetrazolium staining bands by PAGE. The authors' data support and extend the observation of the presence of more than one NOA in *Acholeplasma* and *Mycoplasma* species.

Jinks and Matz (1976a) isolated an NADH dehydrogenase from the membranes of *A. laidlawii* ATCC 14192. The dehydrogenase, assayed as ferricyanide reductase, was prepared by treatment of membranes with 9% ethanol at 43°C for 30 min. The fraction with ferricyanide reductase activity contained, per milligram protein: 1.3 μg iron, 0.19 μg FAD, 0.30 μg FMN, and 642 μg lipid. Activity was observed with a number of electron acceptors: ferricyanide > menadione > DCPIP >> $O_2$ > $Q_{10}$ > cytochrome *c*. Ferricyanide reductase activity was not inhibited by arsenate (0.01 m*M*), rotenone (0.08 m*M*), or azide (10 m*M*). These inhibitors did, however, inhibit membrane-bound NOA, though oligomycin (0.001 m*M*) did not. Oxygen consumption by membranes in the presence of NADH was not or very little altered by the addition of ADP.

The *extracted* ferricyanide reductase activity exhibited a linear function with 0.144 m*M* NADH and a $K_m$ value of $5.7 \times 10^{-4}$ *M* (Fig. 4). A *membrane-bound* ferricyanide reductase activity was also detected and exhibited a biphasic function in Lineweaver–Burk plots, but the Michaelis–Menten values were not reported. These data indicate two reactive sites or two enzymes with ferricyanide reductase activity in *A. laidlawii* membranes. The authors suggest that there is uncoupled respiration at a Site I-like locus in *Acholeplasma laidlawii*.

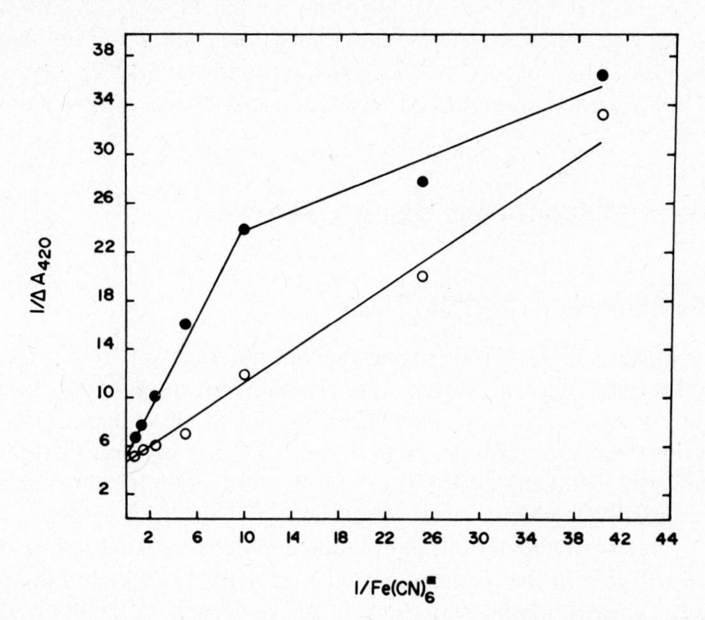


FIGURE 4. Lineweaver-Burk plots of the membrane-bound (closed circles) and ethanol-extracted NADH dehydrogenase (open circles) of *A. laidlawii* at 0.144 m*M* NADH. (From Jinks and Matz, 1976a.)

Stopkie and Weber (1967) also observed that ADP (9 m*M*) inhibited (43%) osmotically prepared membrane-associated NOA of *A. laidlawii* A. Curiously, in the same work the residual NOA found in membrane-free supernatants (105,000 *g* for 6 hr) was not inhibited by 9 m*M* ADP. This residual supernatant NOA appears to be distinct. Membranes with NOA sensitive to 9 m*M* ADP were deliberately sonicated to produce fractions with nonsedimentable fragments. The NOA in these supernatant fractions was found to be sensitive to 9 m*M* ADP as the parent untreated membranes. This suggests that osmotic lysis, but not sonication, separates two NOAs identified by their differential sensitivity to 9 m*M* ADP. The possible different *in situ* localization of these activities was not suggested.

In another study, Jinks and Matz (1976b) found that the purified deoxycholate-treated NADH ferricyanide oxidoreductase, from membranes of *A. laidlawii* ATCC 14192 had one major and two minor bands by PAGE. The specific activity was increased 25- to 35-fold during purification, and the yield was 10–30%. The purified activity was inhibited by $Mg^{2+}$ and $Ca^{2+}$ and had a $K_m$ of $5.10 \times 10^{-4}$ *M* and *V* of 0.236 $\mu$mol/min. Membranes were prepared from cells whose lipid content was modified by growth in media variously supplemented with fatty acids. There was little change in the kinetic constants of either the purified or the membrane-bound activities from these modified membranes compared to unmodified membranes. The authors consider the purified activity to be the "primary" respiratory chain-linked NADH dehydrogenase of *A. laidlawii*.

## III. OTHER ENERGETICALLY USEFUL PATHWAYS

### A. Tricarboxylic Acid (TCA) Cycle

The work of Lynn (1960) suggested that the TCA cycle was present in the nonfermentative mycoplasmas. He showed, as already noted, that certain mycoplasmas were capable of oxidizing short-chain fatty acids. Other investigators (Holmes and Pirie, 1932; Lecce and Morton, 1953; Rodwell and Rodwell, 1954b; Tourtellotte and Jacobs, 1960) working with predominantly fermentative mycoplasmas did not find evidence for the TCA cycle. VanDemark and Smith (1964b) suggested that the TCA cycle functioned only in the oxidation of fatty acids by nonfermentative mycoplasmas. They studied *M. arthritidis* 07. A variety of TCA cycle intermediates were oxidized by acetate-grown cells, as the $Q_{O_2}$(N): succinate > pyruvate > isocitrate > fumarate > malate > $\alpha$-ketoglutarate > acetate. Citrate and glyoxylate were not oxidized. Cell-free extracts possessed TCA cycle enzyme activity, as micromole per hour per milligram enzyme protein: aconitase (as citratase and isocitratase) > succinic dehydrogenase > fumarase > $\alpha$-ketoglutarate dehydrogenase > malic dehydrogenase > isocitric dehydrogenase > acetyl-CoA kinase. Malate synthetase and citrate synthetase activity was also indicated. The presence of the TCA cycle in the nonfermentative *M. arthritidis* 07 appears clear. It was further suggested that the dissimilation of fatty acids, coupled to the TCA cycle and then to oxidative phosphorylation, would be energetically useful. No evidence has been presented that substrate phosphorylation at the succinyl-CoA synthetase locus occurs in this strain or in any other Mollicutes.

Tourtellotte and Jacobs (1960) showed that 11 fermentative strains of mycoplasmas produced lactic, pyruvic, and acetic acids and acetylmethyl carbinol from glucose. These workers indicated that the major source of energy in growing cultures was obtained from the metabolism of carbohydrate to lactate and acetate. The oxidation of TCA cycle intermediates in these fermentative isolates appeared to play a minor role, if any, in the formation of energy. Neimark and Pickett (1960) separated fermentative strains into two groups, the homo- and heterofermentative mycoplasmas, based on the predominance of lactic acid accumulation in the former case and mixed acids in the latter.

## B. Amino Acid Metabolism

### 1. The Production of ATP from Glutamine

Smith (1955, 1957, 1960) investigated whether nonfermentative strains might use amino acids as their sole energy and carbon source. Glutamine at pH 6.0 was shown to undergo cleavage with the production of ATP. This bromsulphalein-sensitive reversible phosphorolytic deamidation of glutamine required ADP, $Mg^{2+}$, and $P_i$ to yield glutamic acid, ATP, and $NH_3$. ATP synthesis was favored (kEq = 4) over the synthetic reaction yielding glutamine (kEq = $2 \times 10^{-2}$). Smith considered that the reaction contributed principally to the synthetic requirements and less to the energy requirements of the cell. Although the role of glutamine in energy conservation or production is considered limited, the paucity of evidence demonstrating oxidative or substrate phosphorylation in the Mollicutes requires that the presence of this reaction be reemphasized.

### 2. The Production of ATP from Arginine

Arginine is also an energy source for nonfermentative mycoplasmas. The formation of ATP proceeds by the arginine dihydrolase pathway (Smith, 1955, 1957, 1960: Schimke and Barile, 1963; Barile *et al.*, 1966, Schimke *et al.*, 1966; Hahn and Kenny, 1974; Fenske and Kenny, 1976; Weickmann and Fahrney, 1977). The first step in the reaction sequence is mediated by arginine deiminase [Eq. (1) ].

$$\text{Arginine} + H_2O \rightarrow \text{citrulline} + NH_3 \tag{1}$$

This enzyme is found almost entirely in prokaryotes; most of the nonfermentative mycoplasma species contain the enzyme. The subsequent reaction sequence involves the phorphorolysis of citrulline to ornithine and carbamoyl phosphate. Carbamoyl phosphate with ADP and $Mg^{2+}$ is cleaved by carbamoyl phosphate synthetase to ATP, $NH_3$, and $CO_2$.

Schimke *et al.* (1966), working with cultures of *M. arthritidis* 07 grown to high densities, indicated that arginine was a major source of energy. Hahn and Kenny (1974) using other nonfermentative mycoplasmas found that arginine was not essential for the initiation of growth in all strains. In experiments using *M. hominis* ATCC 14027 and other mycoplasmas Fenske and Kenny (1976) suggested that the arginine dihydrolase pathway is not the major energy generating system in *M. hominis,* as arginine deiminase was first detectable late in log phase. These authors stated that it was "currently not clear how arginine-utilizing 'mycoplasmata' generate their energy."

### 3. Acetate Kinase Activity

Kahane *et al.* (1978) have demonstrated that ATP is generated in some mycoplasmas from acetyl-CoA, ADP and inorganic phosphate ($P_i$) by the action of phosphate acetyl transferase (PAT) and acetate kinase (AK) [Eq. (2) and (3)].

$$\text{Acetyl-CoA} + P_i \overset{\text{PAT}}{\rightleftharpoons} \text{acetyl phosphate} + \text{CoA} \tag{2}$$

$$\text{Acetyl phosphate} + \text{ADP} \overset{\text{AK}}{\rightleftharpoons} \text{acetate} + \text{ATP} \tag{3}$$

These workers found acetate kinase activity in a cytoplasmic fraction from both *A. laidlawii* (oral) and *M. hominis* (ATCC 15056). They also found phosphate acetyl transferase activity in *M. hominis*. This activity has also been reported in *A. laidlawii* and *M. gallinarum* J by Smith and Henrikson (1965).

Kahane *et al.* (1978) have suggested that ATP is generated in fermentative and non-fermentative mycoplasmas by the activity of acetate kinase. This new concept is a significant contribution to our consideration of how Mollicutes obtain energy.

## IV. COMMENT

The aerobic, nonthermophilic Mollicutes have traditionally and usefully been classified as either fermentative or nonfermentative microorganisms (Rodwell, 1969; Smith, 1971; Razin, 1973). The fermentative mycoplasmas are viewed as obtaining energy from carbohydrates and as lacking oxidative phosphorylation that involves quinones and cytochromes. Flavins may be associated with the final transfer of electrons to the electron sink, which is presumably oxygen, to produce water or peroxide. The presence of NADH–ferricyanide oxidoreductase activity can only suggest, by analogy to mitochondrial or bacterial systems, that there is energy conservation at a Site I-like locus. The presence of the Embden–Meyerhof pathway is likely.

Although the data are not extensive they suggest that the nonfermentative, aerobic, nonthermophilic mycoplasmas obtain energy via the oxidation of fatty acids, the tricarboxylic acid cycle, and the metabolism of amino acids as glutamine and arginine. Electrons derived from these, and probably other sources, are transferred to oxygen via a flavin-containing path similar to that described for fermentative mycoplasmas.

The presence of cytochromes in the nonfermentative *Mycoplasma arthritidis* strain 07 was not confirmed. Cytochromes have not been detected in other strains of the same species, nor in any other available nonthermophilic Mollicutes. Hence, for the time being, the concept of an energy-conserving cytochrome-dependent electron transport system in the nonthermophilic Mollicutes should be deemphasized.

*Thermoplasma acidophilum* most clearly possesses cytochromes.

*Ureaplasma urealyticum* has been reported as nonfermentative, lacking catalase, without either peroxide formation or the ability to reduce tetrazolium or methylene blue (Shepard *et al.*, 1974). No description of the energy-conserving mechanisms of *U. urealyticum*, the *Spiroplasma*, or *Anaeroplasma* have been reported.

In the 25 years since the Rodwells initiated the contemporary study of Mollicutes physiology our metabolic understanding of these ubiquitous organisms has only slowly grown. Lack of taxonomically comprehensive data has not yet permitted a broad understanding of the energy-yielding pathways of the Mollicutes. I hope that there will be continued interest in the metabolic study of these special microbes.

## ACKNOWLEDGMENTS

I wish to thank Dr. R. Holländer for permission to cite his unpublished data, and Ms. M. Warfield and H. Rehn for help in preparing the manuscript.

## REFERENCES

Asano, A., and Brodie, A. F. (1965). *Biochem. Biophys. Res. Commun.* **19,** 121–126.

Barile, M., Schimke, R. T., and Riggs, D. B. (1966). *J. Bacteriol.* **91,** 189–192.

Beinert, H. (1976). *Adv. Exp. Biol. Med.* **74,** 137–149.

Belly, R. T., Bohlool, B. B., and Brock, T. D. (1973). *Ann. N.Y. Acad. Sci.* **225,** 94–107.

Bevers, E. M., Snoek, G. T., Op den Kamp, J. A. F., and van Deenen, L. L. M. (1977). *Biochim. Biophys. Acta* **467,** 346–356.

Bieberstein, E. L., and Gills, M. (1961). *J. Bacteriol.* **81,** 380–384.

Bogin, E., Higashi, T., and Brodie, A. F. (1969). *Arch. Biochem. Biophys.* **129,** 211–220.

Brodie, A. F., and Gutnick, D. L. (1972). *In* "Electron and Coupled Energy Transfer in Biological Systems" (T. E. King and M. Klingenberg, eds.), Vol. 1, Part B, 599–681. Dekker, New York.

Castrejon-Diez, J., Fisher, T. N., and Fisher, E., Jr. (1963). *J. Bacteriol.* **86,** 627–636.

Cherry, J. D., and Taylor-Robinson, D. (1970). *Nature (London)* **228,** 1099–1100.

Cole, B. C., Ward, J. R., and Martin, C. H. (1968). *J. Bacteriol.* **95,** 2022–2030.

Dancey, G. F., and Shapiro, B. M. (1977). *Biochim. Biophys. Acta* **487,** 368–377.

de Kruyff, B., van Dijck, P. W. M., Goldbach, R. W., Demel, R. A., and van Deenen, L. L. M. (1973). *Biochim. Biophys. Acta* **330,** 269–282.

Dolin, M. I. (1961). *In* "The Bacteria" (I. C. Gunsalus and R. Y. Stanier, eds.), Vol. 2, pp. 425–460. Academic, New York.
Edward, D. G. ff. (1954). *J. Gen. Microbiol.* **10,** 27–64.
Fenske, J. D., and Kenny, G. E. (1976). *J. Bacteriol.* **126,** 501–510.
Freundt, E. A. (1958). "The Mycoplasmataceae." Munksgaard, Copenhagen.
Gale, P. H., Erickson, R. E., Page, A. C., Jr., and Folkers, K. (1964). *Arch. Biochem. Biophys.* **104,** 169–172.
Gerwitz, M., and VanDemark, P. J. (1966). *Bacteriol. Proc.* p. 77.
Green, F., III, and Hanson, R. P. (1973). *J. Bacteriol.* **116,** 1011–1018.
Gunsalus, I. C., and Umbreit, W. W. (1945). *J. Bacteriol.* **49,** 347–357.
Haddock, B. A., and Jones, C. W. (1977). *Bacteriol. Rev.* **41,** 47–99.
Hahn, R. G., and Kenny, G. E. (1974). *J. Bacteriol.* **117,** 611–618.
Hatefi, Y. (1976). *Adv. Exp. Biol. Med.* **74,** 150–160.
Holländer, R. (1978). *J. Gen. Microbiol.* **108,** 165–167.
Holländer, R., Wolf, G., and Mannheim, W. (1977). *Antonie van Leeuwenhoek* **43,** 177–185.
Holmes, B. E., and Pirie, A. (1932). *Br. J. Exp. Pathol.* **13,** 364–370.
Jinks, D. C., and Matz, L. L. (1976a). *Biochim. Biophys. Acta* **430,** 71–82.
Jinks, D. C., and Matz, L. L. (1976b). *Biochim. Biophys. Acta* **452,** 30–41.
Johnson, D. W., and Muscoplat, C. C. (1972). *Am. J. Vet. Res.* **33,** 2593–2595.
Kahane, I., Greenstein, S., and Razin, S. (1977). *J. Gen. Microbiol.* **101,** 173–176.
Kahane, I., Razin, S., and Muhlrad, A. (1978). *FEMS Microbiol. Lett.* **3,** 143–145.
Kandler, G., and Kandler, O. (1955). *Zentralbl. Bakteriol., Parasitenkd., Infektionskr. Hyg., Abt. 2* **108,** 383–397.
Langworthy, T. A., Smith, P. F., and Mayberry, W. R. (1972). *J. Bacteriol.* **112,** 1193–1200.
Larraga, V., and Razin, S. (1976). *J. Bacteriol.* **128,** 827–833.
Lecce, J. G., and Morton, H. E. (1953). *J. Bacteriol.* **67,** 62–68.
Low, I. E. (1971). *Infect. Immun.* **3,** 80–86.
Low, I. E., and Eaton, M. D. (1965). *J. Bacteriol.* **89,** 725–728.
Low, I. E., and Zimkus, S. M. (1973). *J. Bacteriol.* **116,** 346–354.
Low, I. E., Eaton, M. D., and Proctor, P. (1968). *J. Bacteriol.* **95,** 1425–1430.
Lynn, R. J. (1960). *Ann. N.Y. Acad. Sci.* **79,** 538–542.
Machinist, J. M., and Singer, T. P. (1965). *J. Biol. Chem.* **240,** 3182–3190.
Maniloff, J. (1972). *Pathog. Mycoplasmas, Ciba Found. Symp. 1972,* pp. 67–87.
Masover, G. K., Razin, S., and Hayflick, L. (1977). *J. Bacteriol.* **130,** 297–302.
Melnick, R. L., Monti, L. G., and Motzkin, S. M. (1976). *Biochem. Biophys. Res. Commun.* **69,** 68–73.
Morowitz, H. J., and Terry, T. M. (1969). *Biochim. Biophys. Acta* **183,** 276–294.
Mudd, J. B., Ittig, M., Roy, B., Latrille, J., and Bové, J. M. (1977). *J. Bacteriol.* **129,** 1250–1256.
Ne'eman, Z., and Razin, S. (1975). *Biochim. Biophys. Acta* **375,** 54–68.
Ne'eman, Z., Kahane, I., and Razin, S. (1971). *Biochim. Biophys. Acta* **249,** 169–176.
Ne'eman, Z., Kahane, I., Kovartovsky, J., and Razin, S. (1972). *Biochim. Biophys. Acta* **266,** 255–268.
Neimark, H. C., and Pickett, M. J. (1960). *Ann. N.Y. Acad. Sci.* **79,** 531–537.
Pinchot, G. B. (1953). *J. Biol. Chem.* **205,** 65–74.
Pirie, A. (1938). *Br. J. Exp. Pathol.* **19,** 9–17.
Pirie, A., and Holmes, B. E. (1933). *Br. J. Exp. Pathol.* **14,** 290.
Plackett, P., and Rodwell, A. W. (1970). *Biochim. Biophys. Acta* **210,** 230–240.
Pollack, J. D. (1975). *Int. J. Syst. Bacteriol.* **25,** 108–113.
Pollack, J. D. (1978). *Int. J. Syst. Bacteriol.* **28,** 425–426.
Pollack, J. D., Razin, S., Pollack, M. E., and Cleverdon, R. C. (1965a). *Life Sci.* **4,** 973–977.

Pollack, J. D., Razin, S., and Cleverdon, R. C. (1965b). *J. Bacteriol.* **90,** 617–622.
Pollack, J. D., Somerson, N. L., and Senterfit, L. B. (1970). *Infect. Immun.* **2,** 326–339.
Ragan, C. I. (1976). *Biochim. Biophys. Acta* **456,** 249–290.
Razin, S. (1973). *Adv. Microb. Physiol.* **10,** 1–80.
Razin, S., and Rottem, S. (1974). *In* "Methods in Enzymology" (S. Fleischer and L. Packer, eds.), Vol. 32, Part B, pp. 459–468. Academic Press, New York.
Ritchey, T. W., and Seeley, H. W., Jr., (1976). *J. Gen. Microbiol.* **93,** 195–203.
Rodwell, A. W. (1967). *Ann. N.Y. Acad. Sci.* **143,** 88–109.
Rodwell, A. W. (1969). *In* "The Mycoplasmatales and L-Phase of Bacteria" (L. Hayflick, ed.), pp. 413–449. Appleton, New York.
Rodwell, A. W., and Rodwell, E. S. (1954a). *Aust. J. Biol. Sci.* **7,** 18–30.
Rodwell, A. W., and Rodwell, E. S. (1954b). *Aust. J. Biol. Sci.* **7,** 31–36.
Rottem, S., and Razin, S. (1967). *J. Gen. Microbiol.* **48,** 53–63.
Ruwart, M. J., and Haug, A. (1975). *Biochemistry* **14,** 860–866.
Schimke, R. T., and Barile, M. (1963). *J. Bacteriol.* **86,** 195–206.
Schimke, R. T., Berlin, C. M., Sweeney, E. W., and Carroll, W. R. (1966). *J. Biol. Chem.* **241,** 2228–2236.
Shepard, M. C., Lunceford, C. D., Ford, D. K., Purcell, R. H., Taylor-Robinson, D., Razin, S., and Black, F. T. (1974). *Int. J. Syst. Bacteriol.* **24,** 160–171.
Singer, T. P. (1963). *In* "The Enzymes" (P. D. Boyer, H. A. Lardy, and K. Myrbäck, eds.), 2nd ed., Vol. 7, pp. 345–381. Academic Press, New York.
Singer, T. P., and Gutman, M. (1971). *Adv. Enzymol.* **34,** 79–153.
Smith, P. F. (1955). *J. Bacteriol.* **70,** 552–556.
Smith, P. F. (1957). *J. Bacteriol.* **74,** 801–806.
Smith, P. F. (1960). *Ann. N.Y. Acad. Sci.* **79,** 543–550.
Smith, P. F. (1971). "The Biology of Mycoplasmas." Academic Press, New York.
Smith, P. F., and Henrikson, C. V. (1965). *J. Bacteriol.* **89,** 146–153.
Smith, S. S. (1963). M.S. Thesis, Cornell University, Ithaca, New York.
Smith, S. L., VanDemark, P. J., and Fabricant, J. (1963). *J. Bacteriol.* **86,** 893–897.
Somerson, N. L., and Morton, H. E. (1953). *J. Bacteriol.* **65,** 245–251.
Somerson, N. L., Walls, B. E., and Chanock, R. M. (1965). *Science* **150,** 226–228.
Stopkie, R. J., and Weber, M. M. (1967). *Biochem. Biophys. Res. Commun.* **28,** 1034–1039.
Tarshis, M. A., Bekkouzjin, A. G., and Ladygina, V. G. (1976). *Arch. Microbiol.* **109,** 295–299.
Thomas, L., and Bitensky, M. W. (1966). *Nature (London)* **210,** 963–964.
Tourtellotte, M. E., and Jacobs, R. E. (1960). *Ann. N.Y. Acad. Sci.* **79,** 521–530.
VanDemark, P. J. (1967). *Ann. N.Y. Acad. Sci.* **143,** 77–84.
VanDemark, P. J. (1969). *In* "The Mycoplasmatales and the L-Phase of Bacteria" (L. Hayflick, ed.), pp. 491–501. Appleton, New York.
VanDemark, P. J., and Smith, P. F. (1964a). *J. Bacteriol.* **88,** 122–129.
VanDemark, P. J., and Smith, P. F. (1964b). *J. Bacteriol.* **88,** 1602–1607.
VanDemark, P. J., and Smith, P. F. (1965). *J. Bacteriol.* **89,** 373–377.
Vinther, O., and Freundt, E. A. (1977). *Acta Pathol. Microbiol. Scand., Sect. B* **85,** 184–188.
Warren, J. (1942). *J. Bacteriol.* **43,** 211–228.
Weibull, C., and Hammarberg, K. (1962). *J. Bacteriol.* **84,** 520–525.
Weickmann, J. L., and Fahrney, D. E. (1977). *J. Biol. Chem.* **252,** 2615–2620.
White, D. C., and Sinclair, P. R. (1971). *Adv. Microb. Physiol.* **5,** 173–211.
Whittenbury, R. (1960). *Nature (London)* **187,** 433–434.
Yarrison, G., Young, D. W., and Choules, G. L. (1972). *J. Bacteriol.* **110,** 494–503.

# 8 / ISOLATION AND CHARACTERIZATION OF MYCOPLASMA MEMBRANES

*Shmuel Razin*



## I. INTRODUCTION

The mycoplasmas, lacking a cell wall and intracytoplasmic membranes, have only one type of membrane, the plasma membrane. The ease with which this membrane can be isolated and the ability to introduce controlled alterations in its composition have made mycoplasma membranes most effective and popular tools in biomembrane research (Razin, 1975). The main aim of this chapter is to discuss and evaluate the different approaches used in the isolation and characterization of mycoplasma membranes. For technical details the reader is referred to Razin and Rottem (1974, 1976).



THE MYCOPLASMAS, VOL. 1

ISBN 0-12-078401-7

## II. MEMBRANE ISOLATION

### A. Osmotic Lysis

Because mycoplasmas have no protective cell walls, they are sensitive to osmotic shock, the simplest and gentlest way to isolate cell membranes. However, this technique does not always work because the organism's sensitivity to osmotic shock decreases most markedly with the aging of the culture (Razin, 1963, 1964) and the presence of even traces of divalent cations in the suspension medium may provide complete protection against lysis (Razin, 1964; Rodwell, 1965). Thus, osmotic lysis is most effective when the organisms are harvested at the logarithmic phase of growth, with the divalent cation concentration in the lysis medium kept to a minimum. Perfect control of the growth rates is absolutely essential and can be obtained only when the strain is well adapted to the growth medium. This is relatively easy to achieve with the fast-growing mycoplasmas, such as *Acholeplasma laidlawii,* or *Mycoplasma mycoides* subsp. *capri,* but may prove extremely difficult with the more exacting, slow-growing mycoplasmas, such as *Mycoplasma pneumoniae* (Pollack *et al.*, 1970).

Preloading the organisms with glycerol increases the sensitivity to osmotic lysis of such strains as *Mycoplasma gallisepticum,* which are relatively resistant to lysis by the regular procedure (Rottem *et al.*, 1968; Gabridge and Murphy, 1971). Glycerol, which penetrates rapidly through the cell membrane of mycoplasmas (McElhaney *et al.*, 1970), intensifies the osmotic shock by increasing the internal osmotic pressure of the cells.

Why is the susceptibility of mycoplasmas to osmotic lysis so dependent on the culture age? We do not have a direct, conclusive answer to this question. The available data indicate that marked changes in the osmotic fragility of the cells can be caused by variations in the cholesterol content (Razin, 1967; Rottem *et al.*, 1973b) or in the ratio of unsaturated to saturated fatty acids in membrane lipids (Razin *et al.*, 1966). Considerable variations in membrane lipid composition, in the lipid-to-protein ratio, and consequently in membrane fluidity are, in fact, known to occur with aging in mycoplasma cultures (Chapter 11, Section IV. C). It is not clear how such changes in membrane composition affect osmotic fragility. Studies by de Kruyff *et al.* (1973) and McElhaney *et al.* (1973) showed that *A. laidlawii* cells suspended in isotonic glycerol or erythritol swelled very rapidly and actually lysed when the temperature dropped to a point where membrane lipids underwent a phase transition from the liquid crystalline to the gel state. It was also found that at this temperature the cells were more susceptible to rupture by the mechanical stress of filtration through membrane filters. More recent work by van Zoelen *et al.* (1975) using *A. laidlawii* supports previous findings by showing that once the membrane lipid bilayer is completely transformed to the gel state, the membrane loses its elasticity so that the cells lyse rather than swell when placed in hypotonic solutions. These findings seem to suggest that the membrane becomes less flexible and more prone to brittle fracture when its lipid domain crystallizes. Since the temperature of this phase transition largely depends on the fatty acid composition of the membrane lipids and on their cholesterol content (Chapter 10), it is not difficult to find the correlation between the membrane lipid composition and osmotic fragility. If this is a valid explanation, then the osmotic fragility of mycoplasmas should increase markedly when the temperature of the medium falls below the phase-transition temperature of membrane lipids. This would fit in well with the extreme sensitivity to osmotic lysis seen with *A. laidlawii* cells whose membranes have been highly enriched with palmitate or stearate (Razin *et al.*, 1966), since at room temperature or even at 37°C all, or nearly all, of their membrane lipids are in the gel state (Steim *et al.*, 1969; McElhaney, 1974). This would also explain the much higher osmotic fragility of the cholesterol-poor *M. mycoides* subsp. *capri* at low temperatures, when compared to the cholesterol-rich native strain (Rottem *et al.*, 1973b). On the other hand, it has long been established that osmotic lysis of mycoplasmas is usually far more effective at 37°C than at 0°C (Razin, 1964). The slowing of the rate of water penetration into the cells due to increased viscosity of the lipid bilayer at low temperatures (e.g., Blok *et al.*, 1976) may be responsible for the decreased sensitivity of the organisms to osmotic lysis in the cold.

In spite of their structural similarity to bacterial protoplasts, mycoplasmas are osmotically less fragile (Razin, 1969). One explanation is that because of the high surface-to-volume ratio of the minute mycoplasma cells, rapid liberation of the internal solutes may occur upon transfer to a hypotonic solution, quickly lowering the internal osmotic pressure so as to cushion the osmotic shock. The marked leakiness of mycoplasmas in nonnutrient solutions is well known. Moreover, filamentous growth, common in mycoplasmas, is likely to augment these effects, since the surface-to-volume ratio is much higher than in spherical cells, and filaments are able to absorb large amounts of water and turn into large spherical bodies without any stretching of the membrane (Rodwell, 1965; Razin and Cosenza, 1966). Hence, if osmotic lysis is to be effective, the cells should be transferred as quickly as possible (e.g., by injection) from the hypertonic medium to deionized water. Special properties of the membrane itself also enhance resistance to osmotic lysis. The mycoplasma membrane was actually found to be far more resistant to fragmentation by sonication than the bacterial protoplast membrane (Razin, 1967). Whether the high cholesterol content is related to the marked tensile strength of the mycoplasma membrane is still a matter of conjecture.

### B. Digitonin-Induced Lysis

Mycoplasmas, in particular the sterol-requiring species, contain considerable quantities of cholesterol in their cell membranes (Chapters 9 and 10) and are therefore sensitive to lysis by digitonin (Smith and Rothblat, 1960; Razin and Argaman, 1963; Razin and Shafer, 1969). Digitonin forms a complex with cholesterol in the membrane, apparently causing considerable rearrangements in the lipid bilayer, leading to an increased permeability and cell lysis (Rottem and Razin, 1972). One of the advantages of digitonin-induced lysis is that it is less dependent on the age of the culture, making it a useful technique for the isolation of membranes from slow-growing mycoplasmas, where the difficulty of determining the appropriate harvesting time constitutes a handicap to osmotic lysis. Another advantage of digitonin-induced lysis is that it can take place in the presence of divalent cations (Rottem and Razin, 1972). Elimination of the need to reduce the ionic strength and divalent cation concentration of the lysis medium may be of importance in view of the finding that some loosely bound membrane proteins may become dissociated and lost under these conditions (Ne'eman *et al.*, 1971; Ne'eman and Razin, 1975). On the other hand, a possible drawback with the membranes obtained by the digitonin technique is that they seem to retain the cholesterol–digitonide complex, so that their organization and composition are altered (Rottem and Razin, 1972). Thus, the high hexose content of *M. mycoides* subsp. *capri* membranes isolated by digitonin may be attributed to the hexose moieties of the digitonin incorporated into the membrane (Archer, 1975). In conclusion, although osmotic lysis is the preferred technique for isolating mycoplasma membranes, lysis by digitonin is gaining popularity (Archer, 1975; Alexander and Kenny; 1977), particularly for such mycoplasmas as *Mycoplasma gallisepticum* and *Ureaplasma urealyticum,* which are relatively resistant to osmotic lysis (Levisohn and Razin, 1973; Masover *et al.*, 1976, 1977; Romano and La Licata, 1978).

### C. Other Techniques

The sensitivity of mycoplasmas to lysis by alkali (Razin and Argaman, 1963) has been applied to isolate membranes from *M. gallisepticum,* which is relatively resistant to osmotic shock (Goel, 1973) and from *Thermoplasma acidophilum* (Ruwart and Haug, 1975), which totally resists osmotic lysis (Belly and Brock, 1972). Lysis at alkaline pH may suffer, however, from the deficiency that some membrane proteins, or even the membranes themselves, may be solubilized at pH values higher than 10 (Ruwart and Haug, 1975; S. Razin, unpublished data). Alternate freezing and thawing (Pollack *et al.*, 1965a; Clyde and Kim, 1967; Williams and Taylor-Robinson, 1967; Hollingdale and Lemcke, 1969), gas cavitation (Hollingdale and Lemcke, 1969), or ultrasonic oscillations (Pollack *et al.*, 1965a; Argaman and Razin, 1969; Hollingdale and Lemcke, 1969; Kahane and Razin, 1969; Smith *et al.*, 1973; Romano and La Licata, 1978) have been utilized to lyse mycoplasmas. Of these, freezing and thawing frequently fails to lyse the bulk of the cells, particularly with mycoplasmas relatively resistant to osmotic lysis. The pliable nature of the enclosing membrane of the mycoplasmas may account for their resistance to the action of freezing and thawing (Hollingdale and Lemcke, 1969). Sonic or ultrasonic oscillators and mechanical presses rupture mycoplasma cells most effectively. Their use is not recommended, however, since the membranes invariably disintegrate into minute particles, and some of these cannot be sedimented even at very high gravitational forces (Pollack *et al.*, 1965a; Kahane and Razin, 1969). Membrane material obtained by these means also loses its characteristic appearance in the electron microscope, forfeiting one of the most useful criteria for checking the purity of a membrane preparation. Nevertheless, in cases where lysis by osmotic shock or by digitonin fails to give satisfactory results, one can resort to ultrasonic treatment. Thus, Romano and La Licata (1978) showed that ultrasonic treatment disrupted all the cells in a *U. urealyticum* cell suspension, whereas osmotic shock only lysed about 50% and digitonin about 70% of the cells. The advantages and disadvantages of the different methods for lysing mycoplasma cells and isolating their membranes are summarized in Table I.

## III. CRITERIA FOR MEMBRANE PURITY

### A. Purification of Membranes

It is the general consensus that our criteria for membrane purity are arbitrary or operational rather than absolute. Some proteins are bound so loosely to the membrane that even slight changes in the ionic strength of the solution used for cell lysis and membrane washing will detach them (Chapter 11, section II B). We should therefore allow for the possibility that some loosely bound proteins and ions may be lost from the mycoplasma membranes during the osmotic lysis of the cells in deionized water

TABLE I. **Methods of Isolating Mycoplasma Membranes**

| Method | Principle | Advantages | Disadvantages |
|---|---|---|---|
| Osmotic lysis | | | |
| a. Regular | Transfer of organisms from media of high to low tonicity (i.e., from 0.25 *M* NaCl to deionized water) | Gentlest technique available with no introduction of foreign substances | Some mycoplasmas relatively resistant; osmotic sensitivity decreases with age of culture; traces of divalent cations interfere with lysis; peripheral membrane proteins may be lost |
| b. Preloading with glycerol | Increase in intracellular osmotic pressure by incubating the organisms in 2 *M* glycerol before osmotic shock | Enables osmotic lysis of mycoplasmas relatively resistant to regular osmotic shock | |
| Digitonin-induced lysis | On exposure to digitonin, a cholesterol–digitonide complex is formed in the membrane, causing damage to membrane permeability | Effective also with mycoplasmas relatively resistant to osmotic shock; lysis not affected by age of culture or by divalent cations | Digitonin remains in isolated membranes; acholeplasmas with low cholesterol are less sensitive to lysis |
| Alkaline pH | Raising the pH of the cell suspension to about pH 9.3 causes cell lysis | A simple technique, effective also with *Thermoplasma* which resists osmotic or digitonin-induced lysis | Some proteins may be dissociated from the membrane at alkaline pH |
| Freezing and thawing | Rapid alternate transfers of cell suspension from −70°C (or −20°C) to 37°C | No introduction of foreign substances; can be performed in media of normal ionic strength | Frequently fails to lyse the bulk of the cells |
| Mechanical means | Disruption of cells by sonic or ultrasonic oscillations, mechanical presses, or gas cavitation | Very effective means for disrupting cells of all mycoplasmas, including the osmotically and digitonin-resistant organisms | Membrane fragments into minute particles, some of which cannot be sedimented by high-speed centrifugation |

and subsequent washings (Ne'eman *et al.*, 1971; Kahane *et al.*, 1973; Ne'eman and Razin, 1975).

As with other microbial membranes, ribosomes and DNA may stick to mycoplasma membranes. Alternate washings of the membranes in deionized water and 0.05 *M* NaCl in 0.01 *M* phosphate buffer, pH 7.5, proved to be quite an effective method for removing them (Rottem *et al.*, 1968). They may also be removed by treating the membranes with ribonuclease and deoxyribonuclease. The addition of nucleolytic enzymes to *A. laidlawii* membranes is usually unnecessary, because the potent endogenous nucleases of this organism, both cytoplasmic and membrane associated, adequately remove the ribosomes and DNA when cell lysis is carried out at 37°C (Pollack *et al.*, 1965b; Ne'eman and Razin, 1975). With several other mycoplasmas, however, treatment with deoxyribonuclease is an essential step in the isolation and purification of their membranes, particularly when large quantities of cells are lysed in relatively small volumes of media. This was shown to be the case with *Spiroplasma citri* lysed by osmotic shock (Razin *et al.*, 1973) and with *M. gallisepticum* lysed by digitonin (Levisohn and Razin, 1973). The poor endogenous deoxyribonuclease activity associated with membranes and cytoplasm of *S. citri* as compared with that of *A. laidlawii* (Kahane *et al.*, 1977) may explain the need for the addition of exogenous deoxyribonuclease during the process of membrane purification.

The demonstration of endogenous peptidases (Choules and Gray, 1971; Pecht *et al.*, 1972) and phospholipases (van Golde *et al.*, 1971; Rottem *et al.*, 1973c) in mycoplasma membranes indicates that some degree of hydrolysis of membrane components is inevitable during membrane isolation and purification and so some modifications in membrane composition can be expected. To minimize that kind of degradation the incubation period in the hypotonic solution used for osmotic lysis can be shortened from the usual 15 min at 37°C (Razin and Rottem, 1976) to less than 5 min, as long as the mycoplasmas tested are highly sensitive to osmotic shock. Furthermore, washing of membranes should be carried out at a low temperature and the membranes should be stored at −70°C rather than at −20°C, because even at −20°C slow activity of the endogenous phospholipase has been detected (Rottem *et al.*, 1973c).

## B. Electron Microscopy

Electron microscopy is one way in which the purity of a membrane preparation is determined. Electron micrographs of thin-sectioned material may show the presence of unlysed cells in the membrane preparation and cytoplasmic contaminants, such as ribosomes, adhering to the membranes (Fig. 1). Thin-sectioned mycoplasma membranes fixed with osmium tetroxide and stained with uranyl acetate show the characteristic triple-layered appearance, with a total thickness of about 100 Å. Some variation in membrane thickness between different mycoplasmas may be expected. Thus, Carstensen *et al.* (1971) claimed on the basis of sections that the membrane of *M. gallisepticum* is 110–120 Å thick, while that of *A. laidlawii* is only 70–80 Å thick. However, one must be aware of the pitfalls encountered while measuring membrane thickness in sections, as slight variations in the lengthy procedure of their preparation may affect membrane thickness. Another interesting point revealed by thin sectioning is the disappearance in isolated membrane preparations of the fuzzy layer frequently seen sticking to the outer surface of the cell membrane in intact cells (Morowitz and Terry, 1969). This finding further supports the argument that isolated membranes may lack certain components found associated with the membrane of the intact cell (Section IV, C).

### C. Density-Gradient Analysis

Another way to check the purity of a membrane preparation is to determine its density. Density-gradient centrifugation has also been useful in the separation of mycoplasma membranes from unlysed cells and from precipitates of the growth medium ingredients (Engelman *et al.*, 1967; Razin and Rottem, 1976; Romano and La Licata, 1978). Since essentially all mycoplasma lipids are localized in their cell membranes, growth in the presence of a radioactive long-chain fatty acid, such as oleic or palmitic acid, specifically labels the cell membrane, facilitating the tracing of membranes or membrane-containing material in gradients (Engelman *et al.*, 1967; Pollack *et al.*, 1970; Romano and La Licata, 1978).

The density of purified mycoplasma membranes was found to vary between 1.14 gm/cm$^3$ and 1.20 gm/cm$^3$, depending on the mycoplasma species and on growth conditions. Thus, *M. gallisepticum* membranes (density 1.199 gm/cm$^3$; Rottem *et al.*, 1968; Levisohn and Razin, 1973) are definitely heavier than those of *Mycoplasma hominis* and *A. laidlawii* (density varying from 1.14 to 1.18 gm/cm$^3$; Kahane and Razin, 1970; Rottem and Razin, 1972), reflecting the higher protein-to-lipid ratio in the former, probably attributable to the protein-rich bleb associated with the *M. gallisepticum* membrane. The density of membranes from several *Mycoplasma* and *Acholeplasma* species has been shown to increase with the age of the culture, reflecting a significant increase in the ratio of membrane protein-to-lipid occurring during aging (Chapter 11, section IV. C). Similar changes were observed when membranes were isolated from cultures of *A. laidlawii* grown at different pH values, a factor which

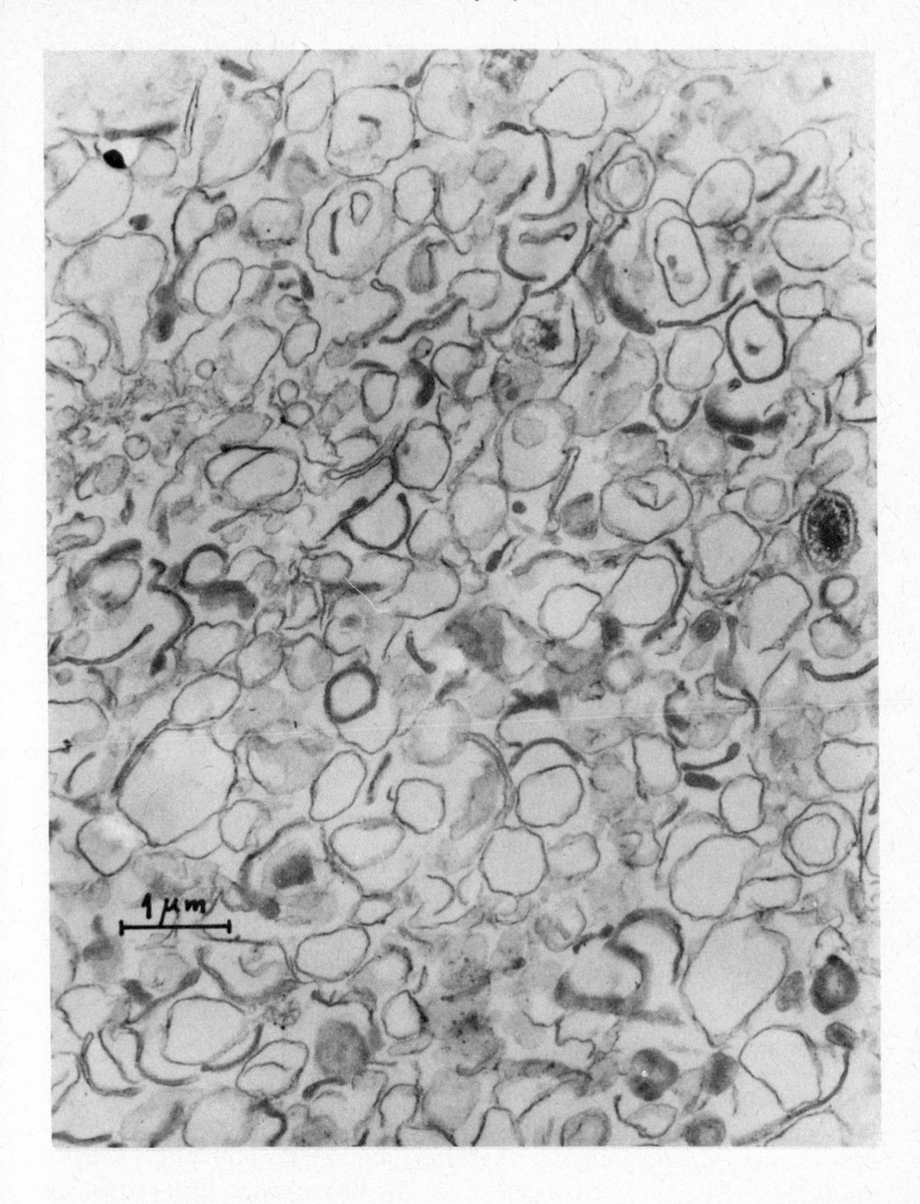


FIGURE 1. Thin sections of *A. laidlawii* membranes isolated by osmotic lysis of the cells. A few unlysed cells can also be seen.

affects the growth rate of the organisms (Kahane and Razin, 1970). It has been suggested that membrane density is also influenced by the fatty acid composition of membrane lipids. *Acholeplasma laidlawii* membrane enriched with oleic acid (Rottem *et al.*, 1970) or with short-chain saturated fatty acids (Huang *et al.*, 1974) exhibited a lower density than did membranes from the same organism enriched with long-chain saturated fatty acids. The lower density was attributed to the looser packing of the lipid bilayer enriched with the unsaturated or short-chain fatty acids (Rottem *et al.*, 1970; Huang *et al.*, 1974). To conclude, membrane density values cannot be regarded as constant for a given species. Nevertheless, one should not forget that the high density of a certain membrane preparation may simply reflect its contamination with nucleic acids.

### D. Enzymatic Activities as Markers

Checking the activity of cytoplasmic and membrane-associated enzymes can aid in determining the purity of mycoplasma membrane preparations. Although enzyme localization studies have not yet covered all mycoplasmas, they have invariably shown that the ATPase activity of the organisms is exclusively associated with the membrane (Chapter 11, Section III. B), whereas the hexokinase activity of fermentative mycoplasmas is restricted to the cytoplasm (Amar *et al.*, 1974). Hence, hexokinase activity in membrane preparations indicates their contamination with cytoplasmic material, while ATPase in the cytoplasmic fluid fraction indicates its contamination with membrane material. The degree of cytoplasmic contamination of ureaplasma membrane preparations can be readily estimated by determining the urease activity in the preparations. Urease is an extremely potent enzyme, which can be easily tested and is localized exclusively in the *U. urealyticum* cytoplasm (Vinther, 1976; Masover *et al.,* 1977; Romano and La Licata, 1978).

### E. Contamination with Noncellular Material

Contamination of mycoplasma membrane preparations with precipitated components of the growth medium may seriously hamper their chemical, enzymatic, and antigenic characterization. This problem is usually encountered in either of two cases: in the fast-growing acid-producing mycoplasmas (such as *M. mycoides* subsp. *capri*) when allowed to grow long enough to decrease the pH of the culture medium to less than pH 6.0, or with mycoplasmas (such as the ureaplasmas) which grow very poorly in the available culture media. In the first case, the low pH causes precipitation of proteins and lipoproteins from the serum component of the growth medium (Chelton *et al.*, 1968; Bradbury and Jordan, 1972; Yaguzhinskaya, 1976), which cosediment with the cells during harvesting and washing, and may consequently cosediment with (Yaguzhinskaya, 1976), or even adsorb to, the isolated membranes (Rottem *et al.*, 1973a). In the case of the ureaplasma, cell yield is so low that most of the pellet obtained by centrifugation of the culture consists of noncellular components which will contaminate the isolated membrane preparations. Thus, Masover *et al.* (1976) reported that the protein content of the "membrane" fraction isolated by digitonin treatment of *U. urealyticum* grown in unfiltered Hayflick's medium, was about 25 times higher than the amount of protein in the soluble cytoplasmic fraction. Reduction of the serum content and prefiltration of the growth medium reduced the "membrane"-to-cytoplasmic protein ratio to about 10 (Masover *et al.*, 1977), a value still much higher than the 0.5–1.0 ratios reported for similar fractions from classical mycoplasmas. Centrifugation on sucrose density gradients has also been suggested as a means for separating the ureaplasma cells from nonspecific precipitates (Rottem *et al.*, 1971). Membranes isolated from cells prepared in this way will presumably be cleaner than those described so far. Clearly, at least for ureaplasmas grown *in vitro*, cell-associated exogenous materials should be accounted for in biologic and biochemical tests. Some recently reported analyses of ureaplasma membrane preparations (Whitescarver *et al.*, 1975, 1976) are therefore questionable, particularly as none of the criteria for checking membrane purity (Razin, 1975; Razin and Rottem, 1976) have been fulfilled.

## IV. CHEMICAL COMPOSITION OF MEMBRANES

### A. Gross Chemical Composition

Mycoplasma membranes, like other biological membranes, consist mainly of proteins and lipids. The protein comprises roughly two-thirds of the mass of the membrane, the balance being mostly lipid (Table II). The detailed chemical composition and other properties of mycoplasma membrane proteins and lipids will be dealt with in Chapters 9 and 11. The discussion here will, therefore, be limited to some of the minor membrane components, such as inorganic ions, and to slime layers which are loosely attached to the outer membrane surface.

### B. Inorganic Ions

Inorganic ions, although often ignored, are important constituents of biological membranes. Our information concerning this area of myco-

TABLE II. **Gross Chemical Composition of Isolated Mycoplasma Membranes**[a]

| Organism | Protein | Lipid | | | Carbohydrate[d] | References |
|---|---|---|---|---|---|---|
| | | Total | Neutral[b] | Polar[c] | | |
| *Acholeplasma laidlawii* | 57 | 32 | 2.5 | 29.5 | 0.5 | Razin (1967) |
| *Mycoplasma bovigenitalium* | 59 | 37 | 11.1 | 25.9 | 2.2 | Razin (1967) |
| *Mycoplasma mycoides* subsp. *capri* | 50 | 40 | 14 | 26 | 2.0 | Razin (1967) |
| *Mycoplasma mycoides* subsp. *mycoides* | 51 | 39 | 15 | 24 | 3.2 | Razin *et al.* (1969) |
| *Mycoplasma hominis* | 57 | 41 | 24.6 | 16.4 | 0.8 | Rottem and Razin (1972) |
| *Mycoplasma fermentans* | 60 | 29 | ND[e] | ND | 9.0[f] | Gabridge and Murphy (1971) |
| *Mycoplasma gallisepticum* | 80 | 19 | 5.7 | 13.3 | 0.2 | Levisohn and Razin (1973) |
| *Spiroplasma citri* | 57 | 34 | 13.6 | 20.4 | 2.2 | Razin *et al.* (1973) |
| *Thermoplasma acidophilum* | 76 | 19 | 3.4 | 15.6 | 5.0 | Ruwart and Haug (1975); Langworthy *et al.* (1972) |

[a] The data (expressed as % dry weight) represent approximate values only, as the ratio of membrane protein to lipid usually varies on aging of cultures, and membrane lipid composition is subject to considerable changes depending on alterations in the lipid composition of the growth medium.

[b] Mostly cholesterol.

[c] Phospholipids and in some mycoplasmas also glycolipids and phosphoglycolipids.

[d] In the lipid-extracted membranes.

[e] ND, not determined.

[f] Not clear whether the determination was carried out before or after removal of lipids.

plasma membrane chemistry is rather scant and mainly comes from a study by Kahane *et al.* (1973). Considerable quantities of $Mg^{2+}$ (up to 2 $\mu$g/mg of membrane protein) and much smaller quantities of $Ca^{2+}$ were detected in membranes of several mycoplasmas. However, since supplementation of the medium with more $Ca^{2+}$ increased its quantity in the membranes, the marked preponderance of $Mg^{2+}$ over $Ca^{2+}$ would appear to reflect the low $Ca^{2+}$ content of the medium rather than a preferential binding of $Mg^{2+}$. The acidic phospholipids of the mycoplasma membrane apparently provide most of the binding sites for the cations. How important the divalent cations are for the structural integrity of mycoplasma membranes is not clear. Treatment with ethylenediaminetetraacetate (EDTA) removed over two-thirds of the $Mg^{2+}$ of the *A. laidlawii* membrane, without causing significant membrane fragmentation or disaggregation (Kahane *et al.*, 1973), although some membrane proteins were lost during this treatment (Ne'eman *et al.*, 1971; Ne'eman and Razin, 1975). It thus appears that divalent cations are not as essential for the integrity of the mycoplasma membrane as they are for that of membranes of the halophilic bacteria, or for the outer membrane of gram-negative bacteria (Brown, 1964; Asbell and Eagon, 1966). It should be stressed, however, that under the most extensive treatment with EDTA, some $Mg^{2+}$ still remained attached to the membrane, and this tightly associated cation entity may play an important role in membrane stabilization. Magnesium, or other divalent cations, is needed for reaggregation of solubilized mycoplasma membrane components to membranous structures (Razin, 1972, 1974). Another possible function of divalent cations is the induction of local and reversible changes in the fluidity of the cell membrane. The binding of the divalent cations to the polar acidic groups of membrane phospholipids results in closer packing of the molecules' headgroups, consequently reducing membrane fluidity by decreasing the freedom of motion of the hydrocarbon chains (Chapter 10).

### C. Slime Layers

Polymeric substances produced and excreted by the cells in the form of a capsule may be lost during membrane isolation. *Mycoplasma mycoides* subsp. *mycoides* was the first mycoplasma shown to be covered by a capsule made of galactan (Gourlay and Thrower, 1968), a polymer possessing toxic properties. The "fuzziness" on the surface of sectioned *A. laidlawii* cells is probably due to a hexosamine polymer consisting of *N*-acetylgalactosamine and *N*-acetylglucosamine (Gilliam and Morowitz, 1972). Of the *Acholeplasma* species, only *A. laidlawii* is capable of producing this polymer (Smith, 1977), but even in this species its quantity

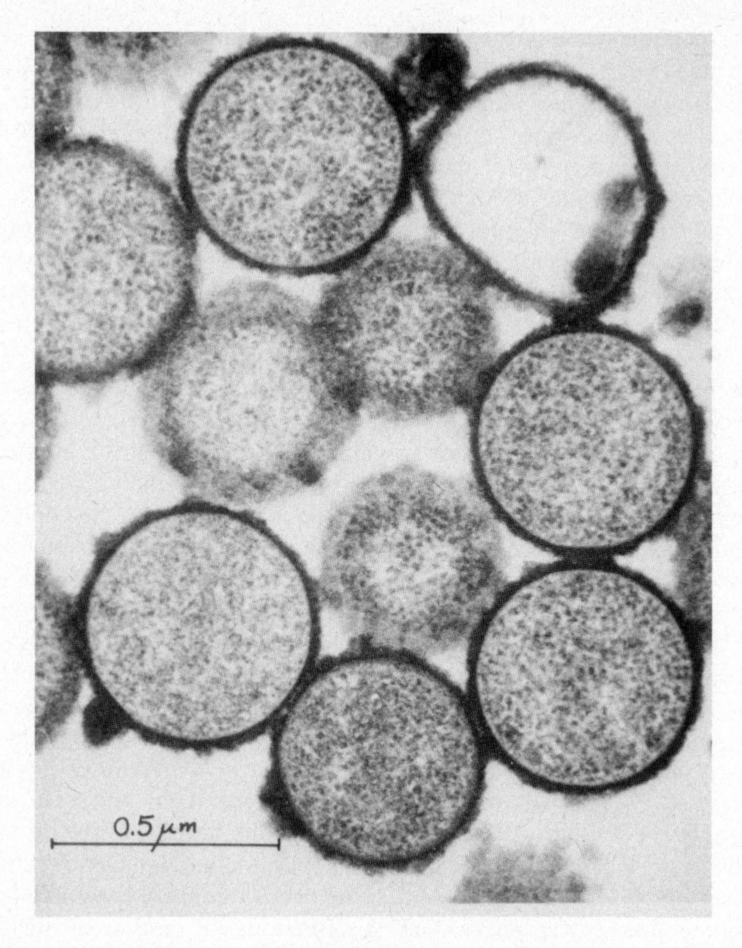


FIGURE 2. Extracellular layer in thin sections of *Ureaplasma urealyticum* cells stained by ruthenium red. (From Robertson and Smook, 1976.)

may vary significantly, depending on the strain and on growth conditions (I. Kahane, personal communication). The hexosamine polymer appears to be tightly bound to the membrane, as even after prolonged washing a considerable part of it remains associated with the membrane. The claim by Terry and Zupnik (1973) that the polymer can be released from the membrane by brief ultrasonic treatment could not be confirmed (J. M. Gilliam and I. Kahane, personal communication). Evidence for capsules in other mycoplasmas is supported solely by electron microscopy. Surface projections and "fuzzy" layers were demonstrated on the cell surface of *Mycoplasma pulmonis* (Hummler *et al.*, 1965), *M. gallisepticum*

(Chu and Horne, 1967), *Spiroplasma* (Horne, 1970), *U. urealyticum* (Black *et al.*, 1972), and *S. citri* (Cole *et al.*, 1973). Capsules staining with ruthenium red (Fig. 2) were shown in *Mycoplasma meleagridis* (Green and Hanson, 1973), *Mycoplasma dispar* (Howard and Gourlay, 1974), *Mycoplasma pneumoniae* (Wilson and Collier, 1976), and *U. urealyticum* (Robertson and Smook, 1976). In none of these cases has the chemical nature of the capsular material been determined. Ruthenium red reacts with a variety of polyanions and has been used to demonstrate the polysaccharide glycocalyxes of eukaryotic cells and capsules of prokaryotic organisms, including the galactan capsule of *M. mycoides* subsp. *mycoides* (Howard and Gourlay, 1974). Hence, it is probable that in mycoplasmas too the extracellular material consists of or contains carbohydrate. In fact, the *U. urealyticum* capsule reacted with concanavalin A iron dextran, indicating the presence of glycosyl residues in it (Robertson and Smook, 1976). The possibility that the extracellular material consists of growth medium components, adsorbed to the cell surface (Schiefer *et al.*, 1974; Robertson and Smook, 1976), cannot be ruled out but does seem unlikely as various *Mycoplasma* species (Howard and Gourlay, 1974) and *U. urealyticum* strains (J. Robertson, personal communication) differ in their ability to produce a ruthenium red-positive capsule when grown under the same conditions.

## REFERENCES

Alexander, A. G., and Kenny, G. E. (1977). *Infect. Immun.* **15,** 313–321.
Amar, A., Rottem, S., and Razin, S. (1974). *Biochim. Biophys. Acta* **352,** 228–244.
Archer, D. B. (1975). *J. Gen. Microbiol.* **88,** 329–338.
Argaman, M., and Razin, S. (1969). *J. Gen. Microbiol.* **55,** 45–58.
Asbell, M. A., and Eagon, R. G. (1966). *Biochem. Biophys. Res. Commun.* **22,** 664–671.
Belly, R. T., and Brock, T. D. (1972). *J. Gen. Microbiol.* **73,** 465–469.
Black, F. T., Andersen, A. B., and Freundt, E. A. (1972). *J. Bacteriol.* **111,** 254–259.
Blok, M. C., van Deenen, L. L. M., and de Gier, J. (1976). *Biochim. Biophys. Acta* **433,** 1–12.
Bradbury, J. M., and Jordan, F. T. W. (1972). *J. Hyg.* **70,** 267–278.
Brown, A. D. (1964). *Bacteriol. Rev.* **28,** 296–329.
Carstensen, E. L., Maniloff, J., and Einolf, C. W., Jr. (1971). *Biophys. J.* **11,** 572–581.
Chelton, E. T. J., Jones, A. S., and Walker, R. T. (1968). *J. Gen. Microbiol.* **50,** 305–312.
Choules, G. L., and Gray, W. R. (1971). *Biochem. Biophys. Res. Commun.* **45,** 849–855.
Chu, H. P., and Horne, R. W. (1967). *Ann. N.Y. Acad. Sci.* **143,** 190–203.
Clyde, W. A., Jr., and Kim, K. S. (1967). *Ann. N.Y. Acad. Sci.* **143,** 425–435.
Cole, R. M., Tully, J. G., and Popkin, T. J. (1973). *Ann. N.Y. Acad. Sci.* **255,** 471–493.
de Kruyff, B., de Greef, W. J., van Eyk, R. V. W., Demel, R. A., and van Deenen, L. L. M. (1973). *Biochim. Biophys. Acta* **298,** 479–499.
Engelman, D. M., Terry, T. M., and Morowitz, H. J. (1967). *Biochim. Biophys. Acta* **135,** 381–390.

Gabridge, M. G., and Murphy, W. H. (1971). *Infect. Immun.* **4,** 678–682.
Gilliam. J. M., and Morowitz, H. J. (1972). *Biochim. Biophys. Acta* **274,** 353–359.
Goel, M. C. (1973). *J. Bacteriol.* **116,** 994–1000.
Gourlay, R. N., and Thrower, K. J. (1968). *J. Gen. Microbiol.* **54,** 155–159.
Green, F., III, and Hanson, R. P. (1973). *J. Bacteriol.* **116,** 1011–1018.
Hollingdale, M. R., and Lemcke, R. M. (1969). *J. Hyg.* **67,** 585–602.
Horne, R. W. (1970). *Micron* **2,** 19–38.
Howard, C. J., and Gourlay, R. N. (1974). *J. Gen. Microbiol.* **83,** 393–398.
Huang, L., Jaquet, D. D., and Haug, A. (1974). *Can. J. Biochem.* **52,** 483–490.
Hummler, K., Tomassini, N., and Hayflick, L. (1965). *J. Bacteriol.* **90,** 517–523.
Kahane, I., and Razin, S. (1969). *J. Bacteriol.* **100,** 187–194.
Kahane, I., and Razin, S. (1970). *FEBS Lett.* **10,** 261–264.
Kahane, I., Ne'eman, Z., and Razin, S. (1973). *J. Bacteriol.* **113,** 666–671.
Kahane, I., Greenstein, S., and Razin, S. (1977). *J. Gen. Microbiol.* **101,** 173–176.
Langworthy, T. A., Smith, P. F., and Mayberry, W. R. (1972). *J. Bacteriol.* **112,** 1193–1200.
Levisohn, S., and Razin, S. (1973). *J. Hyg.* **71,** 725–737.
McElhaney, R. N. (1974). *J. Mol. Biol.* **84,** 145–157.
McElhaney, R. N., de Gier, J., and van Deenen, L. L. M. (1970). *Biochim. Biophys. Acta* **219,** 245–247.
McElhaney, R. N., de Gier, J., and van der Neut-Kok, E. C. M. (1973). *Biochim. Biophys. Acta* **298,** 500–512.
Masover, G. K., Sawyer, J. E., and Hayflick, L. (1976). *J. Bacteriol.* **125,** 581–587.
Masover, G. K., Razin, S., and Hayflick, L. (1977). *J. Bacteriol.* **130,** 297–302.
Morowitz, H. J., and Terry, T. M. (1969). *Biochim. Biophys. Acta* **183,** 276–294.
Ne'eman, Z., and Razin, S. (1975). *Biochim. Biophys. Acta* **375,** 54–68.
Ne'eman, Z., Kahane, I., and Razin, S. (1971). *Biochim. Biophys. Acta* **249,** 169–176.
Pecht, M., Giberman, E., Keysary, A., Yariv, J., and Katchalski, E. (1972). *Biochim. Biophys. Acta* **290,** 267–273.
Pollack, J. D., Razin, S., Pollack, M. E., and Cleverdon, R. C. (1965a). *Life Sci.* **4,** 973–977.
Pollack, J. D., Razin, S., and Cleverdon, R. C. (1965b). *J. Bacteriol.* **90,** 617–622.
Pollack, J. D., Somerson, N. L., and Senterfit, L. B. (1970). *Infect. Immun.* **2,** 326–339.
Razin, S. (1963). *J. Gen. Microbiol.* **33,** 471–475.
Razin, S. (1964). *J. Gen. Microbiol.* **36,** 451–459.
Razin, S. (1967). *Ann. N.Y. Acad. Sci.* **143,** 115–129.
Razin, S. (1969). *In* "The Mycoplasmatales and the L-Phase of Bacteria" (L. Hayflick, ed.), pp. 317–348. Appleton, New York.
Razin, S. (1972). *Biochim. Biophys. Acta* **265,** 241–296.
Razin, S. (1974). *J. Supramol. Struct.* **2,** 670–681.
Razin, S. (1975). *Prog. Surf. Membr. Sci.* **9,** 257–312.
Razin, S., and Argaman, M. (1963). *J. Gen. Microbiol.* **30,** 155–172.
Razin, S., and Cosenza, B. J. (1966). *J. Bacteriol.* **91,** 858–869.
Razin, S., and Rottem, S. (1974). *In* "Methods in Enzymology" (S. Fleischer and L. Packer, eds.), Vol. 32, Part B, pp. 459–468. Academic Press, New York.
Razin, S., and Rottem, S. (1976). *In* "Biochemical Analysis of Membranes" (A. H. Maddy, ed.), pp. 3–26. Chapman & Hall, London.
Razin, S., and Shafer, Z. (1969). *J. Gen. Microbiol.* **58,** 327–339.
Razin, S., Tourtellotte, M. E. McElhaney, R. N., and Pollack, J. D. (1966). *J. Bacteriol.* **91,** 609–616.
Razin, S., Ne'eman, Z., and Ohad, I. (1969). *Biochim. Biophys. Acta* **193,** 277–293.
Razin, S., Hasin, M., Ne'eman, Z., and Rottem, S. (1973). *J. Bacteriol.* **116,** 1421–1435.

Robertson, J., and Smook, E. (1976). *J. Bacteriol.* **128,** 658–660.
Rodwell, A. W. (1965). *J. Gen. Microbiol.* **40,** 227–234.
Romano, N., and La Licata, R. (1978). *J. Bacteriol.* (in press).
Rottem, S., and Razin, S. (1972). *J. Bacteriol.* **110,** 699–705.
Rottem, S., Stein, O., and Razin, S. (1968). *Arch. Biochem. Biophys.* **125,** 46–56.
Rottem, S., Hubbell, W. L., Hayflick, L., and McConnell, H. M. (1970). *Biochim. Biophys. Acta* **219,** 104–113.
Rottem, S., Pfendt, E. A., and Hayflick, L. (1971). *J. Bacteriol.* **105,** 323–330.
Rottem, S., Hasin, M., and Razin, S. (1973a). *Biochim. Biophys. Acta* **298,** 876–886.
Rottem, S., Yashouv, J., Ne'eman, Z., and Razin, S. (1973b). *Biochim. Biophys. Acta* **323,** 495–508.
Rottem, S., Hasin, M., and Razin, S. (1973c). *Biochim. Biophys. Acta* **323,** 520–531.
Ruwart, M. J., and Haug, A. (1975). *Biochemistry* **14,** 860–866.
Schiefer, H.-G., Gerhardt, V., Brunner, H., and Krüpe, M. (1974). *J. Bacteriol.* **120,** 81–88.
Smith, P. F. (1977). *J. Bacteriol.* **130,** 393–398.
Smith, P. F., and Rothblat, G. H. (1960). *J. Bacteriol.* **80,** 842–850.
Smith, P. F., Langworthy, T. A., Mayberry, W. R., and Hougland, A. E. (1973). *J. Bacteriol.* **116,** 1019–1028.
Steim, J. M., Tourtellotte, M. E., Reinart, J. C., McElhaney, R. N., and Rader, R. L. (1969). *Proc. Natl. Acad. Sci. U.S.A.* **63,** 104–109.
Terry, T. M., and Zupnik, J. S. (1973). *Biochim. Biophys. Acta* **291,** 144–148.
van Golde, L. M. G., McElhaney, R. N., and van Deenen, L. L. M. (1971). *Biochim. Biophys. Acta* **231,** 245–249.
van Zoelen, E. E. J., van der Neut-Kok, E. C. M., de Gier, J., and van Deenen, L. L. M. (1975). *Biochim. Biophys. Acta* **394,** 463–469.
Vinther, O. (1976). *Acta Pathol. Microbiol. Scand. Sec. B* **84,** 217–224.
Whitescarver, J., Castillo, F., and Furness, G. (1975). *Proc. Soc. Exp. Biol. Med.* **150,** 20–22.
Whitescarver, J., Trocola, M., Campana, T., Marks, R., and Furness, G. (1976). *Proc. Soc. Exp. Biol. Med.* **151,** 68–71.
Williams, M. H., and Taylor-Robinson, D. (1967). *Nature (London)* **215,** 973–974.
Wilson, M. H., and Collier, A. M. (1976). *J. Bacteriol.* **125,** 332–339.
Yaguzhinskaya, O. E. (1976). *J. Hyg.* **77,** 189–198.

# 9 / THE COMPOSITION OF MEMBRANE LIPIDS AND LIPOPOLYSACCHARIDES

*Paul F. Smith*



The mycoplasmal cell is characterized as being enclosed by a single unit membrane. This membrane is responsible not only for maintenance of cellular integrity but also for carrying out many of the diverse functions required of a growing, reproducing autonomous cell, i.e., transport, energy flow, cellular division, and specific cellular identity. Mycoplasmal membranes are considered to be composed of proteins and lipids, although recently other components have been recognized, e.g., lipopolysaccharides and polysaccharides. This exposition considers only lipids and lipopolysaccharides.

## I. LIPIDS

### A. General Composition

Essentially all of the lipids are found associated with the cytoplasmic membrane. Total lipids of midlogarithmic cells comprise 3–20% of the dry



THE MYCOPLASMAS, VOL. 1

ISBN 0-12-078401-7

weight of whole cells and 25–35% of the dry weight of membranes (Smith, 1968a). Significant quantitative differences are intergeneric, although some variation is observed with age of cultures and exogenous supply of lipids. *Thermoplasma* contains the least, 3% (Langworthy *et al.*, 1972), while *Acholeplasma* contains amounts varying from 8 to 14% (Mayberry *et al.*, 1974); *Anaeroplasma,* 15–16% (Langworthy *et al.*, 1975); and *Mycoplasma* and *Spiroplasma,* 14–23% (Mudd *et al.*, 1977; K. R. Patel, unpublished) of total cellular dry weight. *Spiroplasma* trends toward the highest value but this may reflect lysis of these fragile cells during processing resulting in enrichment of the pellet with membranes. Valid quantitative data are not available for *Ureaplasma* (Romano *et al.*, 1972) because pelleted organisms are contaminated with particulate constituents from the culture medium. In general the contribution of lipid to membranes of all genera is similar.

The distribution of total lipids among the three major classes, neutral lipids, glycolipids, and polar lipids, also varies among different genera, although invariably the percentage by weight of polar lipids equals the combined percentages of neutral lipids and glycolipids. Table I lists values of some selected representative species. These values are representative for cells harvested at the late exponential phase of growth. Some quantitative variation occurs in the growth cycle, particularly at the later stages. Neutral and polar lipids increase relative to the dry weight of cells (Lynn and Smith, 1960; Smith and Koostra, 1967) during the early exponential phase, probably reflecting increased membrane synthesis. Beyond the exponential phase the ratios of polar lipids and neutral lipids to membrane protein decrease (Rottem and Greenberg, 1975; Razin, 1974). This latter phenomenon no doubt is an expression of either a shutdown of lipid synthesis or an increase in membrane protein synthesis, since the ratios of the various classes of lipids to one another remain constant throughout the growth cycle. The glycolipid:dry weight ratio remains constant (Smith and Mayberry, 1968; Shaw *et al.*, 1968). These findings refer to various species of *Mycoplasma* and *Acholeplasma*. No changes in the ratios of neutral or phospholipids to membrane protein have been reported for *Spiroplasma citri* (Mudd *et al.*, 1977), although these data can also be interpreted as a reduction of lipid as the cells age.

The neutral lipid content of both mycoplasmas and acholeplasmas can be modulated up and down by increases or decreases of exogenously supplied cholesterol (Smith and Rothblat, 1960; Rottem and Razin, 1973).

### B. Chemical Structures of Lipids

The mycoplasmas have proven to be a veritable gold mine of new and unusual lipids. Table II presents a compilation of our current knowledge

TABLE I. **Distribution of Total Lipids among Various Classes for Selected Species**

| Organism | Neutral lipids (%) | Glycolipids (%) | Polar lipids (%) | Reference |
|---|---|---|---|---|
| *Acholeplasma modicum* (PG49) | 4.4 | 45.7 | 48.4 | Mayberry *et al*. (1974). |
| *Anaeroplasma bactoclasticum* (7LA) | 38.8 | 22.9 | 38.9 | Langworthy *et al*. (1975) |
| *Mycoplasma arthritidis* (07) | 56.2 | 0 | 43.8 | Smith and Koostra (1967) |
| *Mycoplasma gallinarum* (J) | 29.4 | 21.4 | 48.8 | Smith and Koostra (1967) |
| *Mycoplasma neurolyticum* (PG39) | 18.7 | 37.5 | 43.8 | Smith (1972a) |
| *Spiroplasma citri* (Morocco) | 44.8 | 17.8 | 37.7 | K.R. Patel (unpublished) |
| *Thermoplasma acidophilum* (122–1B2) | 17.5 | 25.1 | 56.6 | Langworthy *et al*. (1972) |

TABLE II. **Lipids of Various Species of Mycoplasmas**[a]

| Organism | Neutral lipids | Glycolipids | Polar lipids | References |
|---|---|---|---|---|
| *Acholeplasma axanthum* (S743) | Long-chain bases of dihydrosphingosine type; β-hydroxy acyl ceramides; acylated ceramides; free fatty acids; carotenoids | Glc*p* (1 $\xrightarrow{\beta}$ 4)-2-*O*-acyl; Glc*p* (1 $\xrightarrow{\beta}$ 3) cholesterol; monoglucosyl cholesterol; Gal*f*(1 $\xrightarrow{\beta}$ 1)-2,3-diglyceride | PG; acyl PG; DPG; lyso PG; ceramide phosphoryl glycerol; acyl ceramide phosphoryl glycerol | Plackett *et al*. (1970); Mayberry *et al*. (1973); Sugiyama *et al*. (1974); Mayberry (1977) |
| *Acholeplasma granularum* (BTS39) | Carotenoids; glycerides | Glc*p*(1 $\xrightarrow{\alpha}$ 2)-Glc*p*(1 $\xrightarrow{\alpha}$ 1)-2,3-diglyceride; Glc*p*(1 $\xrightarrow{\alpha}$ 1)-2,3-diglyceride | PG; DPG; acyl DPG; PA; phosphoglycolipid; aminoacyl PG | P.F. Smith (unpublished) |
| *Acholeplasma laidlawii* (A) | Carotenoids; carotenyl esters; glycerides | MGDG; DGDG; carotenyl-α-D-glucoside | PG; DPG: GPDGDG; GPMGDG | Rothblat and Smith (1961); Smith (1963); Wieslander and Rilfors (1977) |
| *Acholeplasma laidlawii* (B) | Carotenoids; carotenyl esters; glycerides | Carotenyl-β-D-glucoside; Glc*p*(1 $\xrightarrow{\alpha}$ 2)-Glc*p*(1 $\xrightarrow{\alpha}$ 1)-2,3-diglyceride; Glc*p*(1 $\xrightarrow{\alpha}$ 1)-2,3-diglyceride | PG; DPG; D- and L-alanyl PG; glycerophosphoryl Gle*p*(1 $\xrightarrow{\alpha}$ 2)-Glc*p* (1 $\xrightarrow{\alpha}$ 1)-2,3-diglyceride; Phosphatidyl DGDG | Smith and Henrikson (1965a); Smith (1963); Shaw *et al*. (1968, 1972); Koostra and Smith (1969); Smith *et al*., 1965 |
| *Acholeplasma modicum* (PG49) | Carotenoids; glycerides | Glc*p*(1 $\xrightarrow{\alpha}$ 2)-Glcp(1 $\xrightarrow{\alpha}$ 1)-2-3-diglyceride; Glc*p* (1 $\xrightarrow{\alpha}$ 1)-2,3-diglyceride; Gal*p*(1 $\xrightarrow{\alpha}$ 2)-Gal*p*(1 $\xrightarrow{\alpha}$ 3)-D-glycero D-mannohept*p* (1 $\xrightarrow{\beta}$ 3)-Glc*p*(1 $\xrightarrow{\alpha}$ 2)-Glc*p*(1 $\xrightarrow{\alpha}$ 1)-2,3-diglyceride | PG; aminoacyl PG; DPG; acyl DPG; lyso DPG | Mayberry *et al*. (1974, 1976) |

| | | | | |
|---|---|---|---|---|
| *Anaeroplasma abactoclasticum* (6-1) | Cholesterol; unidentified | Unidentified | PG; aminoacyl PG; DPG | Robinson *et al*. (1975); Langworthy *et al*. (1975) |
| *Anaeroplasma bactoclasticum* (7LA) | Unidentified | Unidentified | PG; aminoacyl PG; DPG | Robinson *et al*. (1975); Langworthy *et al*. (1975) |
| *Mycoplasma arthritidis* (07) | Cholesterol; cholesteryl esters | None | PG; aminoacyl PG; DPG | Rothblat and Smith (1961); Smith and Koostra (1967) |
| *Mycoplasma capricolum* (14) | Cholesterol; cholesteryl esters; diglycerides; triglycerides | None | PG; aminoacyl PG; DPG; acyl DPG | P. F. Smith (unpublished) |
| *Mycoplasma gallinarum* (J) | Cholesterol; cholesteryl esters | Cholesteryl-β-D-glucoside; 3,4,6-triacyl β-D-glucose | PGP; lyso PGP; DPG | Rothblat and Smith (1961); Smith and Koostra (1967); Smith and Mayberry (1968) |
| *Mycoplasma gallisepticum* (S-6) | Cholesterol; cholesteryl esters; glycerides | None | PG; DPG | P.F. Smith (unpublished) |
| *Mycoplasma hominis* (ATCC15056) | Cholesterol; cholesteryl esters; glycerides; free fatty acids | None | PG; PA; lyso PG | Rottem and Razin (1973) |
| *Mycoplasma hyorhinis* (BTS7) | Cholesterol; cholesteryl esters; glycerides | None | PG; DPG | P.F. Smith (unpublished) |
| *Mycoplasma mycoides* (V5) | Cholesterol; glycerides | Gal*f*(1 $\xrightarrow{\beta}$ 1)-2,3-diglyceride; unidentified | PG; DPG | Plackett (1967a) |

(*continued*)

TABLE II (*Continued*)

| Organism | Neutral lipids | Glycolipids | Polar lipids | References |
|---|---|---|---|---|
| *Mycoplasma neurolyticum* (PG28 & PG39) | Cholesterol; diglycerides; triglycerides | Glc*p*(1 $\xrightarrow{\beta}$ 6)Glc*p*(1 $\xrightarrow{\beta}$ 1)-2,3-diglyceride; Glc*p*(1 $\xrightarrow{\beta}$ 1)-2,3-diglyceride TGDG; acyl DGDG | PG; aminoacyl PG; PA; DPG; acyl DPG | Smith (1972a); P. F. Smith (unpublished); Sugiyama *et al*. (1974) |
| *Mycoplasma pneumoniae* (FH) | Cholesterol | Diagalactosyl diglyceride; trigalactosyl diglyceride; glucosyl galactosyl diglyceride | PG; DPG | Plackett *et al*. (1969) |
| *Spiroplasma citri* (Morocco) | Cholesterol; cholesteryl esters; triglycerides; diglycerides; free fatty acids | Cholesteryl glucosides | PG; DPG; lyso DPG | Freeman *et al*. (1976); K.R. Patel, and P.F. Smith (unpublished) |
| *Thermoplasma acidophilum* (122-1B2) | Hydrocarbons; diglycerol tetraether; naphthoquinones | Glycosyl diglycerol tetraethers | Glycerophosphoryl glycosyl diglycerol tetraethers | Langworthy *et al*. (1972); Langworthy (1977) |
| *Ureaplasma* sp. (P108) | Cholesterol; cholesteryl esters; free fatty acids; triglycerides; diglycerides | Glucosyl diglycerides | PA; PG; DPG; PE; diaminohydroxy compound with adjacent fatty acid ester and *N*-acyl groups | Rottem *et al*. (1971); Romano *et al*. (1972) |

[a] Abbreviations: PA, phosphatidic acid; PG, phosphatidyl glycerol; DPG, diphosphatidyl glycerol; MGDG, monoglucosyl diglyceride; DGDG, diglucosyl diglyceride; TGDG, triglucosyl diglyceride; GPDGDG, glycerophosphoryl diglucosyl diglyceride; GPMGDG, glycerophosphoryl monoglucosyl diglyceride; PE, phosphatidyl ethanolamine; Glc*p*, glucopyranose; Gal*p*, galactopyranose; Gal*f*, galactofuranose.

of the lipids found in various species. For some the exact chemical structures have been resolved, while for others this information is lacking (see Fig. 1).

### 1. Neutral Lipids

Lipid structures from enough different organisms have been deciphered to warrant some generalizations, albeit the risk of being proved incorrect is ever present. Acholeplasmas are capable of *de novo* synthesis of carotenoids based upon incorporation of $^{14}$C-mevalonic acid and $^{14}$C-acetate (Smith, 1963). With one exception the presence of carotenoids is easily visualized by the yellow coloration of cell pellets and the absorbancy at 468 $\mu$m. The one exception is *Acholeplasma axanthum,* so named because it appears unpigmented (Tully and Razin, 1969). Coloration depends upon the degree of unsaturation and the number of conjugated double bonds in the polyterpene chain. Significant coloration is manifest only when seven conjugated double bonds occur, e.g., $\zeta$-carotene, while its two immediate precursors, phytoene and phytofluene, containing three and four conjugated double bonds, are essentially colorless (Smith, 1971a). Thus *A. axanthum* contains carotenoids based upon $^{14}$C incorporation and spectral properties in the ultraviolet (uv) range. This pitfall of concluding the absence of carotenoids based upon light absorption in the visible range also exists with the attempts to inhibit carotenoid biosynthesis. In many cases only the sequential steps of unsaturation, not the biosynthesis of the polyterpene chain, are inhibited. There is no reason to believe that some, if not all, of the unknown functions of carotenoids in acholeplasmal membranes could not be carried out by the less unsaturated polyterpenes. Nevertheless, *A. axanthum* does not contain the quantity of polyterpenes found in other acholeplasmas (Plackett *et al.*, 1970). Apparently it compensates for this paucity by the biosynthesis of free long-chain bases of the dihydrosphingosine type which, by ionic bonding with the free fatty acids found in its membrane, could functionally replace the polyterpenes. The proof for the polyterpenoid structure of the pigments of acholeplasmas has been reviewed previously (Smith, 1971a). Their exact structural characterization, even the carbon chain length, has yet to be accomplished.

Sterol is found in all species of *Mycoplasma, Spiroplasma,* and *Ureaplasma,* all of which have a growth requirement for sterol. Species of *Acholeplasma* (Smith and Henrikson, 1966) and *Thermoplasma* (Langworthy *et al.*, 1971) incorporate sterol into their membranes when this lipid is supplied in the culture medium. The nature of the sterol found in the membrane is identical to that supplied exogenously; i.e., none of

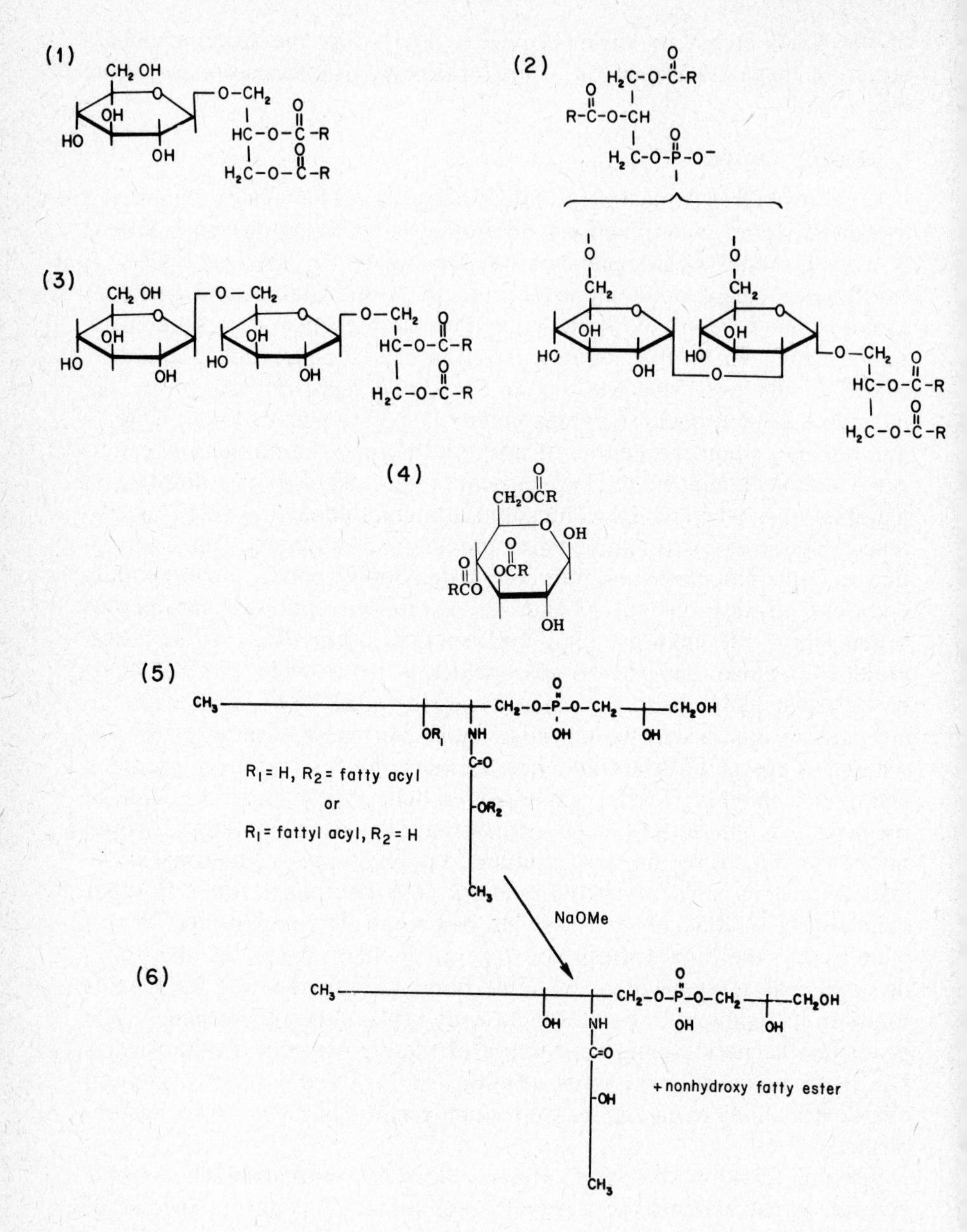


FIGURE 1. (1) Monoglycosyl diglyceride; (2) diglucosyl diglyceride; (3) glycerophosphoryl diglucosyl diglyceride; (4) triacylglucose; (5) *O*-acyl ceramide phosphoryglycerol; (6) ceramide phosphoryl glycerol.

the Mollicutes are capable of any profound biochemical alteration of the sterol molecule (Rothblat and Smith, 1961; Smith, 1962). Although any steroid containing an aliphatic side chain can be incorporated, only planar 3-hydroxy sterols permit growth and, hence, proper fit and function in the membrane (Smith, 1964). Sterols which have been demonstrated to be functionally proper are cholesterol, cholestanol, ergosterol and $\beta$-sitosterol. These sterols can be substituted with appropriate planar 3,3′-dihydroxy polyterpenes, e.g., sarcinaxanthin, lutein, and the carotenol from *Acholeplasma laidlawii*, all of which enter the membrane and remain unaltered (Henrikson and Smith, 1966a). Uptake of sterol or polyterpenol is a first-order physical adsorption process mediated through a micelle composed of sterol or polyterpenol, the protein moiety of an appropriate lipoprotein, and an amphipath, such as a salt of a long-chain unsaturated fatty acid or a phospholipid (Smith and Rothblat, 1960). The energy of activation for uptake has been calculated to be 6 kcal/mol (Gershfeld *et al.*, 1974). Sterol is presumed to be bound by hydrophobic interaction with the apolar regions of phospholipids since only lipid depletion of the membranes interferes with sterol uptake (Smith and Boughton, 1960; Razin *et al.*, 1974). This subject has been reviewed in detail elsewhere (Smith, 1971a). In some species cholesteryl esters are synthesized; in others some uptake of steryl esters occurs, while in yet others no cholesteryl esters exist.

Triglycerides and 1,2- and 1,3-diglycerides are found in small amounts in practically all species. Since they are labeled from $^{14}C$-fatty acids supplied in the culture medium, these glycerides are considered integral components of the membrane lipids. Free fatty acids are found frequently but are considered to be the result of lipolysis during lipid extraction. Exceptions to this generalization are the relatively large amounts of free fatty acids found in *Ureaplasma* (Romano *et al.*, 1972) and *A. axanthum* (Plackett *et al.*, 1970), even when extensive precautions are taken to avoid enzymatic degradation of complex lipids. In *Ureaplasma* free fatty acids have been postulated to control the function and integrity of the membrane by neutralization of the copious quantities of ammonia produced by urea hydrolysis. A possible function of free fatty acids in *A. axanthum* has been mentioned above.

The neutral lipids of *Thermoplasma* are as distinct from those of other genera among the Mollicutes as is the organism's obligate thermoacidophily. Small amounts of straight-chain hydrocarbons varying in chain length from 14 to 32 carbon atoms are found. Free diglycerol tetraether, which serves as the backbone for the complex lipids of this organism, also occurs in the neutral lipid fraction. Vitamin $K_2$-7 has been identified, while several other neutral lipids remain unresolved. A more extensive exposition on the lipids of *Thermoplasma* can be found in Chapter 18 of this volume.

### 2. Glycolipids

Several types of glycolipids have been found distributed among the mycoplasmas; namely, the glycosyl diglyclerides, the polyterpenol glycosides, and the acylated sugars. Glycosyl diglycerides occur most frequently and are found in all *Acholeplasma* species, *Ureaplasma,* and many *Mycoplasma* species. A special type is found in *Thermoplasma* by virtue of its having a diglycerol tetraether rather than diglyceride as the backbone of its complex lipids. The nature and number of sugar residues and their linkages provide for diversity of these lipids. Among those examined in some structural detail, the number of sugars varies from one to five. Usually the monoglycosyl compound predominates, in contrast to bacteria. In *A. laidlawii* B, diglucosyl diglyceride increases to the level of monoglucosyl diglyceride during the exponential phase at the expense of the latter. As cells age, twice as much of the monoglucoside is formed (Shaw *et al.*, 1968). Resting cells exposed to $^{14}$C-glucose synthesize only the diglucosyl lipid (Smith, 1969a). Exogenous cholesterol has no effect on glycolipid content of growing organisms. In *A. laidlawii* A synthesis of monoglucosyl diglyceride is accentuated by addition of a saturated or a trans unsaturated fatty acid to the culture medium. (Wieslander and Rilfors, 1977). Tri- and tetraglycosyl diglycerides are presumed to occur in small amounts based almost solely on color reactions and mobility on thin-layer chromatograms. *Acholeplasma modicum* contains as its predominant glycolipid a pentaglycosyl diglyceride (Mayberry *et al.*, 1974, 1976). Glucose and galactose occur most frequently, the latter sugar being found in both the pyranose and furanose configuration. A heptose, D-glycero-D-mannoheptose, forms part of the oligosaccharide chain in the pentaglycosyl diglyceride of *A. modicum*. All of the sugars have been found to be glycosidically linked but vary as to anomeric configuration and the specific carbon atom to which they are linked. Invariably the internal sugar is bonded to carbon atom 1 of D-glycerol.

The polyterpenol glycosides consist of steryl glycosides and carotenyl glycosides. They have been found to occur in all *Acholeplasma* species (Smith, 1971a; Mayberry, 1977). *Mycoplasma gallinarum* (Rothblat and Smith, 1961), and *Spiroplasma citri* (K. R. Patel and P. F. Smith, unpublished). Carotenyl glucosides are synthesized by acholeplasmas grown in the absence of exogenous sterol. The anomeric configuration of glucose differs in the A and B strains of *A. laidlawii,* being $\alpha$ in the former and $\beta$ in the latter (Smith, 1963). When cholesterol is available in the culture medium cholesteryl glucosides are formed. Glucose has been found only as the $\beta$ anomer in steryl glycosides. Galactose can be utilized for glycoside synthesis in lieu of glucose and any planar 3-hydroxysterol can substitute for cholesterol (Smith, 1971b). A diglucosyl cholesterol is formed by *A. axanthum* and fatty acids can be attached to sugar hydroxyl groups through *O*-acyl bonds (Mayberry, 1977). The cholesteryl glucoside of *M. gallinarum* (Smith and Mayberry, 1968) and the carotenyl glucoside of *A. laidlawii* B (Shaw *et al.*, 1968) vary in concentration with age of the organism, being the greatest during the lag and stationary phases and the least during exponential growth. This behavior suggested turnover of the glucose moiety, which was subsequently demonstrated in pulse-chase experiments with *A. laidlawii* B (Smith, 1969a).

Acylated sugars have been found only in *M. gallinarum,* although they are known to occur also in certain bacteria, e.g., *Corynebacterium* (Weedon, 1971), *Streptococcus* (Welsh *et al.*, 1968), and, of course, *Mycobacterium tuberculosis* (Vilkas *et al.*, 1968) with its cord factor. In contrast to the fluctuation in amount of steryl glucoside, the triacyl glucose content per unit dry mass of *M. gallinarum* remains constant during the growth cycle (Smith and Mayberry, 1968). Oleic acid predominates in the acyl residues of this lipid.

### 3. Polar Lipids

The only common feature of the lipids among all members of the Mollicutes is the existence of acidic glycerophospholipids (Smith, 1973). Phosphatidyl glycerol and diphosphatidyl glycerol are ubiquitous. One exception is *M. gallinarum,* in which phosphatidyl glycerol exists as the phosphate. Lyso and more fully acylated derivatives of these two lipids occur in lesser amounts (Smith and Koostra, 1967). The lyso form probably is the result of partial deacylation during processing of the lipids. Ester lipids exist exclusively in all but *Anaeroplasma* and *Thermoplasma*. The obligately anaerobic *Anaeroplasma* contain significant quantities of plasmalogens restricted to the phospholipids, similar to other anaerobic bacteria (Langworthy *et al.*, 1975). As with its glycolipids, the diglycerol tetraether serves as the backbone for the polar lipids of *Thermoplasma*.

Unique polar lipids are found in *Acholeplasma, Thermoplasma,* and *Ureaplasma*. These consist of ceramides, aminoacyl phosphatidyl glycerols, phosphoglycolipids, and aminophospholipids. An hydroxyceramide phosphoryl glycerol containing an *O*-acyl group is the major polar lipid of *A. axanthum* (Plackett *et al.*, 1970). The length of the long-chain base of the dihydrosphingosine type is dependent upon the nature of fatty acids in the culture medium. Generally the base found is two carbon atoms longer than the supplemented fatty acid, except for fatty acids of chain length greater than 20 carbon atoms, for which no elongation occurs. Unsaturated fatty acids give rise to unsaturated long-chain bases. The fatty acid in *N*-acyl linkage to the amino group of the long-chain base is predominately D(−)-3-hydroxyhexadecanoate but analogs of this typical intermediate of fatty acid biosynthesis are found in smaller quantities (Mayberry *et al.*, 1973). All of these hydroxy fatty acids presumably arise from chain elongation of fatty acids present in the culture medium. The *O*-acyl fatty acid linked to the hydroxyl group of the hydroxy fatty acid is either saturated or unsaturated but nonhydroxylated. The deacylated derivative of this lipid, i.e., lacking the *O*-acyl fatty acid, also occurs.

Aminoacyl phosphatidyl glycerols, in which an amino acid is esterified through its carboxyl group to one of the two free hydroxyl groups of glycerol, are found in many mycoplasmas but the structure has been elucidated only for the alanyl lipid of *A. laidlawii* B. Although the alanyl form predominates, other amino acids, such as glutamic acid, glycine, leucine, lysine, and tyrosine, can be covalently linked to the glycerol. Both the D and L-isomers of alanine are found in the alanyl phosphatidyl glycerol in a nonracemic mixture of approximately 2:1. The alanine is presumed to be linked to carbon atom 3 of the terminal glycerol (Koostra and Smith, 1969).

Phosphoglycolipids have been found in *Acholeplasma granularum, A. laidlawii,* and *Thermoplasma*. The structure of this lipid in *A. granularum* remains unresolved. The lipid from *Thermoplasma* has a glycerophosphate attached to one of the glycerols and a sugar to the other or a glycerophosphoryl glycosyl radical attached to only one of the glycerols of the diglycerol tetraether. Further discussion of this lipid is presented in Chapter 18. The phosphoglycolipid of *A. laidlawii* B occurs both as glycerophosphoryl diglucosyl diglyceride (Shaw *et al.*, 1972) and as phosphatidyl diglucosyl diglyceride (Smith, 1972b). The glycolipid to which the glycerophosphoryl or the phosphatidyl radical is linked is identical to the free diglucosyl diglyceride of the organism. *sn*-Glycerol-3-phosphate or its 1,2-diacyl derivative is attached through a phosphodiester bond to the 6-carbon of the terminal (outside) glucose. Although a similar lipid together with three other phosphoglycolipids have been found in *Streptococcus* (Ambron and Pieringer, 1971; Fischer *et al.*, 1973a,b), the streptococcal lipid contains a *sn*-glycerol-1-phosphate radical on the 6-carbon of the terminal glucose. The streptococcal lipid appears to be involved in covalently linking membrane teichoic acid to the cytoplasmic membrane (Ganfield and Pieringer, 1975). The role in mycoplasmas is unknown, although there is evidence for turnover of both phosphorus and glucose during metabolism of glucose by intact cells. A glycerophosphoryl monoglucosyl diglyceride is suspected in *A. laidlawii*, strains A (Wieslander and Rilfors, 1977) and B (Smith and Henrikson, 1965b).

*Ureaplasma* is the only member of the Mollicutes yet shown to contain the predominant bacterial phospholipid, phosphatidyl ethanolamine (Romano *et al.*, 1972). In addition to this positively charged lipid, *Ureaplasma* also contains an unidentified amino lipid which appears to be a diaminohydroxy compound, possibly derived from an amino acid, with adjacent *O*-acyl and *N*-acyl groups. The uniqueness of the lipids of *Ureaplasma* may reflect both its acidic pH optimum for growth and its copious ammonia production from ureolysis.

## 4. Fatty Acids

The fatty acids found in ester linkage in the complex lipids of mycoplasmas are predominately C14:0, C16:0, and C18:1 (Panos and Rottem, 1970; Smith, 1971a). A variety of other fatty acids, both saturated and unsaturated, of chain length C10–C20 occur in small amounts, usually amounting to only 10% of the total. Few, if any, cyclopropane or highly branched-chain fatty acids are found. Hydroxy fatty acids are limited to *A. axanthum* (Mayberry *et al.*, 1973). The fatty acid composition can be controlled easily by the kinds of the fatty acids supplied in the culture medium. This is possible because most mycoplasmas are incapable of fatty acid biosynthesis; others are limited to synthesis of saturated fatty acids, and a very few are sufficient for total fatty acid biosynthesis. An extreme example has been reported for bovine strain Y, related to *Mycoplasma mycoides,* in which 97% of the total fatty acids was comprised of the elaidic acid (trans 18:1) supplied in the growth medium (Rodwell, 1971). In addition to the exogenous supply of fatty acids, temperature plays a major role in defining fatty acid composition. For example, the ratio of palmitate to oleate incorporated into the lipids of *A. laidlawii* is independent of temperature when incubation is carried out at a temperature below that of the transition of the membrane bilayer. Incorporation becomes temperature dependent when the membrane is above the transition temperature and is fully fluid. In this case the ratio of palmitate to oleate incorporation increases (Melchior and Steim, 1977). Examination of the positional distribution of fatty acids has been limited to the phospholipids of *A. laidlawii.* (Romijn *et al.*, 1972; McElhaney and Tourtellotte, 1969). Unsaturated fatty acids are directed preferentially to the 2 position of glycerol while the saturated acids predominate in the 1 position.

There is little or no distinction in the fatty acid composition among individual lipids. An exception is the phosphatidyl diglucosyl diglyceride of *A. laidlawii* B in which there is a significantly larger amount of short-chain (8–11 carbon atoms) saturated fatty acids in the phosphatidyl radical as compared to the glycosyl diglyceride (Smith, 1972b).

The aldehydes in the plasmalogen phospholipids of *Anaeroplasma* have been found to be predominately the saturated 16- and 18-carbon compounds (Langworthy *et al.*, 1975). In *Thermoplasma* the two glycerol radicals of the diglycerol tetraether are bridged via ether linkages to two isopranoid branched C40 diols (Langworthy, 1977). These diols are homogeneous for a given molecule but may vary from molecule to molecule as to the number of internal rings, e.g., $C_{40}H_{82}$, $C_{40}H_{80}$, $C_{40}H_{78}$ (deRosa *et al.*, 1976).

## C. Biosynthesis of Lipids

### 1. Neutral Lipids

All species of *Acholeplasma* are capable of *de novo* biosynthesis of carotenoids based upon the appearance of pigmentation and the incorporation of $^{14}$C-acetate and $^{14}$C-mevalonate into these pigments. Specific enzymes of the biosynthetic pathway for polyterpenes from the starting compound, acetate, through the five-carbon intermediate, dimethylallyl pyrophosphate, involved in polymerization have been demonstrated in *A. laidlawii* B (Henrikson and Smith, 1966b; Smith and Henrikson, 1965b; Castrejon-Diez *et al.*, 1962). In this organism acetyl-coenzyme A is formed through the mediation of acetate kinase and phosphate acetyltransferase rather than by direct activation of acetate by an acetyl-coenzyme A synthetase. Acetoacetyl-coenzyme A can be synthesized either by acetoacetyl-coenzyme A thiolase or from succinyl-coenzyme A 3-ketoacid-coenzyme A transferase. The pathway for the condensation of the isoprene units and the subsequent oxidation of the polymer to form double bonds and hydroxyl groups remains to be established. A knowledge of the exact structure of the carotenoids is a prior necessity. Hydroxypolyterpenes are vital to the organisms, although their synthesis can be spared by cholesterol (Smith, 1963). The site of cholesterol inhibition in *A. laidlawii* B has been established as isopentenyl pyrophosphate isomerase (Smith and Smith, 1970). Inhibition is not complete since the pathway must operate for the biosynthesis of a lipid intermediate required for lipopolysaccharide formation. Traces of a lipid with the characteristics of bactoprenol phosphate which becomes labeled with $^{14}$C-mevalonate have been detected in all species of *Acholeplasma*. Complete inhibition of enzymes in the pathway for polyterpene biosynthesis prior to the formation of the first polymer results in inhibition of growth, whereas inhibition by diphenylamine of oxidation steps responsible for double-bond formation does not inhibit growth (Smith and Henrikson, 1966). Rather colorless hydroxylated polyterpenes are formed, which appear to carry out the necessary role of carotenoids. A more detailed exposition can be found elsewhere (Smith, 1971a). *Mycoplasma gallinarum,* which requires sterol for growth, is enzymatically sufficient for the biosynthesis of mevalonic acid but lacks the three enzymes necessary for its conversion to isopentenyl pyrophosphate. When this intermediate is supplied *in lieu* of sterol the organism grows and synthesizes a polyterpenol (Smith, 1968b). *Mycoplasma arthritidis,* although capable of formation of acetoacetyl-coenzyme A, lacks all of the subsequent enzymes of polyterpene biosynthesis. However, it appears capable of hydroxylating hydrocarbon polyterpenes.

Esterification of either sterol or carotenol has been demonstrated in *Acholeplasma* (Smith, 1959), a limited number of species of *Mycoplasma,* and *Ureaplasma* (Romano *et al.*, 1972). In most cases this conclusion has been drawn from labeling experiments using $^{14}$C-fatty acids. The polyterpenol esters examined in detail contain short- rather than long-chain fatty acids, e.g., acetic acid in *A. laidlawii* B and *M. gallinarum* and butyric, propionic, and acetic acids in *M. arthritidis* (Rothblat and Smith, 1961; Smith, 1962). Esterification requires the presence of coenzyme A and adenosine triphosphate. Activity decreases as the chain length of the fatty acid is increased. Pulse-labeling experiments with $^{14}$C-glucose result in pulse labeling of the carotenyl ester of *A. laidlawii* B, the fatty acid being exclusively acetate (Smith, 1969a). This behavior is suggestive of some relationship to glucose metabolism and may explain the difficulty in demonstration of esterified polyterpenols in mycoplasmas in general. The short-chain acids can easily be overlooked and the amount of ester varies with the metabolic state.

Long-chain bases of *A. axanthum* are derived from exogenous fatty acids. Usually the base formed contains two more carbon atoms than the precursor but no change in degree of saturation is seen. The upper limit of chain length for a precursor fatty acid is 20 carbon atoms (Plackett *et al.*, 1970).

The mechanisms for biosynthesis of the bactoprenols and the isopranols, diglycerol tetraether, and naphthoquinone of *Thermoplasma* remain to be elucidated. All of these lipids become labeled from $^{14}$C-acetate and $^{14}$C-mevalonate, indicating the participation of the usual isoprene synthetic pathway.

### 2. Glycolipids

Biosynthesis of the monoglucosyl diglyceride of *A. laidlawii* B occurs by the transfer of glucose from uridine-5′-diphosphoglucose to 1,2-diglyceride catalyzed by a membrane-associated $\alpha$-1-glucosyl transferase.

Mixed diglycerides from *A. laidlawii* B or other sources can serve as glucose acceptor but synthetic homogeneous 1,2-diglycerides, such as 1,2-dipalmitin, and 1,3-diglycerides cannot. The diglucosyl diglyceride is formed from uridine-5′-diphosphoglucose and monoglucosyl diglyceride presumably through the mediation of the same enzyme. Monogalactosyl diglyceride cannot substitute for the monoglucosyl lipid. The mechanism for the biosynthesis of 1,2-diglyceride has not been established. The nucleotide sugar intermediate is synthesized from glucose- 1-phosphate, derived from the glycolytic pathway, and uridine triphosphate by the enzyme, UTP: $\alpha$-glucose-1-P uridylyl transferase. $Mg^{2+}$ is required, and sodium dodecyl sulfate and high-ionic strength buffer are stimulatory for optimum synthesis, which occurs at pH 8 (Smith, 1969b). Whether this mechanism serves for the biosynthesis of longer oligosaccharide chains, such as the pentaglycosyl diglyceride of *A. modicum,* is not known.

Cholesteryl glucoside of *M. gallinarum* and presumably the carotenyl glucosides of some acholeplasmas are synthesized by the transfer of glucose from uridine-5′-diphosphoglucose to membrane-bound sterol or carotenol by a membrane-associated enzyme (Smith, 1971b). In *M. gallinarum* galactose can be coupled to cholesterol if uridine-5′-diphosphogalactose serves as a substrate. Only uridine-5′-diphospho sugars function in the reaction. Similar to the synthetic reaction for glucosyl diglycerides, $Mg^{2+}$ is required and the optimum pH is 8.

### 3. Polar Lipids

Mycoplasmas are capable of *de novo* biosynthesis of their phospholipids based upon incorporation of $^{32}P$, $^{14}C$-glycerol, and $^{14}C$-fatty acids (Smith, 1971a). The difficulty in demonstrating $^{14}C$-glycerol incorporation, except for one bovine strain (Y) (Plackett, 1967b), suggests that *sn*-glycerol-3-phosphate and not free glycerol is the substrate for acylation as in bacteria (Ambron and Pieringer, 1973). There is no evidence for the participation of dihydroxyacetone phosphate (Plackett and Rodwell, 1970). Acylation occurs through the intermediary participation of acyl-coenzyme A catalyzed by either a membrane-bound or soluble acyl-coenzyme A: *sn*-glycerol-3-phosphate transacylase (Rottem and Greenberg, 1975). Acyl carrier protein (Rottem *et al.*, 1973) probably also is involved in the acylation of *sn*-glycerol-3-phosphate by *Acholeplasma* and *Ureaplasma*. Studies on the positional distribution of fatty acids in phosphoglycerides suggest the existence of the typical bacterial pathway, first the formation of lysophosphatidic acid and finally phosphatidic acid (Romijn *et al.*, 1972; McElhaney and Tourtellotte, 1970a). The positional distribution of fatty acids in phosphatidic acid and phosphatidyl glycerol are identical. Phosphatidyl glycerol apparently is synthesized from cytidine diphosphate diglyceride and *sn*-glycerol-3-phosphate. Synthesis of the nucleotide intermediate from cytidine triphosphate and phosphatidic acid has been demonstrated in *A. laidlawii* B (P. F. Smith, unpublished). *Mycoplasma hominis* membrane is capable of incorporation of *sn*-glycerol-3-phosphate into phosphatidyl glycerol (Rottem and Greenberg, 1975). The immediate product of the condensation reaction, phosphatidyl glycerophosphate, is rarely found in mycoplasmas with the exception of *M. gallinarum* (Smith and Koostra, 1967), suggesting a very active and specific phosphatidyl glycerophosphatase. Turnover studies with several species indicate that phosphatidyl glycerol serves as immediate precursor for the biosynthesis of diphosphatidyl glycerol (McElhaney and Tourtellotte, 1970b; Smith, 1969a; Plackett, 1967b; Plackett and Rodwell, 1970). The only turnover of phosphatidyl glycerol is coordinated with the formation of diphosphatidyl glycerol and occurs primarily upon aging of the organisms. Presumably two molecules of phosphatidyl glycerol are condensed with the formation of diphosphatidyl glycerol and glycerol. Actual demonstration of these pathways is missing, as is any knowledge of plasmalogen biosynthesis in *Anaeroplasma*.

The mechanisms for the biosynthesis of glycerophosphoryl diglucosyl diglyceride and phosphatidyl diglucosyl diglyceride are unresolved. Extensive attempts to identify the glycerophosphate precursor have failed. In streptococci (Pieringer, 1972) either phosphatidyl glycerol or diphosphatidyl glycerol serves in this capacity. The stimulation of synthesis of the phosphoglucolipid in *A. laidlawii* B by cytidine triphosphate suggests that cytidine diphosphoglycerol may serve as a precursor but the streptococcal system appears more likely.

L-Alanyl phosphatidyl glycerol is synthesized by a membrane-associated enzyme of *A. laidlawii* B from L-alanyl tRNA and phosphatidyl glycerol. Aminoacyl tRNA synthesis required the whole cell lysate. Synthesis of the amino lipid is stereospecific for the L-isomer and is not inhibited by the D-isomer. Synthesis of D-alanyl phosphatidyl glycerol occurs by the transfer of D-alanine from the AMP–D-alanyl–enzyme complex to phosphatidyl glycerol. Like the system for the L-isomer, D-alanyl phosphatidyl glycerol synthesis is stereospecific and is independent of the biosynthesis of the opposite isomer (Koostra and Smith, 1969).

### 4. Fatty Acids

Acholeplasmas are capable of the biosynthesis of saturated fatty acids (Pollack and Tourtellotte, 1967; Herring and Pollack, 1974) and *Ureaplasma* can biosynthesize both saturated and unsaturated fatty acids (Romano *et al.*, 1976) from $^{14}$C-acetate. All species of *Mycoplasma* and *Spiroplasma* so far examined are incapable of fatty acid biosynthesis, while *Anaeroplasma* has not been studied. *Thermoplasma,* lacking fatty acids, synthesizes long-chain isopranoid diols. *De novo* synthesis of saturated fatty acids in *A. laidlawii* A and B proceeds through the malonyl-coenzyme A pathway (Panos and Henrikson, 1969; Panos and Rottem, 1970; Rottem and Panos, 1970). A soluble fatty acid synthetase complex and acyl carrier protein convert acetate primarily into myristic, palmitic, and stearic acids when malonyl-coenzyme A, $NADPH_2$, and $Mg^{2+}$ are present. Only palmitic and myristic acids are formed by intact cells. The acyl carrier protein of *A. laidlawii* can be formed from pantetheine or coenzyme A but not from β-alanine (Rottem *et al.*, 1973). It is a small soluble protein, similar to the acyl carrier protein from *Escherichia coli* except for its increased sensitivity to heat. Cerulenin, an inhibitor of β-ketoacyl-acyl carrier protein, inhibits fatty acid synthesis in *A. laidlawii* (Rottem and Barile, 1976). A block in the pathway of biosynthesis of unsaturated fatty acids by *A. laidlawii* A has been identified at the level of β-hydroxythioester dehydrase. When the organism is supplied with β-hydroxydecanoic acid, the normal precursor at the biosynthetic step where branching of the pathway for formation of unsaturated fatty acids occurs in bacteria, together with β-hydroxythioester dehydrase from *E. coli,* unsaturated fatty acids are synthesized. Although incapable of biosynthesis of unsaturated fatty acids *A. laidlawii* strains are able to elongate exogenously supplied unsaturated acids. Strain A can elongate trans unsaturated acids to $C_{16}$ unsaturated acids but is limited in its ability to elongate to $C_{18}$. This strain also possesses the unique ability to elongate the cyclopropane fatty acid, *cis*-9,10-methylene hexadecanoic acid, to *cis*-11,12-methylene octadecanoic acid (Panos and Leon, 1974). Strain B generally can elongate only cis unsaturated acids, although some isolates may not be as specific. The primary product of unsaturated fatty acid biosynthesis by *Ureaplasma* is palmitoleic acid (Romano *et al.*, 1976).

Thioesterase activity has been demonstrated in both *Acholeplasma* and *Mycoplasma* species (Rottem *et al.*, 1977). The activity is much greater in acholeplasmas and is postulated to regulate fatty acid biosynthesis by controlling either the level of free coenzyme A required for biosynthesis and elongation or the level of acyl-coenzyme A derivatives that might act as feedback inhibitors in fatty acid biosynthesis at the level of acetyl-coenzyme A carboxylase.

## D. Metabolism of Lipids

### 1. Neutral Lipids

A nonspecific lipase capable of the hydrolysis of triglycerides, natural fats, and fatty acid esters (Smith, 1959; Rottem and Razin, 1964) is found in some mycoplasmas. It is a soluble enzyme having no cofactor requirements and it possesses an alkaline pH optimum.

A membrane-associated sterol esterase has been shown to occur in *M. arthritidis, M. gallinarum,* and *A. laidlawii,* although this enzyme is not ubiquitous (Smith, 1959). It catalyzes the hydrolytic or thiolytic cleavage of steryl and carotenyl esters. The enzyme requires a micellar substrate formed in conjunction with an amphipathic compound. It exhibits little specificity toward fatty acids but the greatest activity usually occurs with short-chain fatty acid esters. More specificity resides with the sterol moiety. Esters of cholestanol are more easily hydrolyzed than esters of the $\Delta^5$-cholestenols, indicating that unsaturation in the B ring impedes the esterase. Among sterols containing cis-fused A/B rings, i.e., the coprostanols, esters of the axial 3-hydroxyl group are more readily hydrolyzed than esters of the equatorial 3-hydroxyl group. No such difference is noted with sterol esters containing trans-fused A/B rings, i.e., cholestanols. Increase in the length of the side chain over that found in the cholestane series, e.g., stigmasterol, retards activity. All *trans*-carotenyl esters are easily hydrolyzed by the enzyme (Smith, 1964). Pulse-labeling studies in *A. laidlawii* have demonstrated a turnover of acetate in the carotenyl ester fraction during metabolism of glucose (Smith, 1969a). The functional role of this activity has not been resolved, although speculations have been forwarded (Smith, 1971a).

## 2. Glycolipids

Glucosidases have been detected in *A. laidlawii* A and B and in *M. gallinarum*. The enzyme from *A. laidlawii* A appears specific for $\alpha$-glucosides, while the other two organisms possess a specificity for the opposite anomer (Henrikson and Smith, 1964). In all three organisms there is a preference for an aryl group as the aglycon, which explains their activity with steryl and carotenyl glucosides. Turnover studies using pulse labeling with $^{14}$C-glucose have demonstrated the instability of the glucose moiety during metabolism. Inhibition of glucose fermentation and glucosidase activity depresses the synthesis of the glucoside and completely inhibits the turnover of glucose, suggesting some functional role for these glycolipids. On the other hand, the glycosyl diglycerides and acylated glucose do not exhibit any turnover during metabolism that cannot be accounted for as direct precursor utilization for synthesis of more complex lipids (Smith, 1969a).

## 3. Polar Lipids

Generally mycoplasmas are incapable of degrading polar lipids. With one exception attempts to demonstrate phospholipase activities have given negative results. However, a very active membrane-associated lysophospholipase has been found in *A. laidlawii* B (van Golde *et al.*, 1971). Partial characterization of this enzyme indicated that palmitoyl-*sn*-glycero-3-phosphoryl-1′-glycerol was a far better substrate than either palmitoyl or oleoyl-*sn*-glycero-3-phosphorylcholine. The role of this enzyme is unclear but may reflect a protective device to rid the organism of a very active lytic agent. Turnover studies have failed to show any changes in composition of glycerophospholipids other than a precursor–product relationship, e.g., phosphatidyl glycerol conversion to diphosphatidyl glycerol.

Prelabeling glycerophosphoryl diglucosyl diglyceride with $^{32}P$ followed by metabolism of glucose results in a loss of total radioactivity from this lipid (Smith, 1969a). The most plausible interpretation of this phenomenon is that the glycerophosphoryl moiety is being acylated since the acylated form of this lipid, phosphatidyl diglucosyl diglyceride, accumulates in aging cultures.

#### 4. Fatty Acids

Oxidation of fatty acids of chain length two through ten is carried out by *M. arthritidis* and *M. gallinarum* but not by *A. laidlawii* (Lynn, 1960). The rate of oxidation decreases as chain length increases. The mechanism has been shown for *M. arthritidis* to be the typical β-oxidative pathway. This organism contains an acetyl-coenzyme A synthetase which exhibits some activity toward propionate. Activation of fatty acids other than acetate or propionate occurs through the mediation of a propionate-coenzyme A transferase. Other enzymes of the pathway demonstrated to exist, using butyrate as substrate, are acyl-coenzyme A dehydrogenase, crotonase, an $NADH_2$-requiring β-hydroxylacyl-coenzyme A dehydrogenase, and β-ketothiolase (Vandemark and Smith, 1965). The acetyl-coenzyme A is metabolized further via the tricarboxylic acid cycle (Vandemark and Smith, 1964).

## II. LIPOPOLYSACCHARIDES

### A. Distribution

A new class of lipopolysaccharides, structurally distinct from those occurring in gram-negative bacteria but extractable with hot aqueous phenol, occurs in the membranes of some mycoplasmas. Lipopolysaccharides have been found in four species of *Acholeplasma*, two of *Anaeroplasma*, *Thermoplasma*, and *Mycoplasma neurolyticum*. None has been found in five other species of *Mycoplasma* or in *Spiroplasma* (Mayberry-Carson *et al.*, 1974; Smith *et al.*, 1976). Cytochemical evidence suggests the presence of a glucose polymer on the surface of *Ureaplasma* (Robertson and Smook, 1976), but it is not known whether this polymer is lipopolysaccharide or polysaccharide in nature. Simple polysaccharides, such as glucans and galactans, have been found in *M. mycoides* and related strains (Plackett *et al.*, 1963; Plackett and Buttery, 1964). There is some evidence to suggest that these may contain covalently attached fatty acids in their natural state at the surface of the organisms or in the culture supernatant fluids. The hexosamine polymer of *A. laidlawii* apparently contains no lipid residues but only *N*-acetyl glucosamine and *N*-acetyl galactosamine (Gilliam and Morowitz, 1972). Lipopolysaccharide accounts for about 1% of the dry weight of the acholeplasmas and 2–3% of *Thermoplasma* and *Anaeroplasma*. These compounds represent about 10% of the dry weight of membranes and account for all of the carbohydrate excluding the glycolipids and the hexosamine polymer of *A. laidlawii*.

## B. Chemical Nature of Lipopolysaccharides

Lipopolysaccharides have a tendency to aggregate, when purified, into particle sizes of several million daltons. Disaggregation by deacylation, peracetylation, permethylation, and silylation and the use of detergents has been used successfully to determine the molecular homogeneity or heterogeneity of monomeric units. The polymer from *Thermoplasma* behaves as a monomer with molecular weight of 5300 (after allowance is made for attached radicals) by gel permeation chromatography of trimethyl silyl, acetyl, and methyl derivatives. It is composed of 24 mannose residues, all in the $\alpha$ configuration, seven joined by $1 \rightarrow 3$ linkages, and 17 by $1 \rightarrow 2$ linkages. These mannose units are bonded through a $1 \xrightarrow{\alpha} 3$ linkage to glucose which is glycosidically bonded to diglycerol tetraether. The actual sequence of the mannose residues has not been established (Mayberry-Carson *et al.*, 1974).

The lipopolysaccharide from *A. axanthum* behaves as a single component by permeation chromatography of permethylated and peracetylated derivatives and its water-soluble deacylation product, by polyacrylamide gel electrophoresis in detergent, and by immunodiffusion using antibody prepared against homologous membranes or lipopolysaccharide. Its monomeric size approaches 100,000 as judged both from chemical composition and chromatographic behavior. It is composed of neutral sugars (glucose, galactose), amino sugars (glucosamine, galactosamine, fucosamine, quinovosamine) with the deoxyamino sugars predominating, glycerol, and fatty acids. Only 10% of the fatty acids are in ester linkage, while the remainder are *N*-acyl (Smith *et al.*, 1976; Smith, 1977). Essentially all of the amino groups of the amino sugars are *N*-acylated. The fatty acids mimic those found in the lipids of the organism and include a relatively high concentration of 3-hydroxy acids. Further structural information on this polymer and the lipopolysaccharides from other species of *Acholeplasma, Anaeroplasma,* and *M. neurolyticum* is lacking.

Permeation chromatography of the deacylated derivative and gel electrophoresis suggests two components in the lipopolysaccharide fraction from *A. granularum.* However, these two components have the same chemical composition and behave identically in serologic tests, indicating persistent aggregation rather than heterogeneity. The monomer has a molecular weight of about 20,000 and is composed of neutral sugars (glucose, galactose), amino sugars (glucosamine, fucosamine, and quinovosamine) equally divided between hexosamine and deoxyhexosamine, glycerol, and fatty acids. All of the fatty acids are in ester linkage, yet no free amino groups are detectable, suggesting the presence of *N*-acetyl groups.

The lipopolysaccharide fraction from *A. modicum* behaves as three components on permeation chromatography of the deacylated and peracetylated derivatives. However, using absorbed antisera all three appear antigenically identical. Some variation is seen in the distribution of hexosamine and deoxyhexosamine between the three components. These differences may be real or the three components can be considered to be a monomer with a molecular weight of 36,000, dimer of 72,000, and tetramer of 142,000. In addition to fatty acids in ester linkage and glycerol, the lipopolysaccharide contains neutral sugars (primarily galactose and some glucose) and *N*-acetyl amino sugars (galactosamine, fucosamine, quinovosamine). The neutral sugars predominate, while the amino sugars are equally divided between hexosamine and deoxyhexosamine.

The lipopolysaccharide fraction from *A. laidlawii* is a separable mixture of the hexosamine polymer and lipopolysaccharide. The latter behaves as a homogeneous compound of molecular size 150,000. It is antigenically distinct from the hexosamine polymer and is composed of glucose, mannose, glucosamine, galactosamine, fucosamine, and quinovosamine. Essentially all of the neutral sugar is glucose, and deoxyhexosamines predominate in the amino sugar fraction. Fatty acids in ester linkage and glycerol are found in this compound.

Only compositional data are available for the lipopolysaccharides from *M. neurolyticum* and *Anaeroplasma.* Both toxigenic and nontoxigenic strains of *M. neurolyticum* contain lipopolysaccharide but the neutral sugar content of the toxigenic strain is greater. Both lipopolysaccharides contain the same neutral sugars (galactose, which predominates, and glucose), amino sugars (fucosamine and galactosamine), glycerol, and fatty acid esters. Differences also are seen in the composition of lipopolysaccharides from *Anaeroplasma bactoclasticum* and *Anaeroplasma abactoclasticum.* Both contain glucose, galactose, and mannose but glucose predominates in *A. abactoclasticum.* Only fucosamine and glucosamine are present in the lipopolysaccharide from *A. abactoclasticum,* while quinovosamine and galactosamine are found in addition in the lipopolysaccharide from *A. bactoclasticum.*

The simplicity of composition makes these mycoplasmal lipopolysaccharides different from the typical compounds found in gram-negative bacteria. They lack heptoses, ketodeoxyoctanoate, and phosphoryl ethanolamine. Although the structures have not been defined for any but the lipopolysaccharide from *Thermoplasma,* evidence suggests that for all but the compound from *A. axanthum* they consist of a linear oligosaccharide glycosidically linked to diglyceride. In *A. axanthum* most of the fatty acids are in *N*-acyl linkage to the amino sugars, suggesting a chain of sugars and *N*-acyl amino sugars glycosidically linked to a diglyceride.

### C. Cellular Site of Lipopolysaccharides

The lipopolysaccharides are associated exclusively with the cell membrane since the total content in the organisms can be recovered from purified membrane preparations. Lectin-binding studies suggest that many mycoplasmas contain a carbohydrate surface. The binding site in the acholeplasmas and *M. neurolyticum* may be the oligosaccharide chain of the lipopolysaccharide. The size of this chain in lipopolysaccharides is considerably larger than that of glycolipids and presents a more exposed area for attachment. The specificity of lectin-binding matches well the sugar composition of the lipopolysaccharides (Schiefer *et al.*, 1974; Kahane and Tully, 1976). The histochemical technique of binding concanavalin A, which is specific for α-D-glucosyl or α-D-mannosyl residues, followed by binding of horseradish peroxidase to concanavalin A and visualization with diaminobenzidine has shown the existence of carbohydrate on the surface of *M. mycoides* subsp. *capri* (Schiefer *et al.*, 1975) and on the entire surface of *Thermoplasma* (Mayberry-Carson *et al.*, 1977). Concanavalin A binding and hence the staining reaction can be inhibited with α-D-glucose. These types of studies, including immunologic procedures, should be useful in identifying whether the lipopolysaccharides do face the exterior of the mycoplasmal cell.

The lipopolysaccharides from *Thermoplasma* and *Acholeplasma* appear as fibrous structures when examined by scanning electron microscopy. Transmission electron micrographs of negatively stained preparations exhibit long ribbonlike structures 5 nm in width. In the presence of a detergent, such as sodium dodecyl sulfate, vesicular-shaped particles are formed, a behavior reminiscent of gram-negative bacterial lipopolysaccharides (Mayberry-Carson *et al.*, 1975).

## III. EPILOGUE

Lipids and lipopolysaccharides are important constituents of the cell membranes of mycoplasmas. Aside from their vital functional roles in the living processes of the cell, treated in this volume and elsewhere, the glycolipids and lipopolysaccharides offer additional avenues for speculation. Two species of *Acholeplasma, A. laidlawii* and *A. granularum,* contain large quantities of phosphoglycolipids of structure similar to the lipid involved in covalently linking the membrane to glycerol teichoic acid of streptococci. Since no teichoic acids have ever been detected in mycoplasmas, are these lipids accumulating because of a degenerative wall synthetic system? Sugar residues attached to lipids exhibit a diversity as to type of sugar and nature of linkage, allowing for great variation from species to species. Are these sugar-containing lipoidal structures responsible for the specific identity of a given species or strain? Are they involved in attachment to host cells in the case of pathogens? Are they the cellular components responsible for pathogenesis? Are lipopolysaccharides and glycoproteins exclusive for a given species and are their roles alike? Some preliminary evidence exists bearing on these unanswered questions. Glycolipids, although not immunogenic by themselves, react specifically in serologic tests employing antimembrane sera (Sugiyama *et al.*, 1974; Ryan *et al.*, 1975). Antigenic distinction resides in the number, nature, and linkages of the sugars. Lipopolysaccharides alone are immunogenic. These lipopolysaccharides can bind to sheep erythrocytes and cultured rabbit epithelial cells. Upon exposure of these coated cells to homologous antiserum, agglutination occurs. Addition of complement to the lipopolysaccharide–antibody-coated erythrocytes results in hemolysis (Smith and Lynn, 1976). Some lipopolysaccharides, notably those from *A. axanthum* and *M. neurolyticum,* drastically suppress concanavalin A stimulation of rabbit lymphocytes, an effect opposite that seen with cells of *M. pneumoniae*. Yet *M. neurolyticum* is an accepted pathogen while *A. axanthum* is not considered so in some circles. These and other findings should whet the appetite of the experimentalist in the future.

## REFERENCES

Ambron, R. T., and Pieringer, R. A. (1971). *J. Biol. Chem.* **246,** 4216–4225.

Ambron, R. T., and Pieringer, R. A. (1973). *In* "Form and Function of Phospholipids" (G. B. Ansell, R. M. C. Dawson, and J. N. Hawthorne, eds.), pp. 289–344. Elsevier, Amsterdam.

Castrejon-Diez, T., Fisher, N., and Fisher, E., Jr. (1962). *Biochem. Biophys. Res. Commun.* **9,** 416–420.

deRosa, M., Gambacorta, A., and Bulock, J. D. (1976). *Phytochemistry* **15,** 143–145.

Fischer, W., Ishizuka, I., Landgraf, H. R., and Herrmann, J. (1973a). *Biochim. Biophys. Acta* **296,** 527–545.

Fischer, W., Landgraf, H. R., and Herrmann, J. (1973b). *Biochim. Biophys. Acta* **306,** 353–367.

Freeman, B. A., Sissenstein, R., McManus, T. T., Woodward, J. E., Lee, I. M., and Mudd, J. B. (1976). *J. Bacteriol.* **125,** 946–954.

Ganfield, N. W., and Pieringer, R. A. (1975). *J. Biol. Chem.* **250,** 702–709.

Gershfeld, N. L., Wormser, M., and Razin, S. (1974). *Biochim. Biophys. Acta* **352,** 371–384.

Gilliam, J. M., and Morowitz, H. J. (1972). *Biochim. Biophys. Acta* **274,** 353–363.

Henrikson, C. V., and Smith, P. F. (1964). *J. Gen. Microbiol.* **37,** 73–80.

Henrikson, C. V., and Smith, P. F. (1966a). *J. Gen. Microbiol.* **45,** 73–82.

Henrikson, C. V., and Smith, P. F. (1966b). *J. Bacteriol.* **92,** 701–706.

Herring, P. K., and Pollack, J. D. (1974). *Int. J. Syst. Bacteriol.* **24,** 73–78.

Kahane, I., and Tully, J. G. (1976). *J. Bacteriol.* **128,** 1–7.

Koostra, W. L., and Smith, P. F. (1969). *Biochemistry* **8,** 4794–4806.

Langworthy, T. A. (1977). *Biochim. Biophys. Acta* **487,** 37–50.

Langworthy, T. A., Smith, P. F., and Mayberry, T. A. (1972). *J. Bacteriol.* **112,** 1193–1200.

Langworthy, T. A., Mayberry, W. R., Smith, P. F., and Robinson, I. M. (1975). *J. Bacteriol.* **122,** 785–787.

Lynn, R. J. (1960). *Ann. N.Y. Acad. Sci.* **79,** 538–542.

Lynn, R. J., and Smith, P. F. (1960). *Ann. N.Y. Acad. Sci.* **79,** 493–498.

McElhaney, R. N., and Toutellotte, M. E. (1969). *Science* **164,** 433–434.

McElhaney, R. N., and Tourtellotte, M. E. (1970a). *Biochim. Biophys. Acta* **202,** 120–128.

McElhaney, R. N., and Tourtellotte, M. E. (1970b). *J. Bacteriol.* **101,** 72–76.

Mayberry, W. R. (1977). *Am. Soc. Microbiol., Abstr. Annu. Meet.* p. 193.

Mayberry, W. R., Smith, P. F., Langworthy, T. A., and Plackett, P. (1973). *J. Bacteriol.* **116,** 1091–1095.

Mayberry, W. R., Smith, P. F., and Langworthy, T. A. (1974). *J. Bacteriol.* **118,** 898–904.

Mayberry, W. R., Langworthy, T. A., and Smith, P. F. (1976). *Biochim. Biophys. Acta* **441,** 115–122.

Mayberry-Carson, K. J., Langworthy, T. A., Mayberry, W. R., and Smith, P. F. (1974). *Biochim. Biophys. Acta* **360,** 217–229.

Mayberry-Carson, K. J., Roth, I. L., and Smith, P. F. (1975). *J. Bacteriol.* **121,** 700–703.

Mayberry-Carson, K. J., Jewell, M. J., and Smith, P. F. (1978). *J. Bacteriol.* **133,** 1510–1513.

Melchior, D. L., and Steim, J. M. (1977). *Biochim. Biophys. Acta* **466,** 148–159.

Mudd, J. B., Ittig, M., Roy, B., Latrille, J., and Bové, J. M. (1977). *J. Bacteriol.* **129,** 1250–1256.

Panos, C., and Henrikson, C. V. (1969). *Biochemistry* **8,** 652–656.

Panos, C., and Leon, O. (1974). *J. Gen. Microbiol.* **80,** 93–100.

Panos, C., and Rottem, S. (1970). *Biochemistry* **9,** 407–412.
Pieringer, R. A. (1972). *Biochem. Biophys. Res. Commun.* **49,** 502–507.
Plackett, P. (1967a). *Biochemistry* **6,** 2746–2754.
Plackett, P. (1967b). *Ann. N.Y. Acad. Sci.* **143,** 158–164.
Plackett, P., and Buttery, S. H. (1964). *Biochem. J.* **90,** 201–205.
Plackett, P., and Rodwell, A. W. (1970). *Biochim. Biophys. Acta* **210,** 230–240.
Plackett, P., Buttery, S. H., and Cottew, G. S. (1963). *In* "Recent Progress in Microbiology" (N. E. Gibbons, ed.), pp. 533–547. Univ. of Toronto Press, Toronto.
Plackett, P., Marmion, B. P., Shaw, E. J., and Lemcke, R. M. (1969). *Aust. J. Exp. Biol. Med. Sci.* **47,** 171–195.
Plackett, P., Smith, P. F., and Mayberry, W. R. (1970). *J. Bacteriol.* **104,** 798–807.
Pollack, J. D., and Tourtellotte, M. E. (1967). *J. Bacteriol.* **93,** 636–641.
Razin, S. (1974). *FEBS Lett.* 47, 81–85.
Razin, S., Wormser, M., and Gershfeld, N. L. (1974). *Biochim. Biophys. Acta* **352,** 385–396.
Robertson, J., and Smook, E. (1976). *J. Bacteriol.* **128,** 658–660.
Robinson, I. M., Allison, M. J., and Hartman, P. A. (1975). *Int. J. Syst. Bacteriol.* **25,** 173–181.
Rodwell, A. W. (1971). *J. Gen. Microbiol.* **68,** 167–172.
Romano, N., Smith, P. F., and Mayberry, W. R. (1972). *J. Bacteriol.* **109,** 565–569.
Romano, N., Rottem, S., and Razin, S. (1976). *J. Bacteriol.* **128,** 170–173.
Romijn, J. C., van Golde, L. M. G., McElhaney, R. N., and van Deemen, L. L. M. (1972). *Biochim. Biophys. Acta* **280,** 22–32.
Rothblat, G. H., and Smith, P. F. (1961). *J. Bacteriol.* **82,** 479–491.
Rottem, S., and Barile, M. F. (1976). *Antimicrob. Agents & Chemother.* **9,** 301–307.
Rottem, S., and Greenberg, A. S. (1975). *J. Bacteriol.* **121,** 631–639.
Rottem, S., and Panos, C. (1970). *Biochemistry* **9,** 57–63.
Rottem, S., and Razin, S. (1964). *J. Gen. Microbiol* **37,** 123–134.
Rottem, S., and Razin, S. (1973). *J. Bacteriol.* **113,** 565–571.
Rottem, S., Pfendt, E. A., and Hayflick, L. (1971). *J. Bacteriol.* **105,** 323–330.
Rottem, S., Muhsam-Peled, O., and Razin, S. (1973). *J. Bacteriol.* **113,** 586–591.
Rottem, S., Trotter, S. L., and Barile, M. F. (1977). *J. Bacteriol.* **129,** 707–713.
Ryan, M. D., Noker, P., and Matz, L. L. (1975). *Infect. Immun.* **12,** 799–807.
Schiefer, H. G., Gerhardt, U., Brunner, H., and Krüpe, M. (1974). *J. Bacteriol.* **120,** 81–88.
Schiefer, H. G., Krauss H., Brunner, H., and Gerhardt, U. (1975). *J. Bacteriol.* **124,** 1598–1600.
Shaw, N., Smith, P. F., and Koostra, W. L. (1968). *Biochem. J.* **107,** 329–333.
Shaw, N., Smith, P. F., and Verheij, H. M. (1972). *Biochem. J.* **129,** 167–173.
Smith, P. F. (1959). *J. Bacteriol.* **77,** 682–689.
Smith, P. F. (1962). *J. Bacteriol.* **84,** 534–538.
Smith, P. F. (1963). *J. Gen. Microbiol.* **32,** 307–319.
Smith, P. F. (1964). *J. Lipid Res.* **5,** 121–125.
Smith, P. F. (1968a). *Adv. Lipid Res.* **6,** 69–105.
Smith, P. F. (1968b). *J. Bacteriol.* **95,** 1718–1720.
Smith, P. F. (1969a). *Lipids* **4,** 331–336.
Smith, P. F. (1969b). *J. Bacteriol.* **99,** 480–486.
Smith, P. F. (1971a). "The Biology of Mycoplasmas." Academic Press, New York.
Smith, P. F. (1971b). *J. Bacteriol.* **108,** 986–991.
Smith, P. F. (1972a). *J. Bacteriol.* **112,** 554–558.
Smith, P. F. (1972b). *Biochim. Biophys. Acta* **280,** 375–382.
Smith, P. F. (1973). *Ann. N.Y. Acad. Sci.* **225,** 22–27.

Smith, P. F. (1977). *J. Bacteriol.* **130,** 393–398.
Smith, P. F., and Boughton, J. E. (1960). *J. Bacteriol.* **80,** 851–860.
Smith, P. F., and Henrikson, C. V. (1965a). *J. Lipid Res.* **6,** 106–111.
Smith, P. F., and Henrikson, C. V. (1965b). *J. Bacteriol.* **89,** 146–153.
Smith, P. F., and Henrikson, C. V. (1966). *J. Bacteriol.* **91,** 1854–1858.
Smith, P. F., and Koostra, W. L. (1967). *J. Bacteriol.* **93,** 1853–1862.
Smith, P. F., and Lynn, R. J. (1976). *Proc. Soc. Gen. Microbiol.* **3,** 164.
Smith, P. F., and Mayberry, W. R. (1968). *Biochemistry* **7,** 2706–2710.
Smith, P. F., and Rothblat, G. H. (1960). *J. Bacteriol.* **80,** 842–850.
Smith, P. F., and Smith, M. R. (1970). *J. Bacteriol.* **103,** 27–31.
Smith, P. F., Koostra, W. L., and Henrikson, C. V. (1965). *J. Bacteriol.* **90,** 282–283.
Smith, P. F., Langworthy, T. A., and Mayberry, W. R. (1976). *J. Bacteriol.* **125,** 916–922.
Sugiyama, T., Smith, P. F., Langworthy, T. A., and Mayberry, W. R. (1974). *Infect. Immun.* **10,** 1273–1279.
Tully, J. G., and Razin, S. (1969). *J. Bacteriol.* **98,** 970–978.
Vandemark, P. J., and Smith, P. F. (1964). *J. Bacteriol.* **88,** 1602–1607.
Vandemark, P. J., and Smith, P. F. (1965). *J. Bacteriol.* **89,** 373–377.
van Golde, L. M. G., McElhaney, R. N., and van Deenen, L. L. M. (1971). *Biochim. Biophys. Acta* **231,** 245–249.
Vilkas, E., Adam, A., and Senn, M. (1968). *Chem. Phys. Lipids* **2,** 11–16.
Weedon, B. C. L. (1971). *In* "Carotenoids" (O. Isler, ed.), pp. 29–59. Birkhaueser, Basel.
Welsh, K., Shaw, N., and Baddiley, J. (1968). *Biochem. J.* **107,** 313–314.
Wieslander, A., and Rilfors, L. (1977). *Biochim. Biophys. Acta* **466,** 336–346

# 10 / MOLECULAR ORGANIZATION OF MEMBRANE LIPIDS

*Shlomo Rottem*





THE MYCOPLASMAS, VOL. 1

ISBN 0-12-078401-7

## I. INTRODUCTION

The molecular organization and physical properties of biological membranes have been the subject of intensive studies over the past decade. The cornerstone of these studies was the demonstration that the cytoplasmic membrane of *Acholeplasma laidlawii* underwent a reversible thermal transition similar to that observed in artificial lipid bilayers (Steim *et al.*, 1969; Engelman, 1970), providing evidence that the lipids in a biological membrane are organized in a bimolecular leaflet. Since then mycoplasmas, mainly *A. laidlawii,* have been used as models in research on the biophysical characteristics of biomembranes. Their use stems from the possibility of introducing controlled alterations in the fatty acid composition of their membrane polar lipids, which is due to the total or partial inability of the organisms to synthesize long-chain fatty acids, making them dependent on the fatty acid supply in the growth medium (Razin *et al.*, 1966; Panos and Rottem, 1969; McElhaney and Tourtellotte, 1969). The exogenous fatty acids added to the growth medium are incorporated into *A. laidlawii* membrane lipids (Table I), constituting up to 90 mol% of the total fatty acid residues of membrane lipids, thus affecting the molecular packing of membrane lipids in a controlled and predictable manner. Mycoplasmas are also the only prokaryotes that require cholesterol for growth, making them a unique model for studying the localization of cholesterol in the membrane and its role as a regulator of membrane fluidity (Rottem *et al.*, 1973a,b; Razin and Rottem, 1978).

TABLE I. **Fatty Acid Composition of Membrane Lipids of *A. laidlawii* B Grown with Different Fatty Acids**[a]

| | Major fatty acids found (mol %) | | | | | | |
|---|---|---|---|---|---|---|---|
| Fatty acid added to growth medium | 12:0 | 14:0 | 16:0 | Anteiso 17:0 | 18:0 | Iso 18:0 | 18:1 |
| None | 5.0 | 30.8 | 40.1 | — | 7.7 | — | 1.8 |
| 16:0 | 8.5 | 6.9 | 80.4 | — | 1.9 | — | 1.6 |
| Anteiso 17:0 | 1.3 | 2.6 | 2.2 | 90.7 | 1.0 | — | 1.5 |
| 18:0 | 8.7 | 12.7 | 11.6 | — | 58.4 | — | 4.9 |
| Iso 18:0 | 4.3 | 8.9 | 6.8 | — | Trace | 76.4 | 1.5 |
| 18:1 cis | 2.2 | 6.6 | 22.1 | — | 4.5 | — | 64.4 |
| 18:1 trans | 2.9 | 4.2 | 7.2 | — | Trace | — | 85.4 |

[a] Data from McElhaney (1974b).

## II. FACTORS AFFECTING PHYSICAL PROPERTIES OF MEMBRANE LIPIDS

Intensive physical studies on model systems provided the empirical rules for relating membrane composition to membrane fluidity. The lipid components of a membrane are not rigid structural entities; it is likely that the hydrocarbon chains of the lipid molecules are in constant motion. The freedom of motion of the hydrocarbon chains within the lipid backbone of the membrane will determine the overall fluidity of membrane lipids. In general, fluidity is promoted by higher temperatures. At low temperatures, however, a reversible change in the state of the hydrocarbon chains of membrane phospholipids from a closely packed gel to a fluid state may take place (Steim *et al.*, 1969). Such a temperature-dependent change was variously termed: gel-to-liquid crystalline transition, phase transition, order-to-disorder transition, etc. (Fig. 1).

### A. Fatty Acids

The fluidity of membrane lipids and the temperature range of the gel-to-liquid crystalline phase transition are primarily dependent on the chain length and chemical structure of the hydrocarbon chains of membrane lipids (van Deenen, 1965). The cohesive forces between the hydrocarbon chains of the adjacent phospholipid molecules are predominantly London–van der Waals forces. These forces are additive and increase with a rise in the number of methylene groups in the interacting chains.

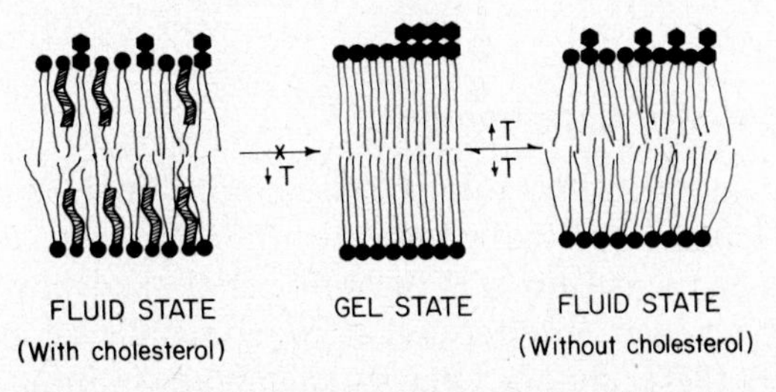


FIGURE 1. Schematic representation of the arrangements of membrane lipids in the gel and liquid crystalline states. The lipid bilayer is built of phospholipids ( ) and glycolipids ( ) with or without cholesterol ( ). Glycolipids are located in the outer half of the bilayer, phospholipids and cholesterol in both outer and inner halves. Decreasing (↓ T) or increasing (↑ T) temperatures result in a reversible gel-to-liquid transition of lipid bilayers but not of those with cholesterol. In the gel state the fatty acyl chains are closely packed, membrane thickness is increased, and the various lipid classes may form separate domains.

Increasing the chain length of the fatty acids increases the attractive forces between the adjacent phospholipids and results in a tightly packed lipid backbone where the freedom of motion of the hydrocarbon chains is low and the temperature of the gel-to-liquid phase transition is high (Ladbrooke and Chapman, 1969). Since the attractive forces between the methylene groups decline rapidly with increasing distance between the groups, unsaturated fatty acids, branched-chain fatty acids, cyclopropane-containing fatty acids, or any fatty acid containing a bulky side group that is incorporated into membrane phospholipids will increase membrane fluidity and decrease the gel-to-liquid crystalline phase transition (Ladbrooke and Chapman, 1969; McElhaney, 1974a; Razin, 1975).

### B. Polar Head Groups

The physical state of membrane lipids is affected also by varying the charge of the polar head groups of membrane lipids. Small changes in the ionic environment (pH, mono- and divalent cations) may induce gross alterations in the structure of lipid bilayers (Träuble and Eibl, 1974). Experiments with artificial lipid bilayers (Träuble *et al.*, 1976; Jähnig, 1976; Forsyth *et al.*, 1977) provide most of the information presently available, showing that with negatively charged phospholipids, a rather small increase in pH increases the charge per polar group and lowers the temperature range of the gel-to-liquid crystalline phase transition. Divalent cations by charge neutralization stabilize the gel state, whereas monovalent cations stabilize the fluid state and decrease transition temperatures.

### C. Cholesterol and Carotenoids

The fluidity and phase transition of membrane lipids is also dependent on the sterol content of the membrane. Sterols must contain a planar nucleus, a free hydroxyl group at the 3-$\beta$ position and a hydrocarbon side chain in order to exert an effect on artificial membrane systems (Demel and de Kruyff, 1976) and mycoplasma membranes (Smith, 1964; Rottem *et al.*, 1971; de Kruyff *et al.*, 1972, 1973a,b, 1974). A sterol such as cholesterol, by virtue of its peculiar molecular shape, will be oriented in the membrane in such a way that its rigid ring system will be aligned parallel to the hydrocarbon chains of membrane phospholipids with the polar hydroxyl group anchoring one end of the cholesterol molecule to the polar surface of the bilayer (Fig. 1). The planar hydrocarbon part of the cholesterol molecule that extends toward the center of the bilayer exerts a condensing effect on lipids at temperatures above the phase transition, but at temperatures below the phase transition, cholesterol will prevent the cooperative crystallization of the hydrocarbon chain and will therefore eliminate the phase transitions (Fig. 1). By eliminating the phase transition at low temperatures and decreasing membrane fluidity at high temperatures, cholesterol creates an intermediate fluid state of the lipid bilayer (Rothman and Engelman, 1972).

The idea that in *Acholeplasma* carotenoids with a planar hydrocarbon structure may fulfill functions analogous to those of cholesterol in the *Mycoplasma* species was put forward by Smith (1963, 1967). This controversial issue was discussed in detail by Razin (1969) and was recently resurrected by Huang and Haug (1974), who showed that the carotenoid pigments of *A. laidlawii* exhibit a condensing effect on membrane lipids in a manner similar to cholesterol, resulting in a decrease in membrane fluidity.

## D. Membrane Proteins

The effect of membrane proteins on the mobility of the hydrocarbon chains of membrane lipids was first noted in spin-labeling studies of *A. laidlawii* membranes (Rottem *et al.*, 1970; Tourtellotte *et al.*, 1970). The fluidity of membranes was lower in membrane preparations than in lipid dispersions prepared from the same membranes, and increasing the protein-to-lipid ratio in reconstituted membranes resulted in decreased fluidity (Rottem *et al.*, 1970). Further studies (Rottem and Samuni, 1973) showed that proteolytic treatment and subsequent binding of cytochrome *c* or lysozyme resulted in an initial increase followed by a decrease in the fluidity of the lipid bilayer.

## E. Age of Culture

During the progression of mycoplasma cultures from the early logarithmic phase to the stationary phase of growth, alterations in the physical properties of the cell membrane were evident with a variety of mycoplasma species (Rottem *et al.*, 1970, 1978; Rottem and Greenberg, 1975; Table III in Chapter 11). Upon aging of *A. laidlawii* cultures a decreased mobility of the hydrocarbon chains of membrane lipids was found (Rottem *et al.*, 1970) and was attributed both to an increase in the ratio of saturated to unsaturated fatty acids and to a decrease in the lipid-to-protein ratio of the membrane. The fatty acid composition of the membrane phospholipids in *Mycoplasma hominis* was, however, unchanged (Rottem and Greenberg, 1975) and palmitic acid predominated throughout the cell growth. The decreased fluidity and increased density of membrane preparations from aged *M. hominis* cultures were a consequence of the steep decline in the rate of phosphatidylglycerol biosynthesis, resulting in a decrease in the lipid-to-protein ratio of the membrane. The pronounced decrease in the phospholipid content of a variety of mycoplasma species upon aging of the cultures resulted in a subsequent decrease in the cholesterol-to-membrane protein ratio but the ratio of cholesterol to phospholipid remained about constant (Razin, 1974). It seems plausible that the changes in the fluidity of membrane lipids on aging may result in a marked decrease in the activity of membrane-associated enzymes and transport systems (Chapters 11 and 12) leading to cell death.

## III. THE ORGANIZATION OF LIPIDS IN MYCOPLASMA MEMBRANES

A wide variety of physical methods is available for analysis of the properties of biological membranes. The procedures applied to mycoplasma membranes have included differential scanning calorimetry (DSC) (Steim *et al.*, 1969; Reinert and Steim, 1970; Melchior *et al.*, 1970; Chapman and Urbina, 1971), differential thermal analysis (McElhaney *et al.*, 1973; McElhaney, 1974b), X-ray diffraction (M. Engelman, 1970, 1971), electron paramagnetic resonance spectroscopy (Rottem *et al.*, 1970; Tourtellotte *et al.*, 1970), nuclear magnetic resonance spectroscopy (Oldfield *et al.*, 1972 ; Metcalfe *et al.*, 1972; Stockton *et al.*, 1975, 1977; de Kruyff *et al.*, 1976), light scattering (Abramson and Pisetsky, 1972), and fluorescence polarization (Rottem *et al.*, 1973b).

### A. Differential Scanning Calorimetry

The most striking event in the membrane is the gel-to-liquid crystalline phase transition. Since melting of most materials is accompanied by heat absorption, a phase change in the membrane will result in uptake of heat. The heat absorption was first observed in *A. laidlawii* membranes by Steim *et al.* (1969), using differential scanning calorimetry (DSC), and later by McElhaney and co-workers (1973; McElhaney, 1974b), using differential thermal analysis. The peak produced by the heat absorption has a shape and position on a temperature scale reflecting the course of the transition, and an area that gives the heat of the transition (Melchior and Steim, 1978). The temperature-base thermograms of *A. laidlawii* membrane lipids (Fig. 2) show that phase transition temperatures decrease as membrane lipids are progressively enriched with fatty acids of

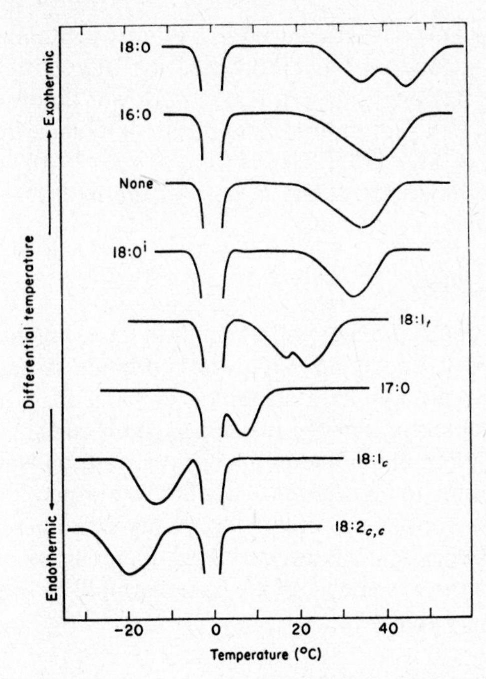


FIGURE 2. Temperature-base thermograms of membranes of *A. laidlawii* cells grown in medium supplemented with different fatty acids. The large endotherm centered around 0°C is due to melting of the ice from excess water associated with the membrane preparations. (From McElhaney, 1974b.)

decreasing melting points. Unlike the narrow and well-defined thermotropic phase transition of dispersions of a specific phospholipid containing uniform fatty acid chains, the phase transition temperature range of isolated membranes or their derived lipid dispersions is broad, and may range up to 30°C (McElhaney, 1974b). This broad range is apparently due to the fact that the bulk of membrane lipids consists of a mixture of lipids with different transition temperatures (Chapman and Urbina, 1971). The similar fatty acid composition and positional distribution of the fatty acids in the various phospho- and glycolipids of *A. laidlawii* (Saito *et al.*, 1977a) suggest that the variations in transition temperatures of the diverse lipid classes of *A. laidlawii* are due to differences in the polar head groups rather than differences in fatty acid composition or distribution. The temperature range of the gel-to-liquid crystalline transition obtained by DSC of *A. laidlawii* membranes was the same as that of membrane lipid dispersions (Steim *et al.*, 1969), and was unaffected by removal of membrane proteins by proteolytic digestion (Melchior *et al.*, 1970). The interpretation of these results in *A. laidlawii* membranes was that over 90%

of the membrane lipids are organized in a bilayer containing hydrocarbon chains which associate with each other rather than with proteins. Spin-labeling studies, however, show a markedly lower fluidity in *A. laidlawii* membranes than in a protein-free lipid dispersion made from these membranes (Rottem *et al.*, 1970; Tourtellotte *et al.*, 1970), suggesting that membrane proteins do affect the fluidity of the hydrocarbon chains.

### B. X-Ray Diffraction

X-Ray diffraction is the most direct method for examining phase transition. The studies of Engelman (1970, 1971) provide strong evidence that the bilayer is the predominant structural element in *A. laidlawii* membranes and demonstrate that the phase transition consists of a change in the lipid hydrocarbon chain within the bilayer, shifting from a liquid state at high temperatures to an ordered array at low temperatures (Engelman, 1971). The phase transitions detected by X-ray diffraction were found to be consistently dependent on the fatty acid composition and were within the same temperature range as those demonstrated by differential scanning calorimetry (Table II).

### C. Electron Paramagnetic Resonance Spectroscopy

The application of electron paramagnetic resonance (epr) spectroscopy to mycoplasma membranes yielded information about the average lipid fluidity, the orientation of lipid components and the phase properties of the membrane. Two types of spin labels have been particularly useful in mycoplasma membrane studies. Both are substituted nitroxides with magnetic properties that result from an unpaired electron, localized primarily on the nitrogen atom (McConnell and McFarland, 1970). The first type consists of spin-labeled fatty acids with nitroxide radicals attached to different carbon atoms on the fatty acid chain. The spin-labeled fatty acids can be incorporated into isolated membranes (Rottem *et al.*, 1970) or can be added to the growth medium and utilized by the organism for the synthesis of spin-labeled complex lipids (Tourtellotte *et al.*, 1970). The ease and reliability of determining the overall membrane fluidity with spin-labeled fatty acids made it a most useful technique for comparative fluidity studies (Rottem *et al.*, 1970, 1978; Tourtellotte *et al.*, 1970; Rottem and Samuni, 1973; Rottem and Greenberg, 1975). Arrhenius plots of the motion parameter ($\tau_0$) versus $°K^{-1}$ of spin-labeled fatty acids incorporated into *A. laidlawii* membranes enriched with different fatty acids

TABLE II. **Some Gel-to-Liquid Crystalline Phase Transitions Observed in *A. laidlawii* Membranes**[a]

| Technique | Temperature range (°C) of phase transitions of membranes from cells grown with: | | | | | References |
|---|---|---|---|---|---|---|
| | Myristate (14:0) | Palmitate (16:0) | Stearate (18:0) | Oleate (18:1 cis) | Elaidate (18:1 trans) | |
| X-Ray diffraction | 34–42° | 28–37° | — | — | 23–30° | Wallace *et al.*, 1976 |
| Calorimetry | — | — | 37–73 | −12 to −22 | — | Steim *et al.*, 1969 |
| | — | 26–44 | — | — | 13–35 | de Kruyff *et al.*, 1973[c] |
| | — | 20–50 | — | −4 to −22 | 5–32 | McElhaney, 1974[b] |
| epr spectroscopy | 36 | — | — | — | 22 | James and Branton, 1973 |
| | — | 35 | — | — | 26 | Rottem *et al.*, 1977 |
| $^3$H-nmr spectroscopy | — | 30–44 | — | — | — | Stockton *et al.*, 1975 |

[a] The mol % of the specific fatty acid incorporated into membrane lipids varied up to 20% of the total fatty acid content among the different studies.

showed inflection points at discrete temperatures that were interpreted as transition temperatures (James and Branton, 1973; Rottem *et al.*, 1977). The inflection points are within the temperature range of the gel-to-liquid crystalline phase transition observed by DSC or X-ray diffraction techniques and were usually located at a temperature close to the lower boundary of the transition (Table II). The second group of spin labels are small nitroxide molecules such as 2,2,6,6,-tetramethylpiperidine-*N*-oxyl (TEMPO) that are soluble in hydrophobic as well as in aqueous media. Partition of these probes between the hydrophobic core of the membrane and the surrounding aqueous media is determined by the physical state of membrane lipids (Shimshick and McConnell, 1973). A break in the TEMPO partitioning in *A. laidlawii* was taken as a measure of phase transition (Metcalfe *et al.*, 1972).

## D. Nuclear Magnetic Resonance Spectroscopy

Nuclear magnetic resonance (nmr) studies on *A. laidlawii* membranes have provided strong evidence for the fluid nature of membrane lipids. The use of selectively perdeuterated fatty acids (Oldfield *et al.*, 1972; Stockton *et al.*, 1975, 1977), $^{13}$C-enriched fatty acids (Metcalfe *et al.*, 1972), or selectively $^{31}$P-enriched phospholipids (de Kruyff *et al.*, 1976) provided a more detailed and quantitative understanding of the structure and organization of the hydrocarbon and polar regions of membrane lipids. Unlike the nitroxide-containing electron-spin resonance probes, most nmr probes do not significantly perturb their environment (Seelig and Seelig, 1974). Deuterium magnetic resonance spectroscopy ($^{2}$H-nmr) of *A. laidlawii* membranes from cells grown with perdeuterated lauric or palmitic acid (Oldfield *et al.*, 1972) or with palmitic acid labeled at the terminal methyl group with deuterium (Stockton *et al.*, 1975, 1977) indicated that only at 44°C, well above the optimal growth temperature, are membrane lipids entirely in the liquid crystalline phase. At 37°C or below, membrane lipids are in both a gel and a liquid crystalline state.

Information on the motion of the phosphate head groups in *A. laidlawii* membranes was obtained by studying the $^{31}$P-nmr spectra of *A. laidlawii* membranes (de Kruyff *et al.*, 1976). The observation that the $^{31}$P-spectrum was insensitive to pronase digestion and subsequent cytochrome *c* binding was taken to indicate that no lipid polar headgroup–protein interactions occur in *A. laidlawii* membranes, or that the lipid–protein "complexes" in the membrane have a fast rotation along an axis perpendicular to the plane of the membrane. The $^{31}$P-nmr spectrum of the membranes was sensitive to the gel-to-liquid crystalline phase transition. The broadening of the $^{31}P$ resonance obtained in the gel state suggests that the motion of the polar head group of membrane phospholipids is more restricted at temperatures below the lipid phase transition temperature.

## E. Other Techniques

Changes in the light-scattering properties of *A. laidlawii* membranes within the temperature range of the phase transition were described by Abramson and Pisetsky (1972) and were attributed to changes in the refractive index accompanying the increase in the hydration of the lipid polar head groups. Fluorescent polarization techniques, extensively employed to determine the fluidity and phase transition of *Escherichia coli* membranes (Cronan and Gelman, 1975) were applied only once to mycoplasma membranes (Rottem *et al.*, 1973b). Fluorescence polarization of diphenylhexatriene in *M. mycoides* subsp. *capri* showed an increased fluidity and a detectable gel-to-liquid phase transition in a cholesterol-depleted strain.

The results obtained by a variety of physical studies firmly established that the bulk of membrane lipids in mycoplasmas constitute a lipid bilayer. However, whereas DSC and X-ray diffraction studies suggested that most lipid hydrocarbon chains associate with each other rather than with proteins, epr spectroscopy (Rottem *et al.*, 1970; Tourtellotte *et al.*, 1970) suggests that some membrane lipids interact with hydrophobic proteins which are partially embedded in, or traverse, the hydrocarbon core of the membrane (Amar *et al.*, 1974). The membrane lipids underwent a reversible gel-to-liquid crystalline phase transition, with the bilayer conformation retained throughout the process. The phase transition is consistently dependent on the chain length and degree of unsaturation of the fatty acid residues of membrane lipids and is undetectable in cholesterol-containing membranes. Because of the heterogeneity of the polar head groups and of the fatty acid chains in the various classes of lipids the phase transition process may range up to 30°C (McElhaney, 1974b). The entire temperature range of the phase transition can be detected calorimetrically or by X-ray diffraction. However, techniques such as epr or fluorescent spectroscopy, where external probes are used, may detect a discrete temperature within the phase change (Table II), which is related to the temperature of a phase transition of specific regions where the probe is concentrated rather than to the transition of the bulk of the membrane lipids (Gaffney and McNamee, 1974). Electron paramagnetic resonance spectroscopy may also detect thermotropic phenomena other than gel-to-liquid crystalline phase transitions, such as cluster formation, liquid–liquid phase separation, transitions in the solvation shells of proteins, etc. (Sackman *et al.*, 1973; Lee *et al.*, 1974; Baldassare *et al.*, 1976; Melchior and Steim, 1978).

## IV. REGULATION OF MEMBRANE FLUIDITY

The physical properties of membrane lipids are vitally important to mycoplasmas. *Acholeplasma laidlawii* was not able to grow at temperatures at which its membrane lipids exist entirely in the gel state (McElhaney, 1974b), confirming the notion that some of the lipid hydrocarbon chains must be in a fluid state to support proper membrane function (Steim *et al.*, 1969). *Acholeplasma laidlawii* cells with up to about half of their membrane lipids in the gel state are able to function as well as cells whose membrane lipids are entirely in the fluid state, but growth rates decrease proportionately with the increase of membrane lipids in the gel state, and growth ceases entirely when more than 90% of membrane lipids are in the gel state (McElhaney, 1974b). The fact that growth and replication of mycoplasmas require a lipid bilayer in a fluid or partially fluid state is due, to a certain extent, to the findings that complete crystallization of membrane lipids results in a marked reduction in the activity of membrane transport systems (Read and McElhaney, 1975; Chapter 7E), activity of membrane-associated enzymes (de Kruyff *et al.*, 1973c; Rottem *et al.*, 1973b; Hsung *et al.*, 1974; Rottem *et al.*, 1977), the permeability of the cells to nonelectrolytes (McElhaney *et al.*, 1970, 1973; de Kruyff *et al.*, 1973a; Rottem *et al.*, 1973b), and elasticity of the membrane (van Zoelen *et al.*, 1975). A regulatory mechanism is therefore necessary to maintain membrane lipids in an appropriate physical state when growth temperature or the fatty acid composition of the medium are changed.

Several types of lipid changes could alter the phase properties of mycoplasma membranes. An increase in the average chain length of the fatty acyl chains or a decrease in the ratio of unsaturated-to-saturated fatty acid moieties would raise the transition temperatures (Steim *et al.*, 1969). Altering the cholesterol content of mycoplasma cell membranes will modify the molecular packing of the phospholipid acyl chains (Rothman and Engelman, 1972). Changes in the distribution of polar head groups of the negatively charged phospholipids of mycoplasma membranes, or allowing these groups to interact with divalent cations, could also markedly change transition temperatures (Träuble and Eibl, 1974).

### A. Regulation of Membrane Fluidity by Varying the Fatty Acid Composition

The regulation of acyl chain length and of the degree of unsaturation seem to play a major role in controlling membrane fluidity of the sterol-nonrequiring *Acholeplasma* species. A regulatory mechanism that senses temperature and maintains a constant membrane fluidity was demonstrated in *A. laidlawii* strain A, where decreasing the growth temperature to 15°C caused a significant increase in the amount of exogenous oleic acid incorporated into membrane lipids (Rottem *et al.*, 1970). This resulted in higher fluidity of membranes from cells grown at 15°C or 28°C than those at 37°C (Rottem *et al.*, 1970; Huang *et al.*, 1974a). Accordingly, a direct downshift in transition temperature at lower growth temperatures was demonstrated by Melchior *et al.* (1970) with the closely related *A. laidlawii* strain B using DSC analyses. When cells were grown at 37°C, membrane lipids were fluid at 37°C but entirely in the gel state at 25°C, whereas when they were grown at 25°C, membrane lipids were mostly fluid even at 25°C. Temperature control of fatty acid composition in bacteria was first noted in *E. coli* by Marr and Ingraham (1962). The phenomenon was intensively investigated in *E. coli* (for review, see Cronan and Gelman, 1975), and the conclusion was reached that the mechanism responsible for the temperature-induced fatty acid alteration operates at the level of both fatty acid biosynthesis, by changing the ratio of unsaturated to saturated fatty acids, and the acyltransferase-mediated incorporation of fatty acids into the glycerol backbone. In *A. laidlawii,* where saturated but not unsaturated fatty acids are synthesized (Rottem and Razin, 1967; Pollack and Tourtellotte, 1967), a thermal regulatory mechanism at the level of fatty acid biosynthesis is expected to affect the chain length of biosynthetic output by decreasing the chain length at lower growth temperatures and increasing it at higher growth temperatures. Yet, changes in the growth temperature produced only minor alterations in the pattern of the saturated fatty acids derived from *de novo* fatty acid biosynthesis (Saito *et al.*, 1977b) or from elongation of exogenous medium-chain fatty acids (Saito *et al.*, 1978). In addition, the temperature range of the thermotropic phase transition of membranes derived from cells grown in fatty acid-free media at various temperatures varied only slightly with growth temperatures (McElhaney, 1974b). It seems, therefore, that the mechanism responsible for temperature-induced fatty acid alterations operates not at the fatty acid biosynthesis level, but at the level of the transacylation reaction, where the membrane-bound transacylase is involved (Rottem and Greenberg, 1975). As the growth temperature is decreased, *A. laidlawii* adjusts the fatty acid composition of its complex lipids by incorporating exogenously supplied fatty acids with progressively lower melting points. Since the requirement for fatty acids with lower melting points can be fulfilled by cis-unsaturated, trans-unsaturated, branched-chain, or cyclopropane-containing fatty acids, it seems that the physical properties rather then the chemical properties and the electronic configuration of the exogenous fatty acid added to the growth medium are important in determining the suitability of the fatty acid for the acyl transfer (Saito and McElhaney, 1977; Melchior and Steim, 1978). It was therefore suggested that the temperature-sensing selection mechanism is thermodynamically determined and is not dependent upon the acyltransferase specificity (Melchior and Steim, 1977). Thus, the control of membrane fluidity resides in the membrane itself, and the membrane-bound acyltransferase is not required to distinguish between different fatty acids. In fact, the pattern of esterification of palmitate and oleate at various temperatures closely parallels the physical state of the membrane bilayer and the physical binding of free fatty acids to liposomes made from membrane lipids (Fig. 3). As the temperature is lowered, an increased amount of oleate relative to palmitate is accepted by the bilayer where it is utilized by the transacylase for complex lipid biosynthesis.

Although the *de novo* fatty acid biosynthesis and the chain elongation systems in *A. laidlawii* are not affected by a shift in the growth temperatures, it seems as if both these mechanisms are involved in regulating the fatty acid composition of membrane lipids in response to variation in the fatty acid composition of the growth medium (Silvius *et al.*, 1977; Saito *et al.*, 1978). Thus, the spectrum of the end products of the *de novo* fatty acid biosynthesis pathway (Silvius *et al.*, 1977) as well as the products of the chain elongation system (Saito *et al.*, 1978) are influenced by exogenous fatty acids added to the growth medium in such a way that exogenous unsaturated or branched-chain fatty acids having low melting points tend to enhance the mean chain length of *de novo* biosynthesized saturated fatty acids and to increase the extent of chain elongation, while exogenous long-chain saturated fatty acids having high melting points tend to exert an opposite influence. Hence, the *de novo* fatty acid biosynthesis and the chain elongation system(s) serve to buffer the physicochemical effect of exogenous fatty acid incorporation by a compensatory shift in the average chain length of the product. A possible explanation for the mechanism of such a regulatory system was proposed by Silvius *et al.* (1977). Since in *A. laidlawii* fatty acids with low melting points are preferentially directed to the 2 position of the glycerol backbone, while fatty acids with high melting points predominate in the 1 position (Romijn *et al.*, 1973; Saito *et al.*, 1977a), an exogenous fatty acid with a low melting point will compete with

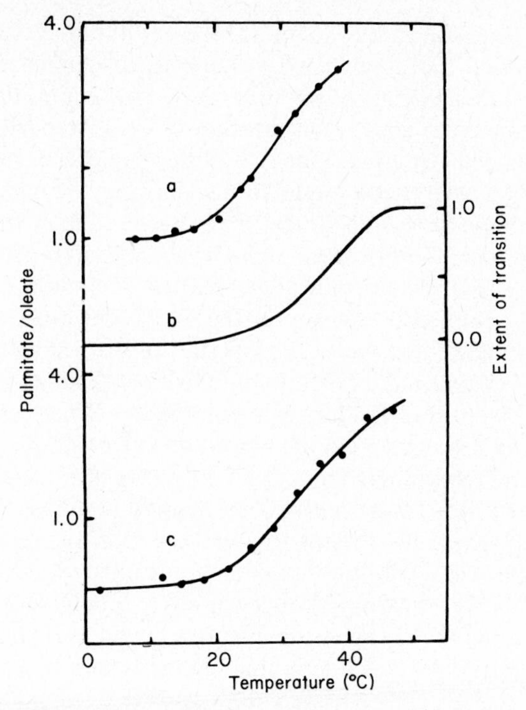


FIGURE 3. Correlation between the palmitate/oleate ratio incorporated into *A. laidlawii* membrane lipids (curve a), the extent of transition in membranes (curve b), and the palmitate/oleate ratio of fatty acids physically bound to bilayers of extracted membrane lipids (curve c). (From Melchior and Steim, 1977.)

the short-chain exogenously synthesized fatty acids for transfer to the 2 position of the glycerol. If the termination of fatty acid chain elongation in *Acholeplasma* is determined by the competition between the fatty acid elongation and the transfer of the acyl groups to the glycerol backbone, an exogenous low-melting fatty acid specific for the 2 position will decrease the rate of acyl transfer of the shorter chain biosynthetic products, enabling continued elongation of the fatty acyl thioesters.

## B. Cholesterol as a Regulator of Membrane Fluidity

The idea that in biological membranes cholesterol regulates the fluidity of membrane phospholipids, forming an intermediate gel state (Rothman and Engelman, 1972), gained strong support from studies with mycoplasmas. These studies showed that the sterols required to promote the growth of *Mycoplasma* (Smith, 1960), *Ureaplasma* (Rottem *et al.*, 1971), and

*Spiroplasma* (Freeman *et al.*, 1976) are those with molecular properties required to induce the condensing effect and to eliminate the thermal phase transition of artificial membranes (de Kruyff *et al.*, 1972, 1973a,b). The ability to alter the cholesterol content of *M. mycoides* subsp. *capri* by adapting the cells to grow with very little cholesterol (Rottem *et al.*, 1973a) provided a most useful model for establishing the regulatory role of cholesterol in biologic systems. The cholesterol content of the adapted strain was less than 3% of the total membrane lipids as against 22–26% in membranes of the native strain. The marked reduction in cholesterol levels produces profound changes in the fatty acid composition, ultrastructure, and biologic and physical properties of the cell membrane of the adapted cells (Rottem *et al.*, 1973a,b) but the most remarkable differences were the increased lipid fluidity in membranes of the cholesterol-poor adapted strain and the detection of a thermotropic gel-to-liquid crystalline phase transition in the adapted but not in the native strain. The differences in the phase behavior of the native and adapted strains is manifested in membrane ultrastructure. Freeze-fracture electron microscopy enables one to compare the hydrophobic membrane core of the native and adapted strains (Fig. 4). Smooth-faced areas are believed to be mainly lipid domains while the particles are apparently of a protein nature (Tourtellotte and Zupnik, 1973). Chilling the adapted strain to 4°C prior to the

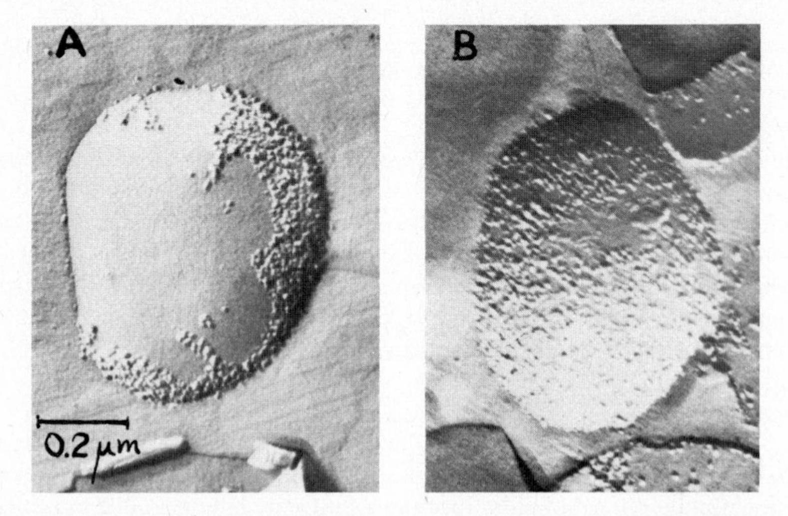


FIGURE 4. Platinum–carbon replica of the convex fracture faces of freeze-fractured *M. mycoides* subsp. *capri* membranes incubated at 4°C prior to deep freezing. (a) Membranes of the cholesterol-depleted adapted strain. (b) Membranes of the native strain. (From Rottem *et al.*, 1973a.)

quick freezing causes the aggregation of particles, leaving over 66% of the fracture faces particle free. No such phenomenon could be demonstrated with the native strain. The aggregation phenomenon is probably due to a thermotropic phase transition of membrane lipids (James and Branton, 1973). The differences in phase properties of the native and adapted strains provided the first evidence to support the notion that cholesterol in a biological membrane functions as a regulator of membrane lipid fluidity by inducing an intermediate fluid state during changes in growth temperature or following alterations in the fatty acid composition of membrane lipids. Cholesterol, by preventing the crystallization of membrane lipids of the native *M. mycoides* subsp. *capri* cells at lower temperatures, keeps membrane lipids in a sufficiently fluid state to support the functions of key membrane enzymes such as the membrane-bound adenosine triphosphatase (ATPase) (Rottem *et al.*, 1973b). Thus, growth of the native cholesterol-rich strain continues at temperatures lower than 37°C, while in the adapted cholesterol-poor strain, growth was almost completely arrested at temperatures where most of the membrane lipids are in a gel state (25°C and below). The low amounts of cholesterol that can be incorporated into *A. laidlawii,* which does not require sterols, were not sufficient to eliminate the phase transition of membrane lipids but only to reduce the energy content of the phase transition (de Kruyff *et al.*, 1972). The incorporation of cholesterol into *A. laidlawii* membranes reduced, however, the permeability of the cells to nonelectrolytes, an observation that can be satisfactorily rationalized by postulating a condensing effect of cholesterol upon the *A. laidlawii* lipid bilayer (McElhaney *et al.*, 1970, 1973; de Kruyff *et al.*, 1972, 1973a). The requirement of mycoplasmas for cholesterol as a regulator that helps to maintain a constant membrane fluidity is probably due to the necessity of overcoming the inability of these cells to operate mechanisms for controlling membrane fluidity at the levels of fatty acid and complex lipid biosynthesis. Such mechanisms operate efficiently in bacteria (Cronan and Gelman, 1975), and to a certain extent in the *Acholeplasma* species that are capable of varying the chain length of the fatty acid synthesized (Silvius *et al.*, 1977) or selectively incorporating exogenous fatty acids from the growth medium (Melchior and Steim, 1977). However, all *Mycoplasma* and *Spiroplasma* species tested to date are incapable of synthesizing or elongating fatty acids (see Chapter 9). Furthermore, it is not clear if the sterol-requiring mycoplasmas possess the ability to adjust incorporation of fatty acids from the growth medium in response to variations in growth conditions, since studies carried out so far with *M. hominis* (Rottem and Razin, 1973) and *Spiroplasma citri* (Freeman *et al.*, 1976) showed that these organisms preferentially incorporated palmitic acid into their membrane phospholipids. The incorporation of large quantities of sterol into the membrane of the cholesterol-requiring parasitic mycoplasmas may be necessary to prevent membrane lipids from crystallization at their optimal growth temperatures (Razin and Rottem, 1978).

## V. LIPID-DEPENDENT STRUCTURAL AND FUNCTIONAL CHANGES IN MYCOPLASMA MEMBRANES

There is little doubt that the lipid phase properties of mycoplasma membranes are of vital importance to the growth and metabolism of the cells. Evidence that pointed to a correlation between the lipid phase properties, membrane structure, and membrane-associated physiologic processes was accumulated during the past few years. The studies on the effect of the physical state of membrane lipids on the permeability and transport processes of mycoplasmas are reviewed in Chapter 12 and, therefore, for the most part, are not dealt with in this chapter.

### A. Distribution of Membrane Proteins

Evidence that the physical properties of mycoplasma membrane lipids may change the distribution of the intramembranous protein particles in the membrane was obtained from freeze-fracture electron microscopy (Verkleij *et al.*, 1972; James and Branton, 1973; Rottem *et al.*, 1973a). In general, *A. laidlawii* membranes frozen from temperatures above the phase transition have a random distribution of intramembranous particles, whereas these particles are aggregated or patched below the transition temperature as a result of "squeezing out" of particles by the lipids as they undergo phase separation (James and Branton, 1973; Rottem *et al.*, 1973a). Data that temperature-induced changes in the physical state of membrane lipids can also cause changes in the location of surface proteins was presented by Wallace *et al.* (1976) using an avidin–ferritin label that was specifically bound to membrane protein. By visualizing the distribution of the complex on the surface of *A. laidlawii* cells, they concluded that in the liquid crystalline or gel phase the labeled sites were dispersed but in membranes labeled at an intermediate temperature, where lipids are in a mixture of the two phases, patches of low- and high-density label were found. It therefore appears that both intramembranous protein particles and surface proteins of mycoplasma are mobile and their relative positions are affected by the physical state of membrane lipids, but whereas intramembranous particles are completely aggregated at the completion of the phase transition and the aggregates are not dispersed as temperature is lowered further (James and Branton, 1973), surface proteins seem to be aggregated at temperatures within the boundaries of the phase transition but dispersed at lower temperatures (Wallace *et al.*, 1976).

## B. Membrane Thickness

The fatty acid composition and physical properties of membrane lipids also have an effect on the thickness of the *A. laidlawii* membranes. X-Ray diffraction studies with *A. laidlawii* membranes enriched with either erucic (C22:1 cis) or palmitic (C16:0) acid showed that the lipid bilayer thickness increased in proportion to the average fatty acyl chain length (Engelman, 1971). Thus, below the phase transition a thickness of 52 Å was measured in the erucate-enriched membranes as against 47 Å in palmitate-enriched membranes. As the chain became more fluid, the dimensions of the bilayer might decrease, and in the fluid phase, the thickness of erucate or palmitate-enriched membranes were 41 and 38 Å respectively. However, membrane thicknesses, measured by thin-section electron microscopy of *A. laidlawii* cells enriched with palmitic or arachidic (C22:0) acid, were found to be identical (Huang *et al.*, 1974b).

## C. Cell Growth

*Acholeplasma laidlawii* shows changes in cell morphology and growth rate depending on the fatty acid supplemented (Razin *et al.*, 1966; McElhaney and Tourtellotte, 1969); thus, when the organisms are grown on oleate, growth rate is high and the cells are filamentous, while poor growth is obtained when they are grown on palmitate, and the cells are spherical. A more precise correlation between growth and the thermotropic transition of membrane lipids revealed that the physical state of membrane lipids has a marked effect on both the temperature range within which *A. laidlawii* cells can grow and on growth rates within the permissible temperature range (McElhaney, 1974b). The minimum growth temperature is defined by the lower boundary of the gel-to-liquid crystalline phase transition of membrane lipids. The optimum and maximum growth temperature are also influenced by the boundaries of the phase transition but to a much smaller extent. The cells do not grow at temperatures below their transition (Steim *et al.*, 1969; McElhaney, 1974b) but some cell growth continues at temperatures within the temperature range of the transition, where most of the membrane lipids are in the gel state. Growth does not stop until only about 10% of membrane lipids remain fluid (McElhaney, 1974b).

### D. Cell Integrity

Lacking a cell wall, the integrity of mycoplasma cells is primarily determined by their osmotic fragility. The precise mechanism of osmotic lysis is still unknown, but there is no doubt that it is determined, at least in part, by the physical state of the membrane lipids (Tourtellotte, 1972). The idea that unsaturated fatty acids may allow for a looser packing of mycoplasma membrane lipids resulting in greater elasticity of the membrane was advanced some time ago (Razin *et al.*, 1966). Furthermore, it was shown (Rottem and Panos, 1969) that the replacement of oleate (C18:1 cis) by elaidate (C:18:1 trans) resulted in a decreased osmotic fragility, suggesting that an optimal fluid state is required to maintain osmotically stable cells. When part of the membrane lipids are in the fluid phase, the elasticity of the cell membrane enables the cells to swell and behave as a good osmometer. However, when membrane lipids are in the gel phase, the cells are unable to swell and will lyse in a slightly hypotonic medium (van Zoelen *et al.*, 1975).

### E. Enzymatic Activity

The marked effect of the physical state of membrane lipids on the growth of *A. laidlawii* suggests that changes in membrane fluidity affect functions that require a fluid lipid environment for optimal activity, such as permeability, transport mechanisms, and enzymatic activities. Direct support for this hypothesis was obtained when the relationship between several membrane-bound enzymes and lipid phase transitions were studied. De Kruyff *et al.* (1973c) examined the rate–temperature profiles for the membrane-bound $Mg^{2+}$-dependent adenosine triphosphatase (ATPase) of *A. laidlawii* from cells supplemented with various fatty acids, and reported inflection points in the slopes of the Arrhenius plots of the activities that represented changes in the activation energy of the enzyme, with lower activation energy at temperatures above the inflection point and higher at temperatures below it. The inflection points were found to be dependent on the fatty acid composition and seem to occur at the lower boundary of the gel-to-fluid phase transition (de Kruyff *et al.*, 1973c; Hsung *et al.*, 1974). Inflection points were also noted in the Arrhenius plot of a membrane-bound thioesterase activity of *A. laidlawii* (Rottem *et al.*, 1977), but not in the Arrhenius plots of the NADH oxidase and *p*-nitrophenylphosphatase activities (de Kruyff *et al.*, 1973c; Rottem *et al.*, 1977), although the membrane lipids underwent a phase transition within the temperature range where enzyme activities were determined. Although these two enzymes are membrane-bound, they do not require lipids for activity (Ne'eman *et al.*, 1972). A lipid dependence of the $Mg^{2+}$–ATPase was previously suggested (Rottem and Razin, 1966;

Ne'eman *et al.*, 1972) and more recently proved to be a specific requirement for phosphatidylglycerol (Bevers *et al.*, 1977b).

The inflection points observed with both $Mg^{2+}$-dependent ATPase and thioesterase were at a single temperature, close to the lower boundary of the lipid phase transition (de Kruyff *et al.*, 1973c; Rottem *et al.*, 1977). Thus, plots of thioesterase activity of elaidate- or palmitate-enriched membranes showed inflection points at 12° and 18°C, respectively. No inflection points were noted in linoleate- or oleate-enriched membranes which do not exhibit a lipid phase transition within the temperature range tested. The possibility brought up by de Kruyff *et al.* (1973c) that the ATPase of *A. laidlawii* is associated with a boundary lipid having a low transition temperature, thus showing inflection points at the lower end of the transition temperature, was recently challenged by Bevers *et al.* (1977b). They demonstrated by reconstitution experiments that the fatty acid composition of both the boundary phospholipids, as well as that of the bulk phospholipids, determine the activation energy and the inflection temperature in the Arrhenius plot of the $Mg^{2+}$-dependent ATPase activity. The pronounced effect of cholesterol on the physical state of membrane lipids is also reflected in the activity of the $Mg^{2+}$-dependent ATPase of mycoplasmas. Arrhenius plots of the ATPase activity of the cholesterol-depleted *M. mycoides* subsp. *capri* adapted strain showed inflection points at temperatures that corresponded well with the thermotropic membrane phase transition (Rottem *et al.*, 1973b; Fig. 5). No inflection

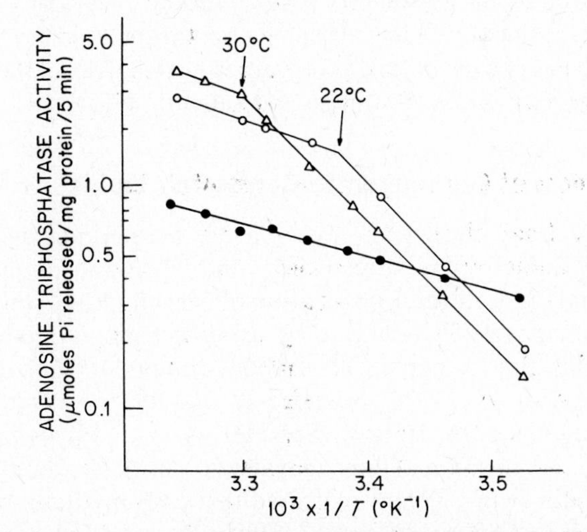


FIGURE 5. Arrhenius plots of ATPase activity of the cholesterol-rich native strain (closed symbols) and the cholesterol-poor adapted strain (open symbols). The adapted strain was grown either with palmitic and oleic acids (○) or with palmitic and elaidic acids (△). (From Rottem *et al.*, 1973b.)

point could be detected, however, in Arrhenius plots of the cholesterol-containing native strain, where a phase transition was eliminated. Cholesterol could also affect the $Mg^{2+}$-dependent ATPase activity of *A. laidlawii* membranes, as the incorporation of cholesterol into these membranes decreases the inflection temperature of the ATPase activity (de Kruyff *et al.*, 1973c). The cholesterol effect was reversed by filipin—a polyene antibiotic which complexes the cholesterol. Temperature-dependent variations in both $V_{max}$ and $K_m$ values of the $Mg^{2+}$-dependent ATPase of *A. laidlawii* that can produce a variety of Arrhenius plot artifacts were recently described (Silvius *et al.*, 1978) suggesting that temperature variations in substrate binding affinity will have to be taken into account when determining the effect of temperature on the rate of a membrane-bound enzyme.

## VI. TRANSBILAYER DISTRIBUTION OF LIPIDS IN MYCOPLASMA MEMBRANES

To gain an insight into the molecular oganization and function of the membrane, information on the transbilayer distribution of components in it is required. The extensive studies of recent years suggest that biological membranes are vectorial structures, that is, their components are asymmetrically distributed between the two surfaces. But whereas for proteins and carbohydrates the asymmetry is absolute and every copy of a polypeptide chain or carbohydrate has the same orientation, lipid asymmetry is not absolute and every type of lipid is present in both halves of the lipid bilayer but in different amounts (Rothman and Lenard, 1977).

### A. Distribution of Carbohydrate-Containing Lipids

The major lipid classes in mycoplasma membranes include carbohydrate-containing lipids (glycolipids and glycophospholipids), phospholipids, and cholesterol. Localization of carbohydrate-containing lipids in mycoplasmas is easily achieved by using specific antibodies or lectins (plant proteins that interact with carbohydrates). In fact, immunologic studies (Razin *et al.*, 1970; Schiefer *et al.*, 1975a, 1977b), agglutination experiments with lectins (Schiefer *et al.*, 1974), and electron microscopic visualization of concanavalin A–surface carbohydrates complexes (Schiefer *et al.*, 1975b) indicated that carbohydrate moieties, presumably of glycolipids and glycophospholipids, are exposed on the external surfaces of several mycoplasma species. The concept of the asymmetrical distribution of the carbohydrate moieties was further promoted by

Kahane and Tully (1976), who showed that in all *Acholeplasma* and *Mycoplasma* species tested the amount of labeled lectins bound to intact cells was almost the same as that bound to isolated membranes, suggesting that the carbohydrate-containing membrane lipids are exposed on the cell surface.

## B. Distribution of Phospholipids

Determination of phospholipid asymmetry in mycoplasmas is more difficult since the methodology developed so far for phospholipid distribution is based to a large extent on the use of agents that react with the primary amines of aminophospholipids and on the use of phospholipases (Rothman and Lenard, 1977). Aminophospholipids, such as phosphatidylethanolamine or phosphatidylserine, are ubiquitous in eukaryotic and most prokaryotic cells but are almost completely absent from mycoplasmas (see Chapter 9). Phospholipases can be used only with organisms whose membrane phospholipids are accessible to the enzyme. Such accessibility is governed initially by the degree of shielding of membrane phospholipid by membrane proteins. It is conceivable that shielding might influence the rate of phospholipid hydrolysis or even completely prevent it. Phosphatidylglycerol, the major phospholipid of *M. hominis* membranes, is such a case, where no hydrolysis by phospholipase C occurred unless membrane proteins were removed by pronase digestion (Rottem *et al.*, 1973c). The masking of *M. hominis* membrane phospholipids by proteins was supported by showing that the phosphatidylglycerol in isolated *M. hominis* membranes fails to interact with its specific antiserum (Schiefer *et al.*, 1975a) or to bind polycationic ferritin (Schiefer *et al.*, 1976), which binds to anionic groups on the membrane surface. The anionic groups appear to be lipid phosphate groups since lipid extraction of the membranes abolished labeling, whereas proteolytic digestion of the membranes by pronase increased labeling (Schiefer *et al.*, 1975a, 1976). In contrast to *M. hominis,* cells of *Mycoplasma pneumoniae, M. mycoides* subsp. *capri, Mycoplasma gallisepticum,* and *A. laidlawii* were labeled by polycationic ferritin (Schiefer *et al.*, 1976) or by ferric oxide (Schiefer *et al.*, 1977a), implying that at least some of the polar head groups of the membrane phospholipids of these organisms are located on the outer half of the bilayer and are not shielded by membrane proteins. Indeed, membrane phospholipids of *A. laidlawii* (Bevers *et al.*, 1977a), *M. gallisepticum,* and *Mycoplasma capricolum* (S. Rottem, unpublished data) are vulnerable to phospholipases that can be used for phospholipid localization studies in these organisms. The localization of phosphatidylglycerol, which constitutes about 30% of *A. laidlawii* membrane lipids, was successfully studied by Bevers *et al.* (1977a) using pancreatic phospholipase $A_2$. Treating intact *A. laidlawii* cells with this enzyme at 5°C led to the hydrolysis of 50% of the phosphatidylglycerol, whereas when isolated membranes of these cells were treated at 5°C, about 70% of the phosphatidylglycerol was hydrolyzed, suggesting the presence of three different phosphatidylglycerol pools: the first (50% of the total) exposed on the external surface; the second (20% of the total) exposed in the inner membrane surface; and the third (30% of the total) protected from the enzyme, probably by interaction with membrane proteins. The complete hydrolysis of the phosphatidylglycerol obtained by phospholipase treatment of intact cells at 37°C was taken to indicate a translocation mechanism ("flip-flop") that enables the phosphatidylglycerol to move from the inner to the outer half of the lipid bilayer. The "flip-flop" rate of phosphatidylglycerol in *A. laidlawii* seems to be much faster than that observed in liposomes (Kornberg and McConnell, 1971; McNamee and McConnell, 1973), or biological membranes (Renooy *et al.*, 1976; Bloj and Zilversmit, 1976). The fast rate may be due to the depletion of the phosphatidylglycerol in the outer half of the bilayer as a result of the phospholipase activity which would trigger the translocation of this compound from the inner half (Bevers *et al.*, 1977a).

### C. Distribution of Cholesterol

The mycoplasmas are most useful tools for studying cholesterol localization and movement in the membrane, being the only prokaryotic cells that require cholesterol for growth. Although cholesterol taken up by mycoplasmas from an exogenous source is first incorporated into the outer half of the lipid bilayer, it is distributed in both halves of the lipid bilayer as indicated by rapid kinetic studies of filipin binding to cholesterol in intact cells and isolated membranes (Bittman and Rottem, 1976). The initial rates of filipin–cholesterol association were significantly lower with intact cells (Fig. 6). The ratio of the second-order rate constants of the filipin–cholesterol association in isolated membranes relative to intact cells pointed to a symmetrical distribution of cholesterol in intact *M. gallisepticum* cells. However, in *M. capricolum* up to about two-thirds of the unesterified cholesterol is localized in the outer half of the lipid bilayer. The approximately symmetrical distribution of cholesterol in the two halves of the lipid bilayer of *M. gallisepticum* has also been established from exchange studies of [$^{14}$C]cholesterol between resting *M. gallisepticum* cells and high-density lipoprotein (Rottem *et al.*, 1978). Cholesterol exists in *M. gallisepticum* cells in two different environments. One, representing about 50% of the total unesterified cholesterol, is readily

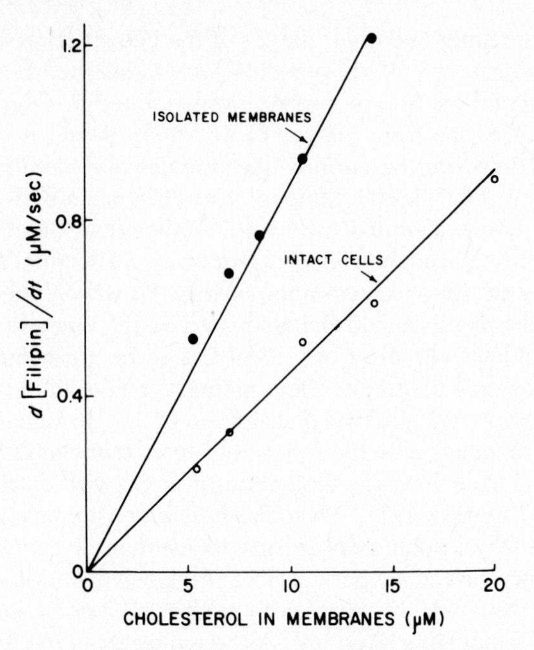


FIGURE 6. Initial rate of filipin binding to *M. gallisepticum* intact cells and isolated membrane preparations. (From Bittman and Rottem, 1976.)

exchanged with exogenous cholesterol, whereas cholesterol in the other environment interchanges at exceedingly slow rates. Since over 90% of the cholesterol in isolated membranes was exchanged rapidly, it is likely that these environments represent the inner and outer halves of the lipid bilayer. Although the exchange studies also suggested that in resting *M. gallisepticum* cells the rate of "flip-flop" of cholesterol from the inner to the outer half of the bilayer is exceedingly slow (Rottem *et al.*, 1978), one must assume that in growing cells, where cholesterol is first incorporated into the outer half of the lipid bilayer, the rate of "flip-flop" to the inner half, initially low in cholesterol, is much faster. Indeed, transfering a cholesterol-depleted *M. capricolum* adapted strain to a cholesterol-rich medium resulted in an almost sixfold increase in the unesterified cholesterol content of the membrane within the first 4 hr of incubation. The transbilayer distribution of cholesterol in the two halves of the bilayer was, however, almost invariant (S. Rottem and R. Bittman, unpublished data), suggesting a very rapid "flip-flop" of cholesterol in growing cells.

The functional significance of lipid asymmetry in mycoplasmas is still obscure. Asymmetry of the polar head groups and variations in the fatty acid constituents of the various lipid classes might result in different fluidities of the inner and outer halves of the bilayer (Bretscher, 1973). In fact, epr spectroscopy of spin-labeled fatty acids incorporated in intact cells and isolated membrane preparations pointed to a higher fluidity of the outer half of the lipid bilayer of *M. hominis* and *A. laidlawii* cells (Rottem, 1975). Since the various lipid species of *A. laidlawii* may differ significantly in their melting temperatures (Chapman and Urbina, 1971), although they have a similar fatty acid composition (Saito *et al.*, 1977a), their asymmetrical transbilayer distribution could account for differences in the fluidity of the two membrane halves. However, the possible contribution of the membrane proteins located on the inner surface of mycoplasma membranes (Amar *et al.*, 1976) has to be considered since membrane proteins may markedly affect membrane lipid fluidity (Rottem and Samuni, 1973). An asymmetric distribution of lipid in specific areas of the cell membrane might give the inner and outer monolayers different surface tensions, leading to the formation of areas with extreme curvature (Rothman and Lenard, 1977). Mycoplasmas, which are devoid of cell walls, might make use of such mechanisms to maintain their filamentous cell shape. A highly curved shape is typical of the cell membrane of the helical *Spiroplasma* cells which inexplicably maintains its cell shape in the absence of any supportive structure. Although electron microscopy studies suggested the presence of fibrils in *Spiroplasma* (Cole *et al.*, 1973; Williamson, 1974) which may be responsible for the organisms' helical shape, the possibility that lipid-asymmetry plays a role in determining the helical shape cannot yet be excluded.

## VII. CONCLUDING REMARKS

The intensive studies on mycoplasma membranes carried out over the past decade have provided information on their thermodynamics and insights into their structure. These studies suggested that the lipids of mycoplasma membranes are organized as a bilayer. The lipid bilayer of mycoplasmas containing little or no cholesterol undergoes a classical gel-to-liquid crystalline phase transition as the temperature is increased. The transbilayer distribution studies of the various lipid components suggest an asymmetric distribution of carbohydrate-containing lipids that are preferentially located in the outer half of the bilayer. Phospholipids and cholesterol, however, are present on both sides of the membrane, though sometimes in different proportions. The transbilayer movement of phospholipids and cholesterol in mycoplasmas seems to occur at a much faster rate than that observed in liposomes, viruses, or red blood cells (Rothman and Lenard, 1977). The rapid movement may imply a catalytic process involving membrane proteins. At the growth temperature membrane lipids are maintained, at least in part, in a fluid state. *Acholeplasma* species which have mechanisms for regulating the degree of unsaturation and the chain length of the fatty acids can maintain membrane lipids in a fluid state at the temperatures of growth. In the *Mycoplasma* species convincing evidence is available that cholesterol acts as a regulator of membrane fluidity. The idea that the physical state of mycoplasma membrane lipids is important for biologic processes is supported by the observed structural–functional relationship between the gel-to-liquid crystalline phase transition of membrane lipids and the activity of membrane-bound proteins, such as membrane enzymes and transport proteins. The mechanisms for controlling membrane fluidity therefore appear to be designed to provide a correct physical state of the lipid microenvironment for optimal activity of these proteins.

## REFERENCES


Abramson, M. B., and Pisetsky, D. (1972). *Biochim. Biophys. Acta* **282,** 80–84.

Amar, A., Rottem, S., and Razin, S. (1974). *Biochim. Biophys. Acta* **352,** 228–244.

Amar, A., Rottem, S., Kahane, I., and Razin, S. (1976). *Biochim. Biophys. Acta* **426,** 258–270.

Baldassare, J. J., Rhinehart, K. B., and Silbert, D. F. (1976). *Biochemistry* **15,** 2989–2996.

Bevers, E. M., Singal, S. A., Op den Kamp, J. A. F., and van Deenen, L. L. M. (1977a). *Biochemistry* **16,** 1290–1294.

Bevers, E. M., Snoek, G. T., Op den Kamp, J. A. F., and van Deenen, L. L. M. (1977b). *Biochim. Biophys. Acta* **467,** 346–356.

Bittman, R., and Rottem, S. (1976). *Biochem. Biophys. Res. Commun.* **71,** 318–324.

Bloj, B., and Zilversmit, D. B. (1976). *Biochemistry* **15,** 1277–1283.

Bretscher, M. S. (1973). *Science* **181,** 622–629.

Chapman, D., and Urbina, J. (1971). *FEBS Lett.* **12,** 169–172.

Cole, R. M., Tully, J. G., and Popkin, T. J. (1973). *Ann. N.Y. Acad. Sci.* **225,** 471–493.

Cronan, J. E., Jr., and Gelman, E. P. (1975). *Bacteriol. Rev.* **39,** 232–256.

de Kruyff, B., Demel, R. A., and van Deenen, L. L. M. (1972). *Biochim. Biophys. Acta* **255,** 331–347.

de Kruyff, B., de Greef, W. J., van Eyk, R. V. W., Demel, R. A., and van Deenen, L. L. M. (1973a). *Biochim. Biophys. Acta* **298,** 479–499.

de Kruyff, B., Demel, R. A., Slotboom, A. G., van Deenen, L. L. M., and Rosenthal, A. F. (1973b). *Biochim. Biophys. Acta* **307,** 1–19.

de Kruyff, B., van Dijck, P. W. M., Goldbach, R. W., Demel, R. A., and van Deenen, L. L. M. (1973c). *Biochim. Biophys. Acta* **330,** 269–282.

de Kruyff, B., Gerritsen, W. J., Oerlemans, A., van Dijck, P. W. M., Demel, R. A., and van Deenen, L. L. M. (1974). *Biochim. Biophys. Acta* **339,** 44–56.

de Kruyff, B., Cullis, P. R., Radda, G. K., and Richards, R. E. (1976). *Biochim. Biophys. Acta* **419,** 411–424.

Demel, R. A., and de Kruyff, B. (1976). *Biochim. Biophys . Acta* **457,** 109–132.

Engelman, D. M. (1970). *J. Mol. Biol.* **47,** 115–117.

Engelman, D. M. (1971). *J. Mol. Biol.* **58,** 153–165.
Forsyth, P. E., Jr., Marčelja, S., Mitchell, D. J., and Ninham, B. W. (1977). *Biochim. Biophys. Acta* **469,** 335–344.
Freeman, B. A., Sissenstein, R., McManus, T. T., Woodward, J. E., Lee, I. M., and Mudd, J. B. (1976). *J. Bacteriol.* **125,** 946–954.
Gaffney, B. J., and McNamee, C. (1974). *In* "Methods in Enzymology" (S. Fleisher and C. Packer, eds.), Vol. 32, pp. 161–198. Academic Press, New York.
Hsung, J. C., Huang, L., Hoy, D. J., and Haug, A. (1974). *Can. J. Biochem.* **52,** 974–980.
Huang, L., and Haug, A. (1974). *Biochim. Biophys. Acta* **352,** 361–370.
Huang, L., Lorch, S. K., Smith, C. S., and Haug, A. (1974a). *FEBS Lett.* **43,** 1–5.
Huang, L., Jaquet, D. D., and Haug, A. (1974b). *Can. J. Biochem.* **52,** 483–490.
Jähnig, F. (1976). *Biophys. Chem.* **4,** 309–318.
James, R., and Branton, D. (1973). *Biochim. Biophys. Acta* **323,** 378–390.
Kahane, I., and Tully, J. G. (1976). *J. Bacteriol.* **128,** 1–7.
Kornberg, R. D., and McConnell, H. M. (1971). *Biochemistry* **10,** 1111–1120.
Ladbrooke, B. D., and Chapman, D. (1969). *Chem. Phys. Lipids* **3,** 304–356.
Lee, A. G., Birdsall, N. J. M., Metcalfe, J. C., Toon, P. A., and Warren, G. B. (1974). *Biochemistry* **13,** 3699–3705.
McConnell, H. M., and McFarland, B. G. (1970). *Rev. Biophys.* **3,** 91–136.
McElhaney, R. N. (1974a). *PAABS Revista* **3,** 753–760.
McElhaney, R. N. (1974b). *J. Mol. Biol.* **84,** 145–157.
McElhaney, R. N., and Tourtellotte, M. E. (1969). *Science* **164,** 433–434.
McElhaney, R. N., de Gier, J., and van Deenen, L. L. M. (1970). *Biochim. Biophys. Acta* **219,** 245–247.
McElhaney, R. N., de Gier, J., and van der Neut-Kok, E. C. M. (1973). *Biochim. Biophys. Acta* **298,** 500–512.
McNamee, M. G., and McConnell, H. N. (1973). *Biochemistry* **12,** 2951–2958.
Marr, A. G., and Ingraham, J. L. (1962). *J. Bacteriol.* **84,** 1260–1267.
Melchior, D. L., and Steim, J. M. (1977). *Biochim. Biophys. Acta* **466,** 148–159.
Melchior, D. L., and Steim, J. M. (1978). *Prog. Surf. Membr. Sci.* **13,** (in press).
Melchior, D. L., Morowitz, H. J., Sturtevant, J. M., and Tsong, T. Y. (1970). *Biochim. Biophys. Acta* **219,** 114–122.
Metcalfe, J. C., Birdsall, N. J. M., and Lee, A. G. (1972). *FEBS Lett.* **21,** 335–340.
Ne'eman, Z., Kahane, I., Kovartovsky, J., and Razin, S. (1972). *Biochim. Biophys. Acta* **266,** 255–268.
Oldfield, E., Chapman, D., and Derbyshire, W. (1972). *Chem. Phys. Lipids* **9,** 69–81.
Panos, C., and Rottem, S. (1969). *Biochemistry* **9,** 407–412.
Pollack, J. D., and Tourtellotte, M. E. (1967). *J. Bacteriol.* **93,** 636–641.
Razin, S. (1969). *In* "The Mycoplasmatales and the L-Phase of Bacteria" (L. Hayflick, ed.), pp. 317–348. Appleton, New York.
Razin, S. (1974). *FEBS Lett.* 47, 81–85.
Razin, S. (1975). *Prog. Surf. Membr. Sci.* **9,** 257–312.
Razin, S., and Rottem, S. (1978). *Trends Biochem. Sci.* **3,** 51–55.
Razin, S., Tourtellotte, M. E., McElhaney, R. N., and Pollack, J. D. (1966). *J. Bacteriol.* **91,** 609–616.
Razin, S., Prescott, B., and Chanock, R. M. (1970). *Proc. Natl. Acad. Sci. U.S.A.* **67,** 590–597.
Read, B. D., and McElhaney, R. N. (1975). *J. Bacteriol.* **123,** 47–55.
Reinert, J. C., and Steim, J. M. (1970). *Science* **168,** 1580–1582.
Renooy, W., van Golde, L. M. G., Zwaal, R. F. A., and van Deenen, L. L. M. (1976). *Eur. J. Biochem.* **61,** 53–58.

Romijn, J. C., van Golde, L. M. G., McElhaney, R. N., and van Deenen, L. L. M. (1972). *Biochim. Biophys. Acta* **280,** 222–232.

Rothman, J. E., and Engelman, D. M. (1972). *Nature (London), New Biol.* **237,** 42–44.

Rothman, J. E., and Lenard, J. (1977). *Science* **195,** 743–753.

Rottem, S. (1975). *Biochem. Biophys. Res. Commun.* **64,** 7–12.

Rottem, S., and Greenberg, A. S. (1975). *J. Bacteriol.* **121,** 631–639.

Rottem, S., and Panos, C. (1969). *J. Gen. Microbiol.* **59,** 317–328.

Rottem, S., and Razin, S. (1966). *J. Bacteriol.* **92,** 714–722.

Rotten, S., and Razin, S. (1967). *J. Gen. Microbiol.* **48,** 53–63.

Rottem, S., and Razin, S. (1973). *J. Bacteriol.* **113,** 565–571.

Rottem, S., and Samuni, A. (1973). *Biochim. Biophys. Acta* **298,** 32–38.

Rottem, S., Hubbell, W. L., Hayflick, L., and McConnell, H. M. (1970). *Biochim. Biophys. Acta* **219,** 104–113.

Rottem, S., Pfendt, E. A., and Hayflick, L. (1971). *J. Bacteriol.* **105,** 323–330.

Rottem, S., Yashouv, J., Ne'eman, Z., and Razin, S. (1973a). *Biochim. Biophys. Acta* **323,** 495–508.

Rottem, S., Cirillo, V. P., de Kruyff, B., Shinitizky, M., and Razin, S. (1973b). *Biochim. Biophys. Acta* **323,** 509–519.

Rottem, S., Hasin, M., and Razin, S. (1973c). *Biochim. Biophys. Acta* **323,** 520–531.

Rottem, S., Trotter, S. L., and Barile, M. F. (1977). *J. Bacteriol.* **129,** 707–713.

Rottem, S., Slutzky, G., and Bittman, R. (1978). *Biochemistry* **17,** 2723–2726.

Sackman, E., Träble, H., Galla, H., and Overath, P. (1973). *Biochemistry* **12,** 5360–5368.

Saito, Y., and McElhaney, R. N. (1977). *J. Bacteriol.* **132,** 485–496.

Saito, Y., Silvius, J. R., and McElhaney, R. N. (1977a). *Arch. Biochem. Biophys.* **182,** 443–454.

Saito, Y., Silvius, J. R., and McElhaney, R. N. (1977b). *J. Bacteriol.* **132,** 497–504.

Saito, Y., Silvius, J. R., and McElhaney, R. N. (1978) *J. Bacteriol.* **133,** 66–74.

Schiefer, H. G., Gerhardt, U., Brunner, H., and Krupe, M. (1974). *J. Bacteriol.* **120,** 81–88.

Schiefer, H. G., Gerhardt, U., and Brunner, H. (1975a). *Hoppe-Seyler's Z. Physiol. Chem.* **356,** 559–565.

Schiefer, H. G., Krauss, H., Brunner, H., and Gerhardt, U. (1975b). *J. Bacteriol.* **124,** 1598–1600.

Schiefer, H. G., Krauss, H., Brunner, H., and Gerhardt, U. (1976). *J. Bacteriol.* **127,** 461–468.

Schiefer, H. G., Krauss, H., Brunner, H., and Gerhardt, U. (1977a). *Zbl. Bakteriol. Parasitenkd., Infektionskr. Hyg., Abt. 1* **237,** 104–110.

Schiefer, H. G., Gerhardt, U., and Brunner, H. (1977b). *Zentralbl. Bakteriol., Parasitenkd., Infektionskr. Hyg., Abt. 1* **239,** 262–269.

Seelig, A., and Seelig, J. (1974). *Biochemistry* **13,** 4839–4846.

Shimshick, E. J., and McConnell, H. M. (1973). *Biochemistry* **12,** 2351–2359.

Silvius, J. R., Saito, Y., and McElhaney, R. N. (1977). *Arch. Biochem. Biophys.* **182,** 455–464.

Silvius, J. R., Read, B. D., and McElhaney, R. N. (1978). *Nature (London)* **272,** 645–647.

Smith, P. F. (1960). *Ann. N.Y. Acad. Sci.* **79,** 508–520.

Smith, P. F. (1963). *Recent Prog. Symp. Int. Congr. Microbiol., 8th 1962,* pp. 518–525.

Smith, P. F. (1964). *J. Lipid Res.* **5,** 121–125.

Smith, P. F. (1967). *Ann. N.Y. Acad. Sci.* **143,** 139–151.

Steim, J. M., Tourtellotte, M. E., Reinert, M. E., and McElhaney, R. N. (1969). *Proc. Natl. Acad. Sci. U.S.A.* **63,** 104–109.

Stockton, G. W., Johnson, K. G., Butler, K. W., Polnaszek, C. F., Cyr, R., and Smith, I. C. P. (1975). *Biochim. Biophys. Acta* **401,** 535–539.

Stockton, G. W., Johnson, K. G., Butler, K. W., Tulloch, A. P., Boulanger, Y., Smith, I. C. P., Davis, J. H., and Bloom, M. (1977). *Nature (London)* **269,** 267–269.

Tourtellotte, M. E. (1972). *In* "Membrane Molecular Biology" (C. F. Fox and A. D. Keith, eds.), pp. 439–470. Sinauer Assoc. Inc., Stamford, Connecticut.

Tourtellotte, M. E., and Zupnik, J. S. (1973). *Science* **179,** 84–86.

Tourtellotte, M. E., Branton, D., and Keith, A. (1970). *Proc. Natl. Acad. Sci. U.S.A.* **66,** 909–916.

Träuble, H., and Eibl, H. (1974). *Proc. Natl. Acad. Sci. U.S.A.* **71,** 214–219.

Träuble, H., Teubner, M., Wooley, P., and Eibl, H. (1976). *Biophys. Chem.* **4,** 319–342.

van Deenen, L. L. M. (1965). *Prog. Chem. Fats Other Lipids* **8,** Part I, 1–47.

van Zoelen, E. J. J., van der Neut-Kok, E. C. M., de Gier, J., and van Deenen, L. L. M. (1975). *Biochim. Biophys. Acta* **394,** 463–469.

Verkleij, A. J., Ververgaert, P. J. H., van Deenen, L. L. M., and Elbers, P. F. (1972). *Biochim. Biophys. Acta* **288,** 326–332.

Wallace, B. A., Richards, F. M., and Engelman, D. M. (1976). *J. Mol. Biol.* **107,** 255–269.

Williamson, D. L. (1974). *J. Bacteriol.* **117,** 904–906.

# 11 / MEMBRANE PROTEINS

*Shmuel Razin*



## I. INTRODUCTION

Most of the mycoplasma membrane is made up of proteins. Apart from their structural and catalytic roles, they also have a major share in the immunologic activities of mycoplasma cells (Razin *et al.*, 1972; this volume, Chapter 13). Some advances have recently been made in the methodology of membrane protein fractionation, so that data on the molecular properties of several isolated mycoplasma membrane proteins are now available. The application of specific labeling agents for proteins has also advanced our knowledge on the transbilayer distribution of



THE MYCOPLASMAS, VOL. 1

ISBN 0-12-078401-7

proteins in mycoplasma membranes, on their lateral mobility within the plane of the membrane, and on their specific interactions with membrane lipids.

## II. MOLECULAR PROPERTIES OF MEMBRANE PROTEINS

### A. Electrophoretic Analysis

The great variety of enzyme and transport activities localized in the single membranous structure of the mycoplasma cell would lead us to predict a high protein content and a great number of different proteins in the membrane. The experimental data support such a prediction, showing that 50–80% of the total dry weight of mycoplasma membranes is protein (Table II in Chapter 8) and that this component can be resolved into several dozen different polypeptide bands by polyacrylamide gel electrophoresis (Rottem and Razin, 1967; Morowitz and Terry, 1969; Hjertén and Johansson, 1972; Kahane and Marchesi, 1973; Amar *et al.*, 1974; Dahl *et al.*, 1977; Fig. 1). There are, undoubtedly, more membrane proteins than are discernible in the gels, since the electrophoretic patterns revealed only the major ones and some may migrate together within the same band. The application of two-dimensional electrophoresis to mycoplasma membranes is expected to disclose many more membrane proteins. Approximately 320 individual proteins were detected by two-dimensional electrophoresis of labeled cell proteins of *Acholeplasma laidlawii*. Of these, about 140 were associated with the cell membrane (Archer *et al.*, 1978). On the other hand, complex membrane proteins tend to dissociate into their component subunits during solubilization and electrophoresis so that each of these complex proteins may be represented by more than one polypeptide band. The electrophoretic patterns of mycoplasma membrane proteins are highly reproducible and species-specific and, since they are usually not significantly affected by variations in growth conditions, they can be used for the identification and classification of mycoplasmas (Rottem and Razin, 1967). The electrophoretic patterns of the *Acholeplasma laidlawii* membrane proteins changed only slightly when the fatty acid composition of the membrane lipids was considerably altered (Pisetsky and Terry, 1972; Amar, 1977; Amar *et al.*, 1978b). Similarly, marked alterations in the cholesterol content of *Mycoplasma mycoides* subsp. *capri* membranes were not accompanied by any significant changes in the electrophoretic pattern of membrane proteins (Rottem *et al.*, 1973b; Archer, 1975). The composition of the major protein components of the mycoplasma membrane may therefore be regarded as a much more stable

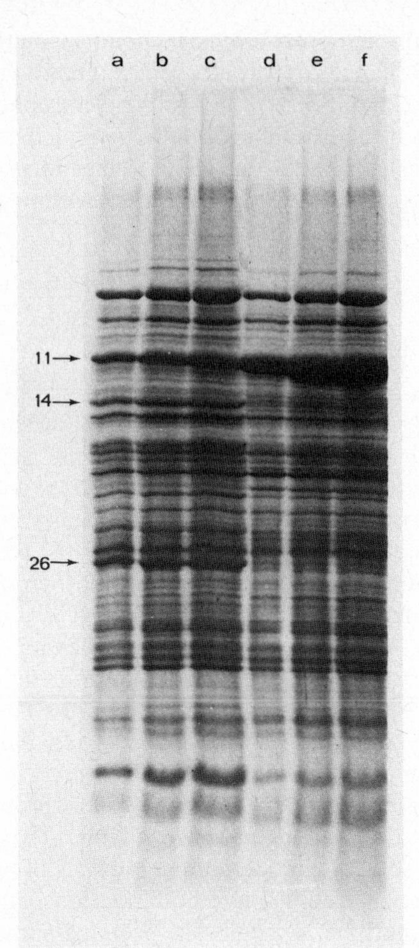


FIGURE 1. Electrophoretic patterns of *A. laidlawii* membrane proteins in slab gels containing 7–15% polyacrylamide with sodium dodecyl sulfate. Patterns a, b, and c were made with increasing amounts of membranes (21, 42, and 63 μg protein) from the wild type of *A. laidlawii* B, and patterns d, e, and f of its mutant 3-28. The high reproducibility of the patterns and the differences in some membrane proteins between the wild type and the mutant are discernible. (From Dahl *et al.*, 1977.)

characteristic of a given species than the composition of its membrane lipids. Nevertheless, one may encounter some changes, usually minor, in membrane protein patterns under certain growth conditions. Thus, two major protein bands were found missing from the pattern of membranes isolated from *A. laidlawii* grown at 25°C instead of 37°C (Amar *et al.*, 1978b), and one protein band became more prominent in the membrane protein profile from aging *Mycoplasma hominis* cultures (Amar *et al.*, 1976). While these changes apparently reflect altered rates of biosynthesis or turnover of some membrane proteins, the electrophoretic patterns may also be influenced by noncellular proteins contaminating the membrane preparations (Armstrong and Yu, 1970; Razin *et al.*, 1970; Rosendal, 1973; Yaguzhinskaya, 1976; also Chapter 8, Section III, E).

## B. Solubilization and Fractionation

### 1. Peripheral Proteins

Solubilization is a prerequisite for the fractionation and characterization of membrane proteins. Under mild nondenaturing conditions, only a minor fraction can be solubilized. These are the peripheral membrane proteins which are not immersed in the lipid bilayer and are held onto the membrane mainly by ionic bonds and salt bridges (Singer and Nicolson, 1972). Changes in the ionic strength, changes in the pH of the suspending medium, or the addition of a chelating agent such as ethylenediamine tetraacetate (EDTA) will detach these proteins in a water-soluble form that is free of lipid. A definitive answer to the question of how much of the mycoplasma membrane protein consists of peripheral proteins cannot be given because the amount seems to depend on the handling of the membrane during its isolation. The isolation of *A. laidlawii* membranes by osmotic lysis of the organisms in deionized water or in very dilute buffer, and subsequent washings of the membranes, most probably cause the loss of some of the peripheral membrane proteins. Two-dimensional electrophoresis of the soluble and membrane proteins of *A. laidlawii* showed many proteins which are common to both fractions, suggesting that these are loosely bound peripheral membrane proteins partially removed from the membrane by the cell lysis and washing procedure (Archer *et al.*, 1978). The addition of $Mg^{2+}$ to the lysis mixture, a practice successfully utilized to minimize peripheral protein loss from bacterial protoplast membranes (Muñoz *et al.*, 1968), could not be applied in this case because divalent cations, even at extremely low concentrations, protect cells from osmotic lysis (Razin, 1964). The application of digitonin-induced lysis may help to solve this problem, since it is also effective in the presence of $Mg^{2+}$ and in media of high ionic strength (Rottem and Razin, 1972). The values of 8–14% (of the total membrane proteins) obtained for peripheral membrane proteins of *A. laidlawii* and *M. hominis* membranes (Ne'eman *et al.*, 1971; Amar, 1977) isolated by osmotic shock and treated with EDTA are apparently minimal values. In fact, a recent reassessment of this problem by Archer *et al.* (1978) indicates that about one-half of the total membrane protein, including about 60 of the 140 membrane protein species, can be released from *A. laidlawii* membranes by exhaustive EDTA treatment. Archer *et al.* (1978) distinguish two groups of peripheral membrane proteins in *A. laidlawii.* One group, which is released by washing the membranes in low-ionic strength buffer without EDTA, apparently consists of proteins bound to the membrane solely by ionic interactions. The other group, released only after EDTA treatment, consists of proteins with hydrophobic regions which are more intimately associated with the membranes.

The peripheral membrane protein fraction released by washing *A. laidlawii* membranes in a low-ionic strength buffer (3 m*M* Tris–HCl) contained about 50% of the total membrane-bound ribonuclease and deoxyribonuclease activities. The ATPase, NADH oxidase, and *p*-nitrophenylphosphatase activities as well as the main protein antigens remained bound to the membrane even when EDTA was added to the wash solutions, and thus appear to belong to the integral membrane protein group (Ne'eman *et al.*, 1971; Ne'eman and Razin, 1975). Since by definition the peripheral membrane proteins are located on the membrane surfaces, it might be expected that these would be the first membrane components to be detached by detergents, as was in fact shown for the ribonuclease activity of *A. laidlawii* membranes (Fig. 2). The high susceptibility of the peripheral membrane proteins to detachment by low detergent concentrations may explain the finding (Ne'eman *et al.*, 1971) that more membrane protein than lipid is solubilized at low detergent concentrations, the reverse being true at high detergent concentrations. Nevertheless, the group of peripheral membrane proteins released from *A. laidlawii* membranes by low-ionic strength buffer without EDTA was poorly soluble in Triton X-100 (Archer *et al.*, 1978); therefore, generalizations cannot be made.

## 2. Integral Proteins

Most membrane proteins are tightly bound to membrane lipids, apparently by a combination of hydrophobic and ionic bonds. To break up these composite bonds, fairly drastic procedures have to be employed. Moreover, in many cases these proteins, named integral membrane proteins (Singer and Nicolson, 1972), remain associated with lipid and usually precipitate in neutral aqueous solutions, when completely lipid-free. The agents most commonly employed for solubilizing the integral proteins are detergents and organic solvents (Razin, 1972). Since the organic solvent itself must be water soluble, the choice is limited. Cold *n*-butanol solubilized *A. laidlawii* membranes efficiently. Some 80–90% of the proteins plus very little lipid separate out in the aqueous phase but apparently are

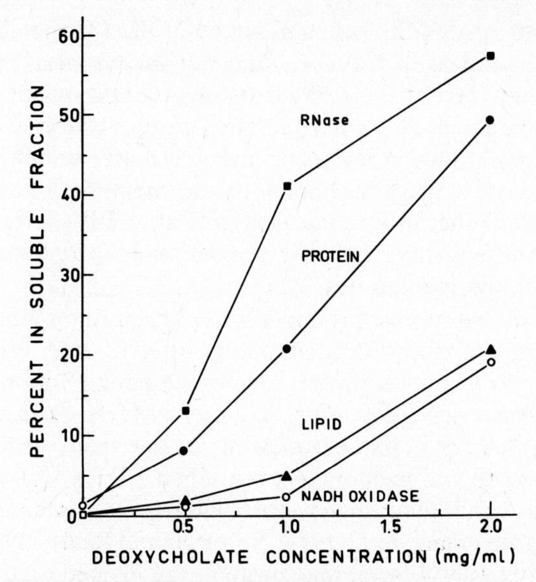


FIGURE 2. Differential solubilization of *A. laidlawii* membrane components by low concentrations of deoxycholate. The peripheral membrane enzyme ribonuclease is solubilized more effectively than the total membrane proteins, while the integral membrane enzyme NADH oxidase is less susceptible to solubilization and is apparently released on the disaggregation of the lipid bilayer. (From Ne'eman and Razin, 1975.)

not in the monomeric form, since they are excluded in the void volume of Sephadex G-200 columns (Rottem *et al.*, 1973a). It must also be pointed out that *n*-butanol extraction is much less effective with *M. mycoides* subsp. *mycoides* and *Mycoplasma gallisepticum* membranes, where less than 50% of the membrane proteins were recovered in the aqueous phase (M. Hasin, unpublished data). In addition, membrane-bound enzymes, such as NADH oxidase, are more susceptible to inactivation by *n*-butanol than by mild detergents (Rodwell *et al.*, 1967), another disadvantage of using organic solvents as membrane solubilizers.

Detergents have proved more effective than organic solvents in solubilizing mycoplasma membranes (Ne'eman *et al.*, 1971); the strongly ionic ones (sodium dodecyl sulfate and cetyltrimethylammonium bromide), not unexpectedly, are more so than the nonionic ones (Triton X-100, Lubrol, or Brij-58), with bile salts occupying an intermediate position (Ne'eman *et al.*, 1971; Ne'eman and Razin, 1975; Wroblewski *et al.*, 1977a,b). The strongly ionic detergents also caused more intensive denaturation of membrane proteins, as indicated by enzyme inactivation (Ne'eman *et al.*, 1971; Ne'eman and Razin, 1975). Not only the membrane proteins and lipids, but also the different membrane protein species were solubilized at different rates, so that there was no random disaggregation of the membrane (Morowitz and Terry, 1969; Ne'eman *et al.*, 1971; Ne'eman and Razin, 1975). The selective solubilization of membrane proteins by the action of mild detergents, such as Tween-20, has been most useful in providing solubilized membrane material enriched with certain membrane proteins for further fractionation (Hjertén and Johansson, 1972; Johansson and Hjertén, 1974; Johansson *et al.*, 1975).

The fractionation, separation, and purification of the integral membrane proteins pose difficult problems. Most available techniques are designed for use with soluble hydrophilic proteins and are not easily adapted to hydrophobic proteins. Attempts to fractionate mycoplasma membrane material, solubilized by detergents, on Sephadex or Sepharose columns devoid of detergents have failed, apparently because the proteins and lipids reaggregated with dilution or removal of the detergent during filtration (Ne'eman *et al.*, 1972). Detergents, therefore, had to be included throughout the fractionation procedure. Where the detergent used for membrane solubilization was also incorporated into the column and the elution buffer, a considerable portion of the enzymatic and antigenic activities of membrane proteins was preserved despite prolonged exposure to detergent (Ne'eman *et al.*, 1972; Razin *et al.*, 1972). Sodium deoxycholate appears to be suitable for this purpose. Its inclusion in the Sephadex G-200 column resulted in the separation of membrane proteins into several reproducible peaks, most of which were devoid of membrane lipid. The fractionation was at least as good as with sodium dodecyl sulfate, and the fractions also retained more of their biological activity (Ne'eman *et al.*, 1972). A highly purified and active preparation of NADH dehydrogenase was recently isolated from *A. laidlawii* membranes in this way (Jinks and Matz, 1976b). The flat bile salt molecule causes fewer conformational changes in membrane proteins than the bar-shaped detergents, probably because it does not penetrate as well into the hydrophobic regions within the protein molecules (Kagawa, 1972).

Considerable advances in the methodology of membrane protein fractionation have been made by a group working in Uppsala. This group applied the mild detergents Tween 20 and deoxycholate to solubilize proteins of *A. laidlawii* and *S. citri* membranes and fractionated the solubilized proteins by gel filtration, agarose suspension electrophoresis, or preparative polyacrylamide or dextran gel electrophoresis (Johansson *et al.*, 1975; Dresdner and Cid-Dresdner, 1976, 1977; Wroblewski *et al.*, 1977b), in the presence of low detergent concentrations or even in the absence of detergent (Johansson *et al.*, 1975; Dresdner and Cid-Dresdner, 1976, 1977). By these procedures, several *A. laidlawii* membrane proteins, including a flavoprotein, an NADH dehydrogenase (Johansson *et al.*, 1975; Dresdner and Cid-Dresdner, 1976; K.-E. Johansson, personal communication), and a major membrane protein from *S. citri,* named spiralin (Wroblewski *et al.*, 1977b) were isolated and partially characterized (Table I). More recently, Johansson (1978) coupled Sepharose beads with monospecific antisera to $T_2$, an *A. laidlawii* membrane protein purified by immunoelectrophoresis. The coupled beads were shown to specifically adsorb protein $T_2$ from a mixture of *A. laidlawii* membrane proteins solubilized in Tween 20. As the adsorbed protein could be eluted from the beads by decreasing the pH to 2.3, this new approach can be employed to isolate relatively large quantities of purified membrane proteins, provided that monospecific antisera are available for these proteins.

### C. Molecular Weight and Amino Acid Composition

The molecular weights of the mycoplasma membrane proteins, assessed by electrophoresis in sodium dodecyl sulfate-containing polyacrylamide gels, range from about 15,000 to over 200,000 (Morowitz and Terry, 1969; Hjertén and Johansson, 1972; Amar *et al.*, 1974; Archer *et al.*, 1978), well within the range of other biological membrane proteins (Guidotti, 1972). The majority of *A. laidlawii* membrane proteins are acidic proteins, having pI values between 4 and 7, as revealed by isoelectric focusing of solubilized membranes (Archer *et al.*, 1978).

Can the hydrophobicity of membrane proteins be explained in terms of a high content of amino acids with hydrophobic side chains, or does it depend on certain amino acid sequences or specific regional adaptations of the polypeptide structure? The latter possibility, suggested by Wallach and Gordon (1968), has found support in recent amino acid sequence analyses of several integral eukaryotic membrane proteins, notably glycophorin, a human erythrocyte membrane protein (Segrest *et al.*, 1973). The early amino acid analyses carried out on the total membrane proteins of *A. laidlawii* revealed only a slightly higher percentage of nonpolar amino acids than in soluble globular proteins (Engelman and Morowitz, 1968; Morowitz and Terry, 1969; Choules and Bjorklund, 1970). Similarly, membrane proteins of *Thermoplasma acidophilum* were found to contain about the same percentage of nonpolar amino acids as the cytoplasmic proteins (Ruwart and Haug, 1975). Hjertén and Johansson (1972) reported, however, that the *A. laidlawii* membrane fraction insoluble in Tween 20 had a higher nonpolar amino acid content than the Tween 20-soluble membrane proteins, pointing to the fact that the various membrane proteins may differ significantly in their amino acid composition. Analysis of isolated membrane proteins would be expected to yield more meaningful data. Unfortunately, very few mycoplasma

TABLE I. **Properties of Some Purified Mycoplasma Membrane Proteins**

| Protein | Organism | Method of isolation | Molecular weight | Amino acid composition | Localization | References |
|---|---|---|---|---|---|---|
| Spiralin | *S. citri* | Solubilization with deoxycholate; fractionation by agarose-suspension electrophoresis | 26,000 | Lacks methionine, histidine, and tryptophan; low in glycine, leucine, tyrosine, and phenylalanine; high threonine, alanine, valine | ND[a] | Wroblewski *et al.* (1977b) |
| Glycoprotein | *M. pneumoniae* | Extraction with lithium diiodosalicylate; phenol–water partition | 60,000 | Very high glycine (52 mol %) and histidine (19.6 mol %); about 7% carbohydrate, mostly glucose, galactose, and glucosamine | External surface | Kahane and Brunner (1977) |
| $D_{12}$ | *A. laidlawii* | Solubilization by deoxycholate; gel filtration | 140,000 | No excess of hydrophobic amino acids; low content of amino sugars, mannose, galactose, and glucose | Inner surface | K.-E. Johansson (personal communication) |
| $T_2$ | *A. laidlawii* | Solubilization by Tween 20; preparative electrophoresis | 52,000 | No excess of hydrophobic amino acids | External surface | Johansson and Hjertén (1974); K.-E. Johansson (personal communication) |
| $T_3$ | | | 110,000 | As for $T_2$ | Inner surface | |
| $T_{4a}$ | | | 34,000 | As for $T_2$; contains flavin | Inner surface | |
| $T_{4b}$ | | | 52,000 | As for $T_2$ | Inner surface | |

[a] ND, not determined.

membrane proteins have been isolated and purified (Table I). The amino acid composition of the purified proteins, though peculiar in the case of the *Mycoplasma pneumoniae* glycoprotein, does not reveal the reason for the hydrophobicity of the molecules. Determination of the amino acid sequence in the polypeptide chains is apparently needed to elucidate this point.

## D. Conformation

The circular dichroism and optical rotatory dispersion spectra of *A. laidlawii* membranes closely resemble those of other biological membranes (Choules and Bjorklund, 1970; Rottem and Hayflick, 1973). The circular dichroism spectra of the membranes in the region of 190–230 nm show some features characteristic of proteins in a partly helical conformation with certain anomalous features, such as low $[\theta]$ values around 190 nm and a red shift above 220 nm. The red shift has been attributed to a number of structural features of the membrane, including interactions between protein helices and lipids. Urry (1972), however, argued that these anomalies are artifacts arising from the particulate, turbid nature of the membrane suspension. The study by Rottem and Hayflick (1973) on *A. laidlawii* membranes supports Urry's view by demonstrating that the spectral distortions could be markedly reduced by ultrasonication of the membranes and completely eliminated by membrane solubilization by sodium dodecyl sulfate. The characteristic red-shift distortion reappeared on the reaggregation of the solubilized membrane material to membranous structures. The findings that the circular dichroism spectra of *A. laidlawii* membranes are unaffected by radical alterations in the fatty acid composition of membrane lipids, or even by the removal of over 90% of the lipids, are further indications that lipid–protein interactions have little or no effect on the secondary structure of the bulk of membrane proteins and therefore are not responsible for the spectral distortions.

Choules and Bjorklund (1970) subjected data from circular dichroism, optical rotatory dispersion, and infrared spectroscopy of *A. laidlawii* membranes to computer analysis in a curve-fitting program. The best fit for the experimental curves was obtained with 56% $\beta$ structure (pleated sheet), 30.7% $\alpha$ helix, and 13.2% random coil. Whereas the value for $\alpha$-helix content agrees with the data for other membranes (Urry, 1972), no significant amount of $\beta$ structure has been reported, except when membrane proteins were extensively denatured (Oldfield *et al.*, 1972). Hence, it is possible that the computer analysis led to an incorrect interpretation of the data.

## E. Glycoproteins

Though very common in membranes of eukaryotes, glycoproteins are rare in prokaryotes (Leaback, 1972). A search for glycoproteins in mycoplasma membranes is especially interesting as mycoplasmas are the only prokaryotes in which the plasma membrane is exposed to the environment. Moreover, many mycoplasmas are known to adhere to epithelial cell surfaces, a phenomenon of great consequence in mycoplasma pathogenicity. There are good indications that in some of these mycoplasmas membrane proteins serve as the specific binding sites (reviewed by Razin, 1978). The possibility that these are glycoproteins, as in viruses, is attractive. The search for glycoproteins in mycoplasma membranes is still in its initial stages. Electrophoresis of *M. pneumoniae* membranes in polyacrylamide gels containing sodium dodecyl sulfate revealed a protein band (MW about 60,000) which stained red with the periodic acid–Schiff reagent, indicative of carbohydrate (Kahane and Marchesi, 1973). This protein, extracted from membranes with lithium diiodosalicylate, consisted of about 80–90% amino acids and about 7% carbohydrate (Table I). Lactoperoxidase-mediated iodination indicated that the glycoprotein is exposed on the external membrane surface, suggesting the possibility that it serves as a binding site in *M. pneumoniae* attachment to epithelial cells (Kahane and Brunner, 1977). Lithium diiodosalicylate was also applied to *M. gallisepticum* membranes to extract a fraction consisting of about two-thirds carbohydrate and one-third protein (Goel and Lemcke, 1975). This fraction blocked the hemagglutinating ability of *M. gallisepticum,* another expression of the pronounced ability of this mycoplasma to adhere to eukaryotic cell surfaces. However, definite identification of this fraction as a glycoprotein could not be obtained, since electrophoresis of the fraction in polyacrylamide gels failed to yield any band which stained with the conventional protein stains and with the periodic acid–Schiff reagent.

An indirect way to look for glycoproteins in mycoplasma membranes is to examine the binding of $^{125}$I-labeled lectins to cells or to isolated membranes, or by testing for agglutination of the cells by lectins. Lectins did bind to every mycoplasma tested (Kahane and Tully, 1976; Kahane *et al.*, 1977; Amar *et al.*, 1978a) but agglutinated only a few (Schiefer *et al.*, 1974); obviously, binding is more sensitive than agglutination. The almost equal binding of labeled lectins to intact cells of a variety of *Mycoplasma* and *Acholeplasma* species and to their isolated membranes suggest that in these mycoplasmas all the carbohydrate-containing components are exposed on the cell surface (Kahane and Tully, 1976; Amar *et al.*, 1978a). To determine whether the carbohydrate groups responsible for the specific binding of lectins are part of glycoprotein molecules, proteolytic digestion or lipid extraction was applied to the membranes. With most *Mycoplasma* and *Acholeplasma* species, proteolytic digestion increased lectin binding, whereas lipid extraction abolished it (Kahane and Tully, 1976; Schiefer *et al.*, 1974), indicating that the binding sites are of a glycolipid or a lipopolysaccharide nature. On the other hand, proteolytic digestion of *M. hominis* (Kahane and Tully, 1976) and *S. citri* (Kahane *et al.*, 1977) membranes diminished their lectin binding capacity and their carbohydrate content, suggesting that in this case the lectin binding sites are parts of glycoprotein molecules. However, glycoproteins could not be detected in either of these two cases by polyacrylamide gel electrophoresis with the periodic acid–Schiff reagent. This failure may be due to a low carbohydrate content and low sensitivity of the staining technique for glycoproteins devoid of sialic acid (Kahane *et al.*, 1977). Protein $D_{12}$, isolated from *A. laidlawii* membranes (Table I), was found to contain about 1.2 mol % amino sugars and a low content of mannose, galactose, and glucose, but failed to stain with the periodic acid–Schiff reagent (K.-E. Johansson, personal communication). Another possibility that cannot be ruled out as yet is that the carbohydrate moieties in *M. hominis* and *S. citri* membranes are part of macromolecules other than glycoproteins, loosely attached to membrane proteins and released following the removal of the proteins (Kahane *et al.*, 1977).

## III. MEMBRANE-BOUND ENZYMES

The relatively easy separation of mycoplasma membranes from cytoplasmic constituents by osmotic lysis or digitonin-induced lysis is a boon to the study of enzyme localization. The susceptibility of almost all mycoplasmas to lysis by these gentle procedures makes it unnecessary to use drastic cell disruption procedures such as mechanical crushing or sonic oscillations. The mechanical means inevitably cause partial disaggregation of the membrane to minute unsedimentable particles which contaminate the "soluble" cytoplasmic fraction, making it unsuitable for enzyme localization studies (Pollack *et al.*, 1965a). Though the mycoplasmas are convenient subjects for enzyme localization studies, their enzyme makeup is still largely unknown. The far from complete list of enzymatic activities detected so far in the *A. laidlawii* membranes (Table II), although steadily increasing, shows that they have not been systematically surveyed even in the most extensively studied mycoplasma membrane.

TABLE II. **Characteristics of Some Membrane-Bound Enzymes in *A. laidlawii***

| Enzymatic activity | Substrates | Products | Integral or peripheral | Location in the membrane | References |
|---|---|---|---|---|---|
| Adenosine triphosphatase | ATP | ADP + $P_i$ | Integral; depends on phosphatidyl glycerol for activity | Inner surface | Rottem and Razin (1966), Ne'eman and Razin (1975), Bevers *et al.* (1977) |
| *p*-Nitrophenylphosphatase | *p*-Nitrophenolphosphate | *p*-Nitrophenol + $P_i$ | Integral; does not depend on lipids | Possibly embedded within the membrane | Ne'eman and Razin (1975) |
| Ribonuclease | Ribonucleic acid | Nucleotides | Mostly peripheral | External surface | Ne'eman and Razin (1975) |
| Deoxyribonuclease | Deoxyribonucleic acid | Nucleotides | Mostly peripheral | External surface | Ne'eman and Razin (1975) |
| Peptidase (aminopeptidase) | Peptides | Amino acids | Integral, probably depends on lipids | External surface | Choules and Gray (1971), Pecht *et al.* (1972) |
| NADH oxidase | NADH | NAD + $H_2O_2$ | Integral; does not depend on lipids | Inner surface | Ne'eman and Razin (1975) |
| Lysophospholipase | Monoacyl phosphoglycerides | Glycerophosphate + fatty acids | Integral | ND[a] | van Golde *et al.* (1971) |
| Acyl-coenzyme A: *sn*-glycerol-3-phosphate transferase | Acyl CoA + glycerol 3-phosphate | Phosphatidic acid | ND[a] | Inner surface | Rottem and Greenberg (1975) |
| Acyl-coenzyme A thioesterase | Thioesters of CoA with long-chain fatty acids | Fatty acids + CoA | Integral; depends on lipids | Inner surface? | Rottem *et al.* (1977) |

[a] ND, not determined.

### A. Electron Transport Enzymes

The electron transport system, a series of enzymes acting in concert, is usually localized in the plasma membrane in prokaryotes (Salton, 1974). Almost all the parasitic mycoplasmas, including *S. citri,* appear not to conform to this rule since their electron transport activity, measured as transfer of electrons from NADH to oxygen, is found in their soluble cytoplasmic fraction (Pollack *et al.*, 1965b; Rodwell, 1969; Low and Zimkus, 1973; Pollack, 1975; Mudd *et al.*, 1977; Kahane *et al.*, 1977). That the NADH oxidase is really cytoplasmic and is not released from the membrane during cell fractionation is supported by the finding (Ne'eman, 1974) that the NADH oxidase activity of *M. mycoides* subsp. *mycoides* was not associated with the membrane even when cell lysis was carried out by digitonin in the presence of $Mg^{2+}$, conditions minimizing the release of loosely associated membrane proteins. Furthermore, histochemical localization of tellurite reduction showed that it takes place in the cytoplasm of *Mycoplasma mycoides* subsp. *mycoides* (Vinther and Freundt, 1977). The finding of a truncate flavin-terminated respiratory system lacking quinones and cytochromes in these mycoplasmas (Holländer *et al.*, 1977; see this volume, Chapter 7) may explain its lack of dependence on any membranous structures for organization.

Unlike *Mycoplasma* and *Spiroplasma* species, the NADH oxidase activity in *Acholeplasma* species is membrane-bound (Pollack, 1975) but, resembling the soluble system of *Mycoplasma* species (Low and Zimkus, 1973), it also appears to consist of only two components (Pollack, 1975; Larraga and Razin, 1976; Jinks and Matz, 1976a,b). The Uppsala group (K.-E. Johansson, personal communication) has recently isolated two proteins from *A. laidlawii* membranes, a flavoprotein (designated $T_{4a}$) and an NADH dehydrogenase, both of which may constitute the NADH oxidase system in this organism. The flavoprotein is a major membrane protein, while the NADH dehydrogenase (estimated molecular weight 140,000) seems to be a minor membrane component. The NADH dehydrogenase of *A. laidlawii* was also purified from an ethanol extract of *A. laidlawii* membranes by deoxycholate and gel filtration (Jinks and Matz, 1976b). The purified enzyme (estimated to be over 90% pure) did not depend on lipids for activity, supporting the results of Ne'eman *et al.* (1972) with native or solubilized *A. laidlawii* membranes and the recent findings that hydrolysis of phosphatidylglycerol in *A. laidlawii* membranes by phospholipase $A_2$ does not affect the NADH oxidase activity though it abolishes the ATPase activity (Bevers *et al.*, 1977). Moreover, unlike ATPase, the NADH oxidase activity of *A. laidlawii* is not influenced by the physical state of membrane lipids (de Kruyff *et al.*, 1973). All this may indicate that the electron transport enzymes in *A. laidlawii* are less integrated with membrane structure than the complex electron transport chains in other bacteria, such as *Micrococcus lysodeikticus* and *Escherichia coli*, where lipids were found to be required for NADH oxidase activity (Nachbar and Salton, 1970; Esfahani *et al.*, 1977). Yet, it must be stressed that NADH oxidase in *A. laidlawii* does not fall into the category of the loosely bound or peripheral membrane proteins, as it cannot be released from the membranes by EDTA and a low ionic environment (Ne'eman *et al.*, 1971).

The different location of the NADH oxidase in the *Mycoplasma* and *Acholeplasma* species, apart from its phylogenetic significance (Pollack, 1975), provides us with an effective tool for comparing the properties of an enzyme system in the soluble and membrane-bound states. The membrane-bound NADH oxidase system of *A. laidlawii*, despite its general resemblance to the soluble system of *M. mycoides* subsp. *capri* (Larraga and Razin, 1976), differed from it in some important properties. In addition to differences in the ratio of oxidase (using $O_2$ as an electron acceptor) to oxidoreductase (using 2,4-dichlorophenol indophenol as an electron acceptor) activities and different sensitivities to several inhibitory agents, the membrane-bound enzyme system was less susceptible than the soluble system to heat inactivation, both in the native and in the detergent-solubilized state. The higher thermostability of the membrane-bound enzyme system might reflect the protective effect of other membrane components closely associated with or adjacent to the enzyme molecules (Larraga and Razin, 1976).

### B. Adenosine Triphosphatase

ATPase has been found to be a component of the cell membrane of every mycoplasma examined so far, including *S. citri* (Mudd *et al.*, 1977; Kahane *et al.*, 1977) and *Ureaplasma urealyticum* (Masover *et al.*, 1977). There can be little doubt that ATPase performs a key role in membrane function, yet we know very little of its molecular properties. Every attempt to release the ATPase from mycoplasma membranes in a soluble and active form has failed so far. The enzyme was not released by the methods employed for the detachment of peripheral membrane proteins (Ne'eman *et al.*, 1971; Ne'eman and Razin, 1975). The marked sensitivity of the *A. laidlawii* ATPase to detergents (Rottem and Razin, 1966; Ne'eman *et al.*, 1971, 1972; Ne'eman and Razin, 1975) caused all attempts to purify it by detergent action to fail. Likewise, this ATPase activity was completely inactivated by *n*-butanol, used to solubilize membrane proteins (S. Razin and L. Gottfried, unpublished data). Ultrasonic treatment liberated about 50% of the enzymatic activity into a fraction that was not sedimentable at 100,000 *g* for 45 min, but it apparently remained bound to minute membrane fragments so that it could not be purified by conventional protein fractionation techniques (Rottem and Razin, 1966). These findings suggested that the mycoplasma ATPase is an integral membrane protein probably dependent on lipids for activity. Conclusive proof became available when the membrane ATPase activity of *A. laidlawii* (de Kruyff *et al.*, 1973; Hsung *et al.*, 1974) and of *M. mycoides* subsp. *capri* (Rottem *et al.*, 1973c) was found to depend on the physical state of membrane lipids. Arrhenius plots of the ATPase activity showed distinct breaks at the lower end of the membrane lipid phase transition, the temperatures at which the breaks occured being dependent on the fatty acid composition of the membrane lipids (Fig. 5 in Chapter 10). Incorporation of large quantities of cholesterol into *M. mycoides* subsp. *capri* membranes abolished both the phase transition of membrane lipids and the break in the Arrhenius plot of the ATPase activity (Rottem *et al.*, 1973c).

Direct proof of the dependence of the *A. laidlawii* ATPase on membrane lipids was recently obtained by Bevers *et al.* (1977), who showed that hydrolysis of over 90% of the membrane phosphatidylglycerol by phospholipase $A_2$ abolished ATPase activity. Activity could be restored by liposomes made of phosphatidylglycerol or other acidic phospholipids, such as phosphatidic acid or phosphatidylserine. The fatty acid composition of the acidic lipid determined the activation energy of the enzyme and the temperature at which the break in the Arrhenius plot occured. It is significant that only about 10% of the membrane phosphatidylglycerol (i.e., less than 3% of the total membrane lipids) is required for ATPase activity, supporting the idea that this lipid forms a "halo" of molecules closely associated with the enzyme protein.

The mycoplasma ATPases, being integral membrane proteins dependent on lipids for activity, bear a measure of resemblance to the ($Na^+$–$K^+$)-ATPase of the eukaryotic plasma membrane (Roelofsen *et al.*, 1971; Bevers *et al.*, 1977). In this respect the mycoplasma ATPases differ from the ATPases characterized so far in other microbial plasma membranes (Salton, 1974). The ATPases of the wall-covered bacteria may be regarded as peripheral membrane proteins since they can be detached as water-soluble, lipid-free proteins by repeated membrane washings using very dilute buffers. However, unlike the ($Na^+$–$K^+$)-activated ATPase of the eukaryotic cell membrane, but as in the microbial and mitochondrial $Mg^{2+}$-dependent ATPases, the mycoplasma enzyme is not activated by sodium and potassium and is insensitive to ouabain (Rottem and Razin, 1966; Cho and Morowitz, 1969). The question of whether the ATPase and *p*-nitrophenylphosphatase activities of *A. laidlawii* membranes are the expression of a single enzyme (Ne'eman *et al.*, 1971) can now be answered negatively. The different sensitivities of the two enzymatic activities to detergents (Ne'eman *et al.*, 1971; Ne'eman and Razin, 1975), organic solvents, and pronase digestion (Ne'eman and Razin, 1975) and the lack of dependence of the *p*-nitrophenylphosphatase on membrane lipids for activity (Bevers *et al.*, 1977) attest to the presence of two different enzymes.

## C. Enzymes of Membrane Biosynthesis

Most of the enzymes participating in membrane lipid synthesis in mycoplasmas appear to be membrane-bound. These include the enzymes responsible for the synthesis of the phospholipids, glucolipids, and phosphoglucolipids in *A. laidlawii* membranes (Smith, 1971). The acyl-CoA:$\alpha$-glycerophosphate transacylase which synthesizes phosphatidic acid, the precursor of membrane phospholipids, was recently located in the membrane of both *A. laidlawii* and *M. hominis* (Rottem and Greenberg, 1975). The activation of fatty acids by an acyl-CoA synthetase was, however, localized in the cytoplasmic fraction of these mycoplasmas. Another membrane-bound enzyme, a long-chain fatty acyl-coenzyme A thioesterase, was recently described in mycoplasmas, its activity being highest in *Acholeplasma* species and very low in most of the *Mycoplasma* species tested, which suggests that it plays a regulatory role in fatty acid biosynthesis (Rottem *et al.*, 1977). The *A. laidlawii* enzyme was shown to be an integral membrane enzyme and, resembling the ATPase, Arrhenius plots of the thioesterase activity showed breaks at temperatures corresponding to the lower end of the membrane lipid phase transition. In the case of the thioesterase, however, it is possible that the effect of the lipid phase transition on the enzyme activity is due to changes in the solubility of the hydrophobic substrate in the lipid bilayer rather than to conformational changes in the enzyme molecule itself (Rottem, *et al.*, 1977).

Membrane lipid synthesis can proceed independently of membrane protein synthesis, and the rate of synthesis may vary considerably under different growth conditions. Thus, membranes of *A. laidlawii* cells grown at high pH had a much lower lipid-to-protein ratio than membranes of cells grown at a low pH value (Kahane and Razin, 1970). Even more striking results can be attained by the addition of chloramphenicol to actively growing mycoplasma cultures. Lipid synthesis continues for several hours after protein synthesis is halted; the membranes of the chloramphenicol-treated cells are consequently richer in lipids and have a lower buoyant density (Kahane and Razin, 1969; Razin, 1974a; Amar *et al.*, 1978b). Hence, membrane lipid synthesis in mycoplasmas is not necessarily synchronized with membrane protein synthesis, so that a stoichiometric relationship between protein and lipid is not essential for membrane function, a conclusion supporting the fluid mosaic model of Singer and Nicolson (1972).

## IV. DISPOSITION OF MEMBRANE PROTEINS

### A. The Transbilayer Distribution of Proteins

The transbilayer asymmetry noted with lipids (Chapter 10, Section VI) is even more pronounced with membrane proteins (Rothman and Lenard, 1977). A number of techniques have been developed to study membrane protein disposition. Most of these procedures are based on the same principle: specific labeling or proteolytic digestion of membrane proteins in intact cells and isolated membranes, assuming that the labeling agent or proteolytic enzyme has access to proteins exposed on both membrane surfaces when isolated membranes are used, but only to proteins on the outer membrane surface when intact cells are used (Fig. 3). The validity of this depends on the fulfillment of several conditions: (a) that the labeling agent or enzyme is incapable of penetrating into the cells, (b) that the isolated membranes do not reseal, and (c) that the disposition of proteins in isolated membranes does not differ significantly from that in membranes of intact cells. Regarding the first condition, the tendency of washed mycoplasmas to lose intracellular components and even lyse on prolonged incubation in nonnutrient media at 37°C (Smith and Sasaki, 1958; Butler and Knight, 1960) could invalidate our major assumption. Indeed, the experiments devised by Amar *et al.* (1974) to test this point

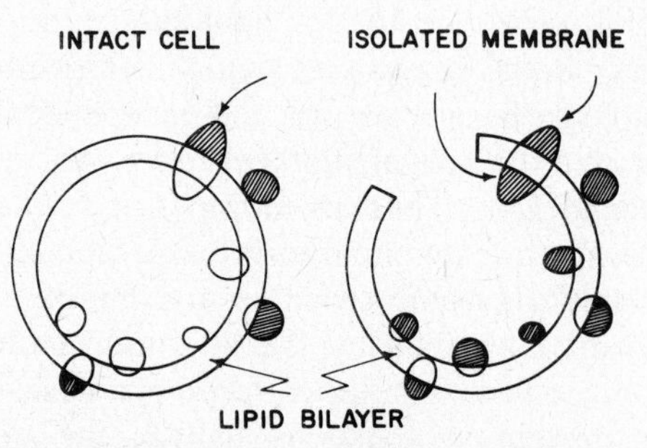


FIGURE 3. Schematic illustration of the principle underlying the techniques used in studies on membrane protein disposition. The shaded parts on the globular protein molecules are those exposed to the labeling agent or the proteolytic enzyme.

showed that some agents, particularly the diazonium salt of sulfanilic acid and pronase, significantly damage the permeability barrier of the cells, restricting the use of these agents in disposition studies on mycoplasma. The second condition appears to apply here since mycoplasma membranes do not usually reseal following their isolation. The large holes produced in the membrane during cell lysis, necessary for the effective release of the highly viscous cellular DNA, are too big to be sealed (see Fig. 4 in Rottem *et al.*, 1968). At this stage it is much more difficult, if not impossible, to state whether the third condition can be fulfilled. It is evident that in isolating cell membranes, even by the gentlest procedure available, we immobilize the membranous structures and place them in a physiologic environment very different from that in the intact functioning cell. Membrane isolation, with its attendant elimination of possible structure-determining substances, such as ions and ATP (Schmidt-Ullrich *et al.*, 1973), or contractile filamentous proteins, such as the actin and spectrin associated with erythrocyte membranes (Sheetz *et al.*, 1976; Yu and Branton, 1976), may leave membranes in a "relaxed" state with many reactive groups available. In the intact cell, the membrane proteins are presumably packed more compactly, with fewer reactive groups exposed. Nevertheless, since spontaneous transmembrane movement ("flip-flop") of both membrane proteins and lipids is thermodynamically unfavorable (Rothman and Lenard, 1977), one would expect the transbilayer asymmetry of membrane components to be retained in isolated membranes, though the faster rate of lateral mobility of the membrane components in the more "relaxed" isolated membrane can be expected to increase the absolute labeling values or the sensitivity of membrane proteins to proteolytic digestion.

### 1. Localization by Means of Labeling Agents and Proteolytic Enzymes

The lactoperoxidase-mediated iodination technique (Hubbard and Cohn, 1976) has proved valuable for examining protein disposition in mycoplasma membranes (Kahane and Marchesi, 1973; Amar *et al.*, 1974, 1976, 1978b; Archer *et al.*, 1978). The high-molecular-weight lactoperoxidase cannot penetrate into mycoplasma cells so that on treatment of intact cells only the tyrosine residues of proteins exposed on the outer membrane surface are labeled by $^{125}I$. Nevertheless a minor fraction of the label (5–10%) was detected in the cytoplasmic protein fraction and in membrane lipids (Amar *et al.*, 1974).

The main conclusion that can be drawn from the iodination studies of Amar *et al.* (1974, 1976, 1978b) is that more proteins are exposed on the

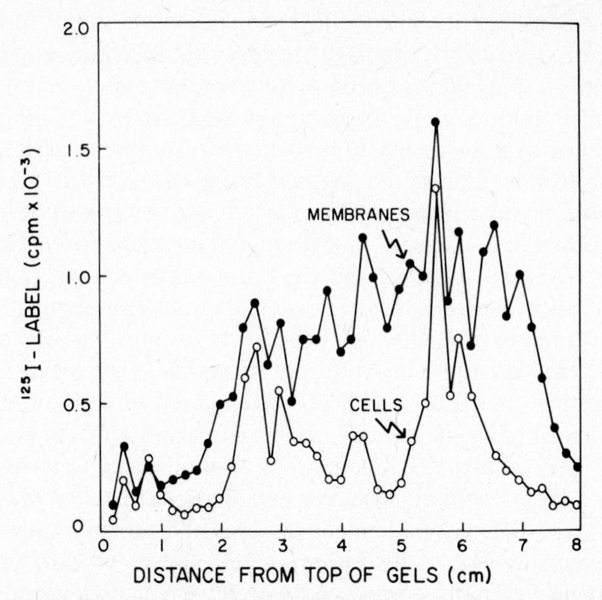


FIGURE 4. Distribution of iodine label in *A. laidlawii* membrane proteins. Isolated membranes or intact cells were subjected to lactoperoxidase-mediated iodination. Membranes were isolated by osmotic lysis of the iodinated cells. The labeled membranes were electrophoresed in polyacrylamide gels containing sodium dodecyl sulfate. The gels were sliced and the iodine label was determined in the slices. (A. Amar, S. Rottem, and S. Razin, unpublished data.)

inner than on the outer membrane surface in mycoplasmas as in other plasma membranes (Rothman and Lenard, 1977). This was inferred from the finding that the labeling intensity of isolated membranes was more than threefold greater than found in membranes of intact *A. laidlawii* cells and more than twice as great for *M. hominis*. Moreover, electrophoretic analysis showed that only some of the membrane protein species had been labeled in intact cells as against almost all the protein bands from isolated membranes (Fig. 4). A more detailed picture has recently been provided by Archer *et al.* (1978), who applied the two-dimensional gel electrophoresis technique to iodinated *A. laidlawii* cells and membranes. Of the 140 membrane proteins detected by this technique, about 90–100 proteins were iodinated on treatment of isolated membranes and a maximum of 40 were labeled after iodination of whole cells. The finding of a significant number of unlabeled proteins suggests that these represent proteins which either are totally buried in the lipid bilayer or lack the iodine-binding tyrosine residues on their polypeptide moieties exposed to the aqueous surroundings.

Pronase appears to be the least suitable of the proteolytic enzymes used in the disposition studies (Morowitz and Terry, 1969; Kahane and Marchesi, 1973; Amar *et al.*, 1974) because it caused the greatest damage to cell permeability. Even at low concentrations and fairly short incubation periods this enzyme caused lysis of some of the cells and hydrolyzed the membrane proteins on the cytoplasmic side as well. Membrane digestion with pronase leads to breakdown of the proteins to small peptides, many of which remain associated with the membrane. Thus, the digested membrane may still retain about 50% of its Lowry-reactive material, while none of the original protein bands can be seen in the polyacrylamide gels (Amar *et al.*, 1974). Furthermore, prolonged pronase treatment results in clear gels, although the membrane may still contain 20% of its Lowry-reactive material. One may conclude that the peptides produced by prolonged pronase treatment are too small to be retained in the gels. That these small fragments of the native membrane proteins may be buried within the lipid matrix of the membrane is supported by their poor labeling by the lactoperoxidase iodination reaction (Amar *et al.*, 1974). Yet, the amino acid composition of this "pronase-resistant" peptide residue was not enriched with hydrophobic amino acids as would be anticipated if it contained hydrophobic "tails" of proteins buried in a lipid interior (Morowitz and Terry, 1969).

Trypsin, being a milder, more selective, and better defined proteolytic enzyme, is preferred over pronase for protein disposition studies. The use of this enzyme corroborated the iodination data by showing that only a few protein bands, mainly of high molecular weight, disappeared on treatment of intact cells, as against the elimination of most of the protein bands on treatment of isolated membranes. The finding that most of the proteins exposed on the outer membrane surface are of high molecular weight might imply that these are proteins which span the membrane, but experimental proof for this suggestion is unavailable.

Labeled lectins can be valuable in localizing carbohydrate-containing membrane components. The almost identical binding of labeled lectins to intact cells of a variety of *Mycoplasma* species and to their isolated membranes (Kahane and Tully, 1976) suggests that all the carbohydrate-containing components, including glycoproteins, are exposed on the cell surface. In fact, the external position of the *M. pneumoniae* glycoprotein was also evinced by the lactoperoxidase-mediated iodination technique (Kahane and Marchesi, 1973).

### 2. Localization by Immunologic Techniques

The crossed immunoelectrophoretic technique, successfully used for the localization of membrane proteins and enzymes of *Micrococcus lysodeikticus* (Owen and Salton, 1975), has also been applied to localize membrane proteins in mycoplasmas. Solubilized membranes or purified membrane proteins serve as antigens, and an antiserum to membranes adsorbed either by intact cells (to remove antibodies to proteins exposed on the outer membrane surface) or by isolated membranes (to remove antibodies to proteins exposed on both membrane surfaces) serve as antisera. A major advantage of using antibodies for localization of membrane components is that they are not expected to penetrate the cells, particularly when the cell–antibody mixtures are kept at 4°C for a short period of time. Moreover, unlike the techniques employing enzymes, membrane components are not modified by the antibodies, minimizing membrane perturbation. Only one of four purified membrane proteins from *A. laidlawii* was detected on the external cell surface by this technique (Johansson and Hjertén, 1974; Fig. 5), while in *M. arginini* one major antigen was found on the outside, and two others were apparently completely immersed in the bilayer since antibodies to them could only be adsorbed by Triton-solubilized membrane material (Alexander and Kenny, 1977).

### 3. Localization by Freeze-Fracturing Electron Microscopy

This technique splits biomembranes and exposes extensive face views of the hydrophobic interior of the lipid bilayer continuum. The bilayer appears to be interrupted by particles interpreted as being the morphologic manifestation of proteins or proteins with tightly bound lipids intercalated into the membrane, possibly traversing it. Freeze-fractured faces of artificial membranes made of phospholipids are smooth and become studded with particles only when certain integral membrane proteins are incorporated into them (Hong and Hubbell, 1972; Yu and Branton, 1976) which is perhaps the best available evidence for the protein nature of the particles. Tourtellotte and Zupnik (1973) substantiated this suggestion by showing that the number of particles in membranes of *A. laidlawii* cells decreased markedly after incubation for several hours in a growth medium either devoid of amino acids or containing puromycin. Under these conditions membrane protein synthesis was essentially abolished within 1 hr, whereas membrane lipid synthesis continued uninterrupted for at least 6 hr.

Predictably, every mycoplasma membrane examined by freeze fracturing so far showed the presence of particles, 50–100 Å in diameter, on its fracture faces. As with other plasma membranes, the distribution of particles on the two fracture faces of the mycoplasma membranes is asymmetrical; more particles are observed on the convex fracture face than on the concave fracture face (Fig. 6; Tillack *et al.*, 1970; Tourtellotte *et al.*, 1970; Maniloff and Morowitz, 1972; Rottem *et al.*, 1973b; Razin *et*

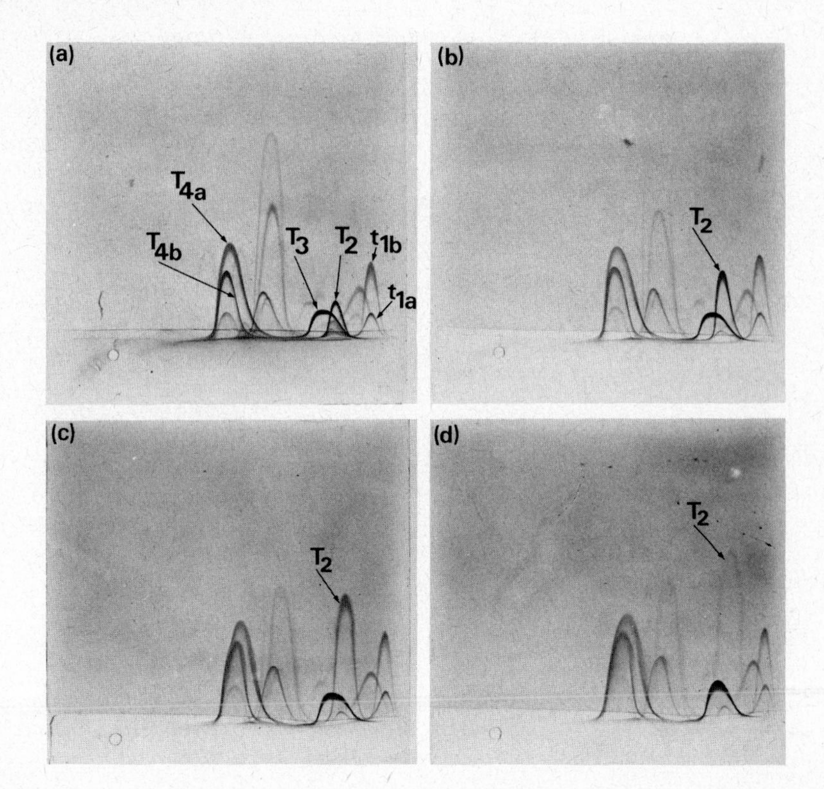


FIGURE 5. Localization of antigenic determinants *in situ* of membrane proteins by crossed immunoelectrophoresis. Antigen, Tween 20-soluble material from *A. laidlawii* membranes (Tween supernatant). Antibodies were produced by injecting washed membranes into rabbits. The concentrations of antibodies against surface proteins were successively diminished by adsorption to an increasing number of intact organisms. After centrifugation, the supernatant was used as antiserum in crossed immunoelectrophoresis. Observe the successive increase of the area under the precipitation line, corresponding to the protein $T_2$, indicating that this protein is exposed on the outside surface of the intact organism. (a) 0 ml cell suspension; (b) 1 ml cell suspension; (c) 2 ml cell suspension; (d) 2.5 ml cell suspension. (From Johansson and Hjertęn, 1974.)

*al.*, 1973; Green and Hanson, 1973). Since the convex fracture face corresponds to the inner half of the membrane lipid bilayer (Branton *et al.*, 1975), the larger number of particles on it supports the conclusion derived from data obtained by other techniques that more membrane proteins face the cytoplasm than the external cell surface.

## 4. Localization of Membrane-Bound Enzymes

That ATPase and NADH oxidase activities are probably localized on the inner cytoplasmic face of the cell membrane in *A. laidlawii* is sup-

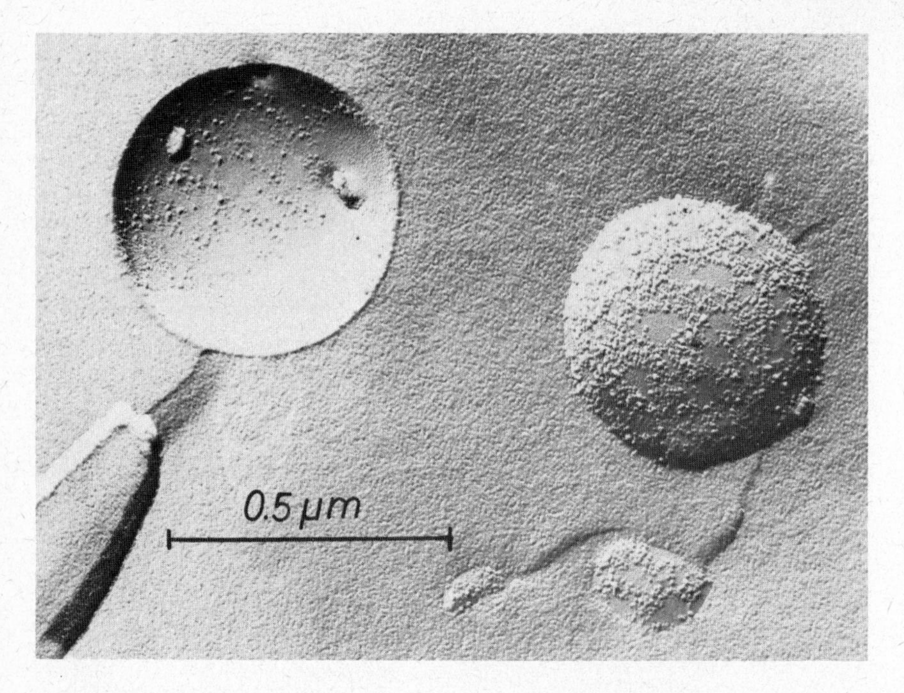


FIGURE 6. Replicas of freeze-cleaved *A. laidlawii* membranes. Many more particles are seen on the convex than on the concave fracture face, demonstrating the asymmetrical transbilayer distribution of membrane proteins. (From Razin, 1974b.)

ported by the inability of intact cells to attack ATP and NADH (Pollack *et al.*, 1965b) and by the resistance, in intact cells, of these enzymes to inactivation by proteases and by antibodies. This contrasts with their high sensitivity to these agents when isolated membranes are treated (Ne'eman and Razin, 1975). ATPase has also been localized histochemically on the inner side of the cell membrane of *M. gallisepticum* (Munkres and Wachtel, 1967). In this respect the mycoplasma ATPase resembles *M. lysodeikticus* ATPase which is similarly located on the inner surface of the plasma membrane (Salton, 1974). The flavoprotein component of the NADH oxidase system of *A. laidlawii* was also localized on the inner membrane face by crossed immunoelectrophoresis (Johansson and Hjertén, 1974). Considering that these enzymes play a key role in respiration and energy coupling, their localization on the cytoplasmic face of the plasma membrane is to be anticipated.

Unlike the NADH oxidase and ATPase, the membrane-bound peptidases and nucleases of *A. laidlawii* are at least partly exposed on the external membrane surface (Table II). Hydrolysis of their substrates can be demonstrated with intact cells (Pecht *et al.*, 1972; Ne'eman and Razin, 1975), supporting the proposed involvement of these enzymes in the provision of transportable nutrients of low molecular weight, such as amino acids and nucleosides, from peptides and nucleic acids supplied in the growth medium. The resistance of the *p*-nitrophenylphosphatase activity of *A. laidlawii* membranes to inactivation by pronase (Ne'eman and Razin, 1975) suggests that this enzyme is embedded within the membrane and in this way is protected from proteolysis.

We have some information on the localization of membrane-bound enzymes of lipid metabolism in *A. laidlawii* (Table II). The acyl-CoA thioesterase probably resides on the inner membrane face since proteolytic treatment of intact cells affected its activity only slightly, whereas treatment of isolated membranes inactivated about 80% of the enzyme activity (Rottem *et al.*, 1977). The acyl-CoA:$\alpha$-glycerophosphate transacylase also appears to be located on the inner membrane surface, since intact *A. laidlawii* and *M. hominis* cells showed no significant enzymatic activity in contrast to isolated membranes (Rottem and Greenberg, 1975).

### B. Distribution of Proteins within the Plane of the Membrane

Lateral diffusion of membrane lipids and proteins is usually much faster than the thermodynamically unfavorable transbilayer movement or "flip-flop" (Rothman and Lenard, 1977). Hence, a homogenous distribution of lipids and proteins in the plane of the membrane would be predicted under normal growth conditions. However, many studies show that this is not always the case, since due to the presence in the membrane of lipids of different melting temperatures, lateral phase separation may take place even within the temperature range permitting normal growth. The differential crystallization of lipids in the bilayer influences the distribution of the proteins partially or wholly immersed in it. Freeze-fracture studies support this notion by showing the aggregation of the intramembranous particles during the progressive crystallization of the lipid bilayer of mycoplasma membranes (Chapter 10, Section V, A). Since in cholesterol-rich membranes phase transition of lipids is averted, aggregation of intramembranous particles could be demonstrated only in *A. laidlawii* (Verkleij *et al.*, 1972; James and Branton, 1973; Wallace and Engelman, 1978), or in the cholesterol-poor *M. mycoides* subsp. *capri* (Rottem *et al.*, 1973b). The distribution of particles on the fracture faces of cholesterol-rich mycoplasma membranes, such as those of *M. gallisepticum* (Maniloff and Morowitz, 1972), *M. mycoides* subsp. *capri* and subsp. *mycoides* (Rottem *et al.*, 1973b; Z. Ne'eman, unpublished data), and *Mycoplasma meleagridis* (Green and Hanson, 1973) was to a large extent homogenous.

While freeze fracturing facilitates the study of the planar distribution and lateral mobility of the proteins immersed within the lipid bilayer, a new technique has recently been employed to study the distribution of proteins partly or entirely exposed on the membrane surface (Wallace *et al.*, 1976; Wallace and Engelman, 1978). The membranes are treated with a biotin–avidin–ferritin complex which specifically links to free amino groups of the exposed membrane proteins. The ferritin component of the complex enables its visualization in the electron microscope. *Acholeplasma laidlawii* membranes labeled above or below the lipid phase transition showed the labeled sites to be relatively dispersed, whereas in membranes labeled at the midtransition, low- and high-density patches of label were found. This patching phenomenon was reversible on changing the temperature. As with freeze fracturing, these results signify that lateral mobility of membrane proteins is affected by the physical state of membrane lipids. However, in contrast to the results of freeze fracturing, where the intramembranous particles remained aggregated below the phase transition, the labeling data indicate that lateral mobility of membrane proteins may also take place when membrane lipid is entirely in the crystalline state, suggesting that this state is not as solid and resistant to diffusion as is widely supposed. To eliminate the possibility that the freeze-fracture or ferritin-labeling techniques were producing artifacts which resulted in the different distribution patterns or that variations in experimental details of the sample preparations were responsible, Wallace and Engelman (1978) tested the behavior of surface sites and intramembranous particles in a single sample and obtained the same results as previously.

To explain their data, Wallace and Engelman (1978) propose that the two techniques focus on two classes of membrane proteins which are differently influenced in their distribution by the physical state of the membrane lipids. The first class of protein consists of all integral membrane proteins having primary amino groups exposed at the cytoplasmic membrane surface. This is the set of proteins observed by the avidin–ferritin labeling and may include all the proteins in the membrane. The second class consists of proteins which penetrate deeply into the bilayer and are seen as particles in freeze fracture. This is a subset of the total set of proteins and possibly a subset of the first class. Accordingly, the intramembranous particles are influenced by the lipids on both the cytoplasmic and the exoplasmic halves of the bilayer, whereas the remainder of the surface proteins detected by the avidin–ferritin are influenced primarily by the lipids on the cytoplasmic side. As the lipids may have an asymmetrical distribution, the behavior of lipids in both leaflets of the bilayer may not be identical.

Restricted mobility of some proteins in membranes of intact cells is a well-established phenomenon in erythrocytes. The intramembranous particles in human erythrocytes are immobile in the membrane of intact cells, apparently due to their association with a network made of spectrin and actin on the cytoplasmic face of the membrane (Yu and Branton, 1976; Sheetz *et al.*, 1976). Other membrane proteins are free to move under the same conditions (Fowler and Branton, 1977). In view of the possible presence of contractile proteins in mycoplasmas (Neimark, 1977; Kahane and Muhlrad, 1977; Ghosh *et al.*, 1978; Searcy *et al.*, 1978), it would be pertinent to test whether the intramembranous particles are free to move in membranes of intact mycoplasma cells. Most of the studies using freeze fracturing have been carried out on isolated mycoplasma membranes, where actin-like proteins, if present, might have dissociated from the membrane. The study of Copps *et al.* (1976) on the aggregation and redistribution of the intramembranous particles in *A. laidlawii* membranes as a function of pH, was apparently performed on cells, but data on the intactness of the cells were not provided. Clearly, this point requires more rigorous clarification.

### C. Effects of Culture Age and Membrane Potential on Protein Disposition

The process of aging in mycoplasma cultures is accompanied by a marked decrease in the activity of membrane-associated enzymes and transport systems and quite frequently culminates in lysis of the wall-less organisms. There is expanding evidence to indicate that these manifestations of aging are associated with alterations in the composition and physical properties of the cell membrane. Thus, the phospholipid and cholesterol content of membranes from different mycoplasma species were found to decrease most markedly on aging of cultures, resulting in a significantly higher ratio of protein-to-lipid, a higher density, and reduced fluidity (Table III). The finding that membranes from aging mycoplasma cultures become richer in protein raises two questions: (a) Does the membrane protein composition remain constant during aging, and (b) are there any changes in the disposition of the various membrane proteins during aging? As previously mentioned in this chapter (Section II, A) the overall profile of the major membrane proteins, detectable by polyacrylamide gel electrophoresis, is hardly altered by aging (Amar *et al.*, 1976, 1978b). However, significant changes in the vertical disposition of the proteins were indicated by experiments utilizing the lactoperoxidase-

TABLE III. **Changes in Membrane Composition and Properties in Aging *M. hominis* Cultures**

| Absorbance of culture at 640 nm | Viable organisms (CFU/ml) | Membrane lipid-to-protein ratio | | Membrane density[b] ($gm/cm^3$) | Lectin binding[c] (μg concanavalin A per mg protein) | ATPase activity[d] | Membrane fluidity[e] (hyperfine splitting in gauss) |
|---|---|---|---|---|---|---|---|
| | | μg lipid phosphorus per mg protein[a] | μg cholesterol per mg protein[a] | | | | |
| 0.10 | $7.5 \times 10^8$ | 9.5 | 226 | 1.162 | 2.3 | 3.07 | 53.9 |
| 0.30 | $1.2 \times 10^9$ | 7.5 | 158 | 1.172 | 1.3 | 2.00 | 55.2 |
| 0.40 | $1.7 \times 10^9$ | 6.3 | 121 | 1.183 | 1.1 | 1.40 | 56.0 |

[a] From Razin (1975a).
[b] From Amar *et al.* (1976).
[c] From Amar *et al.* *(1978a)*.
[d] ATPase activity is expressed in micromoles of inorganic phosphate released from ATP per milligram protein in 30 min.
[e] From Rottem and Greenberg (1975).

mediated iodination system and proteolytic enzymes. These experiments suggested that aging in *M. hominis* cultures is accompanied by a continuous increase in the packing density of the protein molecules on the inner surface of the cell membrane (Amar *et al.*, 1976). In *A. laidlawii* cultures the number of protein sites on the cell surface available for iodination and exposed to proteolytic digestion decreased considerably with age (Amar *et al.*, 1978b).

The question of whether membrane fluidity influences the vertical disposition of mycoplasma membrane proteins was recently tackled in our laboratory (Amar *et al.*, 1978b). Alterations in the fluidity of *A. laidlawii* membranes were introduced by varying the age or the growth temperature of the culture, by modifying the fatty acid composition of the membranes, by changing the iodination temperature, or by modulating the lipid-to-protein ratio in the membrane by inhibiting protein synthesis by chloramphenicol. No consistent correlation could be found between altered membrane fluidity and degree of exposure of membrane proteins to lactoperoxidase-mediated iodination. Nevertheless, the iodination values of intact cells regularly decreased on the exposure of the organisms to the suboptimal growth conditions employed to introduce changes in membrane fluidity. The iodination values of isolated membranes, on the other hand, remained unchanged regardless of growth conditions. These findings suggested that membrane potential or the energized membrane state that are affected by growth conditions might influence membrane protein disposition. Supportive evidence for this hypothesis is presented in Table IV. The exposure of *A. laidlawii* cells to the ionophore valinomycin, which dissipates the $K^+$ gradient, or to carbonylcyanide *m*-chlorophenylhydrazone (CCCP), which causes the collapse of the proton gradient, resulted in a rapid drop in the iodination values of the proteins exposed on the cell surface. Conformational changes in membrane proteins or enzymes resulting from changes in the proton gradient have been described for several microbial (Rudnick *et al.*, 1976), mitochondrial (Pfaff and Klinberger, 1968), and chloroplast (Jagendorf, 1975) membrane systems. It can thus be proposed that the decreased availability of iodine binding sites on the mycoplasma cells surface may reflect conformational changes in membrane proteins triggered by short-circuiting the membrane potential.

## V. CONCLUSIONS

The first strides toward the molecular characterization of the mycoplasma membrane proteins have already been made following the de-

TABLE IV. **Effects of Valinomycin and Carbonylcyanide *m*-Chlorophenylhydrazone (CCCP) on the Lactoperoxidase-Mediated Iodination of Intact Cells and Isolated Membranes of *A. laidlawii***

| Inhibitor[a] | $^{125}$I label ($10^5$ cpm/mg membrane protein) | | labeling ratio (membranes/cells) |
|---|---|---|---|
| | Isolated membranes | Intact cells | |
| No inhibitor | 30.9 | 7.5 | 4.2 |
| Valinomycin ($10^{-5}$ *M*) | 30.5 | 5.8 | 5.3 |
| CCCP ($10^{-5}$ *M*) | 30.4 | 5.3 | 5.7 |
| Valinomycin ($10^{-5}$ *M*) + CCCP ($10^{-5}$ *M*) | 30.8 | 2.7 | 11.2 |

[a] Washed cells (10mg cell protein/ml) in 0.15 *M* NaCl containing 0.05 sodium phosphate buffer, pH 7.5 were treated with inhibitors for 10 min at 37°C. Part of the cells were osmotically lysed and the isolated membranes as well as the intact cells were subjected to lactoperoxidase-mediated iodination (From Amar *et al.* 1978c).

velopment of methods for their solubilization and fractionation. Some mycoplasma membrane proteins have been purified to a degree enabling the meaningful analysis of their amino acid composition. Determination of the amino acid sequence in these proteins will apparently be the next step leading to a better understanding of their hydrophobic properties. The finding of carbohydrates associated with some of these membrane proteins is of particular interest in view of the rarity of glycoproteins in prokaryotes, and the possibility that these externally located components participate in the specific binding of mycoplasmas to their host cells.

Specific interactions of membrane proteins with membrane lipids do occur, as indicated by the requirement of the *A. laidlawii* ATPase for phosphatidyl glycerol. Better characterization of the mycoplasmal ATPase is warranted, as it differs from the ATPase of other prokaryotes in its dependence on phospholipids. The membrane-bound NADH oxidase of *A. laidlawii* does not depend on lipids for activity and seems to consist of two enzyme proteins which may soon be characterized. Although many of the enzymes involved in the biosynthesis of membrane components are located in the membrane, only a few have been studied so far and none has been purified. The finding that at least some of these enzymes depend on membrane lipids for activity is an obstacle to their isolation in an active state.

The reassembly of detergent-solubilized mycoplasma membrane proteins and lipids into membranous structures, though boosting research on membrane reconstitution (reviewed by Razin, 1972, 1974b), failed to provide correct answers to questions concerning the mode of organization of proteins in native mycoplasma membranes. The recent application of labeling agents and proteolytic enzymes has been more successful by indicating an asymmetrical transbilayer distribution of membrane proteins, with more proteins facing the cytoplasm than the external cell surface. The findings that culture age, adverse growth conditions, and ionophores affect the vertical disposition of mycoplasma membrane proteins exposed on the cell surface are of great biological interest and deserve further investigation.

## REFERENCES


Alexander, A. G., and Kenny, G. E. (1977). *Infect. Immun.* **15,** 313–321.

Amar, A. (1977). Ph.D. Thesis, Hebrew University, Jerusalem.

Amar, A., Rottem, S., and Razin, S. (1974). *Biochim. Biophys. Acta* **352,** 228–244.

Amar, A., Rottem, S., Kahane, I., and Razin, S. (1976). *Biochim. Biophys. Acta* **426,** 258–270.

Amar, A., Kahane, I., Rottem, S., and Razin, S. (1978a). *J. Gen. Microbiol.* (in press).

Amar, A., Rottem, S., and Razin, S. (1978b). *Biochim. Biophys. Acta* (in press).

Amar, A., Rottem, S., and Razin, S. (1978c). *Biochem. Biophys. Res. Commun.* **84,** 306–312.

Archer, D. B. (1975). *J. Gen. Microbiol.* **88,** 329–338.

Archer, D. B., Rodwell, A. W., and Rodwell, E. S. (1978). *Proc. Austral. Biochem. Soc.* **11,** 109.

Armstrong, D., and Yu, B. (1970). *J. Bacteriol.* **104,** 295–299.

Bevers, E. M., Snoek, G. T., Op den Kamp, J. A. F., and van Deenen, L. L. M. (1977). *Biochim. Biophys. Acta* **467,** 346–356.

Branton, D., Bullivant, S., Gilula, N. B., Karnovsky, M. J., Moore, H., Mühlethaler, K., Northcote, D. H., Packer, L., Satir, B., Satir, P., Speth, K., Staehlin, L. A., Steere, R. L., and Weinstein, R. S. (1975). *Science* **190,** 54–56.

Butler, M., and Knight, B. C. J. G. (1960). *J. Gen. Microbiol.* **22,** 470–477.

Cho, H. W., and Morowitz, H. J. (1969). *Biochim. Biophys. Acta* **183,** 295–303.

Choules, G. L., and Bjorklund, R. F. (1970). *Biochemistry* **9,** 4759–4767.

Choules, G. L., and Gray, W. R. (1971). *Biochem. Biophys. Res. Commun.* **45,** 849–855.

Copps, T. P., Chelack, W. S., and Petkau, A. (1976). *J. Ultrastruct. Res.* **55,** 1–3.

Dahl, J. S., Hellewell, S. B., and Levine, R. P. (1977). *J. Immunol.* **119,** 1419–1426.

de Kruyff, B., van Dijck, P. W. M., Goldbach, R. W., Demel, R. A., and van Deenen, L. L. M. (1973). *Biochim. Biophys. Acta* **330,** 269–282.

Dresdner, G., and Cid-Dresdner, H. (1976). *FEBS Lett.* **72,** 243–246.

Dresdner, G., and Cid-Dresdner, H. (1977). *Anal. Biochem.* **78,** 171–181.

Engelman, D. M., and Morowitz, H. J. (1968). *Biochim. Biophys. Acta* **150,** 376–384.

Esfahani, M., Rudkin, B. B., Cutler, C. J., and Waldron, P. E. (1977). *J. Biol. Chem.* **252,** 3194–3198.

Fowler, V., and Branton, D. (1977). *Nature (London)* **268,** 23–26.

Ghosh, A., Maniloff, J., and Gerling, D. A. (1978). *Cell* **13,** 57–64.

Goel, M. C., and Lemcke, R. M. (1975). *Ann. Microbiol. (Paris)* **126b,** 299–312.

Green, F., III, and Hanson, R. P. (1973). *J. Bacteriol.* **116,** 1011–1018.

Guidotti, G. (1972). *Annu. Rev. Biochem.* **41,** 731–752.

Hjertén, S., and Johansson, K.-E. (1972). *Biochim. Biophys. Acta* **288,** 312–325.

Holländer, R., Wolf, G., and Mannheim, W. (1977). *Antonie van Leeuwenhoek J. Microbiol. Serol.* **43,** 177–185.
Hong, K., and Hubbell, W. L. (1972). *Proc. Natl. Acad, Sci. U.S.A.* **69,** 2617–2621.
Hsung, J.-C., Huang, L., Hoy, D. J., and Haug, A. (1974). *Can. J. Biochem.* **52,** 974–980.
Hubbard, A. L., and Cohn, Z. A. (1976). *In* "Biochemical Analysis of Membranes" (A. H. Maddy, ed.). pp. 427–501. Chapman & Hall, London.
Jagendorf, A. T. (1975). *In* "Bioenergetics of Photosynthesis" (R. Govindjee, ed.), pp. 414–492. Academic Press, New York.
James, R., and Branton, D. (1973). *Biochim. Biophys. Acta* **323,** 378–390.
Jinks, D. C., and Matz, L. L. (1976a). *Biochim. Biophys. Acta* **430,** 71–82.
Jinks, D. C., and Matz, L. L. (1976b). *Biochim. Biophys. Acta* **452,** 30–41.
Johansson, K.-E. (1978). *Zentralbl. Bakteriol., Parasitenkd., Infektionskr. Hyg., Abt. 1: Orig., Reihe A* **241,** 199–200.
Johansson, K.-E., and Hjertén, S. (1974). *J. Mol. Biol.* **86,** 341–348.
Johansson, K.-E., Blomquist, I., and Hjertén, S. (1975). *J. Biol. Chem.* **250,** 2463–2469.
Kagawa, Y. (1972). *Biochim. Biophys. Acta* **265,** 297–338.
Kahane, I., and Brunner, H. (1977). *Infect. Immun.* **18,** 273–277.
Kahane, I., and Marchesi, V. T. (1973). *Ann. N.Y. Acad. Sci.* **225,** 38–45.
Kahane, I., and Muhlrad, A. (1977). *Israel J. Med. Sci.* **13,** 956.
Kahane, I., and Razin, S. (1969). *Biochim. Biophys. Acta* **183,** 78–89.
Kahane, I., and Razin, S. (1970). *FEBS Lett.* **10,** 261–264.
Kahane, I., and Tully, J. G. (1976). *J. Bacteriol.* **128,** 1–7.
Kahane, I., Greenstein, S., and Razin, S. (1977). *J. Gen. Microbiol.* **101,** 173–176.
Larraga, V., and Razin, S. (1976). *J. Bacteriol.* **128,** 827–833.
Leaback, D. H. (1972). *Biochem. J.* **128,** 127p.
Low, I. E., and Zimkus, S. M. (1973). *J. Bacteriol.* **116,** 346–354.
Maniloff, J., and Morowitz, H. J. (1972). *Bacteriol. Rev.* **36,** 263–290.
Masover, G. K., Razin, S., and Hayflick, L. (1977). *J. Bacteriol.* **130,** 297–302.
Morowitz, H. J., and Terry, T. M. (1969). *Biochim. Biophys. Acta* **183,** 276–294.
Mudd, J. B., Ittig, M., Roy, B., Latrille, J., and Bové, J. M. (1977). *J. Bacteriol.* **129,** 1250–1256.
Munkres, M., and Wachtel, A. (1967). *J. Bacteriol.* **93,** 1096–1103.
Muñoz, E., Nachbar, M. S., Schor, M. T., and Salton, M. R. J. (1968). *Biochem. Biophys. Res. Commun.* **32,** 539–546.
Nachbar, M. S., and Salton, M. R. J. (1970). *Biochim. Biophys. Acta* **223,** 309–320.
Ne'eman, Z. (1974). Ph.D. Thesis, Hebrew University, Jerusalem.
Ne'eman, Z., and Razin, S. (1975). *Biochim. Biophys. Acta* **375,** 54–68.
Ne'eman, Z., Kahane, I., and Razin, S. (1971). *Biochim. Biophys. Acta* **249,** 169–176.
Ne'eman, Z., Kahane, I., Kovartovsky, J., and Razin, S. (1972). *Biochim. Biophys. Acta* **266,** 255–268.
Neimark, H. C. (1977). *Proc. Natl. Acad. Sci. U.S.A.* **74,** 4041–4045.
Oldfield, E., Chapman, D., and Derbyshire, W. (1972). *Chem. Phys. Lipids* **9,** 68–91.
Owen, P., and Salton, M. R. J. (1975). *Proc. Natl. Acad. Sci. U.S.A.* **72,** 3711–3715.
Pecht, M., Giberman, E., Keysary, A., Yariv, J., and Katchalski, E. (1972). *Biochim. Biophys. Acta* **290,** 267–273.
Pfaff, E., and Klinberger, M. (1968). *Eur. J. Biochem.* **6,** 66–79.
Pisetsky, D., and Terry, T. M. (1972). *Biochim. Biophys. Acta* **274,** 95–104.
Pollack, J. D. (1975). *Int. J. Syst. Bacteriol.* **25,** 108–113.
Pollack, J. D., Razin, S., Pollack, M. E., and Cleverdon, R. C. (1965a). *Life Sci.* **4,** 973–977.
Pollack, J. D., Razin, S., and Cleverdon, R. C. (1965b). *J. Bacteriol.* **90,** 617–622.

Razin, S. (1964). *J. Gen. Microbiol.* **33,** 471–475.
Razin, S. (1972). *Biochim. Biophys. Acta* **265,** 241–296.
Razin, S. (1974a). *FEBS Lett.* **47,** 81–85.
Razin, S. (1974b). *J. Supramol. Struct.* **2,** 670–681.
Razin, S. (1978). *Microbiol. Rev.* **42,** 414–470.
Razin, S., Valdesuso, J., Purcell, R. H., and Chanock, R. M. (1970). *J. Bacteriol.* **103,** 702–706.
Razin, S., Kahane, I., and Kovartovsky, J. (1972). *Pathog. Mycoplasmas, Ciba Found. Symp., 1972* pp. 93–122.
Razin, S., Hasin, M., Ne'eman, Z., and Rottem, S. (1973). *J. Bacteriol.* **116,** 1421–1435.
Rodwell, A. W. (1969). *In* "The Mycoplasmatales and the L-Phase of Bacteria" (L. Hayflick, ed.), pp. 413–450. Appleton, New York.
Rodwell, A. W., Razin, S., Rottem, S., and Argaman, M. (1967). *Arch. Biochem. Biophys.* **122,** 621–628.
Roelofsen, B., Zwaal, R. F. A., and van Deenen, L. L. M. (1971). *In* "Membrane Bound Enzymes" (G. Porcellati and F. DiJeso, eds.), pp. 209–228. Plenum, New York.
Rosendal, S. (1973). *Acta Pathol. Microbiol. Scand.* **81,** 273–281.
Rothman, J. E., and Lenard, J. (1977). *Science* **195,** 743–753.
Rottem, S., and Greenberg, A. S. (1975). *J. Bacteriol.* **121,** 631–639.
Rottem, S., and Hayflick, L. (1973). *Can. J. Biochem.* **51,** 632–636.
Rottem, S., and Razin, S. (1966). *J. Bacteriol.* **92,** 714–722.
Rottem, S., and Razin, S. (1967). *J. Bacteriol.* **94,** 359–364.
Rottem, S., and Razin, S. (1972). *J. Bacteriol.* **110,** 699–705.
Rottem, S., Stein, O., and Razin, S. (1968). *Arch. Biochem. Biophys.* **125,** 46–56.
Rottem, S., Hasin, M., and Razin, S. (1973a). *Biochim. Biophys. Acta* **298,** 876–886.
Rottem, S., Yashouv, J., Ne'eman, Z., and Razin, S. (1973b). *Biochim. Biophys. Acta* **323,** 495–508.
Rottem, S., Cirillo, V. P., de Kruyff, B., Shinitzky, M., and Razin, S. (1973c). *Biochim. Biophys. Acta* **323,** 509–519.
Rottem, S., Trotter, S. L., and Barile, M. F. (1977). *J. Bacteriol.* **129,** 707–713.
Rudnick,, G., Schuldiner, S., and Kaback, R. H. (1976). *Biochemistry* **15,** 5126–5132.
Ruwart, M. J., and Haug, A. (1975). *Biochemistry* **14,** 860–866.
Salton, M. R. J. (1974). *Adv. Microb. Physiol.* **11,** 213–283.
Schiefer, H.-G., Gerhardt, U., Brunner, H., and Krupe, M. (1974). *J. Bacteriol.* **120,** 81–88.
Schmidt-Ullrich, R., Knüfermann, H., and Wallach, D. F. H. (1973). *Biochim. Biophys. Acta* **307,** 353–365.
Searcy, D. G., Stein, D. B., and Green, G. R. (1978). *Bio Systems* **10,** 19–28.
Segrest, J. P., Kahane, I., Jackson, R. L., and Marchesi, V. T. (1973). *Arch. Biochem. Biophys.* **155,** 167–183.
Sheetz, M. P., Painter, R. G., and Singer, S. J. (1976). *Biochemistry* **15,** 4486–4492.
Singer, S. J., and Nicolson, G. L. (1972). *Science* **175,** 720–731.
Smith, P. F. (1971). "The Biology of Mycoplasmas." Academic Press, New York.
Smith, P. F., and Sasaki, S. (1958). *Appl. Microbiol.* **6,** 184–189.
Tillack, T. W., Carter, R., and Razin, S. (1970). *Biochim. Biophys. Acta* **219,** 123–130.
Tourtellotte, M. E., and Zupnik, J. S. (1973). *Science* **179,** 84–86.
Tourtellotte, M. E., Branton, D., and Keith, A. (1970). *Proc. Natl. Acad. Sci. U.S.A.* **66,** 909–916.
Urry, D. W. (1962). *Biochim. Biophys. Acta* **265,** 115–168.
van Golde, L. M. G., McElhaney, R. N., and van Deenen, L. L. M. (1971). *Biochim. Biophys. Acta* **231,** 245–249.

Verkleij, A. J., Ververgaert, P. H. J., van Deenen, L. L. M., and Elbers, P. F. (1972). *Biochim. Biophys. Acta* **288,** 326–332.

Vinther, O., and Freundt, E. A. (1977). *Acta Pathol. Microbiol. Scand., Sect. B* **85,** 184–188.

Wallace, B. A., and Engelman, D. M. (1978). *Biochim. Biophys. Acta* **508,** 431–449.

Wallace, B. A., Richards, F. M., and Engelman, D. M. (1976). *J. Mol. Biol.* **107,** 255–269.

Wallach, D. F. H., and Gordon, A. (1968). *Fed. Proc., Fed. Am. Soc. Exp. Biol.* **27,** 1263–1268.

Wroblewski, H., Johansson, K.-E., and Burlot, R. (1977a). *Int. J. Syst. Bacteriol.* **27,** 97–103.

Wroblewski, H., Johansson, K.-E., and Hjertén, S. (1977b). *Biochim. Biophys. Acta* **465,** 275–289.

Yaguzhinskaya, O. E. (1976). *J. Hyrg.* **77,** 189–198.

Yu, J., and Branton, D. (1976). *Proc. Natl. Acad. Sci. U.S.A.* **73,** 3891–3895.

# 12 / TRANSPORT SYSTEMS

*Vincent P. Cirillo*



## I. INTRODUCTION

It is now universally accepted that the lipids of cell membranes are arranged in bilayers which constitute the primary barriers to the movement of solutes across the membrane (Stein, 1967; Singer and Nicolson, 1972). This concept applies equally to mycoplasmas as to other prokaryotes (Razin, 1975; Smith, 1971; van Deenen *et al.*, 1975).

In the absence of specific conductors inserted into the lipid bilayer, transport of the solutes necessary for cell activity, growth, and survival could not take place or would occur at rates many orders of magnitude lower than they do. The natural conductors are specifically designed membrane proteins which complex with their substrates. The characteristics of the solute–conductor complex determines whether it serves merely



THE MYCOPLASMAS, VOL. 1

ISBN 0-12-078401-7

to equilibrate the solute across the cell membrane in accordance with its electrochemical gradient or results in the transport of the solute against its electrochemical gradient. These alternative features are referred to as *facilitated diffusion* and *active transport,* respectively (Cirillo, 1961; Stein, 1967). In the latter case, the work of active transport must be the result of coupling between the energy-requiring (endergonic) transport process and energy-releasing (exergonic) processes. In both facilitated diffusion and active transport, the solute molecules are not chemically altered and the membrane proteins are referred to as *carriers*. However, the transport of some solutes is accompanied by their chemical modification, i.e., phosphorylation of sugars, phosphoribosylation of purines, CoA thioesterification of fatty acids or hydrolysis of saccharides and peptides. If this chemical modification is obligatory, the transport process is called *group translocation* (or vectorial metabolism) and involves membrane-bound enzymes (Simoni, 1972; Kaback, 1974; Roseman, 1977).

Transport processes mediated either by carriers or enzymes differ from unmediated *simple diffusion* processes by the additional feature that they show stereo and isomeric substrate specificity (i.e., $\alpha$ versus $\beta$ or D versus L isomers of their substrates) and saturation kinetics. These latter features reflect the specificity of the interaction between the solutes and the membrane carriers and enzymes.

The kinetic and thermodynamic characteristics which differentiate mediated transport processes from simple diffusion are summarized in Fig. 1. The graphs at the top of the figure compare the kinetics of transport for a hypothetical pair of isomers (D and L) across a lipid bilayer. In the unmediated processes, the rate of transport ($v$) for both isomers is the same and obeys Fick's laws at all substrate concentrations (Stein, 1967). By contrast, the rate of mediated transport is different for the D and L isomer (i.e., D-versus L-glucose) and exhibits saturation kinetics with a characteristic $K_m$ and $V_{max}$. The three types of mediated transport are distinguished on the basis of whether the solutes are chemically modified (group translocation) or unmodified and whether the unmodified solute is transported uphill against its electrochemical gradient (active transport) or downhill with its electrochemical gradient (facilitated diffusion).

The transport processes studied in mycoplasmas are listed in Table I. As can be seen, mycoplasmas present examples of transport by simple diffusion (glycol, glycerol, and erythritol), group translocation (sugars), and active transport ($K^+$, amino acids, and sugars). The specific characteristics of each system are presented in the remaining sections of this chapter.

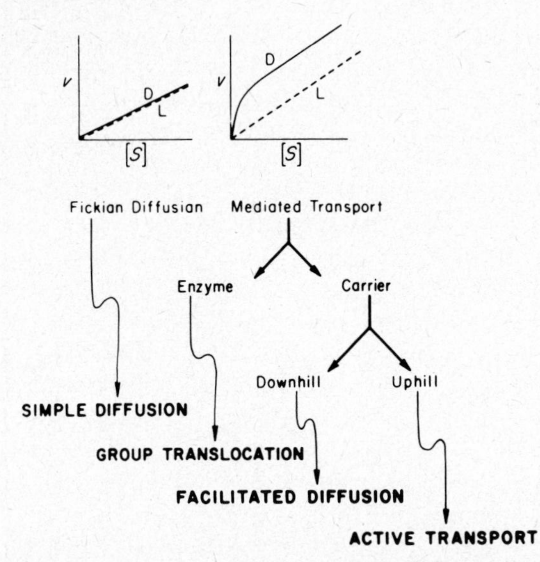


FIGURE 1. A key to transport systems. Graphs show dependence of transport rate (*v*) on substrate concentration [*S*] for Fickian diffusion and mediated transport for a hypothetical pair of isomers, D and L. Mediated transport is further classified according to whether the substrate is enzymatically modified (enzyme) or is transported in an unmodified form (carrier). Facilitated diffusion and active transport are distinguished on the basis of whether substrates are transported uphill or downhill with respect to their electrochemical gradients.

## II. SIMPLE DIFFUSION AND PERMEABILITY

### A. The Energy Barrier to Diffusion across Lipid Bilayers

Studies of the unmediated uptake of the homolgous series of polyols, glycol, glycerol, and erythritol, by *Acholeplasma laidlawii* and liposomes derived from *A. laidlawii* lipids have helped to confirm some fundamental principles of membrane permeability (McElhaney *et al.*, 1970, 1973; de Gier *et al.*, 1971; de Kruyff *et al.*, 1972; Romijn *et al.*, 1972; Read and McElhaney, 1975). These principles are based on the absolute rate theory (Johnson *et al.* 1954; de Gier *et al.*, 1971), which proposes that the rate of a molecular process, whether it is a chemical reaction or simple diffusion, is determined by the energy required by the reacting molecules to reach the energy level of the "transition intermediate." The higher the energy level of the transition intermediate, the fewer molecules in a given population will have the requisite energy to react and the lower will be the rate of reaction. For the diffusion of nonelectrolytes such as polyols from one aqueous phase to another across a lipid bilayer, the transition state is

TABLE I. **Transport Processes Studied in Mycoplasmas**

| Transport process | | | |
|---|---|---|---|
| Solute | Type | Organism | References[a] |
| A. Ions | | | |
| $K^+$ | Active transport | *A. laidlawii* | 1–3 |
| B. Amino acids | | | |
| 1-Histidine | Active transport | *M. fermentans* | 4 |
| 1-Methionine | | *M. hominis* | 4 |
| C. Polyhydric alcohols | | | |
| Glycol, glycerol, erythritol | Simple diffusion | *A. laidlawii* | 5–10 |
| D. Sugars | | | |
| Glucose ($\alpha$-MG)[b] | Group translocation | *M. gallisepticum*, *M.* sp. strain Y, *M. mycoides* subsp. *mycoides*, *M. mycoides* subsp. *capri*, *M. capricolum*, *S. citri* | 11–17 |
| Glucose, 3-*O*-methylglucose, 2-deoxyglucose, 6-deoxyglucose | Active transport | *A. laidlawii* | 10, 18–26 |
| Fructose | Group translocation | *M. mycoides* subsp. *capri* | 12, 16 |
| Fructose | Active transport | *A. laidlawii* | 19, 20, 25 |
| Mannose | Group translocation | *M. mycoides* subsp. *capri* | 12 |
| Maltose | Active transport | *A. laidlawii* | 19, 20 |

[a] References: (1) Rottem and Razin (1966), (2) Cho and Morowitz (1969), (3) Cho and Morowitz (1972), (4) Razin *et al.* (1968), (5) de Gier *et al.* (1971), (6) McElhaney *et al.* (1970), (7) deKruyff *et al.* (1972), (8) Romijn *et al.* (1972), (9) McElhaney *et al.* (1973), (10) Read and McElhaney (1975), (11) Rottem and Razin (1969), (12) Cirillo and Razin (1973), (13) Van Demark and Plackett (1972), (14) Jaffor Ullah and Cirillo (1976), (15) Jaffor Ullah and Cirillo (1977), (16) Mugharbil and Cirillo (1978), (17) A. H. Jaffor Ullah and V. P. Cirillo (unpublished), (18) Tarshis *et al.* (1972), (19) Tarshis *et al.* (1973), (20) Panchenko *et al.* (1973), (21) Fedotov *et al.* (1974), (22) Fedotov *et al.* (1975a), (23) Fedotov *et al.* (1975b), (24) Panchenko *et al.* (1975), (25) Tarshis *et al.* (1976a), (26) Tarshis *et al.* (1976b).

[b] $\alpha$-Methyl-D-glucopyranoside ($\alpha$-MG) serves as a nonmetabolized glucose analog for group translocation.

proposed to be the molecular species which must separate itself from the aqueous phase and penetrate the lipid bilayer.

The fundamental theorem of absolute rate theory applied to diffusion is that the diffusion constant $D$ of Fick's diffusion equation [Eq. (1)] is determined by the $\Delta G^*$ of activation by Eq. (2).

$$v = -AD\ ds/dx \tag{1}$$

in which $A$ is the unit area of flux, $D$ is the diffusion constant, and $ds/dx$ is the concentration gradient in the direction of the flux.

$$D = \lambda^2 \frac{KT}{h} e^{-\Delta G^*/RT} \tag{2}$$

in which $\lambda$ is the lattice parameter, the distance between energy minima, $K$ is the Boltzmann constant; $h$ is Plank's constant; $R$ is the gas constant; and $T$ is the absolute temperature. From the relationship between $\Delta G^*$ and its components, the enthalpy (or heat) of activation $\Delta H^*$, and the entropy of activation $\Delta S^*$, [Eq. (3)], the definition of the diffusion constant is expressed by Eq. (4).

$$\Delta G^* = \Delta H^* - T\Delta S^* \tag{3}$$

$$D = \lambda^2 \frac{KT}{h} e^{-(\Delta H^*/RT)} e^{(\Delta S^*/R)} \tag{4}$$

Conversion of Eq. (4) to its logarithmic form (to the base 10) produces Eq. (5), which allows determination of $\Delta H^*$ from the slope of the plot of log $D$ against the reciprocal of the absolute temperature ($1/T$).

$$\log D = [\log \frac{\lambda^2 K}{h} + \Delta S^*/R] + \log T - \Delta H^*/RT \tag{5}$$

This plot is equivalent to an Arrhenius plot in which the slope is $E/2.3R$. The values of $E$ and $\Delta H^*$ differ by only 600 cal/mole at 25°–30°C as seen from the relationship given in Eq. (6).

$$E = RT + \Delta H^* \tag{6}$$

## B. Unmediated Transport of Polyhydric Alcohols

From an analysis of the rate of uptake of various solutes according to these principles, one can determine the effect of various changes in cell membrane composition on the $\Delta H^*$ and $\Delta S^*$ of activation for the uptake of various solutes and reach conclusions about their role in the uptake process. De Gier *et al.* (1971) have applied such an analysis to the unmediated uptake of the homologous series of polyhydric alcohols, glycol, glycerol, and erythritol, by *A. laidlawii* containing a wide range of lipid compositions. The ability to dramatically alter the lipid composition of *A. laidlawii* membranes (McElhaney and Tourtellotte, 1969) makes these organisms ideal for such studies. As a verification that the changes in permeability are the result of the lipid composition of the cell, the permeability of liposomes prepared from the lipids of *A. laidlawii* were also studied. The permeability of both the whole cells and liposomes was measured from the rate of swelling when the cells and liposomes are suspended in isosmotic (i.e., 200 m*M*) solutions; the cells and liposomes behave as nearly ideal osmometers at this concentration (McElhaney *et al.*, 1973). Figure 2, taken from the latter study, compares the rate of glycerol uptake by cells grown in the presence of different combinations of fatty acids with and without cholesterol. The remarkable similarity between the permeability behavior of the liposomes and the whole cells shows that the characteristics of unmediated transport of the cell mem-

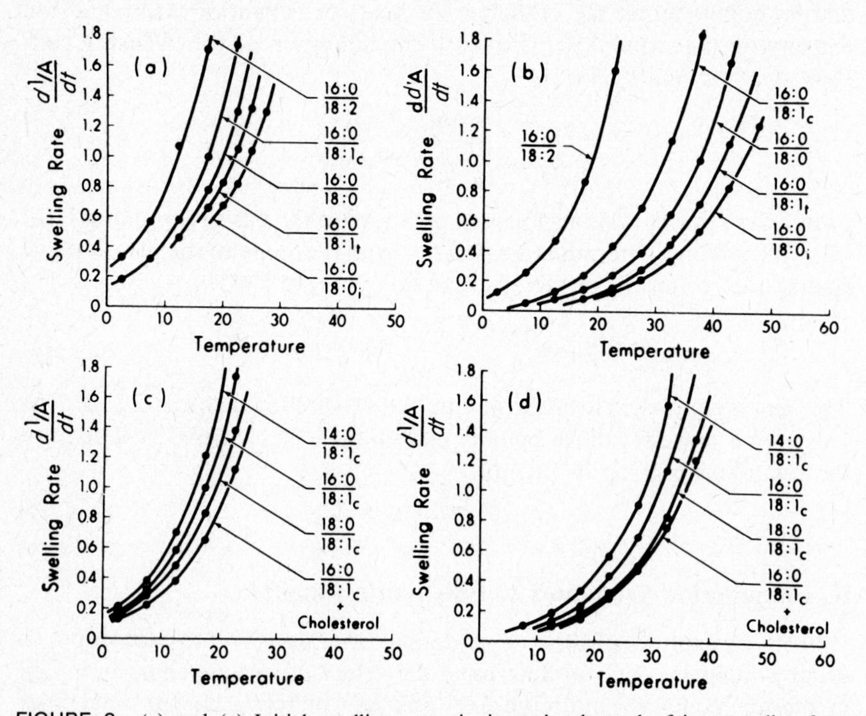


FIGURE 2. (a) and (c) Initial swelling rates in isotonic glycerol of intact cells of *A. laidlawii* B grown in the presence of different combinations of fatty acids, with or without cholesterol, as a function of temperature. (b) and (d) Initial swelling rates in isotonic glycerol of liposomes, prepared from the total membrane lipids of *A. laidlawii* B, as a function of temperature. (From McElhaney *et al.*, 1973.)

brane is determined primarily by the membrane lipids. Arrhenius plots of these data showed the remarkable fact that the activation energy (i.e., $E = RT + \Delta H^*$) is essentially the same for a given substrate (in both cells and liposomes) irrespective of the lipid composition. From these data it is possible to conclude that: (1) the rate of uptake is correlated with the relative fluidity of the membrane lipids; transport at any given temperature is greater the more unsaturated the membrane lipids; (2) cholesterol further decreases membrane permeability in accordance with its known decrease in membrane fluidity (Rottem *et al.*, 1973); and (3) the energy of activation is independent of the lipid composition. The significance of these results can be understood by examination of Eq. (4). It can be seen that a wide variation of $v$, the rate of uptake, with a change in lipid composition in the absence of a change of $\Delta H^*$ must mean that the lipid composition affects primarily the entropy for forming the transition state. Since the entropy of activation can be viewed as the result of increased packing of the lipid side chains upon penetration of the alcohol into the bilayer, the higher rate of uptake by fluid (i.e., unsaturated lipid) membranes may be viewed as resulting in less lipid packing than occurs in viscous (i.e., saturated lipid) membranes. The constancy of $\Delta H^*$, irrespective of lipid composition, is interpreted to mean that $\Delta H^*$ reflects a characteristic of the solute rather than the membrane. Following a similar analysis by Stein (1967), de Gier *et al.* (1971) indentify the $\Delta H^*$ for membrane penetration with the energy expected for complete dehydration of the permeant molecules (i.e., breaking the hydrogen bonds between the alcohol groups and water). The $\Delta H^*$ of activation for glycol, glycerol, and erythritol uptake are 14.3, 19.4, and 20.8 kcal/mole, respectively. Using a value of 5 kcals/mole needed to break one hydrogen bond (i.e., 10 kcal/mole per pair) and allowing for intramolecular hydrogen bonds to be formed in the dehydrated molecules, the $\Delta H^*$ for dehydration for glycol, glycerol, and erythritol would be 15, 20, and 25 kcal/mole, respectively. The closeness between these calculated values with the experimental ones lends support to the proposal. This proposal is further supported by the observation that the $\Delta H^*$ for the uptake of the homologous family of diols from ethane-1,2-diol to pentane-1,5-diol varies only from 12.5 to 14 kcal/mole. However, the rate of uptake of glycol, glycerol, and erythritol is not determined only by the $\Delta H^*$ since their relative rates of uptake would be ca. 100,000, 100, and 1, respectively. Actually their relative rates of uptake are ca. 50, 10, and 1 (de Gier *et al.*, 1971). As in the interpretation of the variation of the rate of transport of these molecules in membranes of different lipid composition, the lack of correspondence between $\Delta H^*$ and the rate of transport highlights the entropic factors which affect the penetration of hydrophilic molecules into the hydrophobic core of the lipid bilayer.

The analysis of unmediated transport of polyhydric alcohols with *A. laidlawii* cells and liposomes provides a perspective for understanding the role of carriers in cell membranes. As catalysts of the transport process, they must reduce the $\Delta G^*$ of activation [Eq. (2)]; this can be accomplished [see Eq. (4)] either by decreasing the $\Delta H^*$ or by increasing $\Delta S^*$, or both. The mediated transport of glucose by *A. laidlawii* cells (Read and McElhaney, 1975) exhibits a $\Delta H^*$ of activation of about 16.5 kcal/mole; the value for unmediated glucose transport is 27 (Chen *et al.*, 1971). In mediated transport, therefore, the membrane is not a silent partner in the dehydration process and reduces the enthalpy of activation for dehydration.

Undoubtedly, carriers also contribute to an increase in the $\Delta S^*$ of activation probably by providing an entropically more favorable bypass to the inhospitable lipid barrier. In the subsequent sections the characteristics of mediated transport are reviewed in detail.

## III. ACTIVE TRANSPORT

Unlike the lower polyhydric alcohols, glucose is taken up by a mediated process. Two different mechanisms are known to be involved in glucose transport in the Mollicutes: active transport in *A. laidlawii* and group translocation in *Mycoplasma* species.

### A. Sugar Transport in *Acholeplasma laidlawii*

#### 1. Characteristics of Carrier-Mediated Transport

The active transport of glucose and its analogs by *A. laidlawii* has been established by Tarshis and his colleagues in the Soviet Union in a series of papers beginning in 1972. Group translocation mediated by the PEP:sugar phosphotransferase system (PTS) could be excluded easily by the fact that whole cells do not take up α-methyl-D-glucoside (α-MG), the glucose analog for this transport process, and cell extracts lack the PEP dependent: sugar phosphotransferase system (PTS) (Tarshis *et al.*, 1972; Cirillo and Razin, 1973). However, the results from the early papers of the series (Tarshis *et al.*, 1972, 1973; Panchenko *et al.*, 1973) did not distinguish between facilitated diffusion and active transport because the studies involved metabolized sugars, namely glucose, fructose, and maltose. The same reservation applied to the excellent study by Read and McElhaney

(1975) on the dependence of glucose transport on the physical state of membrane lipids. However, the recent studies on glucose uptake by membrane vesicles (Fedotov *et al.*, 1975a,b; Panchenko *et al.*, 1975) and 3-*O*-methylglucose (3-*O*-MG) uptake by whole cells (Tarshis *et al.*, 1976a,b) has confirmed the earlier characterization of this process as a true active transport as defined above and in Fig. 1.

The kinetic characteristics of sugar transport by *A. laidlawii* cells and vesicles allow the following conclusions. (1) The 100-fold greater uptake rate of glucose by cells than by liposomes (Read and McElhaney, 1975) shows that the process is mediated. (2) This is confirmed by the strict substrate specificity which distinguishes between D- and L-glucose (see Fig. 1) as well as among other sugars which would not be distinguished by liposomes; for example, the uptake by cells and vesicles of glucose, D-fructose, and maltose but the exclusion of the pentoses (D-ribose, D-xylose, and L-arabinose), αMG, and lactose (Tarshis *et al.*, 1973; Read and McElhaney, 1975; Fedotov *et al.*, 1975a). (3) Saturation kinetics with a $K_m$ of 4.6 $\mu M$ for 3-*O*-MG uptake by whole cells (Tarshis *et al.*, 1976a) and 21.2 $\mu M$ for glucose uptake by vesicles (Panchenko *et al.*, 1975) suggest that there are a limited number of carriers in the cell membrane with a very high affinity. (4) The temperature characteristics show: (a) that as for unmediated transport, the rate of carrier transport is dependent on lipid composition and that the rate depends on the relative fluidity of the membrane lipids (Read and McElhaney, 1975). In this respect, carrier-mediated sugar transport in *A. laidlawii* is similar to that of other cells (Overath *et al.*, 1970; van Deenen *et al.*, 1975) in which there is a correlation between the effects of temperature on transport and the phase transition of membrane lipids. (b) The remarkable decrease in the enthalpy of activation for glucose uptake by cells compared to liposomes (16,500 cal/mole for cells compared with 27,000 cal/mole for liposomes) reflects the fact that part of the catalytic action of the carrier on transport is to reduce the enthalpy of activation (Chen *et al.*, 1971; Read and McElhaney, 1975). The independence of the enthalpy of activation on membrane lipid composition (Read and McElhaney, 1975) in spite of the dependence of the rate of uptake on lipid composition, was also encountered in unmediated transport in liposomes. As for unmediated transport, this represents the entropic factor of transport. (c) The denaturation between 40° and 45°C (Read and McElhaney, 1975; Tarshis *et al.*, 1976a) is expected for a protein carrier. (5) Counterflow was demonstrated in this case by measuring the uptake of $^3$H-labeled 3-*O*-MG at an extracellular concentration of 2.6 $\mu M$ into carbonyl cyanide-*m*-chlorophenyl hydrazone (CCCP)-treated washed cells preloaded with 1 m$M$ unlabeled 3-*O*-MG. In unloaded cells, the intracellular concentration of labeled sugar never exceeded that of the external concentration; in loaded cells the intracellular concentration of labeled sugar "overshoots" that of the external medium (Tarshis *et al.*, 1976a). Counterflow, in energy-depleted cells, is commonly held to be a confirmation of carrier activity (Cirillo, 1961; Stein, 1967; Winkler, 1973). (6) Inhibition by thiol and amino blocking reagents (Tarshis *et al.*, 1973, 1976; Read and McElhaney, 1975; Fedotov *et al.*, 1975b) adds further evidence for the protein nature of the sugar carriers. (7) The inhibition of glucose utilization by phlorizin and its aglycone, phloretin, was reported by Smith (1969) and Read and McElhaney (1975). These agents are remarkably specific inhibitors of sugar transport in vertebrates (LeFevre, 1961; Cirillo, 1972) and invertebrate (Fisher and Read, 1971) tissues. LeFevre (1961) has presented extensive evidence that phlorizin and phloretin owe their activity to their steroidal structure which allows them to become incorporated into the lipid bilayer in the manner of cholesterol and, at a critical concentration, to interfere with sugar carrier interaction. In mammalian tissues the relative activity of phlorizin and phloretin differ depending on the mechanism of sugar transport. In facilitated diffusion systems phloretin is active at micromolar concentrations, whereas phlorizin is active only at millimolar concentrations; in active transport systems the converse is true. In *A. laidlawii*, phloretin is more active than phlorizin when used at 0.2 m*M* concentration (Smith, 1969; Read and McElhaney, 1975). While this would suggest that sugar transport in *A. laidlawii* is a facilitated diffusion process, the accumulation data presented below suggest an active transport process. (8) The inhibitory action of the steroids deoxycortisone and estradiol on glucose utilization (Smith, 1969) is probably analogous to that of phlorizin and phloretin; namely, their incorporation into the cell membrane with attendant interference with sugar carrier interaction. (9) Finally, the inhibitory action of uranyl ion on sugar transport (Smith, 1969) was first described by Rothstein for yeast cells. Rothstein's studies suggest that the uranyl ions combine primarily with phosphoryl groups in the cell membrane with some interaction with less reactive carboxyl groups (Rothstein and Van Steveninck, 1966). Wilkins and O'Kane (1964) showed that uranyl ions also inhibit sugar transport in bacteria.

### 2. Evidence for Chemiosmotic Coupling

The kinetics of sugar transport, including the absence of a PTS system and $\alpha$-MG uptake, and the action of inhibitors on the transport process clearly establish that sugar transport is a carrier-mediated process, but they do not allow a distinction to be made between facilitated diffusion and active transport. Such a distinction must come from information on whether accumulation takes place against an electrochemical gradient. The data available provide this information. Glucose uptake by vesicles and 3-*O*-MG uptake by whole cells result in the accumulation of the unmodified sugars against significant concentration gradients (Fedotov *et al.*, 1975b; Tarshis *et al.*, 1976a). Furthermore, their accumulation is inhibited by inhibitors which dissipate a proton motive force [dinitrophenol (DNP), carbonyl cyanide *m*-chlorophenyl hydrazone (CCCP), pentachlorophenol, and tetrachlorotrifluoromethylbenzimidazole] (Fedotov *et al.*, 1975b; Tarshis *et al.*, 1976a,b). These inhibitory effects strongly suggest that the active transport of sugars in *A. laidlawii* is energized by a proton motive force.

In bacteria, as in mitochondria, a proton motive force is generated either by electron transport through the respiratory chain or by ATP hydrolysis by the membrane-bound, magnesium-dependent ATPase complex (Harold, 1972; Konings, 1977). The relationship between these two pumps is diagrammed in Fig. 3. The independent formation of a proton motive force by either the respiratory chain or the ATPase and the reversible nature of the ATPase reaction forms the basis for the Mitchell chemiosmotic theory of oxidative phosphorylation (Mitchell, 1966). In either case, protons are translocated across the membrane to produce a proton motive force, $\Delta\,\bar{\mu}_{H^+}$,which consists of two components, a proton concentration difference, ΔpH, and a membrane potential, $\Delta\psi$, related by Eq. (7).

$$\Delta\bar{\mu}_{H^+} = \Delta\psi - 60\text{ mV }\Delta\text{pH} \qquad (7)$$

The energy of the proton motive force may be used secondarily to drive the transport of other solutes as diagrammed in Fig. 4. Cations such as $K^+$ may respond directly to the membrane potential, interior negative, lead-

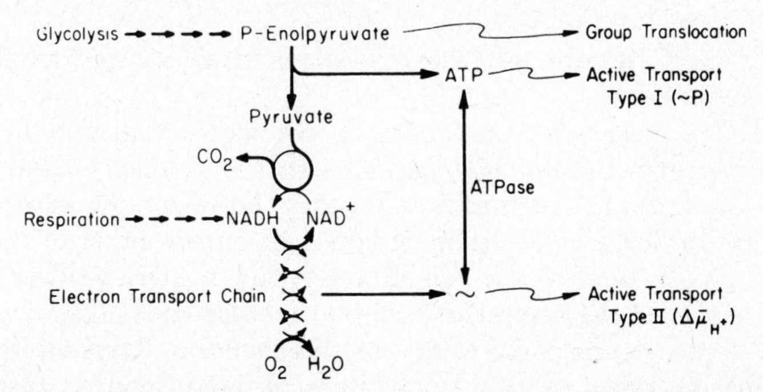


FIGURE 3. A scheme for the energy sources for active transport and group translocation: Active transport may be mediated directly by ATP (type I) or by a proton motive force (type II). PEP is the energy source for sugar group translocation. (Modified from Berger, 1973.)

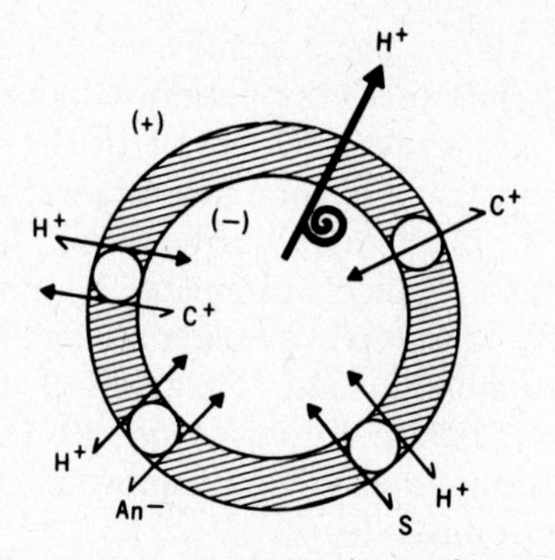


FIGURE 4. Secondary coupling of transport carriers to the proton circulation according to chemiosmotic theory. The primary proton pump (mechanism unspecified) is vectorial; the secondary carriers are reversible. Metabolites: S, uncharged; $C^+$, cationic; $An^-$, anionic. (From Harold, 1977.)

ing to a concentration gradient as described by the Nernst equation (Cirillo, 1966):

$$\Delta\psi = -2.3\ RT/F\ \log K^+_i/K^+_o \tag{8}$$

Uncharged solutes, such as sugars or amino acids, may cross the membrane coupled with protons; such a linkage (called *symport*) will result in accumulation of the solute driven by the $\Delta\bar{\mu}_{H^+}$. Alternatively, proton uptake in response to the proton motive force may occur in exchange (i.e., *antiport*) for an intracellular cation, thus resulting in the extrusion of an intracellular cation. Weak acids cross the membrane in the uncharged form and accumulate in response to the ΔpH, interior alkaline, as described by the Jacobs equation (Jacobs, 1940):

$$\frac{C_i}{C_o} = \frac{1 + 10^{pH_i - pK}}{1 + 10^{pH_o - pK}} \tag{9}$$

where $C_i$ and $C_o$ are the total concentrations inside and outside the cell, respectively.

The action of proton conductors on the active transport of sugars demonstrates that a proton motive force exists in *A. laidlawii* and may be used to drive the active transport process; however, the origin of the proton motive force is uncertain. It could arise from either of two independent proton pumps, the membrane-bound NADH oxidase or the membrane-bound ATPase. The respiratory chain in *A. laidlawii*, however, is truncated compared to that of mitochondria. Respiration is only marginally sensitive to site I inhibitors of mitochondrial respiration (amytal, rotenone, and phenanthrolene) and is completely insensitive to site II [antimycin A and 2-heptyl-4-hydroxyquinoline-*N*-oxide (HOQNO)] and site III (NaCN and $NaN_3$) inhibitors (Tarshis *et al.*, 1976a). Consistent with these inhibitor effects is the reported absence of cytochromes and quinones in *A. laidlawii* (Tarshis *et al.*, 1976a,b). It seems likely, therefore, that the NADH oxidase is flavin linked to oxygen through a nonheme iron intermediate (Van Demark, 1969). It is uncertain whether this truncated respiratory chain can generate a proton motive force since *A. laidlawii* does not carry out oxidative phosphorylation (Van Demark, 1969; this volume, Chapter 7) although there is an active membrane-bound ATPase (Ne'eman and Razin, 1975; Bevers *et al.*, 1977). This would suggest that the proton motive force arises primarily, if not exclusively, from ATP hydrolysis. The effect of inhibitors on sugar transport in whole cells supports this suggestion. However, the relative role of NADH oxidase versus ATPase in producing the proton motive force for active sugar transport is reported to be different in whole cells and membrane vesicles. Thus, glucose accumulation by vesicles is inhibited by proton conductors and partially inhibited by site I inhibitors but is not inhibited by arsenate or the ATPase inhibitor, dicyclohexylcarbodiimide (DCCD) (Fedotov *et al.*, 1975b). This would suggest that the proton motive force in vesicles arises from respiration and not from ATP hydrolysis. The authors could not identify the energy source in vesicles. Glucose accumulation was supported by an endogenous energy source and was not stimulated by any exogeneous energy source tested. Extensive washing of the vesicles depleted their ability to accumulate glucose, presumably by depleting them of the endogenous energy source. On the other hand, active transport of 3-*O*-MG in whole cells is significantly inhibited by arsenate and DCCD in addition to proton conductors and site I inhibitors (Tarshis *et al.*, 1976a,b). These results suggest that the proton motive force for active transport by whole cells can be produced either by respiration or by ATP hydrolysis. However, Tarshis *et al.*(1976b) question the specificity of site I inhibitors on respiration and propose that the proton motive force arises only from ATP hydrolysis. While this proposal would explain the data for active transport by whole cells and accommodate the fact that there is no demonstrable oxidative phosphorylation in *A. laidlawii* (see Chapter 7), it leaves unexplained the data on membrane vesicles. Resolution of this apparent inconsistency awaits further experiments.

The sensitivity of active transport by whole cells to both arsenate and DCCD excludes direct ATP coupling to active transport (type I in Fig. 3). Direct ATP coupling is characteristic of binding protein-linked active transport processes in *Escherichia coli* (Berger, 1973; Berger and Heppel, 1974; Ferenci *et al.*, 1977) and is resistant to ATPase inhibition by DCCD. Type I ($\sim$P) active transport is also characterized by transinhibition by the internal substrate on substrate uptake, exchange, and efflux so that addition of uncouplers to preloaded cells does not result in the loss of internal substrate. Such a loss is characteristic of a type II ($\sim\Delta\,\bar{\mu}_{H^+}$) active transport process. Consistent with type II active transport, uncouplers cause the release of accumulated sugars in *A. laidlawii* whole cells and vesicles (Fedotov *et al.*, 1975b; Tarshis *et al.*, 1976a).

## B. Amino Acid Transport

Amino acid transport has been studied in only two *Mycoplasma* species, namely *l*-histidine in *Mycoplasma fermentans* and *l*-methionine in *Mycoplasma hominis* (Razin *et al.*, 1968).

The *l*-histidine transport process in *M. fermentans* meets all the criteria of an active transport system for basic amino acids: (1) high affinity ($K_m$ 80 $\mu M$) and high capacity (17 nmol/mg protein/min); (2) high substrate specificity; (3) high accumulation ratio of the amino acid into a free amino acid pool; (4) temperature characteristics typical of other active transport processes; (5) inhibition by *p*-chloromercuribenzoate (*p*CMB) at a modest concentration ($10^{-4}$ *M*). Energy for transport is provided by endogenous metabolism and was not stimulated by glucose or any other exogenous energy source tested. The nature of the energy coupling mechanism is not obvious from the effect of inhibitors since inhibitors of glycolysis (i.e., iodoacetate and fluoride) as well as respiration (cyanide and azide) at 1 m*M* concentrations result in only modest inhibition. Most significant, however, is the insensitivity to dinitrophenol. Insensitivity to proton conductors by a transport system capable of a very high accumulation ($C_i/C_o = 1000$) suggests a type I active transport (Fig. 3) energized directly by ATP rather than by a proton motive force. If this were a type I active transport process it should also be insensitive to ATPase inhibitors (i.e., DCCD) when the energy is provided by glycolysis. Furthermore, type I coupling should be abolished by arsenate, but 1-histidine transport was only inhibited by ca. 40% at 1 m*M* Na arsenate. The lack of information on the energy source involved in this transport, the significant resistance to arsenate, and the lack of information on the sensitivity to DCCD make it unwise to conclude that a type I coupling mechanism is involved. Finally, the efflux data are inconsistent with such a mechanism since, in bacteria, type I active transport shows little efflux of the accumulated substrate because of strong transinhibition of carrier activity (Harold, 1977). In the *l*-histidine system, however, efflux is significant when cells are placed in an amino acid-free medium and accelerated fivefold when an exogenous amino acid is present. Thus, there are some features of both types of active transport mechanisms. However, the paucity of data does not allow reaching any conclusions except that more experiments are needed and other uncouplers should be tested.

The insensitivity to thallium acetate deserves a special comment. Thallium acetate has recently been shown to uncouple mammalian mitochondria (Melnick *et al.*, 1976). The use of thallium acetate in mycoplasma growth medium, to prevent the growth of bacteria and fungi (Hayflick, 1969), attests to the fact that thallium acetate is tolerated by mycoplasmas. Whether this means that thallium acetate does not act as an uncoupler or that mycoplasmas are insensitive is an interesting question. The data of Tarshis *et al.* (1976a,b) suggest that *A. laidlawii* is in fact sensitive to uncouplers. However, the possibility that mycoplasmas in complex media are not sensitive to proton conducting uncouplers is interesting in the context of some recent experiments performed by Harold and Van Brunt (1977) in *E. coli.* They showed that *E. coli* could grow in the presence of uncouplers which abolish pH and electrical potential differences across the membrane in a rich medium, with a slightly alkaline pH and a minimum potassium concentration of 0.1 *M*. These conditions are met by standard mycoplasma growth media except for a somewhat lower potassium concentration (Rodwell, 1969). This intriguing question is, therefore, open for future investigation.

The *l*-methionine transport system of *M. hominis* shares many of the features of the *l*-histidine system in *M. fermentans:* (1) There is a relatively high affinity but low capacity for *l*-methionine transport. (2) The substrate specificity is very high since none of 15 amino acids tested inhibited *l*-methionine uptake even when they were present at a 1000-fold excess. (3) Accumulation capacity is very low ($C_i/C_o = 30$) in the presence of chloramphenicol and less in the absence of chloramphenicol. (4) Susceptibility to inhibitors is the same as for *M. fermentans* including the insensitivity to dinitrophenol. (5) In these cells, amino acid accumulation was stimulated by a mixture of acetate, butyrate, and arginine. Acetate and butyrate stimulate only slightly. Arginine stimulates transport some eight-fold. Arginine presumably produces ATP by the arginine dihydrolase pathway (Barile *et al.*, 1966). In this case, the combination of insensitivity to dinitrophenol and stimulation by ATP suggests a type I active transport processes; however, the low capacity of this system makes further interpretation difficult. No information on efflux was presented.

The results of this single study on amino acid transport by Razin *et al.* (1968) are highly provocative. They raise a number of interesting and important questions about the mechanism of coupling, suggesting the possibility that amino acid transport may be energized directly by ATP instead of via a proton motive force. However, these interesting suggestions require more information for their confirmation or refutation.

### C. $K^+$ Transport in *Acholeplasma laidlawii*

Our knowledge of the characteristics of $K^+$ transport in *A. laidlawii* comes from only a few publications (Rottem and Razin, 1966; Cho and Morowitz, 1969, 1972; van der Neut-Kok *et al.*, 1974).

Rottem and Razin (1966) established that $K^+$ is accumulated against a concentration gradient by an energy-dependent process. When washed cells were suspended in a buffered solution containing glucose and 2.6 m*M* $K^+$, the cells accumulated $K^+$ to an intracellular concentration of 60–65 m*M*. The accumulation depended on glucose as an energy source and was inhibited by iodoacetate, presumably by inhibiting glucose metabolism. Cho and Morowitz (1969) confirmed and extended these observations. They showed that the capacity to accumulate $K^+$ could be increased by adapting the cells to a growth medium containing 0.068 m*M* $K^+$. They found that the intracellular $K^+$ concentration remained essentially constant at ca. 70 m*M*, even though its concentration in the medium was reduced from 5.4 m*M* to 0.068 m*M*; this represents an increase in the intracellular accumulation ratio from 13 to 1000. Thus the cells adapt to a lowered external $K^+$ concentration by increasing their $K^+$ accumulation capacity.

The effects of inhibitors indicate that the energy of glycolysis is required for accumulation but that respiration, as in the case of 3-*O*-MG accumulation, is only marginally effective in $K^+$ accumulation. The inactivity of dinitrophenol is surprising inasmuch as the ionophores, gramicidin S and D, are remarkably active. The gramicidins and dinitrophenol act to dissipate the proton motive force, although by different mechanisms (Gomez-Puyou and Gomez-Lojero, 1977). The inhibition by valinomycin is expected since this ionophore leads to the leakage of $K^+$ from *A. laidlawii* cells (van der Neut-Kok *et al.*, 1974; van Deenen *et al.*, 1975). The inactivity of ouabain is not surprising since this cardiac glycoside acts specifically on eukaryotic, $Na^+$–$K^+$ ATPases (i.e., the sodium pump ATPase). The *A. laidlawii* $K^+$ transport system does not require $Na^+$ ions to be present in the uptake medium (Rottem and Razin, 1966).

The partial inhibition of $K^+$ accumulation by phospholipase A (Cho and Morowitz, 1969) could mean that phospholipids are needed to maintain bilayer integrity or are required for the activity either of the $K^+$ carrier itself or of the energy linking mechanism. In this regard it is significant that greater than 90% hydrolysis of phosphatidyl glycerol by phospholipases $A_2$ and C strongly reduced the activity of $Mg^{2+}$-ATPase of *A.*

*laidlawii* (Bevers *et al.*, 1977). If ATP hydrolysis generates a proton motive force and $K^+$ accumulation is driven by such a proton motive force, as suggested by the inhibition by the gramicidins, phospholipase A inhibition of $K^+$ accumulation could be due to an inhibition of an ATPase proton pump. It would be interesting to test the effect of phospholipase D since this enzyme did not affect ATPase activity even when 99% of the phosphatidyl glycerol was converted to phosphatidic acid (Bevers *et al.*, 1977).

The effects of temperature shifts (Cho and Morowitz, 1972) suggest asymmetric effects on influx and efflux of $K^+$ in *A. laidlawii*. A greater stimulation of efflux than influx by an increase in temperature would explain the decrease in accumulation and, conversely, a decrease in temperature would increase accumulation. The effect of inhibitors on $K^+$ efflux (Cho and Morowitz, 1972) demonstrates that $K^+$ efflux is regulated by membrane proteins (i.e., inhibition by the thiol blocking reagents). The marked stimulation of $K^+$ efflux by gramicidin D and valinomycin (van der Newt-Kok *et al.*, 1974) are expected since these agents are $K^+$ ionophores (Pressman, 1976). The stimulation of $K^+$ efflux by the presence of extracellular $K^+$ (Cho and Morowitz, 1969) is characteristic of type II ($\Delta \bar{\mu}_{H^+}$) active transport processes. Finally, $K^+$ efflux exhibits a high enthalpy of activation which is consistent with the suggestion made above that the temperature shift effects are probably due to asymmetrically greater effect on efflux.

## D. Concluding Remarks on Active Transport in Mycoplasmas

The studies of sugar, amino acid, and $K^+$ transport reviewed above provide evidence that all these processes are carrier-mediated, active transport processes. The kinetics, the substrate specificity, and the effect of protein-directed reagents clearly show that all of these systems are protein-mediated. The accumulation of the unmodified substrate, especially in the case of the glucose analog 3-*O*-MG and, of course, $K^+$, shows that these are carrier-mediated, active transport processes and not group translocations. Not so clear, however, is the distinction between type I and type II active transport (Fig. 3). The evidence for 3-*O*-MG transport by *A. laidlawii* strongly suggests an active transport process driven by a proton motive force. The insensitivity to respiratory inhibitors, the apparent absence of oxidative phosphorylation, the sensitivity to inhibitors of glycolysis, and, finally, the sensitivity to DCCD and uncouplers all point to a proton motive force generated by the membrane ATPase. The evidence for amino acid transport in the two *Mycoplasma* species and $K^+$

transport in *A. laidlawii* is uneven. Particularly puzzling is the insensitivity to dinitrophenol. Since in the case of $K^+$ transport in *A. laidlawii* gramicidin D was a very active inhibitor, one might be justified in concluding that the conditions of the experiment prevented dinitrophenol inhibition and, therefore, that this transport system is also driven by a proton motive force. The situation for the dinitrophenol-insensitivie amino acid transport system in *M. fermentans* and *M. hominis* is more difficult to evaluate. If one emphasizes the fact that efflux is not internally trans-inhibited, one could tentatively conclude that this system is also driven by a proton motive force. However, for both the amino acid systems and $K^+$ more experiments must be done. One must be aware that the conclusions reached thus far are based on a relatively small literature. One hopes that the necessary experiments to resolve these questions will be carried out soon.

## IV. GROUP TRANSLOCATION

### A. The PEP: Sugar Phosphotransferase System (PTS)

In contrast to the mechanism of sugar transport in *Acholeplasma,* sugar transport by the glucose fermenting species of the genus *Mycoplasma* is mediated by the PTS discovered in 1964 by Roseman and his colleagues (Kundig *et al.*, 1964). The details of this system have been described in a number of recent reviews (Postma and Roseman, 1976; Cordaro, 1976; Hengstenberg, 1977; Roseman, 1977). By this mechanism, which is aptly called "vectorial phosphorylation," the transport of the sugar across the cell membrane is obligatorily coupled to its phosphorylation. The phosphoryl donor for sugar phosphorylation is a phosphoprotein which is linked to the ultimate phosphoryl donor, PEP, by a phosphoryl transfer chain. In *E. coli* the PTS consists of four proteins: two general, nonspecific proteins (enzyme I and HPr) which are constitutive, "cytoplasmic" components and two sugar-specific proteins, which may be inducible, both of which may be membrane-bound (IIA and IIB) or may consist of one cytoplasmic (III) and one membrane-bound component (IIB′). The phosphoryl transfer chain ending in either IIA/IIB or III/IIB′ is shown in Fig. 5. Sugar phosphorylation is mediated by a family of sugar-specific IIB (or IIB′) components which are also responsible for the sugar translocation across the cell membrane.

#### 1. Distribution of the PTS among Mollicutes

The PTS is unique to prokaryotes. Among bacteria, it is present in both gram-positive and gram-negative organisms. Romano *et al.*, in a survey of

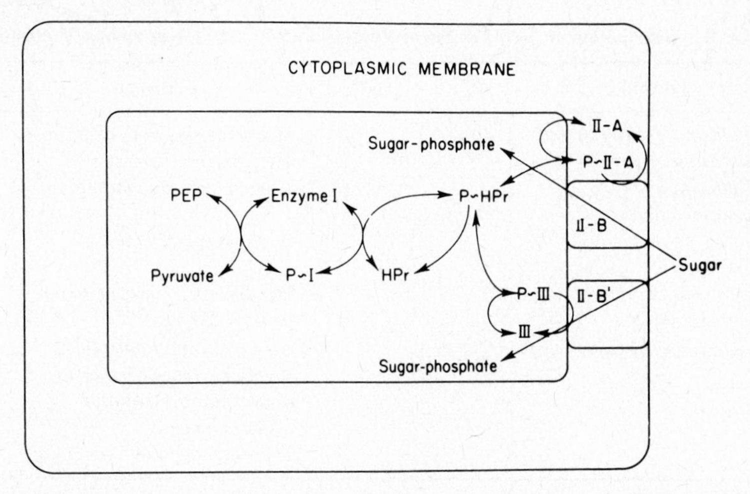


FIGURE 5. The phosphoenol pyruvate phosphotransferase system in *Escherichia coli*. PEP indicates phosphoenol pyruvate; HPr, histidine-containing phosphocarrier protein; II-A, enzyme II-A; II-B, enzyme II-B, and III, factor III. (From Kundig, 1976.)

the literature as of 1970 and based on examination of representative organisms, concluded that among bacteria the PTS is present exclusively in facultative and obligate anaerobes. Few exceptions to this rule have been reported. In one apparent exception, the occurrence of a PTS for fructose in the obligate aerobe, *Arthrobacter pyridinolis* (Wolfson and Krulwich, 1974), the PTS system is induced in aerobically grown cells late in the growth of the culture when the large cell population reduces the oxygen tension of the medium and renders it more "anaerobic" (T. A. Krulwich, personal communication). The occurrence of a PTS in mycoplasmas was reported independently by Van Demark and Plackett (1972) for *Mycoplasma* sp. strain Y and by Cirillo and Razin (1973) for *Mycoplasma gallisepticum* and *Mycoplasma mycoides* subsp. *mycoides* and *capri*. In retrospect, the occurrence of the PTS in *M. gallsepticum* was suggested by the observation by Rottem and Razin (1969) that these organisms accumulate α-methyl-D-glucopyronoside (α-MG). In bacteria this substrate is phosphorylated by the glucose-specific PTS (Romano *et al.*, 1970; Kornberg, 1976) and its uptake is often a clue to the presence of a PTS.

A survey of several Mollicutes (Table II) has shown that a PTS is present also in two strains of the glucose fermenter, *Mycoplasma capricolum* (Kid and M14) and in *Spiroplasma citri* (Jaffor Ullah and Cirillo, 1976, 1977; V. P. Cirillo, unpublished). The PTS is absent, however, in all the glucose-fermenting *Acholeplasma* species tested (Cirillo and Razin, 1973; Tarshis *et al.*, 1972). It is also absent in species of *Mycoplasma*

TABLE II. **Distribution of the PEP: Sugar Phosphotransferase System among Mollicutes**

| Present | Absent |
|---|---|
| *Mycoplasma gallisepticum*[a] | *Mycoplasma bovigenitalium*[c] |
| *Mycoplasma* sp. *strain Y*[b] | *Mycoplasma hominis*[c] |
| *Mycoplasma mycoides* subsp. *mycoides*[c] | *Mycoplasma agalactiae*[c] |
| *Mycoplasma mycoides* subsp. *capri*[c] | *Mycoplasma flocculare*[c] |
| *Mycoplasma capricolum* (Kid and M14)[d] | *Mycoplasma hypopneumoniae*[e] |
| *Spiroplasma citri* (Morroco)[e] | *Acholeplasma laidlawii*[c,f] |
| | *Acholeplasma granularum*[c] |
| | *Acholeplasma axanthum* (410 and 743)[c] |
| | *Thermoplasma acidophilum*[e] |

[a] Rottem and Razin (1969).
[b] Van Demark and Plackett (1972).
[c] Cirillo and Razin (1973).
[d] Jaffor Ullah and Cirillo (1976, 1977).
[e] V. P. Cirillo (unpublished).
[f] Tarshis *et al.* (1972).

which are only marginal or absolute nonfermenters (Cirillo and Razin, 1973; J. G. Tully and V. P. Cirillo, unpublished) and it is absent in *Thermoplasma acidophilum* (V. P. Cirillo unpublished) which appears to be a chemolithoautotroph (P. F. Smith, personal communication).

The distribution of the PTS among the Mollicutes extends the "rule" of its occurrence in facultative anaerobes. The Mollicutes among which it is distributed are lactic acid fermenters. Hawthorne and Van Demark (1977) have recently presented evidence that *S. citri* is a homolactic fermenter. The absence of the PTS in acholeplasmas is intriguing for the mycoplasmologist ever on the lookout for clues to the phylogeny of mycoplasmas; however, it is difficult at this stage to interpret the significance of this fact.

## 2. Characteristics of PTS-Mediated Sugar Uptake

The characteristics of PTS-mediated sugar uptake by whole cells mimic those of carrier-mediated active transport when the transport is measured by the use of radioactively labeled, nonmetabolized analogs such as $\alpha$-MG (Hoffee and Englesberg, 1962; Rottem and Razin, 1969; Winkler, 1971). Uptake shows the same kinetic characteristics of any mediated transport process; the $\alpha$-MG reaches a maximum intracellular concentration in which the label is "accumulated against a gradient"; the internal sugar is rapidly exchanged with sugar in the external medium (i.e., internal, labeled sugar is chased by the addition of unlabeled sugar) and there is net loss of internal sugar if energy metabolism is blocked by appropriate inhibitors (i.e., iodoacetate and fluoride). These are essentially the characteristics described by Rottem and Razin (1969) for α-MG uptake by *M. gallisepticum*. However, these characteristics result not from the action of a permease pump and leak system (Cirillo, 1961), but from a series of chemical modifications of the sugar substrate; namely, its phosphorylation associated with transport by the PTS, its hydrolysis by an intracellular phosphatase, and the diffusion of the free sugar back out into the medium (Winkler, 1971). A characteristic feature of PTS-mediated sugar transport is its insensitivity to uncouplers. [Instead of inhibition, dinitrophenol stimulates α-MG uptake in *E. coli* (Hoffee and Englesberg, 1962).] Van Demark and Plackett (1972) confirmed that α-MG uptake by *Mycoplasma* sp. strain Y is severely inhibited by iodoacetate and fluoride but is insensitive to dinitrophenol. Unequivocal evidence for the role of the PTS in sugar transport requires proof that the sugar phosphorylation is a necessary part of the transport process and the demonstration of the existence of the requisite enzymatic activity. Among the Mollicutes, the PTS enzymes have been found in all glucose-fermenting species of the genus *Mycoplasma* and in *S. citri* (Table II). The requirement for PTS for α-MG uptake has been demonstrated by the correlation between α-MG uptake and the presence of the enzyme system (Cirillo and Razin, 1973) and by the loss of the ability to transport α-MG in mutants defective in PTS activity (Mugharbil and Cirillo, 1978).

### 3. Enzymology of the PTS of *Mycoplasma capricolum*

Among the organisms listed in Table II which have a PTS, the *M. capricolum* system has been studied in some detail (Jaffor Ullah and Cirillo, 1976, 1977). Only two cytoplasmic components have been identified so far, enzyme I and the phosphocarrier protein, HPr. These components have been purified to homogeneity and are the only protein components together with the membrane required to phosphorylate glucose and fructose. This suggests that the *M. capricolum* membrane contains the IIA/IIB components for these sugars. However, one cannot exclude the possibility that a factor III/IIB′ system exists for these or other sugars since in *E. coli* both a IIA/IIB and a III/IIB′ system exist for glucose (Fig. 5) (Postma and Roseman, 1976; Cordaro, 1976).

**a. HPr.** The *M. capricolum* HPr has been purified to homogeneity and its amino acid composition determined (Jaffor Ullah and Cirillo, 1976). The similarities among the HPrs of *M. capricolum, E. coli* and *Staphylococcus aureus* are striking. Its molecular weight (9506) is very similar to that of *E. coli* (9537) and *S. aureus* (8630). It is similarly heat stable and has a single histidine whose imidazole side chain is presumed to be phosphorylated in P-HPr because its activity is destroyed by photooxidation in the presence of Rose Bengal dye (Jaffor Ullah and Cirillo, 1976). It differs significantly from the other HPrs in havinq two cysteine residues. In its P-HPr form it serves equally as well as *E. coli* P-HPr with *E. coli* membranes for $\alpha$-MG phosphorylation. This confirms the earlier observation that the mycoplasma cytoplasm complements *E. coli* membrane for sugar phosphorylation (Cirillo, 1973). However, the *M. capricolum* HPr is a poor substrate for *E. coli* enzyme I and it does not cross-react with antibodies produced in goats against *E. coli* HPr. The HPr from wild-type cells grown either on glucose or fructose is not sugar-specific as a phosphoryl donor for either glucose or fructose which are phosphorylated by different enzymes II, one constitutive and the other inducible, respectively (Mugharbil and Cirillo, 1978). However, in a fructose grown, $\alpha$-MG-resistant mutant, the general HPr is replaced by a fructose-specific HPr, $HPr^{fru}$, which is 10 times more active with the fructose-specific enzyme II complex than with the glucose-specific enzyme II.

The *S. aureus* HPr has now been sequenced (Hengstenberg, 1977) and the *E. coli* sequence will soon be completed (S. Roseman, personal communication). The homologies which these sequences will reveal are awaited with considerable expectation. The mycoplasma HPr sequence, alas, is still to be started (V. P. Cirillo, personal confession).

**b. Enzyme I.** The *M. capricolum* enzyme I has also been purified to homogeneity (Jaffor Ullah and Cirillo, 1977). Unlike the enzymes I of *E. coli* and *S. aureus* (Postma and Roseman, 1976; Hengstenberg, 1977), which have been difficult to purify and consist of a single polypeptide of molecular weight between 70,000 and 90,000 daltons, the *M. capricolum* enzyme I is stable and has a molecular weight of 220,000 daltons and consists of four subunits: two $\alpha$ subunits (44,500 daltons), one $\beta$ (62,000 daltons), and one $\gamma$ subunit (64,500 daltons). The high stability of the mycoplasma enzyme may be due to its tetrameric structure. It is not possible at this time to assign the three *N*-terminal amino acids (glycine, alanine, and tyrosine) to specific subunits. It is not known whether all four subunits are required for enzymatic activity. It is tempting to suppose that one or more of the subunits is involved in interaction with the cell membrane and/or participate in regulatory activities of the PTS (see Section IV).

The mycoplasma enzyme I phosphorylates the *E. coli* HPr but does so with reduced activity (ca. 25%) compared with its activity with mycoplasma HPr.

**c. Enzyme II.** Neither enzyme II activity nor its components has been separated from other cell membrane components, although a very preliminary experiment suggests that activity may be reconstituted by dialysis of cholate extracts (S. C. Kozinn and V. P. Cirillo, unpublished). It is presumed by analogy with the *E. coli* system that the mycoplasma enzyme II consists of a IIA/IIB complex. The ability of the *E. coli* P-HPr to react with the mycoplasma membrane, even though at only 10% the level of a homologous combination with mycoplasma P-HPr and the full activity of the mycoplasma P-HPr with *E. coli* membranes suggest that the mycoplasma enzyme II is not too different from that of *E. coli*. However, the observed structural differences between the mycoplasma and *E. coli* enzymes I, in spite of significant cross-reaction with heterologous HPrs, serves as a warning against reaching conclusions about structural similarities based only on enzyme activities.

The *M. capricolum* enzyme II for fructose is inducible (Mugharbil and Cirillo, 1978); however, in *M. mycoides* subsp. *capri* the enzymes II for glucose, fructose, and mannose are all constitutive (Cirillo and Razin, 1973). The glucose enzyme II activity of several *M. capricolum* mutants selected for $\alpha$-MG resistance have all been found to be defective (Mugharbil and Cirillo, 1978; V. P. Cirillo, unpublished). Finally, enzyme II activity has been found to be required for regulation of intracellular levels of cyclic AMP by exogenous sugars (see below).

### 4. Regulation of Intracellular Cyclic AMP

One of the reasons for choosing to study the PTS of mycoplasmas was the hope that the PTS of these small genome organisms might prove to be simpler than that in "higher bacteria." However, it is clear that the PTS in *M. capricolum* and in other mycoplasmas is proving to be as complex as that of *E. coli*. In fact, the PTS seems too complex even for *E. coli*. Saier and Feucht (1975) recently observed for the *E. coli* system, "The complexity of the phosphoenolpyruvate:sugar phosphotransferase system is bewildering, and one wonders what evolutionary advantages an organism might derive from the presence of so intricate a system." For organisms with only 20% the genome of *E. coli,* this complexity seems even more extreme. The complexity of the PTS system is now being interpreted to reflect its role in many regulatory processes, including: (1) regulation of the intracellular level of cyclic AMP (Peterkovsky and Gazdar, 1975; Saier and Feucht, 1975; Peterkovsky, 1976; Harwood *et al.*, 1976), (2) inducer exclusion (i.e., inhibition of the permeases of substrates used as inducible energy sources) in gram-negative bacteria (Saier and Roseman, 1976; Postma and Roseman, 1976; Cordaro, 1976), (3) regulation of carbohydrate uptake in gram-positive bacteria (Saier and Simoni, 1976) and (4) chemoreceptors for certain sugars in *E. coli* chemotaxis (Adler and Epstein, 1974; Adler, 1975).

Thus far a regulatory role for PTS in mycoplasmas has only been established for control of the intracellular concentration of cyclic AMP (Mugharbil and Cirillo, 1978). The intracellular level of cyclic AMP in *M. capricolum* is decreased by the presence of exogenous glucose or fructose as described by Makman and Sutherland (1965) for *E. coli*. For a sugar to regulate the intracellular level of cyclic AMP, however, there must be an intact PTS for that sugar. Thus, in *M. capricolum* mutants with a defective enzyme II, glucose is no longer able to reduce the intracellular concentration of cyclic AMP; however, fructose, for which there is a specific enzyme II, continues to control the concentration of intracellular cyclic AMP as in wild-type cells. Thus, in mycoplasmas, the complexity of the PTS will undoubtedly be related to its role as a regulator of cell growth and enzyme induction. It is tempting to predict that the occurrence of the PTS in *S. citri* suggests the occurrence of chemotaxis in these motile organisms.

While we marvel at the many functions of the PTS in bacteria, and now in mycoplasmas, it must not be forgotten that many organisms (i.e., the acholeplasmas among the Mollicutes) do not have this system. This naturally raises the question of what mechanisms are used by these other organisms to serve these regulatory functions.

## V. CONCLUSIONS

The transport characteristics of the few systems studied so far among the Mollicutes appear as complex and diversified as among other prokaryotes. On the one hand, this seems surprising in organisms of such a small genome size. On the other hand, their limited biosynthetic capacity places on these organisms an even greater need for transport systems than organisms with greater biosynthetic abilities. Irrespective of our expectations, the few systems studied so far emphasize complexity and similarity with bacterial transport systems rather than simplification. However, it must be emphasized that very few systems have been studied so far and any conclusions about complexity versus simplicity must await additional and more detailed studies.

## REFERENCES

Adler, J. (1975). *Annu. Rev. Biochem.* **44,** 534–544.
Adler, J., and Epstein, W. (1974). *Proc. Natl. Acad. Sci. U.S.A.* **71,** 2985–2899.
Barile, M. F., Schimke, R. I., and Riggs, D. B. (1966). *J. Bacteriol.* **91,** 189–192.
Berger, E. A. (1973). *Proc. Natl. Acad. Sci. U.S.A.* **70,** 1514–1518.

Berger, E. A., and Heppel, L. A. (1974). *J. Biol. Chem.* **249,** 7747–7755.
Bevers, E. M., Snoek, G. T., Op den Kamp, J. A. F., and van Deenen, L. L. M. (1977). *Biochim. Biophys. Acta* **467,** 346–356.
Chen, L.-F., Lund, D. B., and Richardson, T. (1971). *Biochim. Biophys. Acta* **225,** 89–95.
Cho, H. W., and Morowitz, H. J. (1969). *Biochim. Biophys. Acta* **183,** 295–303.
Cho, H. W., and Morowitz, H. J. (1972). *Biochim. Biophys. Acta* **274,** 105–110.
Cirillo, V. P. (1961). *Annu. Rev. Microbiol.* **15,** 197–218.
Cirillo, V. P. (1966). *Bacteriol. Rev.* **30,** 68–79.
Cirillo, V. P. (1972). *Metab. Inhibitors* **3,** 47–67.
Cirillo, V. P. (1973). *Abstr. Meet. Int. Assoc. Microbiol. Soc.* Vol. I, p. 57.
Cirillo, V. P., and Razin, S. (1973). *J. Bacteriol.* **113,** 212–217.
Cordaro, C. (1976). *Annu. Rev. Genet.* **10,** 341–359.
de Gier, J., Mandersloot, J. G., Hupkes, J. V., McElhaney, R. N., and van Beck, W. P. (1971). *Biochim. Biophys. Acta* **233,** 610–618.
de Kruyff, B., Demel, R. A., and van Deenen, L. L. M. (1972). *Biochim. Biophys. Acta* **255,** 331–347.
Fedotov, N. S., Panchenko, L. F., and Tarshis, M. A. (1974). *Mikrobiologia* **43,** 543–545.
Fedotov, N. S., Panchenko, L. F., and Tarshis, M. A. (1975a). *Folia Microbiol. (Prague)* **20,** 470–479.
Fedotov, N. S., Panchenko, L. F., and Tarshis, M. A. (1975b). *Folia Microbiol. (Prague)* **20,** 488–495.
Ferenci, T., Boos, W., Schwartz, M., and Szmelcman, S. (1977). *Eur. J. Biochem.* **75,** 187–193.
Fisher, F. M., and Read, C. P. (1971). *Biol. Bull. (Woods Hole, Mass.)* **140,** 46–47.
Gomez-Puyou, A., and Gomez-Lojero, C. (1977). *Curr. Top. Bioenerg.* **6,** 221–257.
Harold, F. M. (1972). *Bacteriol. Rev.* **36,** 172–230.
Harold, F. M. (1977). *Curr. Top. Bioenerg.* **6,** 83–149.
Harold, F. M., and Van Brunt, J. (1977). *Science* **197,** 372–373.
Harwood, J. P., Gazdar, C., Prasad, C., and Peterkovsky, A. (1976). *J. Biol. Chem.* **251,** 2462–2468.
Hawthorne, J. D., and Van Demark, P. J. (1977). *Abstr. Annu. Meet. Am. Soc. Microbiol.* No. G17, p. 133.
Hayflick, L. (1969). *In* "The Mycoplasmatales and the L-Phase of Bacteria" (L. Hayflick, ed.), pp. 15–47. Appleton, New York.
Hengstenberg, W. (1977). *Curr. Top. Microbiol. Immunol.* **77,** 97–126.
Hoffee, P., and Englesberg, E. (1962). *Proc. Natl. Acad. Sci. U.S.A.* **48,** 1759–1767.
Jacobs, M. H. (1940). *Cold Spring Harbor Symp. Quant. Biol.* **8,** 30–35.
Jaffor Ullah, A. H., and Cirillo, V. P. (1976). *J. Bacteriol.* **127,** 1298–1306.
Jaffor Ullah, A. H., and Cirillo, V. P. (1977). *J. Bacteriol.* **131,** 988–996.
Johnson, F. H., Eyring, H., and Polissar, M. J. (1954). "The Kinetic Basis of Molecular Biology." Wiley, New York.
Kaback, H. R. (1974). *Science* **186,** 882–892.
Konings, W. N. (1977). *Adv. Microb. Physiol.* **15,** 175–251.
Kornberg, H. L. (1976). *J. Gen. Microbiol.* **96,** 1–16.
Kundig, W. (1976). *Enzymes Biol. Membr.* **3,** 31–55.
Kundig, W., Ghosh, S., and Roseman, S. (1964). *Proc. Natl. Acad. Sci. U.S.A.* **52,** 1067–1074.
LeFevre, P. G. (1961). *Pharmacol. Rev.* **13,** 39–70.
McElhaney, R. N., and Tourtellotte, M. (1969). *Science* **164,** 433–434.
McElhaney, R. N., de Gier, J., and van Deenen, L. L. M. (1970). *Biochim. Biophys. Acta* **219,** 245–247.

McElhaney, R. N., de Gier, J., van Deenen, L. L. M., and van der Neut-Kok, E. C. M. (1973). *Biochim. Biophy-. Acta* **298,** 500–512.

Makman, R. S., and Sutherland, E. W. (1965). *J. Biol. Chem.* **240,** 1309–1314.

Melnick, R. L., Monti, L. G., and Motzkin, S. M. (1976). *Biochem. Biophys. Res. Commun.* **69,** 68–73.

Mitchell, P. (1966). *Biol. Rev. Cambridge Philos. Soc.* **41,** 445–502.

Mugharbil, U., and Cirillo, V. P. (1978). *J. Bacteriol.* **133,** 203–209.

Ne'eman, Z., and Razin, S. (1975). *Biochim. Biophys. Acta* **375,** 54–68.

Overath, P., Schairer, H. U., and Stoffel, W. (1970). *Proc. Natl. Acad. Sci. U.S.A.* **67,** 606–612.

Panchenko, L. F., Migushina, V. L., Fedotov, N. S., and Tarshis, M. A. (1973). *Dokl. Akad. Nauk SSSR* **209,** 213–216; *Doklady Biol. Sci. (Eng. Transl.)* **208,** 135–137 (1973).

Panchenko, L. F., Fedotov, N. S., and Tarshis, M. A. (1975). *Folia Microbiol. (Prague)* **20,** 480–487.

Peterkovsky, A. (1976). *Adv. Cyclic Nucleotide Res.* **7,** 1–48.

Peterkovsky, A., and Gazdar, C. (1975). *Proc. Natl. Acad. Sci. U.S.A.* **72,** 2920–2924.

Postma, P. W., and Roseman, S. (1976). *Biochim. Biophys. Acta* **457,** 213–257.

Pressman, B. (1976). *Annu. Rev. Biochem.* **45,** 501–530.

Razin, S. (1975). *Prog. Surf. Membr. Sci.* **9,** 135–216.

Razin, S., Gottfried, L., and Rottem. S. (1968). *J. Bacteriol.* **95,** 1685–1691.

Read, B. D., and McElhaney, R. N. (1975). *J. Bacteriol.* **123,** 47–55.

Rodwell, A. (1969). *In* "The Mycoplasmatales and the L-Phase of Bacteria" (L. Hayflick, ed.), pp. 413–449. Appleton, New York.

Romano, A. H., Eberhard, S. J., Dingle, S. L., and McDowell, T. D. (1970). *J. Bacteriol.* **104,** 808–813.

Romijn, J. C., van Golde, L. M. G., McElhaney, R. N., and van Deenen, L. L. M. (1972). *Biochim. Biophys. Acta* **280,** 22–32.

Roseman, S. (1977). *FEBS-Symp.* **42,** 582–597.

Rothstein, A., and Van Steveninck, J. (1966). *Ann. N.Y. Acad. Sci.* **137,** 606–623.

Rottem, S., and Razin, S. (1966). *J. Bacteriol.* **92,** 714–722.

Rottem, S., and Razin, S. (1969). *J. Bacteriol.* **97,** 787–792.

Rottem, S., Cirillo, V. P., de Kruyff, B., Shinitzky, M., and Razin, S. (1973). *Biochim. Biophys. Acta* **323,** 509–519.

Saier, M. H., Jr., and Feucht, B. U. (1975). *J. Biol. Chem.* **250,** 7078–7080.

Saier, M. H., Jr., and Roseman, S. (1976). *J. Biol. Chem.* **251,** 6606–6615.

Saier, M. H., Jr., and Simoni, R. D. (1976). *J. Biol. Chem.* **251,** 893–894.

Simoni, R. D. (1972). *In* "Membrane Molecular Biology" (C. F. Fox and A. D. Keith, eds.), pp. 289–322. Sinauer Assoc., Inc., Stamford, Connecticut.

Singer, S. J., and Nicolson, G. L. (1972). *Science* **175,** 720–731.

Smith, P. F. (1969). *Lipids* **4,** 331–336.

Smith, P. F. (1971). "The Biology of Mycoplasmas." Academic Press, New York.

Stein, W. D. (1967). "The Movement of Molecules across Cell Membranes." Academic Press, New York.

Tarshis, M. A., Panchenko, L. F., Migoushina, V. L., and Bourd, G. I. (1972). *Biokhimiya* **37,** 930–935.

Tarshis, M. A., Migoushina, V. L., Panchenko, L. F., Fedotov, N. S., and Bourd, H. I. (1973). *Eur. J. Biochem.* **40,** 171–175.

Tarshis, M. A., Bekkouzjin, A. G., and Ladygina, V. G. (1976a). *Arch. Microbiol.* **109,** 295–299.

Tarshis, M. A., Bekkouzjin, A. G., Ladygina, V. G., and Panchenko, L. F. (1976b). *J. Bacteriol.* **125,** 1–7.

van Deenen, L. L. M., de Gier, J., Demel, R. A., de Kruyff, B., Blok, M. C., van der Neut-Kok, E. C. M., Haest, C. W. M., Ververgaert, P. H. J. T., and Verkleij, A. J. (1975). *Ann. N.Y. Acad. Sci.* **264,** 124–141.

Van Demark, P. J. (1969). *In* "The Mycoplasmatales and the L-Phase of Bacteria" (L. Hayflick, ed.), pp. 491–501. Appleton, New York.

Van Demark, P. J., and Plackett, P. (1972). *J. Bacteriol.* **111,** 454–458.

van der Neut-Kok, E. C. M., de Gier, J., Middelbeek, E. J., and van Deenen, L. L. M. (1974). *Biochim. Biophys. Acta* **332,** 97–103.

Wilkins, P. O., and O'Kane, D. J. (1964). *J. Gen. Microbiol.* **34,** 389–399.

Winkler, H. H. (1971). *J. Bacteriol.* **106,** 362–368.

Winkler, H. M. (1973). *J. Bacteriol.* **116,** 203–209.

Wolfson, E. B., and Krulwich, T. A. (1974). *Proc. Natl. Acad. Sci. U.S.A.* **71,** 1739–1742.

# 13 / ANTIGENIC DETERMINANTS

*George E. Kenny*



## I. INTRODUCTION

Studies on the antigens of the Mycoplasmatales have been limited by the small amounts of material which can be obtained from these organisms. Nevertheless substantial progress has been achieved in recent years. It is the purpose of this review to define the current status of the antigens of the Mycoplasmatales with particular emphasis on advances made since the last review by this author, which had a literature search current to 1973 (Kenny, 1975).



THE MYCOPLASMAS, VOL. 1

ISBN 0-12-078401-7

## II. PREPARATION OF ANTIGENS AND IMMUNOGENS

Particular attention is paid in this section to problems with contamination of pellets of Mycoplasmatales with medium components. Presence of such nonmycoplasmic materials will compromise studies on the chemical nature of specific antigens of cellular components to the extent of uselessness unless adequately controlled. Such contamination poses a significant problem to evaluation of past results in the literature and its significance cannot be overemphasized presently.

### A. Purity

The major problems in preparing antigens of the organisms in the Mycoplasmatales are the poor yield of organisms and the contamination of the antigenic preparation by medium components. The yield of organisms from broth cultures is about 1–100 mg protein/liter. In comparison to this small yield, the amount of potentially contaminating material in the medium is huge: medium supplemented with 10% serum contains 7000 mg of serum protein/liter as well as yeast extract components and peptones (20–30 gm). Although washing of the pellets extensively has been relied upon to eliminate medium components, additional methods are needed to verify that the antigens being studied are in fact derived from the organisms and not from medium components. Antigens prepared by centrifugation from dialysate broth supplemented with 10% agamma horse serum showed clear evidence of contamination with horse serum when tested serologically (Kenny, 1969). Although it has been claimed that contamination of antigens might be a result of absorption of serum proteins to the organisms (Bradbury and Jordan, 1972; Smith *et al.*, 1966), Yaguzhinskaya (1976) has produced clear evidence that specific serum components coprecipitate with the pellet during centrifugation and that these components cannot be removed by washing pellets of *Acholeplasma laidlawii* and *Mycoplasma arthritidis* but may be removed by sucrose gradient techniques. Washing procedures were sufficient to remove highly soluble components, such as albumin, but other components with slower electrophoretic mobilities were concentrated with the pellet as judged by two-dimensional immunoelectrophoresis (Fenske and Kenny, 1976). The problem of inclusion of medium components in washed pellets appears most marked with *Ureaplasma urealyticum* because of its small yield in culture; Masover *et al.* (1976) found 25 times as much protein in the membrane fraction as in the cytoplasmic fraction and concluded that the excess material was likely medium components. Razin *et al.* (1977) have pointed out that the study of such materials cannot give reasonable results from a standpoint of chemical composition. Some information as to the purity of the pellet of organisms can be obtained by determining the protein content per colony forming unit (CFU) since this value will be inflated by presence of medium components (Kenny and Cartwright, 1977). Purification of the antigenic preparation as suggested by Yaguzhinskaya (1976) is clearly indicated for analyses which might be compromised by medium components. Proof that the particular component was in fact synthesized by the organism and is not a medium component contaminant can be obtained by labeling the organism with specific precursors as has been done for glycolipids of both *Mycoplasma pneumoniae* (Plackett *et al.*, 1969) and *U. urealyticum* (Romano *et al.*, 1972) as well as proteins of *A. laidlawii* and *M. arthritidis* (Yaguzhinskaya, 1976). Immunologic methods have great utility for proving specificity of given components to the mycoplasma cell and such methods will be discussed in detail in Section III.

The viability of the organisms in the immunogen is an additional and important criterion of purity. Cultures harvested for preparation of antigen should be in log phase or late log phase since aging cultures show substantial degradation of structure (Boatman and Kenny, 1970) and biochemical changes in lipids and proteins (Rottem and Greenberg, 1975; Amar *et al.*, 1976). Control of pH for organisms which ferment glucose or utilize arginine or urea is important since immunogenicity was impaired by low pH for *M. pneumoniae* (Pollack *et al.*, 1969) a problem which could be reversed by the use of *N*-2-hydroxyethylpiperazine-*N'*-2-ethanesulfonic acid (HEPES) buffer. *N*-Tris(hydroxymethyl)methyl-2-amino ethanesulfonic acid (TES) buffer is suitable for a wide variety of mycoplasma species at 10 m*M* concentration (Kenny, 1969, 1977) and 2-(*N*-morpholino) ethanesulfonic acid (MES) buffer is excellent for controlling the pH of cultures of *U. urealyticum* (Kenny and Cartwright, 1977). Control of both pH and incubation time was found important to the serologic activity of *Mycoplasma gallisepticum* (Bradbury and Jordan, 1971).

## B. Media

The medium for cultivation of the Mycoplasmatales ordinarily consists of a peptone, some form of yeast extract, and animal serum at a concentration of 10–20%. Defined media are suitable for only some *Mycoplasma* and *Acholeplasma* species and the yields of culture may be small (Rodwell, 1969). The peptones and yeast extract may precipitate at low pH and give the appearance of excellent growth of the organisms when the organisms cultivated produce acid. One solution used for this problem is to prepare a dialysate from soy peptone and yeast extract (Pollock and Bonner,

1969a) or fresh yeast (Kenny, 1967), a method similar to that used for pneumococci 40 years ago (Brown and Robinson, 1938). Little precipitation of the dialyzable fraction is observed at the extremes of pH encountered during cultivation of the organisms, in contrast to extensive precipitation of medium components in medium prepared with whole peptones and filtered yeast extract. Adequate buffering will eliminate part of this problem. Filtered and autoclaved fresh yeast extract is particularly troublesome, since the pellets from cultures will frequently be black in color, indicating massive contamination. This component can be prepared separately as a dialysate (Kenny, 1969) which both speeds preparation and lessens contamination problems. The soy peptone–fresh yeast dialysate medium has been used successfully on a great variety of *Mycoplasma, Acholeplasma*, and *Ureaplasma* species (Kenny, 1967, 1969, 1975, 1977; Kenny and Cartwright, 1977). Yields of organisms in dialysate broth supplemented with 10% serum range from 0.6 mg protein per liter for *U. urealyticum* to 5–20 mg/liter for most *Mycoplasma* and *Acholeplasma* species. Higher yields might be achieved by rapid serial passage in the medium, but that approach is neither practical nor desirable for antigenic analysis.

The serum component poses an even larger problem because of its ready denaturation in medium at 37°C, particularly when cultures are agitated. In this laboratory, this problem has been circumvented to a degree by the use of commercially available "agamma sera" (serum chemically fractionated to reduce gammaglobulin content.) Immunoglobulins may be particularly vulnerable to denaturation (Yaguzhinskaya, 1976). Even agamma sera need to be pretested by incubation in medium to determine whether or not they yield a significant precipitate. Some lots are satisfactory and others are not. Another method used to limit medium precipitates is to incubate the complete medium and then filter it to remove insoluble material.

### C. Immunogens

Several approaches have been used to the immunologic solution of the medium component problem. Rabbits have been immunized with nonantigenic medium: Taylor-Robinson *et al.* (1963) used as immunogens for rabbits organisms grown in rabbit infusion broth supplemented with rabbit serum and cholesterol. Serologic test antigens were grown in the same medium. Kenny (1967, 1969) used organisms cultivated in dialysate broth supplemented with 10% "agamma" calf serum as immunogens and serologic test antigens were grown in dialysate broth supplemented with

10% "agamma" horse serum. The cross-reaction between horse and calf serum is small enough that the medium component antigens did not interfere with the specificity of the serologic tests. Either of these methods permits antigenic analysis of the organisms without interference with medium components, but chemical analysis of the antigens may be considerably in error because of the inclusion of medium materials in the pellets. "Agamma" rabbit serum is a suitable supplement for many *Mycoplasma* species but is not readily available commercially (G. E. Kenny, unpublished data).

### D. Preparation of Antisera

The choice of methods of preparing antiserum to a given immunogen depends to a large extent upon the purpose for which the antiserum is to be used. For example, if the antiserum is to be used for the purpose of detecting as many antigens as possible in a given organism; then intensive immunization with large amounts of organisms is warranted. In contrast, satisfactory reagents (or possibly superior for the purpose) might be produced with less intensive immunization (Lemcke, 1972) in cases where the antigen to be tested is highly immunogenic or the serologic testing system is highly sensitive (Table I). It is well recognized that intensive immunization is required for production of antibody suitable for growth inhibition on agar studies using the method of Clyde (1964). Such antibodies in the case of *M. pneumoniae* are produced late in immunization and require higher immunogen doses than the induction of complement-fixing antibodies to the lipid antigen (Kenny, 1971a). For production of highly specific antisera (antibodies restricted to a few antigenic specificities) both lower immunogen doses and shorter immunization schemes are warranted. Such sera might be used for typing of organisms by the fluorescent antibody method.

The method used for immunization in this laboratory is designed to produce antisera suitable for double immunodiffusion and two-dimensional immunoelectrophoresis as well as for growth inhibition typing of organisms. Accordingly, relatively intensive immunization is required. Briefly, animals are immunized intramuscularly with incomplete adjuvant, followed 3 weeks later by a series of four biweekly injections of increasing doses of fluid immunogen intravenously (Kenny, 1971a), a method based on that of Lemcke (1964). Animals are bled 1 week after final immunization without the necessity of trial bleedings. Best results were obtained when rabbits were injected with a total dose of 10 mg or more (protein) in the case of *M. pneumoniae*. This method has been

TABLE I. **Specificity and Sensitivity of Serologic Tests for Antigenic Analysis of the Mycoplasmatales**[a]

| Test | Specificity | Sensitivity | | Analytical[b] abilities | Antigens[c] |
|---|---|---|---|---|---|
| | | Antibody | Antigen | | |
| Growth inhibition (Clyde) | Species | + | ++++ | + | M |
| Metabolic inhibition | Species/subspecies | +++ | ++++ | + | M,S |
| Complement-mediated killing | Species/subspecies | ++++ | ++++ | + | M,S |
| Complement fixation | Genus/species/subspecies | ++ | ++ | + | M,C |
| Fluorescent antibody | Species/subspecies | ++++ | ++++ | + | M,S? |
| Gel diffusion | Genus/species/subspecies | ++ | ++ | ++[d] | M,C |
| Two-dimensional immunoelectrophoresis | Genus/species/subspecies, negatively charged antigens | ++ | +++ | ++++[d] | M,C,S |

[a] Adapted from Kenny (1977).
[b] Analytical abilities for directly differentiating antigens.
[c] M, membrane; S, surface; C, cytoplasm (soluble fraction).
[d] Broadly reactive antisera required for maximal analytical results.

successful in producing antisera to a variety of *Mycoplasma*, *Acholeplasma*, and *Ureaplasma* species (Kenny, 1977). For small doses of purified immunogens, complete adjuvant may be necessary (Alexander and Kenny, 1977); however, it is necessary to prepare antiserum to the complete adjuvant alone to demonstrate that the antibodies generated are not in response to mycobacteria. The overall purpose of this immunization scheme is to produce antibodies against as many determinants as possible with affinities suitable to the precipitin reaction. For production of high-affinity antibodies, smaller immunogen doses and a longer time interval between injections would be required as indicated by Lemcke (1973). Immunosuppression may be associated with some *Mycoplasma* species and may prevent preparation of adequate antibody for antigenic analysis. Immunosuppression has been associated with *M. arthritidis* (Kaklamanis and Pavlotis, 1972; Bergquist *et al.*, 1974). The antisera prepared to *M. arthritidis* were the weakest of those prepared to various arginine-utilizing species even though several attempts were made to prepare better sera (Thirkill and Kenny, 1974). This problem is probably restricted to only a few species.

## III. SEROLOGIC TESTS

Three general classes of serologic tests can be employed for serotyping and antigenic analysis of the Mycoplasmatales. Detection of inhibition of growth or killing of the organisms is carried out by growth inhibition on agar, metabolic inhibition, and the complement mediated killing test. Classical serologic tests for measurement of antibody to either whole organisms or purified fractions include: fluorescent antibody (primarily whole organisms), complement fixation, precipitin reaction, passive hemagglutination, agglutination, and other tests. Analytical procedures which can directly differentiate antigens from solubilized whole organisms or fractions thereof include gel diffusion and immunoelectrophoresis. The relative sensitivities and analytical abilities of these procedures are presented in Table I.

### A. Inhibition and Killing Tests

#### 1. Growth Inhibition

Growth inhibition on agar as a means of speciating the Mycoplasmatales was described by Edward and Fitzgerald (1954). The method using antibody-impregnated paper disks was first described by Huijsmans-

Evers and Ruys (1956) and further refined by Clyde (1964). The mechanism of the test is obscure, but the antibodies active in the test are induced by immunization with membrane preparations (Kahane and Razin, 1969; Levisohn and Razin, 1973; Hollingdale and Lemcke, 1969; Goel, 1973). The results of the method are usually considered to be influenced little by antigenic variations between strains of a species. However, extensive variation has been observed in *M. hominis* strains (Lin *et al.*, 1975). This did not appear to be true for *Mycoplasma arginini:* strain G-230 contains a prominent strain-specific membrane antigen on its surface which is not shared by strains 23243 or leonis (Alexander and Kenny, 1977), but these strains show growth inhibition with G-230 antiserum (Hahn and Kenny, 1974). The three strains of *M. arginini* do share one surface membrane antigen (Alexander and Kenny, 1977).

### 2. Metabolic Inhibition Test

The metabolic inhibition test measures inhibition of growth of organisms in fluid medium. In these tests, growth of the organism is assessed by the formation of a metabolic product which can be detected by a color reaction: acid from glucose, ammonia from urea and arginine, or the reduction of tetrazolium (Purcell *et al.*, 1966a,b; Taylor-Robinson *et al.*, 1966; Senterfit and Jensen, 1966). Specific antibody inhibits growth and hence the formation of the metabolic product; the end point can be measured by titration of the antibody against a fixed number of "color changing" units of organism. The antibodies measured are induced by immunization with membrane preparations (Williams and Taylor-Robinson, 1967; Pollack *et al.*, 1970; Kahane and Razin, 1969; Hollingdale and Lemcke, 1969; Levisohn and Razin, 1973; Goel, 1973). Glycolipids are the determinant(s) for this reaction in the case of *M. pneumoniae* since antibody is induced by immunization with *M. pneumoniae* glycolipids reaggregated with *A. laidlawii* membranes (Razin *et al.*, 1970b, 1971). In contrast, the determinant(s) for metabolic inhibition of *A. laidlawii* appears to be membrane proteins (Dörner *et al.*, 1976; Ne'eman *et al.*, 1972) as is also true for *M. hominis* (Hollingdale and Lemcke, 1972). In general, metabolic inhibition reactions are more prone than growth inhibition tests to show strain variation within a species; extensive variation in antibody titers to homologous and heterologous strains has been shown with *Mycoplasma pulmonis* strains (Haller *et al.*, 1973; Forshaw and Fallon, 1972) and *M. hominis* (Hollingdale and Lemcke, 1970; Lin and Kass, 1974; Razin *et al.*, 1972) and with a number of other species (Purcell *et al.*, 1967, 1969). Since the metabolic inhibition test requires far less antibody than the growth inhibition on agar technique, the failure of the latter technique to ordinarily demonstrate intraspecies heterogeneity is understandable since end points can be measured only by zone diameters and not by serum dilution (Clyde, 1964). A reverse system of metabolic inhibition testing has been used: Fernald *et al.* (1967) used as the end point the number of organisms inhibited since high-titered sera could inhibit more organisms than low-titered serum when tested at the same dilution. They also demonstrated that a heat-labile factor greatly enhanced the test. The role of complement in the metabolic inhibition test has been controversial; in general, titers are at least stabilized by the addition of a source of complement. One of the reasons for this controversy is that the medium used for the reaction may be anticomplementary by itself or it becomes anticomplementary by the addition of medium component antibody in the antiserum; this may cause the formation of heterologous antigen–antibody complexes which fix complement (Gale and Kenny, 1970). Thus, the addition of heat-labile factors may have little influence on the test because they are inactivated by the medium in the course of the reaction. Additionally, early antisera have a larger requirement for complement than late sera (Coleman *et al.*, 1974). For measurement of human serum antibody, the test may be compromised by the presence of antimicrobials in the serum (Smith and Herrmann, 1971), a problem which can be avoided by the use of antibiotic-resistant strains (Niitu *et al.*, 1974). A previous review (Purcell *et al.*, 1969) is recommended for additional information.

### 3. Mycoplasmacidal Reactions

Killing of *M. gallisepticum* in broth cultures was demonstrated by Barker and Patt (1967). The reaction was dependent upon a heat-labile factor in immune serum for which fresh guinea pig serum could be substituted. The heat-labile factor had a number of characteristics of complement: inhibition by chelating agents, absorption by unrelated antigen–antibody complexes, temperature dependence, and the factor in guinea pig serum could not be removed by absorption with *M. gallisepticum*.

Gale and Kenny (1970) showed that *M. pneumoniae* was killed when suspensions of organisms were incubated with antibody and complement, diluted, and plated on agar. Killing was rapid and the rate of killing was more dependent upon the concentration of fresh guinea pig serum than on the concentration of antibody. Accordingly, the reaction could be stopped by dilution, which indicated that the organisms were killed and not merely inhibited. Similar evidence for the factor being complement was obtained as for *M. gallisepticum*. In addition, the reaction was profoundly temperature dependent, with no killing of *M. pneumoniae* at 4°C, and only occurred over narrow ranges of ionic strength and pH (Gale and Kenny, 1970). Considerable killing was observed by guinea pig serum alone, an effect which appears to be a result of the complement alone (Bredt *et al.*, 1977). Organisms treated with complement alone showed rounding (Bredt and Bitter-Suermann, 1975); the rounding activity was dependent upon $Mg^{2+}$ but not $Ca^{2+}$, whereas killing by guinea pig serum required both ions. Antibody apparently is not involved in killing by nonimmune guinea pig serum and inactivation of the organism is dependent upon a means of activating complement other than antibody. Complement-mediated killing of *M. pneumoniae* has been difficult to quantitate. The first-order reaction constant $K$ was used by Gale and Kenny (1970). Brunner *et al.* (1972) transformed the S-shaped curve obtained upon plotting the log of the surviving fraction against the dilution of human serum used in the test into a straight line. That dose of antibody which killed 90% of the organisms was used as the end point. Titers by the mycoplasmacidal test were far higher than those obtained by other tests, including the sensitive immunofluorescence test.

*Ureaplasma urealyticum* is killed rapidly by antibody and complement (Lin and Kass, 1970) and a serotyping scheme has been proposed using the mycoplasmacidal assay (Lin and Kass, 1973; Lin *et al.*, 1972). A similar assay was devised for *M. hominis* strains (Lin and Kass, 1975) which shows large antigenic heterogeneity among strains. In both assays, the generation of ammonia from either urea or arginine was proposed as inhibiting the reaction by blocking complement (Lin and Kass, 1970, 1975). However, the amounts of arginine or urea required for inhibition were large (229 m*M* for arginine and 40–200 m*M* for urea) and within the range of concentration where osmotic effects and ionic strength may block complement-mediated reactions (Gale and Kenny, 1970; Levine *et al.*, 1961); Masover *et al.* (1975) found that 5 m*M* $NH_4Cl$ had no effect on complement-mediated killing of *U. urealyticum*.

*Acholeplasma laidlawii* is killed in a complement-mediated reaction, but the determinant is apparently membrane protein (Dörner *et al.*, 1976), in contrast to the situation with *M. pneumoniae,* where killing is obtained with antibody to glycolipids (Brunner *et al.*, 1971). The mechanism of killing of both *M. pneumoniae* and *A. laidlawii* appears to be by specific damage to the membrane (Brunner *et al.*, 1971, 1976). A mutant of *A. laidlawii,* derived by mutagenesis of the parent strain, was found resistant to lysis by antibody (against the parent strain) and complement, whereas the parent readily lysed (Dahl *et al.*, 1977). The interesting question is whether the mutant is resistant because of an antigenic change (it is unknown whether the mutant could have been killed by homologous antibody and complement) or because of alterations in membrane properties unrelated to antibody fixation but related to structural stability.

Complement-mediated killing of the Mycoplasmatales appears to be a general phenomenon because a number of species are killed in such reactions: *Mycoplasma canis* (Tachibana *et al.*, 1970), *M. arthritidis* (Cole and Ward, 1973), and *Mycoplasma meleagridis* (Matsumoto and Yamamoto, 1973).

## B. Classical Serologic Tests

A wide variety of classical serologic tests are available in the immunology literature for measurement of antibodies and antigens. In general, these are readily adaptable for the Mycoplasmatales and require no special comment. However, certain features of the tests are particularly relevant to the Mycoplasmatales and will be discussed.

### 1. Complement Fixation

The complement-fixation test has a major advantage in that the antigens tested need not be soluble as required in most other serologic tests; thus, it is highly useful in measuring antibodies to glycolipids or other antigens which are poorly water-soluble. Although the test, as ordinarily carried out with whole organisms or sonicates, yields titers which are a summation of whatever antigens and antibodies are present in the reaction mixture, it has been possible to denote the major complement-fixing antigens for several organisms: lipid for *M. fermentans* and *M. pneumoniae* (Kenny, 1971b; Kenny and Grayston, 1965), heat-stable proteins for *M. pulmonis* (Deeb and Kenny, 1967b), and carbohydrate-containing antigens for *M. mycoides* (Buttery, 1972). When comparing antigens, or determining major antigens, it is important to report both antigen and antibody titers (Deeb and Kenny, 1967b; Lemcke *et al.*, 1967; Kenny, 1969; Sugiyama *et al.*, 1974). This is particularly important if the antibody titer is little affected by changes in antigen concentration.

### 2. Agglutination

Agglutination reactions have been used to localize surface antigens in the Mycoplasmatales. Antibody prepared against *M. pneumoniae* glycolipids (reaggregated with *A. laidlawii* membranes) agglutinated *M. pneumoniae,* thus verifying the surface location of the glycolipids (Razin *et al.*, 1970b). Antibodies to *M. hominis* membranes, as well as to delipidized membranes, agglutinated native *M. hominis* cells, whereas antibodies to phosphatidylglycerol did not (Schiefer *et al.*, 1975), indicating that proteins were the surface antigens. However, pronase treatment of the cells rendered them agglutinable by antibody to phosphatidylglycerol, suggesting that this phospholipid was covered by protein. A novel growth–agglutination test has been proposed by Lin and Kass (1975) for measurement of antibody to *M. hominis*. Briefly, microtiter wells containing organisms, antibody, and complement are examined microscopically for the presence of clumps of cells in the bottom of the well which occur at high dilutions of specific antibody in the presence of complement.

### 3. Fluorescent Antibody

The importance of the fluorescent antibody method for detection of small amounts of organisms and antibody is emphasized by its role in the discovery of the role of the Eaton agent (*M. pneumoniae*) in primary atypical pneumonia (Liu, 1957). The test is species specific (Del Giudice *et al.*, 1971, 1974; Barile *et al.*, 1972), suggesting that the antigens measured are surface components. The effect of fixatives such as organic solvents (acetone is commonly used) on specific antigens has not been determined; however, one would suspect that glycolipids would be removed by such treatment. The indirect method of fluorescent antibody testing for serodiagnosis is advantageous because antibodies to specific immunoglobulin classes can be determined (Biberfeld and Sterner, 1971).

### 4. Radioimmunoassay and ELISA

Brunner and Chanock (1973) devised a radioimmunoassay for detection of antibody to *M. pneumoniae*. Organisms were grown with [$^{14}$C]oleic and palmitic acids and used as antigen. The organism–antibody complexes were precipitated with antiimmunoglobulin. The sensitivity of the test was equal to or superior to the mycoplasmacidal assay. The glycolipids of *M. pneumoniae* blocked some but not all the reactivity of the serum, suggesting that other components might also participate in the reaction. The assay was adapted to measure specific immunoglobulins (Brunner *et al.*, 1973). A further modification of this test was the use of staphylococci (which contain protein A, which specifically binds IgG but not other immunoglobulins) to precipitate the antigen–antibody complexes (Brunner *et al.*, 1977). This test was also highly sensitive.

The enzyme-linked immunoabsorbent assay (ELISA) described by Engvall and Perlmann (1971) appears to be a highly useful new method for detection of antibody and antigens as can be judged from the extensive literature developing on this assay (Sever and Madden, 1977). The assay has been used for *M. pneumoniae* (Voller *et al.*, 1976) and for *Mycoplasma suipneumoniae*, where specificity and sensitivity were excellent (Bruggmann *et al.*, 1977). The advantages of this assay are large: Performance of the test is simple, radioisotopes are avoided, sensitivity is as high as that in radioimmunoassay, and conjugated reagents are available which permit detection of antibody to specific immunoglobulin classes. However, best use of ELISA (and radioimmunoassay) requires the preparation of purified antigens so that antibodies to specific components may be detected. This may be feasible because of the small amounts of antigen required.

An additional and novel technique is the immune complex-induced platelet aggregation and [$^3$H]serotonin release used for measurement of antigens, antibodies, and complexes (Patscheke *et al.*, 1977). This technique is apparently as sensitive as radioimmunoassay. Biberfeld and Norberg (1974) used a platelet aggregation technique (Penttinen and Myllala, 1968) to detect immune complexes in sera of patients with *M. pneumoniae* infection.

#### 5. Controls

In contrast to the growth inhibition, metabolic inhibition, and mycoplasmacidal assays, which are little affected by medium component antibody, the classic serologic tests are extraordinarily vulnerable to such antibodies. Accordingly, it is most important to prepare antisera to the serum component of the medium (or the whole medium if a dialysate is not used) for immunization and to use this serum to demonstrate that the antigen tested does not contain antibodies to medium components. Since it is clear that only certain serum components contaminate the pellets and, in fact, are enriched in the pellet (Section II, A), both the testing of the serologic test antigen for presence of medium component antibodies and the testing of the mycoplasmic antiserum against whole medium may not serve as adequate controls. A superior control is to show both that the antimycoplasma serum does not react with a heterologous species (known not to cross-react) and that the antigen does not contain components which can be detected by antibody to the heterologous species (Kenny, 1969). In the future, it would appear incumbent upon mycoplasmologists not only to prove serologic specificity of the test system, but also to prepare antigens free of medium components.

### C. Analytical Methods for Antigenic Analysis

Three major methods of antigenic analysis may be described: (1) cross-absorption of antisera can be used to prepare monospecific reagents for identification of specific antigens on cell surfaces, as has been done for a variety of bacteria; (2) organisms can be fractionated, the purity of the fractions assessed biochemically (as discussed in a number of chapters in this volume) and their activity assessed against antisera (such purified antigens would be excellent for radioimmunoassay and ELISA);

and (3) antigens can be identified immunologically by use of analytical methods which can distinguish a variety of antigens in complex mixtures. This section focuses on the descriptions of analytical methods which have proved useful for the Mycoplasmatales: double immunodiffusion (Ouchterlony), immunoelectrophoresis, and two-dimensional (crossed) immunoelectrophoresis.

### 1. Double Immunodiffusion

The principal uses of the double-immunodiffusion test are to detect antigens and antibodies as well as to compare antigens (Crowle, 1973). The test is fairly insensitive, requiring both concentrated antigens and strong antisera for detection of antigens and antibodies to the Mycoplasmatales (Kenny, 1971a). Antigenic comparison have been made for a wide variety of species, of which only a few will be cited: human species (Taylor-Robinson *et al.*, 1963), *Thermoplasma* (Bohlool and Brock, 1974), *Anaeroplasma* (Robinson and Rhoades, 1977), and a variety of other species (Lemcke, 1965; Kenny, 1969, 1977). The major problem for this assay with the Mycoplasmatales lies in the fact that many of the antigens are located in the membrane and hence poorly soluble in aqueous solvents. The use of nonionic detergents such as Triton X-100 was found to greatly enhance the number of precipitin lines which could be observed with membranes of *M. hominis, A. laidlawii*, or *M. gallisepticum* (Hollingdale, and Lemcke, 1969; Kahane and Razin, 1969). Solubilization is apparently not required for the "soluble" (cytoplasmic) fraction. Best visualization is obtained by washing the preparations and then staining with a sensitive stain such as Coomassie blue. However, precipitin lines formed between *M. pneumoniae* glycolipids (with auxiliary lipids) and antibody have not proved stable to washing and must be photographed dirctly (G.E. Kenny, unpublished data). It should also be noted that the practice of refilling wells to enhance precipitin lines is hazardous since additional lines may appear for the same antigen–antibody system (Crowle, 1972). In my experience, the use of both a plastic matrix (Sharpless and LoGrippo, 1965) with a short diffusion distance of 4 mm and 0.5% agarose in a thin film (0.6 mm) gives excellent results (Kenny, 1969). Even in this system, the ability to discriminate between numerous lines (five to ten) is difficult, particularly for assessing "identity" and "nonidentity."

### 2. Immunoelectrophoresis

Resolving power is improved when an electrophoretic step is used to separate the antigens before the pattern is developed by application of antibody. Although the method has been used in a number of studies (Fowler *et al.*, 1967; Argaman and Razin, 1969; Stone and Razin, 1973;

Thirkill and Kenny, 1974), it was not nearly as popular as double immunodiffusion. Several reasons are apparent: the method requires large amounts of antibody; the insolubility of membranes is an even greater disadvantage (Hollingdale and Lemcke, 1969); and it is difficult to compare antigens. A variation on this technique is to carry out the electrophoretic separation in polyacrylamide rather than agar and then to imbed the polyacrylamide strip (or sections thereof) in agar and develop the precipitin lines with antibody (Argaman and Razin, 1969). A related technique, counterimmunoelectrophoresis, has been used for detection of antibody in human infections with *M. pneumoniae* (Goldschmidt *et al.*, 1976; Menonna *et al.*, 1977) and identification of *Mycoplasma* species (Cho and Langford, 1974). Counterimmunoelectrophoresis is highly sensitive for detection of antigen but requires strong antisera, in contrast to immunoelectrophoresis, which requires both large amounts of antigen and antibody.

### 3. Two-Dimensional (Crossed) Immunoelectrophoresis

Two-dimensional (crossed) immunoelectrophoresis belongs to a family of techniques by which antigens can be not only compared, but also quantitated. Rocket electrophoresis is the simplest of the methods. Briefly, antigen (from a point source) is electrophoresed into a agarose bed containing a fixed amount of antibody. A precipitin peak in the form of a "rocket" results, the area (height may be used if the bases of the peaks are the same as is usually true) of which is directly proportional to the antigen concentration and inversely proportional to the antibody concentration. This method has been used for quantitating antigens of *M. hominis* (Fenske and Kenny, 1976) and for determining the distribution of cytoplasmic and membrane components in fractions of *M. arginini* cells (Alexander and Kenny, 1977). Fractions of organisms may be compared by "fused" rocket electrophoresis (Svendsen, 1973); briefly, fractions are placed in wells, allowed to diffuse for 1–2 hr to produce overlapping rockets, and then electrophoresed into antibody. This technique was used for immunologic characterization of fractions of membranes of *Spiroplasma citri* obtained by agarose suspension electrophoresis (Wroblewski *et al.*, 1977b).

Two-dimensional (crossed) immunoelectrophoresis involves two steps: (1) separation of antigenic components electrophoretically in agarose and (2) development of the separated components as rocket peaks in the second phase (Fig. 1). The method has significantly greater resolving power than double immunodiffusion; as many as 20 precipitin peaks have been resolved for solubilized whole cells of *M. arginini* (Thirkill and Kenny, 1974). Most important, antigens can be quantitated since the areas

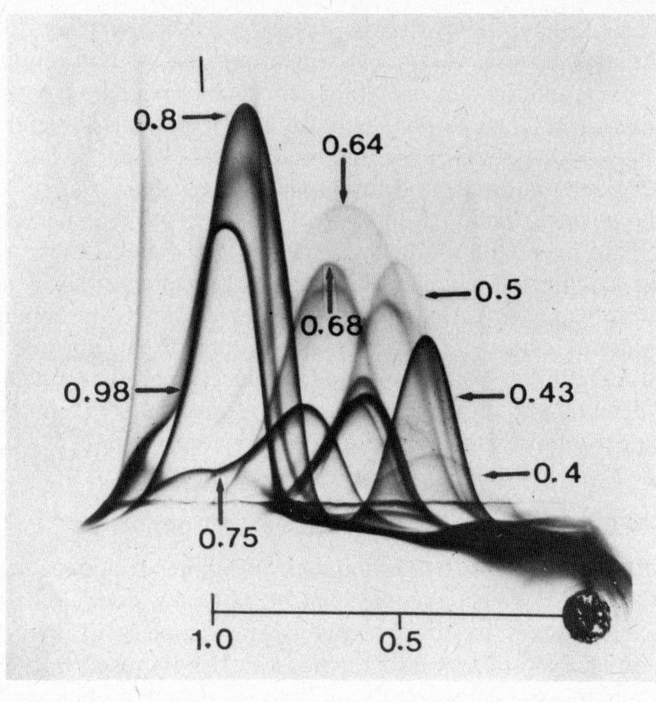


FIGURE 1. Two-dimensional immunoelectrophoresis profile of *M. arginini* strain G-230. The slide was covered with agarose first. Fifty micrograms (protein) of antigen was placed in the well (dark circle at lower right). Electrophoresis was carried out with the anode to the left of the figure until the bovine albumin marker at 6 V/cm (included with the antigen) was judged to have reached 1.0 (on the bottom scale). The agar was removed from the slide above the antigen track (a straight line can be seen traversing the base of the peaks) and replaced with agarose containing antibody to strain G-230. Electrophoresis was now carried out with the anode to the top of the figure until peaks had reached their maximum heights (6 hr). The scale at the bottom of the figure indicates the mobilities of the peaks relative to bovine albumin, which peak is identified by the vertical bar at the top left of the figure. (From Thirkill and Kenny, 1975.)

of peaks are directly proportional to antigen concentration (peak heights cannot be used because the base length of the peak usually increases with antigen concentration). Solubilization of membrane components is required: Not only was it necessary to solubilize organisms in Triton X-100 but also (and more important) Triton X-100 was required to be present at 0.5% concentration in both the first phase and the second phase of the electrophoresis to produce reproducible peaks with most species in the arginine-utilizing group of organisms (Thirkill and Kenny, 1974). Similarly, *A. laidlawii* membranes were effectively solubilized by a variety of detergents (deoxycholate, Triton X-100, and Tween 20), whereas solubilization of membranes of *S. citri* required deoxycholate (Wroblewski *et al.*, 1977a). Mobility of amphiphilic, but not hydrophilic, proteins is enhanced by the addition of a charged detergent to proteins solubilized in a nonionic detergent (charge-shift electrophoresis; Helenius and Simons, 1977). When the first phase of two-dimensional immunoelectrophoresis was run in a mixture of 0.1% deoxycholate (an anionic detergent) and 0.5% Triton X-100, striking anodic shifts of membrane antigens were observed compared to profiles in Triton X-100 alone, whereas cytoplasmic antigens (presumably hydrophilic) did not shift (Alexander and Kenny, 1978). Thus, charge-shift conditions may be useful for presumptive determination of the membrane or cytoplasmic nature of antigens in solubilized whole organism preparations of the Mycoplasmatales.

The high resolving power of two-dimensional immunoelectrophoresis has been used to advantage to compare species in the Mycoplasmatales. Clear evidence for common antigens was observed in five arginine-utilizing species (Thirkill and Kenny, 1974). Wroblewski and Ratanasavanh (1976) demonstrated one common membrane antigen between *A. laidlawii* and *Acholeplasma granularum* (the homologous reactions showed 12–15 components), but membranes of either *Acholeplasma* species showed no cross-reaction with those of *M. fermentans*. Clear evidence for subspecies-specific antigens was obtained in studies of three *M. arginini* strains (Thirkill and Kenny, 1975). The quantitative nature of the technique was particularly useful for comparison of strains since it was possible to establish that the major strain-specific antigens were unique to the strain and if present in other strains, their concentration was minute. A variety of techniques are available for showing antigenic identity and nonidentity (Axelsen and Bock, 1972). Two-dimensional immunoelectrophoresis has further advantages in that not only can antigenic identity be established, but also the antigen can be characterized by its electrophoretic mobility relative to a standard such as bovine albumin (Thirkill and Kenny, 1974, 1975). Surface antigens can be identified by absorption of the antisera with whole organisms: The removal of even 50% of the antibody will be revealed by a doubling of the peak height of affected antigens, whereas internal antigen peak heights will be unaffected by the absorption (Johansson and Hjerten, 1974; Alexander and Kenny, 1977).

Two-dimensional immunoelectrophoresis can be used as a preparative method for preparing immunogens for induction of monospecific antisera to individual components. Precipitin peak material is excised from specific peaks (taking care that the material excised does not overlap with other peaks), emulsified with complete adjuvant, and used as immunogen.

This technique was employed for preparation of monospecific antiserum to the major external membrane component of *M. arginini* (Alexander and Kenny, 1977). This technique appears to have general application (Vestergard, 1975; Caldwell *et al.*, 1975) and the monospecific antiserum can be used in an immunoabsorbent column for fractionation of specific antigens (Caldwell and Kuo, 1977).

In conclusion, two-dimensional immunoelectrophoresis appears to have great utility for antigenic analysis of the Mycoplasmatales. The technique is applicable to a number of *Mycoplasma* species (Thirkill and Kenny, 1974; Wroblewski, 1975; Wroblewski and Ratanasavanh, 1976). It has the advantages of both high sensitivity and high resolution for detection of antigens, but the disadvantage of requiring large amounts of antibody. The method is restricted to antigens which are soluble or can be made soluble by detergent and which have an electrophoretic charge. Antibodies to medium components are a problem as for the other analytical techniques. However, precipitin peaks formed by such components not only can be identified more easily but also can be eliminated from the profiles by the suppression technique (Thirkill and Kenny, 1974, 1975). Although two-dimensional immunoelectrophoresis appears complex, it is actually not difficult and can readily be performed in the laboratory. The technology is rapidly developing and an extensive literature is available with a number of recent review articles (Crowle, 1973; Axelsen *et al.*, 1973; Axelsen, 1975; Verbruggen, 1975).

## IV. ANTIGENIC HETEROGENEITY OF GENUS *Mycoplasma*

Genus *Mycoplasma* is a collection of species which vary markedly: (1) some species ferment glucose, whereas others hydrolyze arginine, properties usually mutally exclusive; (2) the guanine plus cytosine (% G + C) content in their DNA ranges from 25% to 40%; and (3) strong morphologic differences can be observed (these characteristics are discussed in detail in Chapters 1, 2, 3 and 6 of this volume). Separation of species has been carried out by serologic methods, with growth inhibition on agar (Clyde, 1964) being a useful and popular method. Analytical methods have not been used for taxonomic separation except to confirm results obtained by less sensitive but more species-specific methods.

### A. Antigenic Heterogeneity between *Mycoplasma* Species

Striking differences were observed between *Mycoplasma* and *Acholeplasma* species when antigens and antisera were cross-tested by double immunodiffusion (Taylor-Robinson *et al.*, 1963, 1965; Lemcke, 1965) and no clear evidence of common antigens was observed except with the arginine-utilizing strains. I surveyed ten glycolytic strains in an attempt to locate common antigens between species using a microdiffusion test, antigens which contained 5–12 mg protein per milliliter, and antisera which were capable of producing 5–11 precipitin lines in the homologous system (Kenny, 1969). No common components were observed in the entire group, but four groups of serologically related organisms were observed which were related also as judged by a comparison of their % G + C contents; i.e., organisms which had significantly different % G + C were in separate groups, whereas organisms with similar % G + C were usually, but not necessarily, related. The study has been repeated with a total of 23 species, all of which were compared using two separate lots of antigens (Kenny, 1977; Table II). The original four groups have increased to six or seven groups with properties indicated in Table III. The general serologic relationships shown in Table II have been amplified by other workers in double-immunodiffusion studies. The homogeneity of the strains related to *M. mycoides* (group 1) and their lack of relationship to other groups was demonstrated by Argaman and Razin (1969). Common antigens between the arginine-utilizing species were clearly illustrated by Cole *et al.* (1967).

Group 2 organisms have been reclassified as *Acholeplasma* on the basis of their *lack* of a requirement for cholesterol (Edward and Freundt, 1969, 1970). Had Acholeplasmataceae been separated from Mycoplasmataceae on their *lack* of serologic cross-reactivity, a precedent would have been set for further division of the Mycoplasmataceae on an immunologic basis. It is interesting to note that if this were done from the data in Table II, only group 1 organisms would retain the genus name *Mycoplasma*. Species which are closely related serologically, such as *A. laidlawii* and *A. granularum,* also proved to be related when tested by DNA–DNA hybridization (Pollock and Bonner, 1969b; Neimark, 1970), whereas serologically unrelated strains showed no homology. Additional closely related organisms which would be worth testing for DNA–DNA homology would include (1) *Mycoplasma canis, Mycoplasma felis, Mycoplasma edwardii*; (2) *Mycoplasma hyorhinis, Mycoplasma pulmonis*; and (3) *M. arginini, Mycoplasma gateae*. The last pair is of particular interest because of the strong cross-reactivity shown on two-dimensional immunoelectrophoresis (Thirkill and Kenny, 1974). Such experiments would provide an interesting test of the ability of analytical serology to recapitulate genetic relationships. A combination of serologic data and other data would probably permit the separation of the organisms presently known as *M. pneumoniae, M. gallisepticum*, and the arginine-utilizing, non-

TABLE II. **Serologic Comparison of 23 Species in the Mycoplasmatales by Double Immunodiffusion**[a]

| Antigen | Antisera[b] | | | | | | | | | | | | | | | | | | | | | | | | | Strain | Serol group[c] |
|---|---|---|---|---|---|---|---|---|---|---|---|---|---|---|---|---|---|---|---|---|---|---|---|---|---|---|---|
| | M | C | 1 | 2 | 3 | 4 | 5 | 6 | 7 | 8 | 9 | 10 | 11 | 12 | 13 | 14 | 15 | 16 | 17 | 18 | 19 | 20 | 21 | 22 | 23 | | |
| 1. Bovine group 7[d] | 5[e] | 5 | *8* | 3 | | | | | | | | | | | | | | | | | | | | | | ATCC 19884 | 1 |
| 2. *M. putrefaciens* | 3 | 2 | 3 | *6* | | | | | | | | | | | | | | | | | | | | | | ATCC 15718 | 1 |
| 3. *A. laidlawii* | | | | | *7* | 4 | 1 | | | | | | | | | | | | | | | | | | | ATCC 14089 | 2 |
| 4. *A. granularum* | | | | | 4 | *5* | | | | | | | | | | | | | | | | | | | | Friend | 2 |
| 5. *A. axanthum* | | | | | 2 | 2 | *6* | | | | | | | | | | | | | | | | | | | S-743 | 2 |
| 6. *M. bovigenitalium* | | | | | | | | *7* | | | | | | | | | | 1 | | | | | | | | ATCC 19852 | 3 (?) |
| 7. *M. gallisepticum* | | | | | | | | | *5* | | | | | | | | | | | | | | | | | ATCC 15302 | 4 |
| 8. *M. pneumoniae* | | | | | | | 1 | | | *6* | | | | | | | | | | | | | | | | AP-164 | 5 |
| 9. *M. anatis* | | | | | | | | | | | *8* | 3 | | | 3 | | 2 | 1 | | | | | | | | 1340 | 6 |
| 10. *M. bovimastitidis* | | 1 | | | | | | 2 | | | | *6* | | | 1 | 1 | 1 | 1 | 1 | | | | | | | ATCC 25025 | 6 (?) |
| 11. *M. canis* | | | | | | | | | | | 4 | | *5* | 3 | 3 | | 1 | 1 | | | | | | | | ATCC 19525 | 6 |
| 12. *M. edwardii* | | | | | | | | | | | | | 3 | *4* | 4 | 2 | 2 | 1 | 2 | | | | | | | ATCC 23462 | 6 |
| 13. *M. felis* | | | | | | | | | | | 3 | | | 3 | *5* | 1 | | | | | | | | | | B2 | 6 |
| 14. *M. fermentans* | | | | | | | | | | | 2 | 1 | 1 | | 2 | *6* | | | | | | | | | | PG 18 | 6 |
| 15. *M. hyorhinis* | | | | | 1 | | | | | | 1 | 1 | | | | 3 | *8* | 3 | | | | | | | | ATCC 17981 | 6 |
| 16. *M. pulmonis* | | | | | 1 | | | | | | 2 | 1 | 1 | | 1 | | 3 | *9* | 3 | | | | | | | 63 | 6 |
| 17. *M. neurolyticum* | | | | | | | | | | | 1 | | | 2 | 1 | | 2 | | *7* | 1 | | | | | | ATCC 15049 | 6 |
| 18. *M. synoviae* | | | | | | | | | | | 1 | 1 | 1 | 1 | 2 | | 1 | | | *3* | | | | | | WVU 1853 | 6 |
| 19. *M. arginini* | | | | | | | | | | | | | | | | | | | | | *9* | 4 | 1 | 2 | 2 | G-230 | 7 |
| 20. *M. gateae* | | | | | | | | | | | | | | | | | | | | | 3 | *11* | 4 | 4 | 2 | Siam | 7 |
| 21. *M. gallinarum* | | | | | | | | | | | | | | | | | | | | | | 1 | *8* | 2 | 1 | ATCC 15315 | 7 |
| 22. *M. hominis* | | | | | | | | | | | | | | | | | | | | | | 3 | | *6* | | ATCC 14027 | 7 |
| 23. *M. salivarium* | | | | | | | | | | | | | | | | | | | | | 2 | | | 1 | *9* | ATCC 23064 | 7 |

[a] Adapted from Kenny (1977).

[b] M = antisera to M. *mycoides* subsp. *mycoides*, C = antisera to *M. mycoides* subsp. *capri* (both from the National Institutes of Health Reference Reagents Branch). Numbers of antisera correspond to antigen numbers.

[c] Serologic group numbers are provisional and not intended as a final designation.

[d] The strain used for these experiments was obtained by cloning ATCC 19884 (*M. bovirhinis*). The clone was a contaminant and was related, if not identical, to bovine group 7 as represented by strain N-29B.

[e] Number of precipitin lines; homologous reactions are in italics.

TABLE III. **Antigenic Groupings of the Mycoplasmatales**

| Group | Number of species[a] | Representative species | % G + C[b] | Features |
|---|---|---|---|---|
| 1 | 2[c] | *M. mycoides* related | 25–28 | Glycolytic |
| 2 | 3 | *A. laidlawii* | 30–32 | Glycolytic, do not require cholesterol, genome $10^9$ daltons |
| 3 | 1 | *M. bovigenitalium* | 30 | Possibly belongs with group 6 |
| 4 | 1 | *M. gallisepticum* | 32 | Glycolytic, specific morphology |
| 5 | 1 | *M. pneumoniae* | 40 | Glycolytic |
| 6 | 10 | *M. felis*[d]<br>*M. hyorhinis*[d] | 25–28 | Glycolytic[e] |
| 7 | 5 | *M. hominis* | 27–29 | Nonglycolytic, arginine-utilizing |

[a] Number of species falling into group of the 23 compared in Table II.
[b] Data from Neimark (1970), Askaa *et al*. (1973), Freundt (1974).
[c] Does not include *M. mycoides* since antigen was not tested.
[d] Strains which yielded most cross-reactive antisera within group.
[e] *Mycoplasma fermentans* is both glycolytic and arginine-utilizing.

glycolytic organisms from genus *Mycoplasma* at the present moment. Such a division would be consistent with considerations put forth by Neimark (1970) when he compared physiologic and biochemical characteristics of the organisms. The biochemical uniqueness of serologic group 7 is illustrated further not only by their ability to utilize arginine, but also by their failure to incorporate nucleosides into acid-insoluble material, whereas species from other serologic groups readily incorporate nucleosides (McIvor and Kenny, 1977, 1978). *Mycoplasma fermentans*, which utilizes both arginine and glucose, did not cross-react with group 7 (Table II); however, Lemcke (1965, 1973) demonstrated common lines between the group 7 organisms and *M. fermentans*. The differences between these results are not readily explainable; however, Lemcke (1973) suggested that the cross-reactions between *M. fermentans* and group 7 might be due to common arginine deiminase enzymatic components. Antibody to arginine deiminase has been demonstrated (Gill and Pan, 1970) but only some of our antisera had such antibodies (Hahn and Kenny, 1974).

The relationships of other groups in the Mollicutes, *Spiroplasma, Anaeroplasma, Thermoplasma,* and *Ureaplasma,* to *Mycoplasma* and *Acholeplasma* is presently unknown. Some evidence is available for serologic heterogeneity of *Anaeroplasma* (Robinson and Rhoades, 1977) and *Thermoplasma* (Bohlool and Brock, 1974). The difficulties in propagating *U. urealyticum* have made antigenic analysis troublesome because of both poor yield and contamination with medium components (Masover *et al.*, 1975). However, improvements in growth of the organism (Kenny and Cartwright, 1977) have permitted some limited experiments. Two prominent common antigens were observed between types I, II, and VIII and at least one type-specific antigen was observed which showed little electrophoretic mobility in two-dimensional immunoelectrophoresis (G. E. Kenny, unpublished data). The type-specific components showed an anodic charge shift on addition of deoxycholate to Triton X-100-solubilized cells, suggesting that they might be membrane-associated, whereas the common components did not shift, suggesting that they were of cytoplasmic origin if the conclusions from the studies with charge-shift conditions with *M. arginini* membranes are relevant (Section III, C, 3).

Clearly, analytical serology is a highly useful method for further classification of the Mycoplasmatales. However, it will be necessary to identify biochemically specific common antigens between groups so that these can be used as criteria for classification, a prospect which appears likely because of the excellent analytical abilities of two-dimensional immunoelectrophoresis. A major drawback in analytical serologic methods which makes comparison of data difficult is the fact that the ability to detect components is solely dependent upon both the strength of the antiserum, and the number of components which it can recognize, factors which make comparison of data extremely difficult. This problem can be mitigated by the ability to produce monospecific antiserum to given components (Alexander and Kenny, 1977).

Analytical serology appears more sensitive for the detection of relationships than the more stringent DNA–DNA homology technique. However, phylogenetic distances between unrelated species (unrelated by both techniques) are unknown, although it is possible that certain species, now classified in genus *Mycoplasma,* are as distant from each other as *Escherichia* is distant from *Mycobacterium*.

## B. Antigenic Heterogeneity within *Mycoplasma* Species

It is to be expected that strains of a microbial species show antigenic variation. Heterogeneity between strains of *M. pulmonis* is well documented (Fallon and Jackson, 1967; Deeb and Kenny, 1967a,b; Forshaw and Fallon, 1972; Haller *et al.*, 1973). A subspecies-specific antigen, which was heat-stable and protein, was identified by Deeb and Kenny (1967b). This finding is somewhat controversial because Forshaw and

Fallon (1972) confirmed the existence of a heat-stable antigen but could only demonstrate its specific nature in several instances. This controversy is illustrative of a major problem in comparing serologic results: the number of components resolved is solely dependent upon the power and breadth of the antiserum. The antisera used by Forshaw and Fallon (1972) recognized fewer components in double immunodiffusion than those used by Deeb and Kenny (1967b), though their sera had similar complement fixing titers. The antigenic heterogeneity within *M. pulmonis* is further illustrated by the fact that a strain of *M. pulmonis* (*Mycoplasma histotropicum*) could be differentiated from other strains of the species by fluorescent antibody methods (Tully and Ruchman, 1964), but the strain was readily identified as *M. pulmonis* by growth inhibition on agar and double immunodiffusion (Kenny, 1969; Lemcke *et al.*, 1969).

Strains of *M. hominis* also show striking antigenic differences (Purcell *et al.*, 1967; Taylor-Robinson *et al.*, 1963; Hollingdale and Lemcke, 1970; Lin and Kass, 1974, 1975). Specific differences were shown between membrane preparations of various strains but not between soluble fractions (Hollingdale and Lemcke, 1970). Rather startling antigenic heterogeneity was observed between three strains of *M. arginini:* each strain was shown to possess one or more strain-specific antigens and some antigens were shared by only two of the three strains (Thirkill and Kenny, 1975). Five membrane antigens were characterized for strain G-230, three of which were strain specific, whereas four of the five cytoplasmic antigens were common to the other strains (Alexander and Kenny, 1977).

The above data suggest that antigenic heterogeneity will be observed to a greater or lesser degree within all species of the Mycoplasmatales. Clearly, criteria will need to be developed to apply the results of antigenic analysis to serotyping of strains of a species.

## V. ANTIGENIC STRUCTURE

Compared to other areas in the biology of the Mycoplasmatales, progress in defining specific antigens in chemical terms has been slow. The great cost of growing the Mycoplasmatales and their poor yield in culture are major responsible factors. This discussion will be divided into two parts: membrane antigens and cytoplasmic antigens.

### A. Membrane Antigens

Research into membrane antigens has been stimulated by two factors: (1) Logically, surface antigens must play the most important role in the initial response to a pathogen; and (2) membranes are relatively easily separated from the whole organism (this volume, Chapter 8). For most studies, membranes have been prepared by lysing the cells and sedimenting the lysate at relatively low force to remove "unbroken cells," followed by centrifugation at a higher speed to pellet the membrane components. Although such preparations have been labeled as "purified membranes," purity must be tested as recommended by Razin (1975) because medium component antigens appear to cosediment with the membrane pellet (Razin *et al.*, 1977). The number of antigens which have been recognized in double immunodiffusion range from two to five for a number of species: *M. pneumoniae* (Pollack *et al.*, 1970), *M. hominis* (Hollingdale and Lemcke, 1970, 1972; Lemcke, 1973), *T. acidophilum* (Bohlool and Brock, 1974), and *A. laidlawii, M. mycoides,* and *M. gallisepticum* (Argaman and Razin, 1969; Kahane and Razin, 1969; Ne'eman *et al.*, 1972). This number of precipitin lines is strikingly less than the number of proteins which can be demonstrated on polyacrylamide gel electrophoresis of membrane proteins (Hollingdale and Lemcke, 1970; Ne'eman *et al.*, 1972). However, studies by two-dimensional immunoelectrophoresis reveal 5–20 components (Johansson and Hjerten, 1974; Wroblewski, 1975; Wroblewski and Ratanasavanh, 1976; Alexander and Kenny, 1977, 1978), values which give greater hope that membrane proteins can be analyzed adequately antigenically.

### 1. Membrane Protein Antigens

Some characterization has been carried out on membrane protein antigens. Hollingdale and Lemcke (1972) identified three membrane antigens in *M. hominis*: two components were relatively heat-labile, whereas the third component was stable to 100°C. This latter component was also relatively stable to both pronase and trypsin. Antisera prepared to the three components (isolated by polyacrylamide electrophoresis) were active in the metabolic inhibition test. Ne'eman *et al.* (1972) fractionated membranes of *A. laidlawii* by gel filtration (solubility was maintained by running the column in deoxycholate). Immunization with one specific fraction induced antibody active by both growth inhibition on agar and metabolic inhibition testing, whereas immunization with the other three fractions induced complement-fixing antibodies only. Wroblewski *et al.* (1977b) fractionated membranes of *S. citri* by agarose suspension electrophoresis and purified one major antigen, spiralin. This component had a molecular weight of 26,000, possibly composed as much as 20% of the membrane protein, and its amino acid composition was different from that of the whole membrane.

The surface nature of membrane components has been studied. One of the Tween-20-soluble proteins in *A. laidlawii* membranes appeared to be external because its antibodies could be absorbed with whole cells (Johansson and Hjerten, 1974). Three other components appeared to be exposed on the inside of the membrane. An electrophoretically fast component was identified among the membrane antigens of *M. arginini* (Alexander and Kenny, 1977). This component was specific to strain G-230 and was not found in two other *M. arginini* strains. This component and two other antigens were located on the surface of the organism (as judged by absorption of antiserum with whole cells), whereas two other antigens were not so exposed.

Thus, important beginnings have been made at the antigenic characterization of membrane antigens and surface antigens of the Mycoplasmatales. The greatest differences between strains of *M. hominis* appears to be in the membrane antigens and not in the soluble antigens (Hollingdale and Lemcke, 1970). A similar conclusion for *M. arginini* strains appears valid (Alexander and Kenny, 1977).

The possible role of Mycoplasmatales viruses of *A. laidlawii* (MVL) in the antigenic structure of membranes can only be answered in part. Four of the peptides of MVL-2 have been analyzed immunologically and have been found to be specific to the virus and not to the host *A. laidlawii*, suggesting that the synthesis of these peptides is virally directed (Watkins, 1977).

### 2. Lipid Antigens

Consideration of lipid antigens is included with the membrane antigens because lipids are an integral part of membranes. The major antigen of *M. pneumoniae*, as measured by both rabbit immune and human convalescent serum, is found in the lipid fraction of the organism (Kenny and Grayston, 1965). Serologically active lipid fractions were separated by silicic acid column chromatography and found to contain glucose and galactose, though all fractions contained more than one glycolipid (Beckman and Kenny, 1968). A family of glyceroglycolipids was separated, including diglycosyl and triglycosyl components with glucose and/or galactose residues (Plackett *et al.*, 1969). Clear evidence for synthesis of the glycolipids by the organism was obtained because labeled glucose was incorporated into the compounds. Seven distinct serologically active components were recovered, not all of which contained carbohydrate (Plackett *et al.*, 1969). A cross-reaction was observed with spinach digalactosyl diglyceride. Further characterization of the fractions has been hindered by the extraordinarily small amounts of material and the difficulties of separating the glycolipids. At least three diglycosyl components have been recognized with two or three triglycosyl components when *M. pneumoniae* crude glycolipids were deacylated and tested by gas chromatography (G. E. Kenny, unpublished data). The diglycosyl compounds accounted for 55–60% of the total glycolipid, whereas the triglycosyl fraction accounted for most of the rest with a trace of a tetraglycosyl fraction. A striking cross-reaction with a triglycosyl fraction of spinach chloroplast glycolipids was observed (Kenny and Newton, 1973). This fraction contained a triglycosyl fraction (over 95% galactose) with the same retention time (as tested by gas chromatography of deacylated glycolipids) as one of the triglycosyl components of *M. pneumoniae* (G. E. Kenny, unpublished data). However, the serologic activity of the deacylated fragments could not be assessed because such components are not serologically active and efforts to reacylate the component were not successful. The difficulties in fractionation of glycolipids by silicic acid chromatography are large: the spinach trigalactosyl fraction similar to that which appeared to be homogeneous on thin-layer chromatography (Kenny and Newton, 1973) when deacylated was found to contain a substantial amount (60%) of a digalactosyl component which apparently represented monacyl digalactosyl diglyceride which had comigrated with the diacyl triglycosyl fraction, which fraction in turn had two triglycosyl components (G. E. Kenny, unpublished data). The triglycosyl fraction of spinach is of considerable interest because this component provides for excellent serodiagnosis of *M. pneumoniae* infections, with antibody titers closely paralleling those obtained with *M. pneumoniae* crude lipids (Kenny and Newton, 1973). We have produced antibody to the spinach triglycosyl fraction by immunizing rabbits with glycolipid complexed with lecithin, cholesterol, and methylated bovine albumin by the method of Kataoka and Nojima (1970). This antibody reacted with both the spinach trigalactosyl fraction and crude *M. pneumoniae* lipids in double immunodiffusion (G. E. Kenny, unpublished data). As evidenced above, progress with characterization of *M. pneumoniae* glycolipids has been hampered not only by the small amounts of material, but also by the presence of a large number of glycolipids which are difficult to separate completely from each other. It should be recognized that trace amounts of a glycolipid in an ostensibly pure preparation of another lipid could contribute to the serologic activity observed (Plackett *et al.*, 1969). Clearly, evaluation of purity by thin-layer chromatography (of fractions purified by that technique) is inadequate to demonstration of freedom from such components since gas chromatography of deacylated fractions frequently shows heterogeneity. Furthermore, certain lipids, such as phosphatidyl glycerol, are excellent auxiliary lipids and may potentiate the activities of trace amounts of glycolipids (Razin *et al.*, 1970a).

In addition to the striking cross-reaction with spinach glycolipids, a cross-reaction was observed with lipids of *Mycoplasma neurolyticum* (Kenny, 1971b), but no components were shared in double-immunodiffusion tests of whole organism antigens (Kenny, 1969). This cross-reaction is of considerable interest because *M. neurolyticum* lipids contain only glucose (Smith, 1972). A cross-reaction with *M. pneumoniae* glycolipids was not surprising because our preparations contained 75% galactose and 25% glucose. What was surprising was the fact that rabbits immunized with *M. neurolyticum* produced antibodies measurable with the spinach trigalactosyl fraction. Although this fraction, as prepared in our laboratory, does contain 3% glucose, the cross-reaction was startling. However, the lipopolysaccharides of *M. neurolyticum* contain 94% galactose in the neutral sugar fraction (Smith *et al.,* 1976) which might have induced the antibodies detected. A number of remarkable antibody reactions accompany *M. pneumoniae* infections in humans. Antibodies are produced to brain (Biberfeld, 1971), to smooth muscle (Biberfeld and Sterner, 1976), and on occasion to the glycopeptide antigen of *Micropolyspora faeni* (Davies *et al.,* 1975). Some of these antibodies may be a result of antibodies to the *M. pneumoniae* glycolipids.

Serologically active lipids have been recovered from other species: *A. laidlawii, A. granularum,* and *M. fermentans* (Kenny, 1971b); *Acholeplasma modicum* and *Acholeplasma axanthum* (Sugiyama *et al.*, 1974); and *U. urealyticum* (Romano and Scarlata, 1974). However, the presence of serologically active lipids is not a generality; many species do not have significant serologic activity in their lipid fractions: *M. canis, M. felis, M. gallisepticum, M. hyorhinis* (Kenny, 1971b), and *M. pulmonis* (Deeb and Kenny, 1967b). Only slight activity has been observed with lipid fractions of *M. hominis* (Kenny, 1967; Hollingdale and Lemcke, 1969); however, Schiefer *et al.* (1975) have demonstrated that phosphatidyl glycerol not only is a major component of the membrane phospholipids of this organism but also is both an antigen and an immunogen, though proteins appear to be the major antigens (Hollingdale and Lemcke, 1969). Diglucosyl diglyceride purified from *A. laidlawii* is serologically active and cross-reacts with the same compound obtained from *Streptococcus* MG (Plackett and Shaw, 1967). Ryan *et al.* (1975) fractionated glycolipids and phospholipids from *A. laidlawii* by thin-layer chromatography and tested both their immunogenicity and their antigenicity. Antigenic activity was demonstrated with fractions described as monoglucosyl diglyceride and diglucosyl diglyceride as well as cardiolipin and phosphatidyl glycerol. Glycerylphosphoryl diglucosyl diglyceride was anticomplementary, an effect not observed in the study of Sugiyama *et al.* (1974). The anticomplementary effects of *A. granularum* and *A. laidlawii* antigens were found to be a result of antibody in some guinea pig sera (Kenny, 1971b).

### 3. Polysaccharide Antigens

The consideration of polysaccharide antigens is included with the membrane section because the polysaccharides, if not intimately associated with membranes, are likely membrane products. The galactan of *M. mycoides* is well known (Plackett and Buttery, 1958). The galactose residues are present in furanose configuration (Plackett *et al.*, 1963). The galactan appears to be in the form of a capsule since treatment of "thread phase" cultures with antiserum gives a precipitate outlining the threads (Gourlay and Thrower, 1968). The galactan cross-reacts strongly with a galactan from normal bovine lung (Shifrine and Gourlay, 1965). Glucans have been recovered from bovine arthritis strains (Plackett *et al.*, 1963) also known as bovine group 7. A number of lipopolysaccharides have been recovered from various species in the Mycoplasmatales by first removing lipids from membrane preparations with organic solvents and then extracting the lipopolysaccharide with aqueous phenol (Smith *et al.*, 1976). Polysaccharides containing glycerol and fatty acids were recovered from *Thermoplasma acidophilum, A. laidlawii, A. modicum, A. granularum, M. neurolyticum*, and *Anaeroplasma bactoclasticum*. Interestingly, lipopolysaccharides were not recovered from *S. citri, M. capricolum, M. gallisepticum, Mycoplasma gallinarum, M. arthritidis*, and *M. hyorhinis*, supporting the fundamental heterogeneity of the Mycoplasmatales. Glucose and galactose were the predominant neutral sugars and fucosamine was also detected. The lipopolysaccharides of *A. modicum, A. laidlawii, A. axanthum, T. acidophilum*, and *M. neurolyticum* were tested by complement fixation and found to be antigenic (Sugiyama *et al.*, 1974). *Mycoplasma pneumoniae* organisms were extracted with phenol after extensive delipidation (G. E. Kenny, unpublished data). A minute amount of polysaccharide was recovered which contained glucose, galactose, and mannose. This fraction reacted with some human sera by gel diffusion and the presence or absence of antibody correlated with antilipid complement-fixing antibody. It reacted with both antiserum to whole organisms and with antiserum to the spinach triglycosyl fraction, indicating that it had similar determinants to some *M. pneumoniae* lipids. This result would confirm the findings of Sobeslavsky *et al.* (1966), who recovered a polysaccharide which was reactive with rabbit serum but not human serum by complement-fixation. A glycoprotein has also been reported from *M. penumoniae* membranes (Kahane and Marchesi, 1973). A polyhexosamine, which is apparently antigenic, has been associated with *A. laid-*

*lawii* membrane (Terry and Zupnik, 1973). Circumstantial evidence exists for a polysaccharide antigen in *M. pulmonis* in that a periodiate-labile, pronase-stable antigen has been reported (Deeb and Kenny, 1967b).

### B. Cytoplasmic Antigens

Studies on cytoplasmic antigens are little advanced for several reasons: (1) Cytoplasmic preparations are frequently extensively contaminated with membrane fragments, particularly if sonication is used for breaking the cells (Razin *et al.*, 1972; Alexander and Kenny, 1977), and (2) most attention has been focused on membrane antigens. A soluble fraction of *M. pneumoniae* was prepared by osmotic lysis and used as immunogen for rabbits (Pollack *et al.*, 1970). The resulting antisera did not have metabolic inhibiting antibody or antilipid CF antibody but did have antibody to the phenol-treated CF antigen of Chanock *et al.* (1962), suggesting that the determinants in that antigen might be cytoplasmic. The fact that only small amounts of whole organisms are required to induce antilipid antibody (Kenny, 1971a) suggests that their soluble fraction must have been quite free of membrane. The soluble fractions of *M. hominis* gave clear reactions of identity between strains, whereas the membrane fractions showed striking heterogeneity (Hollingdale and Lemcke, 1970). Four of five cytoplasmic antigens of three *M. arginini* strains were common, whereas greater heterogeneity was observed with membrane antigens (Alexander and Kenny, 1977). Thus, cytoplasmic antigens may be less variable and a case might be made for the use of cytoplasmic antigens for grouping *Mycoplasma* species. Hollingdale and Lemcke (1970) pointed out from their experience with *M. hominis* that serologic surveys by double immunodiffusion using sonicated or freeze–thawed antigens might well measure primarily soluble (cytoplasmic) antigens. Such an approach has been made to taxonomy by using specific types of lactate dehydrogenases which are cytoplasmic components (Neimark and Lemcke, 1972; Neimark, 1973; see also Chapter 12, this volume). Indeed, the four organisms that produced D(−)lactate (Neimark, 1973) all happened to fall in a serologic cluster; group 6, Table II. However, *M. neurolyticum,* which is also in this group, did not fit the pattern and produced a different lactate (Neimark and Lemcke, 1972). A serologic comparison of arginine deiminase enzymes showed that the enzymes of *M. arginini* and *M. hominis* were related, but the enzyme of *M. gallinarum* was serologically different (Hahn and Kenny, 1974). Although *M. gallinarum* was related to the other arginine-utilizing organisms, it appeared to be the most serologically different of the group (Thirkill and

Kenny, 1974). Accordingly, the ability to identify cytoplasmic antigenic components specifically might prove very helpful in grouping *Mycoplasma* species into families.

## ACKNOWLEDGMENTS

Some of the studies cited in this chapter were supported in part by grants AI-06720, AI-HD-12005, and AI-10695 from the National Institute of Allergy and Infectious Diseases, National Institutes of Health.

## REFERENCES

Alexander, A. G., and Kenny, G. E. (1977). *Infect. Immun.* **15,** 313–321.

Alexander, A. G., and Kenny, G. E. (1978). *Infect. Immun.* **20,** 861–863.

Amar, A., Rottem, S., Kahane, I., and Razin, S. (1976). *Biochim. Biophys. Acta* **426,** 258–270.

Argaman, M., and Razin, S. (1969). *J. Gen. Microbiol.* **55,** 45–46.

Askaa, G., Christiansen, C., and Ernø, H. (1973). *J. Gen. Microbiol.* **75,** 283–286.

Axelsen, N. H., ed. (1975). "Quantitative Immunoelectrophoresis: New Developments and Applications." Universitetsforlaget, Oslo, Norway.

Axelsen, N. H., and Bock, E. (1972). *J. Immunol. Methods* **1,** 109–121.

Axelsen, N. H., Kroll, J., and Weeke, B. (1973). "A Manual of Quantitative Immunoelectrophoresis: Methods and Applications." Universitetsforlaget, Oslo, Norway.

Barile, M. F., Del Giudice, R. A., and Tully, J. G. (1972). *Infect. Immun.* **5,** 70–76.

Barker, L. F., and Patt, J. K. (1967). *J. Bacteriol.* **94,** 403–408.

Beckman, B. L., and Kenny, G. E. (1968). *J. Bacteriol.* **96,** 1171–1180.

Bergquist, L. M., Lau, B. H. S., and Winter, C. E. (1974). *Infect. Immun.* **9,** 410–415.

Biberfeld, G. (1971). *Clin. Exp. Immunol.* **8,** 319–333.

Biberfeld, G., and Norberg, R. (1974). *J. Immunol.* **112,** 413–415.

Biberfeld, G., and Sterner, G. (1971). *Acta Pathol. Microbiol. Scand.* **79,** 599–605.

Biberfeld, G., and Sterner, G. (1976). *Clin. Exp. Immunol.* **24,** 287–291.

Boatman, E. S., and Kenny, G. E. (1970). *J. Bacteriol.* **101,** 262–277.

Bohlool, B. B., and Brock, T. D. (1974). *Infect. Immun.* **10,** 280–281.

Bradbury, J. M., and Jordan, F. T. W. (1971). *J. Hyg.* **69,** 593–606.

Bradbury, J. M., and Jordan, F. T. W. (1972). *J. Hyg.* **70,** 267–278.

Bredt, W., and Bitter-Suermann, D. (1975). *Infect. Immun.* **11,** 497–504.

Bredt, W., Wellek, B., Brunner, H., and Loos, M. (1977). *Infect. Immun.* **15,** 7–12.

Brown, R., and Robinson, L. K. (1938). *J. Immunol.* **34,** 61–62.

Bruggmann, S., Keller, H., Bertschinger, H. U., and Engberg, B. (1977). *Vet. Rec.* **101,** 109–111.

Brunner, H., and Chanock, R. M. (1973). *Proc. Soc. Exptl. Biol. Med.* **143,** 97–105.

Brunner, H., Razin, S., Kalica, A. R., and Chanock, R. M. (1971). *J. Immunol.* **106,** 907–916.

Brunner, H., James, W. D., Horswood, R. L., and Chanock, R. M. (1972). *J. Immunol.* **108,** 1491–1498.

Brunner, H., Greenberg, H. B., James, W. D., Horswood, R. L., Couch, R. B., and Chanock, R. M. (1973). *Infect. Immun.* **8,** 612–620.

Brunner, H., Dörner, I., Schiefer, H.-G., Krauss, H., and Wellensiek, H.-J. (1976). *Infect. Immun.* **13,** 1671–1677.

Brunner, H., Schaeg, W., Brück, U., Schummer, U., and Schiefer, H.-G. (1977). *Med. Microbiol. Immunol.* **163,** 25–35.

Buttery, S. H. (1972). *Aust. J. Exp. Biol. Med. Sci.* **50,** 567–576.

Caldwell, H. D., and Kuo, C.-C. (1977). *J. Immunol.* **118,** 437–441.

Caldwell, H. D., Kuo, C.-C., and Kenny, G. E. (1975). *J. Immunol.* **115,** 969–975.

Chanock, R. M., James, W. D., Fox, H. H., Turner, H. C., Mufson, M. A., and Hayflick, L. (1962). *Proc. Soc. Exp. Biol. Med.* **110,** 884–889.

Cho, H. J., and Langford, E. V. (1974). *Appl. Microbiol.* **28,** 897–899.

Clyde, W. A. (1964). *J. Immunol.* **92,** 958–965.

Cole, B. C., and Ward, J. R. (1973). *Infect. Immun.* **8,** 199–207.

Cole, B. C., Golightly, L., and Ward, J. R. (1967). *J. Bacteriol.* **94,** 1451–1458.

Coleman, L. H., Lynn, R. J., and Patrick, R. A. (1974). *Antonie van Leeuwenhoek* **40,** 401–407.

Crowle, A. J. (1973). "Immunodiffusion," 2nd ed. Academic Press, New York.

Dahl, J. S., Hellewell, S. B., and Levine, R. P. (1977). *J. Immunol.* **119,** 1419–1426.

Davies, B. H., Edwards, J. H., and Seaton, A. (1975). *Clin. Allergy* **5,** 217–224.

Deeb, B. J., and Kenny, G. E. (1967a). *J. Bacteriol.* **93,** 1416–1424.

Deeb, B. J., and Kenny, G. E. (1967b). *J. Bacteriol.* **93,** 1425–1429.

Del Giudice, R. A., Carski, T. R., Barile, M. F., Lemcke, R. M., and Tully, J. G. (1971). *J. Bacteriol.* **108,** 439–445.

Del Giudice, R. A., Purcell, R. H., Carski, T. R., and Chanock, R. M. (1974). *Int. J. Syst. Bacteriol.* **24,** 147–153.

Dörner, I., Brunner, H., Shiefer, H.-G., and Wellensiek, H.-J. (1976). *Infect. Immun.* **13,** 1663–1670.

Edward, D. G. ff., and Fitzgerald, W. A. (1954). *J. Pathol. Bacteriol.* **68,** 23–30.

Edward, D. G. ff., and Freundt, E. A. (1969). *J. Gen. Microbiol.* **57,** 391–395.

Edward, D. G. ff., and Freundt, E. A. (1970). *J. Gen. Microbiol.* **62,** 1–2.

Engvall, E., and Perlmann, P. (1971). *Immunochemistry* **8,** 871–874.

Fallon, R. J., and Jackson, D. K. (1967). *Lab. Anim.* **1,** 55–64.

Fenske, J. D., and Kenny, G. E. (1976). *J. Bacteriol.* **126,** 501–510.

Fernald, G. W., Clyde, W. A., and Denny, F. W. (1967). *Proc. Soc. Exp. Biol. Med.* **126,** 161–166.

Forshaw, K. A., and Fallon, R. J. (1972). *J. Gen. Microbiol.* **72,** 501–510.

Fowler, R. C., Coble, D. W., Kramer, N. C., Pai, R. R., Serrano, B. A., and Brown, T. M. (1967). *Ann. N.Y. Acad. Sci.* **143,** 641–653.

Freundt, E. A. (1974). *In* "Bergey's Manual of Determinative Bacteriology" (R. E. Buchanan and N. E. Gibbons, eds.), 8th ed., pp. 929–955. Williams & Wilkins, Baltimore, Maryland.

Gale, J. L., and Kenny, G. E. (1970). *J. Immunol.* **104,** 1175–1183.

Gill, P., and Pan, J. (1970). *Can J. Microbiol.* **16,** 415–419.

Goel, M. C. (1973). *J. Bacteriol.* **116,** 994–1000.

Goldschmidt, B. L., Menonna, J. P., Dowling, P. C., and Cook, S. D. (1976). *J. Immunol.* **117,** 1054–1055.

Gourlay, R. N., and Thrower, K. J. (1968). *J. Gen. Microbiol.* **54,** 155–159.

Hahn, R. G., and Kenny, G. E. (1974). *J. Bacteriol.* **117,** 611–618.

Haller, G. J., Boiarski, K. W., and Somerson, N. L. (1973). *J. Infect. Dis.* **127,** 538–542.

Helenius, A., and Simons, K. (1977). *Proc. Natl. Acad. Sci. U.S.A.* **74,** 529–532.

Hollingdale, M. R., and Lemcke, R. M. (1969). *J. Hyg.* **67,** 585–602.

Hollingdale, M. R., and Lemcke, R. M. (1970). *J. Hyg.* **68,** 469–477.
Hollingdale, M. R., and Lemcke, R. M. (1972). *J. Hyg.* **70,** 85–98.
Huijmans-Evers, A. G. M., and Ruys, A. C. (1956). *Antonie van Leewenhoek* **22,** 377–384.
Johansson, K.-E., and Hjerten, S. (1974). *J. Mol. Biol.* **86,** 341–348.
Kahane, I., and Marchesi, V. T. (1973). *Ann. N.Y. Acad. Sci.* **225,** 38–45.
Kahane, I., and Razin, S. (1969). *J. Bacteriol.* **100,** 187–194.
Kaklamanis, E., and Pavlotos, M. (1972). *Immunology* **22,** 695–702.
Kataoka, T., and Nojima, S. (1970). *J. Immunol.* **105,** 502–511.
Kenny, G. E. (1967). *Ann. N.Y. Acad. Sci.* **143,** 676–681.
Kenny, G. E. (1969). *J. Bacteriol.* **98,** 1044–1055.
Kenny, G. E. (1971a). *Infect. Immun.* **3,** 510–515.
Kenny, G. E. (1971b). *Infect. Immun.* **4,** 149–153.
Kenny, G. E. (1975). *In* "The Antigens" (M. Sela, ed.), Vol. 3, p. 449. Academic Press, New York.
Kenny, G. E. (1977). *In* "Non-gonococcal Urethritis and Related Infections" (D. Hobson and K. K. Holmes, eds.), p. 376. Am. Soc. Microbiol., Washington, D.C.
Kenny, G. E., and Cartwright, F. D. (1977). *J. Bacteriol.* **132,** 144–150.
Kenny, G. E., and Grayston, J. T. (1965). *J. Immunol.* **95,** 19–25.
Kenny, G. E., and Newton, R. M. (1973). *Ann. N.Y. Acad. Sci.* **225,** 54–61.
Lemcke, R. M. (1964). *J. Hyg.* **62,** 199–219.
Lemcke, R. M. (1965). *J. Gen. Microbiol.* **38,** 91–100.
Lemcke, R. M. (1973). *Ann. N.Y. Acad. Sci.* **225,** 46–53.
Lemcke, R. M., Marmion, B. P., and Plackett, P. (1967). *Ann. N.Y. Acad. Sci.* **143,** 691–702.
Lemcke, R. M., Forshaw, K. A., and Fallon, R. J. (1969). *J. Gen Microbiol.* **58,** 95–98.
Levine, L., Wasserman, E., and Mills, S. (1961). *J. Immunol.* **86,** 675–680.
Levisohn, S., and Razin, S. (1973). *J. Hyg.* **71,** 725–737.
Lin, J.-S., Kass, E. H. (1970). *J. Infect. Dis.* **122,** 93–95.
Lin, J.-S., and Kass, E. H. (1973). *Infect Immun.* **7,** 499–500.
Lin, J.-S., and Kass, E. H. (1974). *Infect. Immun.* **10,** 535–540.
Lin, J.-S., and Kass, E. H. (1975). *J. Med. Microbiol.* **8,** 397–404.
Lin, J.-S., Kendrick, M. I., and Kass, E. H. (1972). *J. Infect. Dis.* **126,** 658–663.
Lin, J.-S. L., Alpert, S., and Radnay, K. M. (1975). *J. Infect. Dis.* **131,** 727–730.
Liu, C. (1957). *J. Exp. Med.* **106,** 455–466.
McIvor, R. S., and Kenny, G. E. (1977). *Abstr. Annu. Meet., Am. Soc. Microbiol.* p. 133.
McIvor, R. S., and Kenny, G. E. (1978). *J. Bacteriol.* **135,** 483–489.
Masover, G. K., Mischak, R. P., and Hayflick, L. (1975). *Infect. Immun.* **11,** 530–539.
Masover, G. K., Sawyer, J. E., and Hayflick, L. (1976). *J. Bacteriol.* **125,** 581–587.
Matsumoto, M., and Yamamoto, R. (1973). *J. Infect. Dis.* **127,** 543–551.
Menonna, J., Chmel, H., Menegus, M., Dowling, P., and Cook, S. (1977). *J. Clin. Microbiol.* **5,** 610–612.
Ne'eman, Z., Kahane, I., Kovartovsky, J., and Razin, S. (1972). *Biochim. Biophys. Acta* **266,** 255–268.
Neimark, H. (1973). *Ann. N.Y. Acad. Sci.* **225,** 14–21.
Neimark, H., and Lemcke, R. M. (1972). *J. Bacteriol.* **111,** 633–640.
Neimark, H. C. (1970). *J. Gen. Microbiol.* **63,** 249–263.
Niitu, Y., Hasegawa, S., and Kubota, H. (1974). *Antimicrob. Agents & Chemother.* **5,** 111–113.
Patscheke, H., Breinl, M., and Schafer, E. (1977). *J. Immunol. Methods* **16,** 31–38.
Penttinen, K., and Myllala, G. (1968). *Ann. Med. Exp. Biol. Fenn.* **46,** 188–192.
Plackett, P., and Buttery, S. H. (1958). *Nature (London)* **182,** 1236–1237.

Plackett, P., and Shaw, E. J. (1967). *Biochem. J.* **104,** 61C–62C.

Plackett, P., Buttery, S. H., and Cottew, G. S. (1963). *Recent Microbiol. Symp. Int. Congr. Microbiol., 8th, 1962* p. 535.

Plackett, P., Marmion, B. P., Shaw, E. J., and Lemcke, R. M. (1969). *Aust. J. Exp. Biol. Med. Sci.* **47,** 171–195.

Pollack, J. D., Somerson, N. L., and Senterfit, L. B. (1969). *J. Bacteriol.* **97,** 612–619.

Pollack, J. D., Somerson, N. L., and Senterfit, L. B. (1970). *Infect. Immun.* **2,** 326–339.

Pollock, M. E., and Bonner, S. V. (1969a). *J. Bacteriol.* **97,** 522–525.

Pollock, M. E., and Bonner, S. V. (1969b). *Bacteriol. Proc.* p. 32.

Purcell, R. H., Taylor-Robinson, D., Wong, D., and Chanock, R. M. (1966a). *J. Bacteriol.* **92,** 6–12.

Purcell, R. H., Taylor-Robinson, D., Wong, D. C., and Chanock, R. M. (1966b). *Am. J. Epidemiol.* **84,** 51–66.

Purcell, R. H., Wong, D., Chanock, R. M., Taylor-Robinson, D., Canchola, J., and Valdesuso, J. (1967). *Ann. N.Y. Acad. Sci.* **143,** 664–675.

Purcell, R. H., Chanock, R. M., and Taylor-Robinson, D. (1969). *In* "The Mycoplasmatales and L-phase of Bacteria" (L. Hayflick, ed.), p. 221.

Razin, S. (1975). *Prog. Surf. Membr. Sci.* **9,** 257.

Razin, S., Prescott, B., Caldes, G., James, W. D., and Chanock, R. M. (1970a). *Infect. Immun.* **1,** 408–416.

Razin, S., Prescott, B., and Chanock, R. M. (1970b). *Proc. Natl. Acad. Sci. U.S.A.* **67,** 590–597.

Razin, S., Prescott, B., James, W. D., Caldes, G., Valdesuso, J., and Chanock, R. M. (1971). *Infect. Immun.* **3,** 420–423.

Razin, S., Kahane, I., and Kovartovsky, J. (1972). *Pathog. Mycoplasmas, Ciba Found. Symp., 1972* pp. 93–122.

Razin, S., Masover, G. K., and Hayflick, L. (1977). *In* "Non-gonococcal Urethritis and Related Infections" (D. Hobson and K. K. Holmes, eds.), p. 358. Am. Soc. Microbiol., Washington, D.C.

Robinson, I. M., and Rhoades, K. R. (1977). *Int. J. Syst. Bacteriol.* **27,** 200–203.

Rodwell, A. (1969). *In* "The Mycoplasmatales and L-phase of Bacteria" (L. Hayflick, ed.), p. 413. Appleton, New York.

Romano, N., and Scarlata, G. (1974). *Infect. Immun.* **9,** 1062–1065.

Romano, N., Smith, P. F., and Mayberry, W. R. (1972). *J. Bacteriol.* **109,** 565–569.

Rottem, S., and Greenberg, A. S. (1975). *J. Bacteriol.* **121,** 631–639.

Ryan, M. D., Noker, P., and Matz, L. L. (1975). *Infect. Immun.* **12.** 799–807.

Schiefer, H.-G., Gerhardt, U., and Brunner, H. (1975). *Hoppe-Seyler's Z. Physiol. Chem.* **356,** 559–565.

Senterfit, L. B., and Jensen, K. E. (1966). *Proc. Soc. Exp. Biol. Med.* **122,** 786–790.

Sever, J. L., and Madden, D. L., eds. (1977). "Enzyme-linked Immunoabsorbent Assay (ELISA) for Infectious Agents," J. Infect. Dis., Vol. 136, Suppl., p. S258. Univ. of Chicago Press, Chicago, Illinois.

Sharpless, N. S., and LoGrippo, G. A. (1965). *Henry Ford Hosp. Med. Bull.* **13,** 55–77.

Shifrine, M., and Gourlay, R. N. (1965). *Nature (London)* **208,** 498–499.

Smith, P. F. (1972). *J. Bacteriol.* **112,** 554–558.

Smith, P. F., Langworthy, T. A., and Mayberry, W. R. (1976). *J. Bacteriol.* **125,** 916–922.

Smith, S. C., Dunlop, W. R., and Strout, R. G. (1966). *Avian Dis.* **10,** 173–176.

Smith, T. F., and Herrmann, E. C. (1971). *Appl. Microbiol.* **21,** 160–161.

Sobeslavsky, O., Prescott, B., James, W. D., and Chanock, R. M. (1966). *J. Bacteriol.* **91,** 2126–2138.

Stone, S. S., and Razin, S. (1973). *Infect. Immun.* **7,** 922–930.

Sugiyama, T., Smith, P. F., Langworthy, T. A., and Mayberry, W. R. (1974). *Infect. Immun.* **10,** 1273–1279.

Svendsen, P. J. (1973). *Scand. J. Immunol.* **2,** Suppl. 1, 69–70.

Tachibana, D. K., Hayflick, L., and Rosenberg, L. T. (1970). *J. Infect. Dis.* **121,** 541–544.

Taylor-Robinson, D., Somerson, N. L., Turner, H. C., and Chanock, R. M. (1963). *J. Bacteriol.* **85,** 1261–1273.

Taylor-Robinson, D., Sobeslavsky, O., and Chanock, R. M. (1965). *J. Bacteriol.* **90.,** 1432–1437.

Taylor-Robinson, D., Purcell, R. H., Wong, D. C., and Chanock, R. M. (1966). *J. Hyg.* **64,** 91–104.

Terry, T. M., and Zupnik, J. S. (1973). *Biochim. Biophys. Acta* **291,** 144–148.

Thirkill, C. E., and Kenny, G. E. (1974). *Infect. Immun.* **10,** 624–632.

Thirkill, C. E., and Kenny, G. E. (1975). *J. Immunol.* **114,** 1107–1111.

Tully, J. G., and Ruchman, I. (1964). *Proc. Soc. Exp. Biol. Med.* **115,** 554–558.

Verbruggen, R. (1975). *Clin. Chem.* **21,** 5–43.

Vestergard, B. F. (1975). *In* "Quantitative Immunoelectrophoresis: New Developments and Applications" (N. H. Axelsen, ed.), p. 203. Universitetsforlaget, Oslo, Norway.

Voller, A., Bidwell, D., and Bartlett, A. (1976). *In* "Manual of Clinical Immunology" (N. R. Rose and H. Friedman, eds.), p. 506. Am. Soc. Microbiol., Washington, D.C.

Watkins, A. L. (1977). *Abstr. Annu. Meet. Am. Soc. Microbiol.* p. 134.

Williams, M. H., and Taylor-Robinson, D. (1967). *Nature (London)* **215,** 973–974.

Wroblewski, H. (1975). *Biochimie* **57,** 1095–1098.

Wroblewski, H., and Ratanasavanh, D. (1976). *Can. J. Microbiol.* **22,** 1048–1053.

Wroblewski, H., Johansson, K.-E., and Burlot, R. (1977a). *Int. J. Syst. Bacteriol.* **27,** 97–103.

Wroblewski, H., Johansson, K.-E., and Hjerten, S. (1977b). *Biochim. Biophys. Acta* **465,** 275–289.

Yaguzhinskaya, O. E. (1976). *J. Hyg.* **77,** 189–198.

# 14 / MYCOPLASMA AND SPIROPLASMA VIRUSES: ULTRASTRUCTURE

*Roger M. Cole*



## I. INTRODUCTION

The first isolation of a virus from a mycoplasma was made only a few years ago (Gourlay, 1970), although some electron microscopic evidence of viruslike particles in mycoplasmas was reported earlier (Edwards and Fogh, 1960; Swartzendruber *et al.*, 1967). The original isolation was rapidly confirmed and extended (Liss and Maniloff, 1971), and subsequently several viruses of different morphologies were also isolated (Gourlay, 1971; Liska, 1972; Gourlay and Wyld, 1973). All of these, with some few exceptions (Liss and Maniloff, 1971), were found to occur in and to infect indicator strains of *Acholeplasma laidlawii*. They comprise three classes of viruses which are conveniently referred to as groups L1, L2, and L3 (Liss and Maniloff, 1973a; Maniloff and Liss, 1974); morphologically, the viruses in each group are, respectively, rods, spheres, and polyhedrons with short tails.

Viruslike particles were also detected, by electron microscopy only, in *Spiroplasma citri* (Cole *et al.*, 1973a,b, 1974) and have been found since in other spiroplasmas (Cole *et al.*, 1976; Cole, 1977). Some were previously recognized in the sex-ratio organism of *Drosophila*, at a time when the microorganism was described as a spirochete (Oishi and Poulson, 1970; Oishi, 1971) instead of a spiroplasma (Williamson and Whitcomb, 1974;



THE MYCOPLASMAS, VOL. 1

ISBN 0-12-078401-7

Williamson *et al.*, 1977). The particles which, in analogy with the acholeplasma viruses, can be placed in groups C1, C2, and C3, are morphologically rods, polyhedrons with long tails, and polyhedrons with short tails, respectively. Only representatives of the third group have been propagated and characterized (Cole *et al.*, 1977). The relationships, if any, of these spiroplasma viruses to the acholeplasma viruses have not been studied—although it is clear that groups L1 and C1 and groups L3 and C3 share morphologic features.

Other viruslike particles detected only by electron microscopy in, or in association with, mycoplasmas or mycoplasma-like organisms were rodlike (Ploaie, 1971; Allen, 1971; Gourret *et al.*, 1973; Cadilhac and Giannotti, 1975) or apparently spherical (Edwards and Fogh, 1960; Swartzendruber *et al.*, 1967; Robertson *et al.*, 1972).

For details of the occurrence, characteristics, and molecular biology of mycoplasma viruses, the reader is referred to Chapter 15 and to several published reviews (Gourlay, 1972, 1973; Maniloff, 1972; Maniloff and Liss, 1973, 1974; Gourlay, 1974; Cole, 1977, 1978; Maniloff *et al.*, 1977b; Williamson *et al.*, 1977). This chapter considers the morphology and ultrastructure of the viruses and stages of their intracellular development as seen by transmission electron microscopy of negatively strained preparations and ultrathin sections.

## II. ULTRASTRUCTURAL CHARACTERIZATIONS BY SHAPE, DIMENSIONS, CAPSID SYMMETRY, INTRACELLULAR AND INCOMPLETE FORMS, AND MODE OF VIRUS RELEASE

### A. Rod-Shaped Viruses

#### 1. Group L1

The prototype of this group is Mycoplasmatales virus *laidlawii* 1 (MVL1), which was the first mycoplasma virus isolated (Gourlay, 1970). The naked rod- or bullet-shaped particles are assembled at and protrude from the surface of the infected cell (Figs. 1–3). Apparent errors of assembly frequently lead to the simultaneous production of long forms with a variety of diameters (Figs. 3 and 4) (Bruce *et al.*, 1972; Milne *et al.*, (1972). In purified preparations (Fig. 5), the normal infectious particles measure 13–16 by 80–90 nm, although some of double length (not shown) are occasionally seen. The virion was shown, by optical diffraction methods, to possess a helical symmetry (Fig. 6) consisting of subunits hexagonally arranged to form the helix (Bruce *et al.*, 1972). The distal end

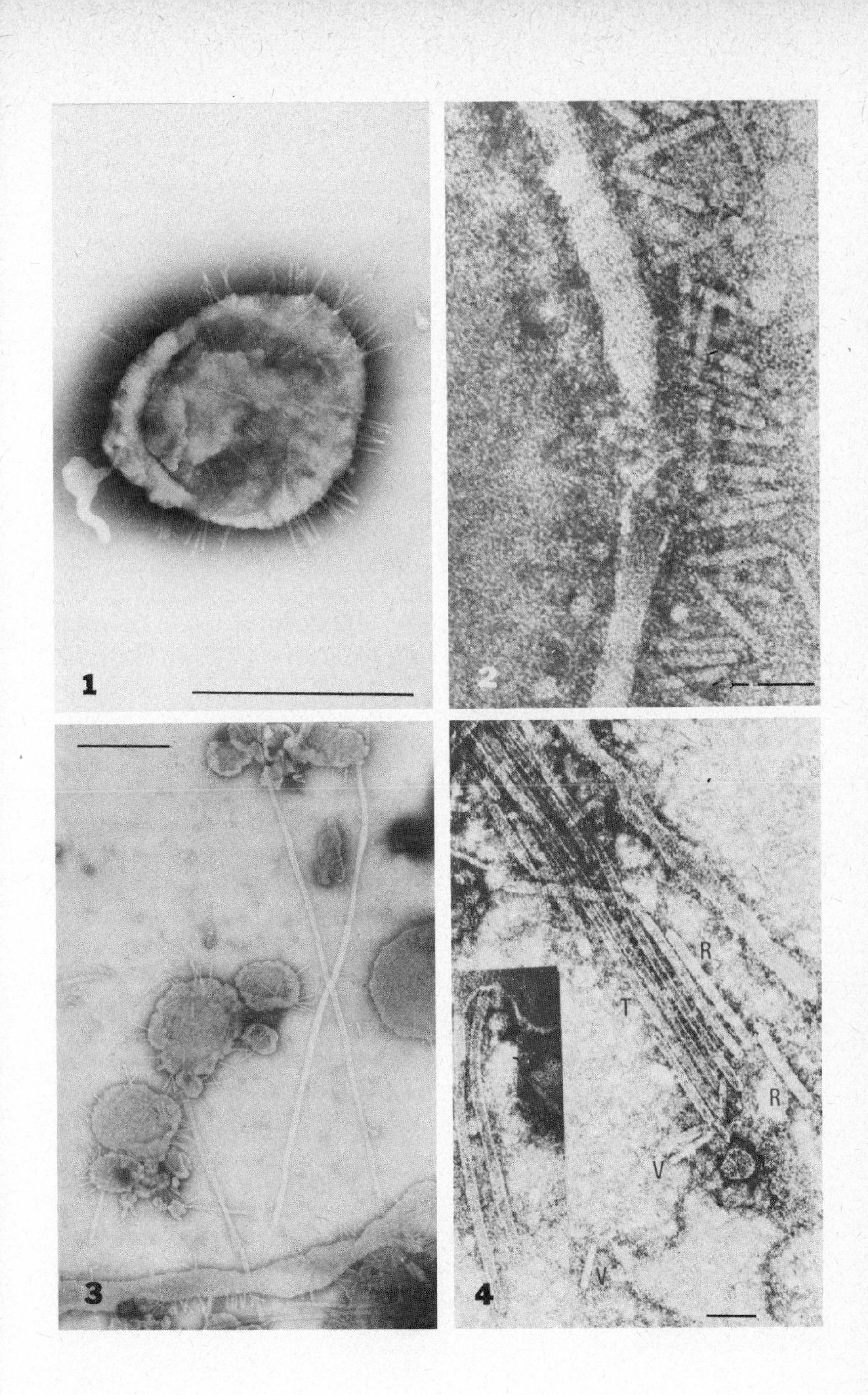
1
2
3
4
R
T
R
V
V

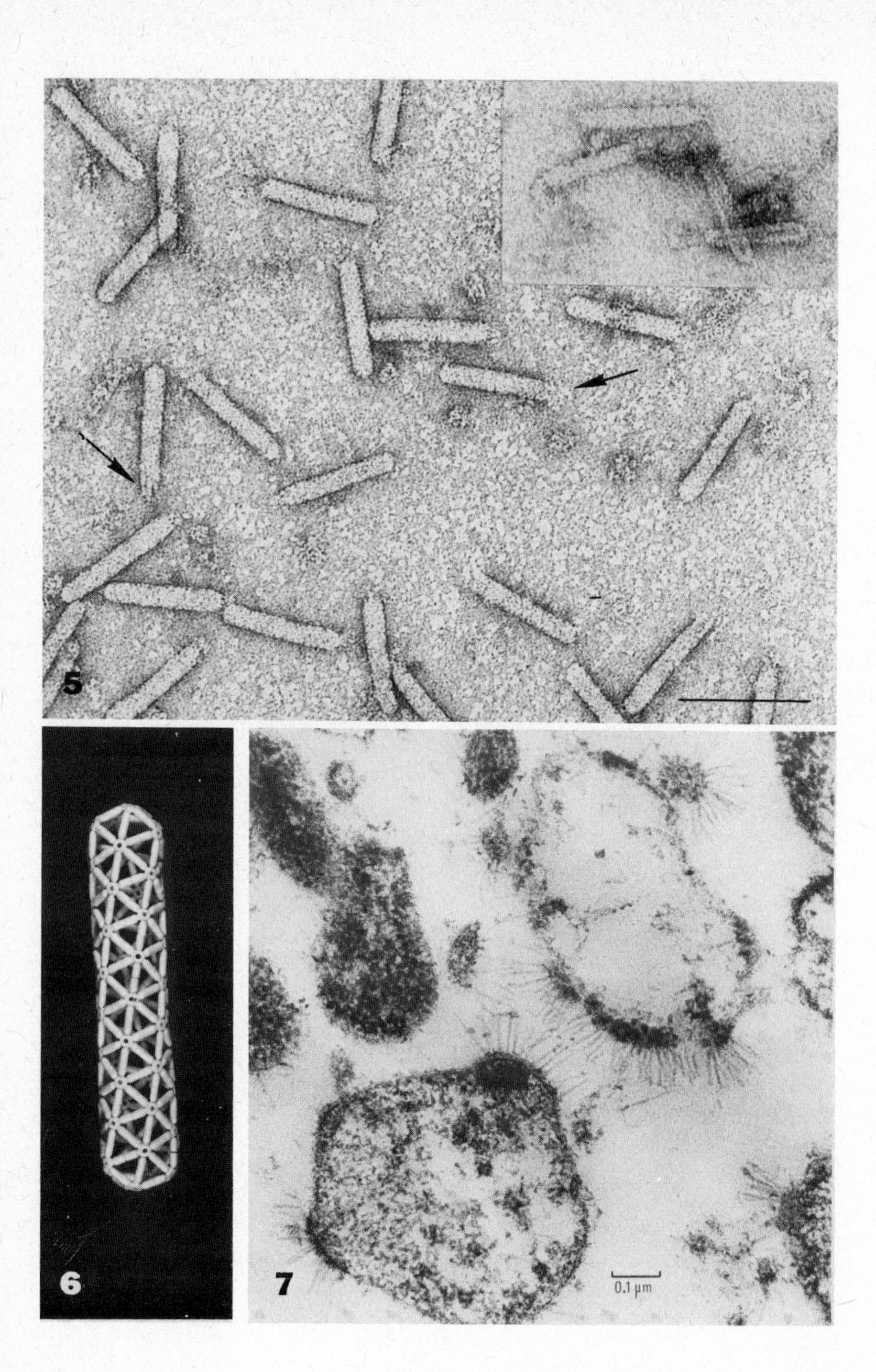
5
6
7
0.1 μm

is rounded and the other appears irregular to degraded (Fig. 5). Some particles appear empty (Fig. 5, insert); the thickness of the capsid "wall" was reported to be 5.2–6.2 nm. (Bruce *et al.*, 1972).

Another isolate, designated as MVL51 (Liss and Maniloff, 1972), was reported to consist of slightly smaller particles measuring 11.7 ± 1.6 by 71 ± 4 nm and contained ring structures of the same diameter (Liss and Maniloff, 1973b). Rods of double length were seen, as well as the long, slightly sinuous filaments (Fig. 3) also seen in MVL1. The filaments may be as long as 1.5 μm. In MVL1, although often hollow and of the same diameter as the bullet-shaped viroids, some filaments were of nearly twice the diameter and had "wall" thicknesses of 7.5 nm (Bruce *et al.*, 1972); others, though not hollow, may be of various lengths and diameters (Milne *et al.*, 1972). In reexamining MVL51 (suppled by Dr. Jack Maniloff), I found the "bullet" to be 11.2–12 by 66–80 nm; there were also solid rods of approximately 20 nm diameter by 160–240 nm in length and hollow tubes of 14.4–20 nm diameter by 435–1067 nm in length, with "walls" 3.6–4.7 nm in thickness. While the reasons for the consistent appearance of long forms and the nature of their relation to the normal particles are not known, the observations suggest that all isolates of group L1 "normal" virions are not identical in dimensions; serologic differences have also been suggested by some studies (Maniloff and Liss, 1974).

The first sections of cells infected with MVL1, in 72 hr cultures, showed the rod-shaped virions radiating from the surface of degenerating cells over densely staining areas (Fig. 7), but no intracellular particles or recognizable precursors have been seen (Milne *et al.*, 1972; Liss and

---

FIGURE 1. MVL1 protruding from the surface of infected *A. laidlawii* JA-1, negatively stained with ammonium molybdate. × 68,000; bar = 0.5 μm.

FIGURE 2. MVL1 at surface of infected cell, negatively stained with sodium phosphotungstate. × 133,000; bar = 100 nm. (From Milne *et al.*, 1972, with permission of Springer-Verlag.)

FIGURE 3. Emerging short rods and long forms in culture of *A. laidlawii* JA-1 infected with MVL 51. Negatively stained with ammonium molybdate. × 27,000; bar = 0.5 μm.

FIGURE 4. Long forms or tubular structures (T) of MVL1, associated with virions of normal size (V) and solid rods (R) of various lengths. Negatively stained with sodium phosphotungstate. × 70,000; bar = 100 nm. (From Milne *et al.*, 1972, with permission of Springer-Verlag.)

FIGURE 5. Purified virions of MVL1, negatively stained with uranyl acetate. One end is rounded and the other incomplete or degraded (arrows). × 200,000; bar = 100 nm. (From Bruce *et al.*, 1972, with permission of Cambridge University Press.) Insert: MVL51, negatively stained with uranyl acetate, showing two hollow forms. × 241,500; bar = 100 nm.

FIGURE 6. Model of helical structure of MVL1. (From Bruce *et al.*, 1972, with permission of Cambridge University Press.)

FIGURE 7. Section of culture of *A. laidlawii* infected for 72 hr with MVL1, showing virions radiating from surface over densely stained areas of degenerating cells. × 69,000; bar = 100 nm. (From Milne *et al.*, 1972, with permission of Springer-Verlag.)

Maniloff, 1973b). The concentrated dense areas are not seen in infected cultures examined at earlier times, but the virions clearly originate at the membrane (Figs. 8 and 9). Other evidence indicates that virions assemble at this level and the infection is nonlytic (Liss and Maniloff, 1973b; Maniloff and Liss, 1974; Maniloff *et al.*, 1977a). The nucleic acid that is enclosed, which results from a series of intracellular replicative intermediate forms (Das and Maniloff, 1975), is single-stranded, covalently closed circular DNA of molecular weight $2 \times 10^6$ daltons (Liss and Maniloff, 1973a). The virions are constituted of four proteins (Maniloff and Das, 1975).

## 2. Group C1

Rod-shaped particles resembling L1 viruses, but longer, were first seen in *Spiroplasma citri* cultures as clusters on the cell surfaces (Figs. 10 and

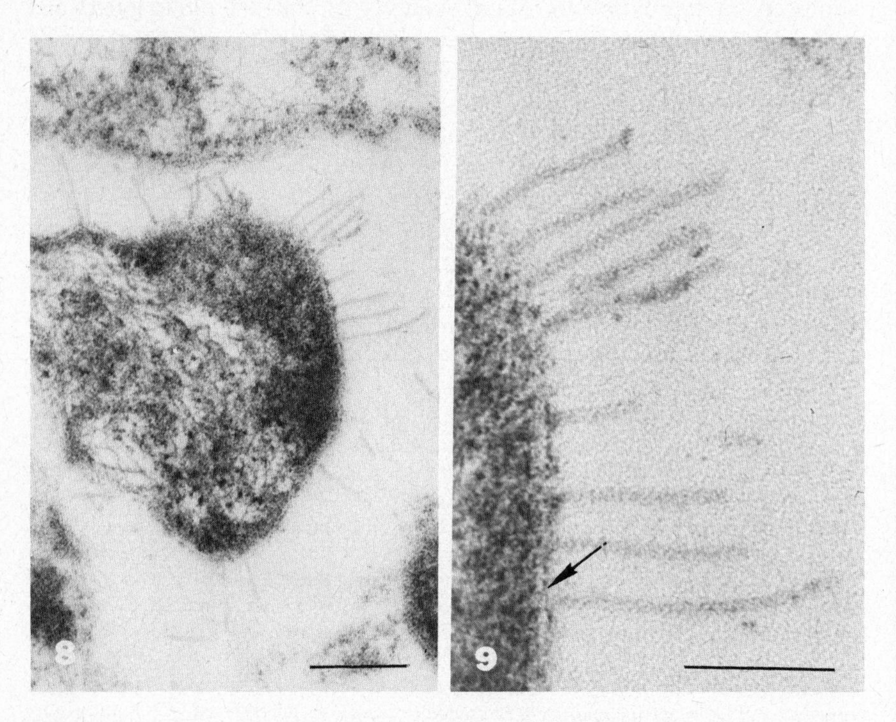


FIGURE 8. Section of culture of *A. laidlawii* JA-1 infected for 48 hr with MVL51. Concentrated areas of unusual density are not present beneath emerging virions. × 131,000; bar = 100 nm.

FIGURE 9. Enlargement of portion of Fig. 8. Virions arise at and may be extended through (arrow) the membrane. × 393,600; bar = 50 nm.

11) (Cole *et al.*, 1974; Cole, 1977). The individual particles (Fig. 12) measure 10–15 by 230–280 nm. As in L1, the distal ends are rounded; the others have a flat plate or fragment that may be a remnant of prior attachment to host cell membrane (Fig. 11). Empty particles (Fig. 11) and filaments of double (Fig. 10) or greater length can also be seen. The particles appear composed of subunits, but studies to confirm a possible helical arrangement have not been made. Sections (Fig. 13) show origin from the membrane and do not reveal intracellular particles or precursors, but details of the rods are not clear. These particles have not been propagated and the nature of their nucleic acid and other characteristics are not known. I have found particles of this morphologic type in suckling mouse cataract agent (Tully *et al.*, 1976), in spiroplasma 277F (Brinton and Burgdorfer, 1976), in the corn stunt agent (Chen and Liao, 1975; Williamson and Whitcomb, 1975), and in the sex-ratio organism (Williamson *et al.*, 1977).

### 3. Other Rodlike Particles

In several yellowing diseases of plants, sections of infected tissues have consistently revealed mycoplasma-like bodies which have not yet been cultured for definitive identification or proof of etiology. In association with some of these bodies, rod-shaped particles resembling viruses—often attached to or protruding from the membranous surface of the body—have been occasionally seen. Those occurring in sections of periwinkle (*Vinca rosea* L.) infected with clover dwarf agent measured 31–33 by 85–88 nm (Fig. 14) (Ploaie, 1971). In sections of clover afflicted with clover phyllody, the particles were 27 ± 3 nm in diameter and 50–90 nm in length (Figs. 15–17); they were found associated with the mycoplasma-like bodies in both the diseased plant and the insect vector (Gourret *et al.*, 1973). In the phloem of *Vinca rosea* infected with stolbur, rods of 30–33 nm diameter were seen (Figs. 18 and 19); some were straight and approximately 200–250 nm long and were sometimes present in bundles (Fig. 20), whereas others were helical and as long as 500 nm (Figs. 18 and 19) (Cadilhac and Giannotti, 1975). Rods (not shown) similarly surrounding or emerging from mycoplasma-like bodies or arranged in bundles were also described in infections with aster yellows agent (Allen, 1971); these measured 70 nm in length and 24 nm in diameter. All particles, in cross-sections, showed a dense inner, ringlike core surrounding a central canal (Figs. 16 and 18, insert), but none has been examined by negative staining and no other details of their ultrastructure have been reported.

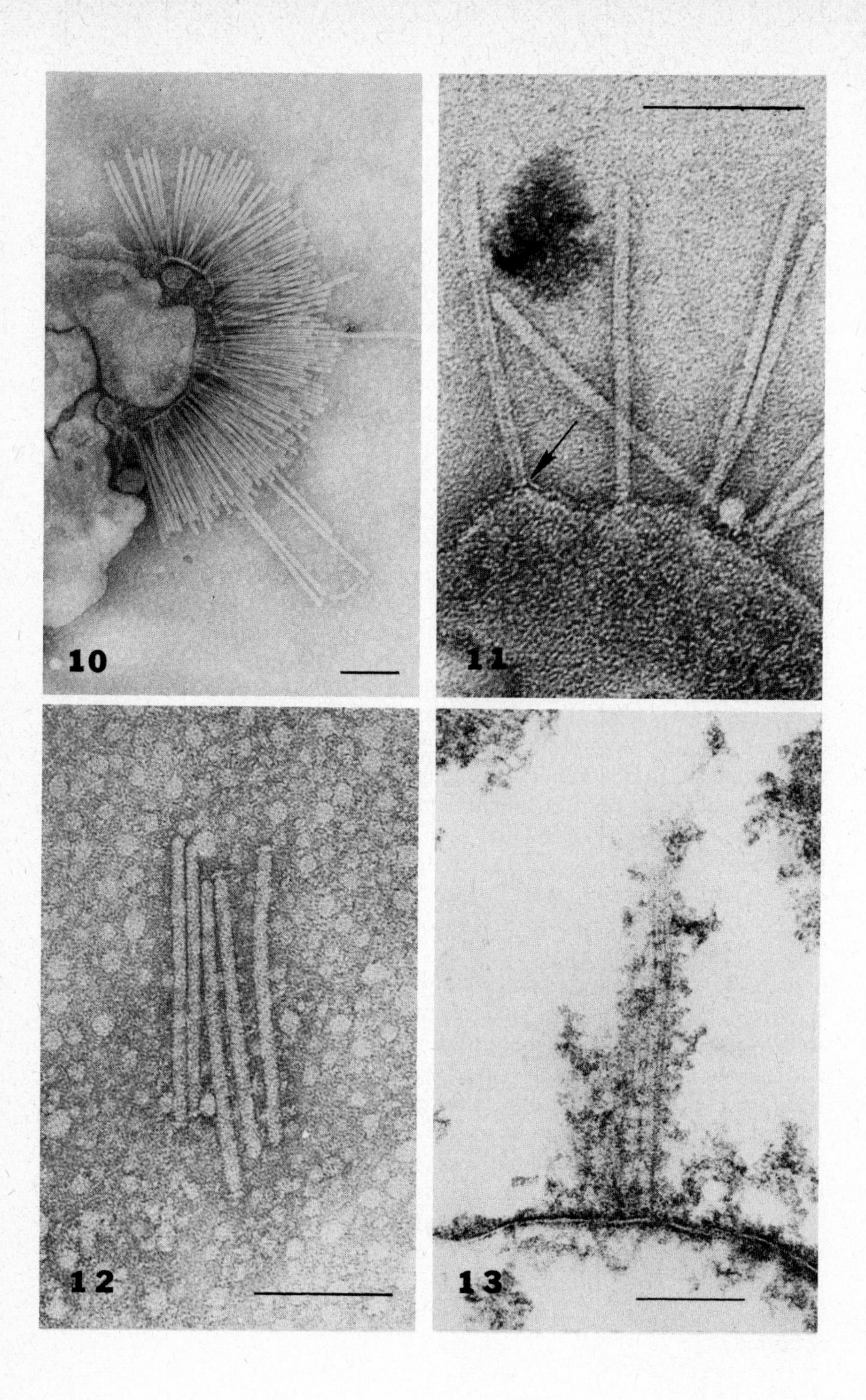
10
11
12
13

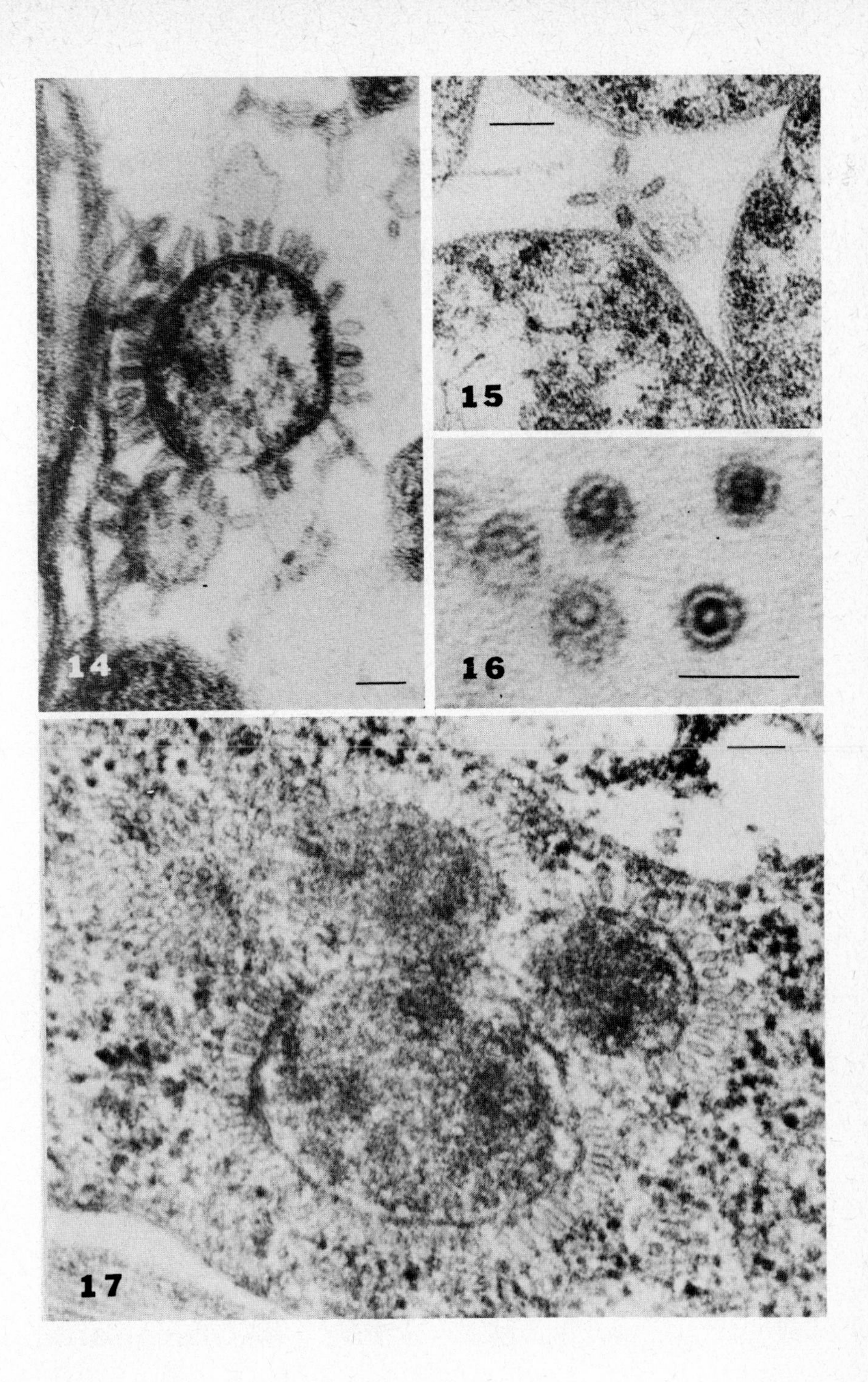

14
15
16
17

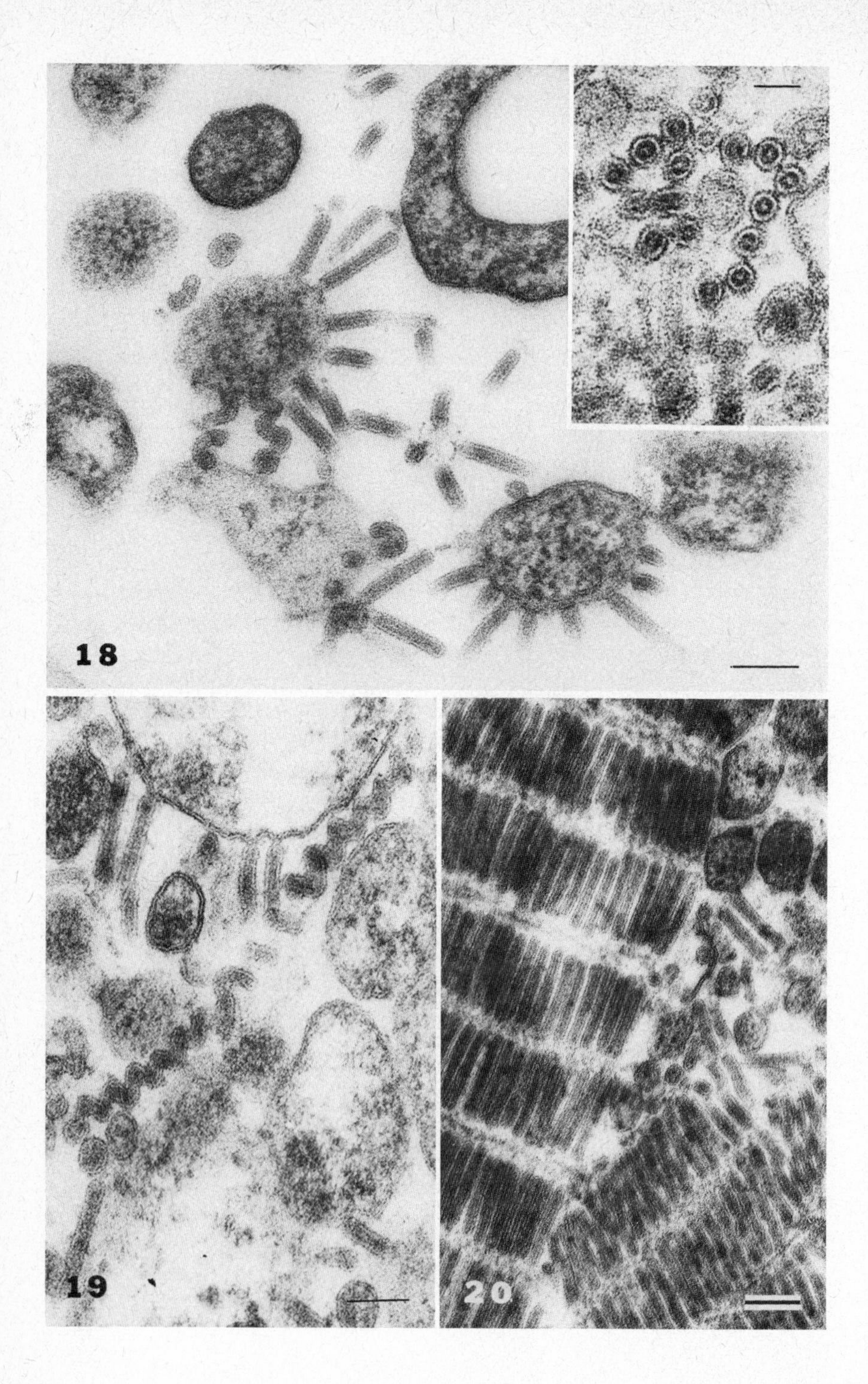
18
19
20

## B. Spherical Viruses

### 1. Enveloped Particles (Group L2)

In 1971, a second agent from *A. laidlawii* that produced plaques on lawns of other strains and was uninhibited by antiserum against MVL1 was isolated and named Mycoplasmatales virus *laidlawii* 2, or MVL2 (Gourlay, 1971). It is sensitive to detergents, ether, and chloroform, in contrast to MVL1 and all other mycoplasma viruses tested to date. The roughly spherical particles are seen surrounding, attached to, or apparently emerging from cells of the infected culture (Figs. 21 and 22) (Gourlay, 1971; Gourlay *et al.*, 1973). In sections the particles are seen to

---

FIGURE 10. Cluster of SVC1 on surface of *S. citri* 551, negatively stained with ammonium molybdate. × 81,800; bar = 100 nm. Some particles of double length are seen. (From Cole, 1977, with permission of Academic Press.)

FIGURE 11. Detail of SVC1 at surface of *S. citri* 551, negatively stained with ammonium molybdate. Note subunit structure and hollow particle showing flat base at membrane attachment point (arrow). × 225,000; bar = 100 nm. (From Cole, 1977, with permission of Academic Press.)

FIGURE 12. Free virions of SVC1, negatively stained with ammonium molybdate. One end is rounded and one has a flat plate or disk. × 198,000; bar = 100 nm. (From Cole, 1977, with permission of Academic Press.)

FIGURE 13. Section of SVC1 arising from membrane of *S. citri* cell. × 160,000; bar = 100 nm. (From Cole *et al.*, 1974, with permission of Department de Colloques et des Publications, l'Institut National de la Santé et de la Recherche Medicale, Paris.)

FIGURE 14. Section of rod-shaped viruslike particles attached to membrane of mycoplasma-like organism in periwinkle infected with clover dwarf agent. × 58,000; bar = 100 nm. (From Ploaie, 1971, with permission of the Institute of Plant Protection and the Publishing House of the Academy of the Socialist Republic of Romania, Bucharest.)

FIGURE 15. Rod-shaped viruslike particles among mycoplasma-like organisms in section of salivary cells of a leafhopper transmitting clover phyllody. × 96,000; bar = 100 nm. (From Gourret *et al.*, 1973, with permission of Cambridge University Press.)

FIGURE 16. Cross section of particles like those in Fig. 15, showing dense hollow core, electron-lucent ring, and dense outer ring. × 350,000; bar = 50 nm. (From Gourret *et al.*, 1973, with permission of Cambridge University Press.)

FIGURE 17. Rod-shaped viruslike particles attached to or surrounding mycoplasma-like organism in phloem cell of clover infected with clover phyllody. × 88,000; bar = 100 nm. (From Gourret *et al.*, 1973, with permission of Cambridge University Press.)

FIGURES 18–20. Viruslike particles associated with mycoplasma-like organisms in sections of periwinkle infected with Stolbur. [From Cadilhac and Giannotti, 1975, with permission of l'Academie des Sciences (Paris) and Gauthier-Villars et Cie.]

FIGURE 18. Rod-shaped and helical particles arising from one MLO. × 92,000; bar = 100 nm. Insert: cross sections of particles showing internal structure like that in Fig. 16. × 154,000; bar = 50 nm.

FIGURE 19. A long helical particle and several rods that clearly originate from the membrane of a lysed mycoplasma-like particle. × 95,000; bar = 100 nm.

FIGURE 20. Bundles of rod-shaped particles in phloem of stolbur-infected periwinkle. × 79,000; bar = 100 nm.

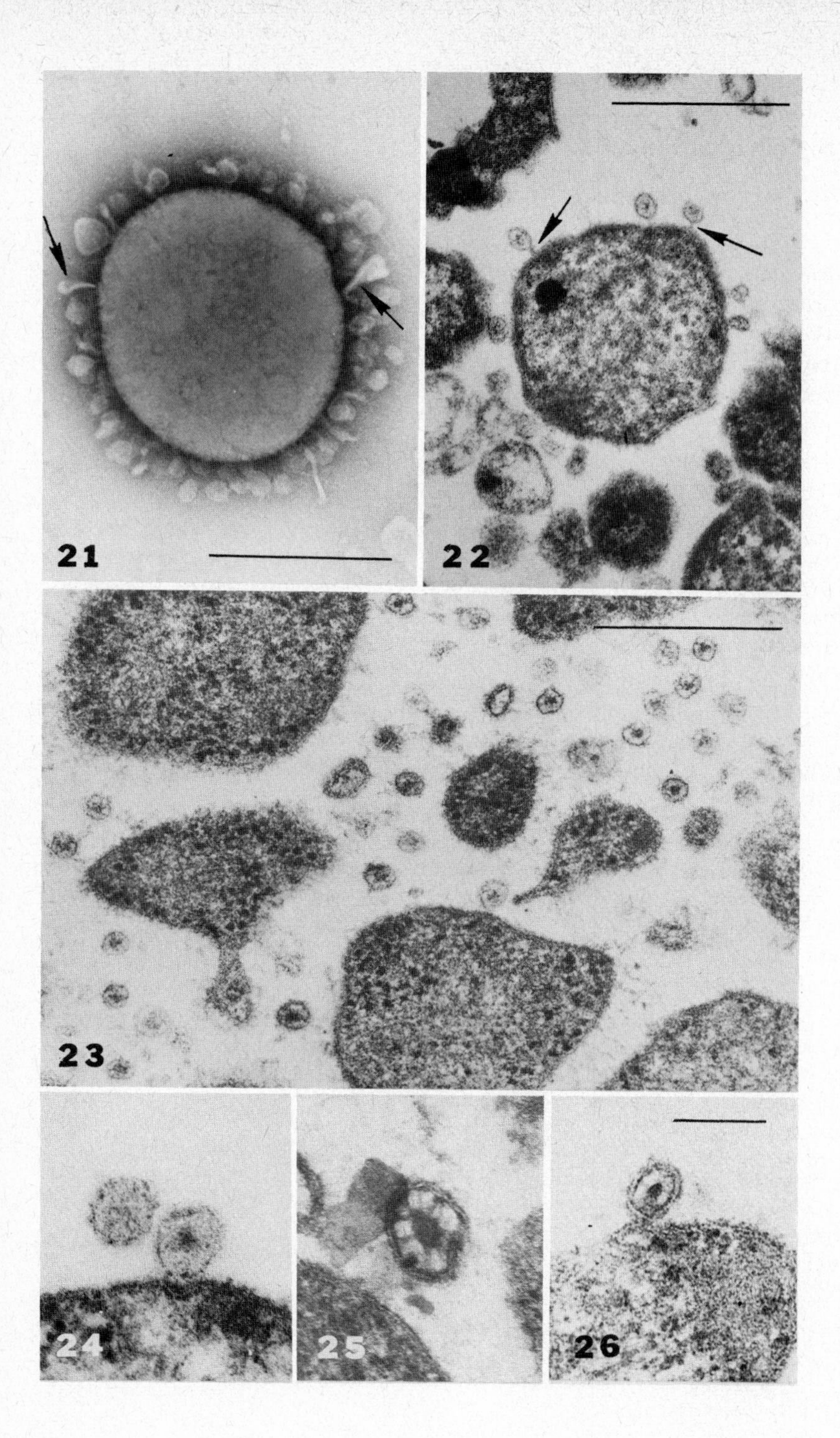
21
22
23
24
25
26

contain dense centers resembling condensed DNA (perhaps a fixation artifact) and to be surrounded by a unit membrane (Figs. 22–26). Unequivocal stages of budding were not observed originally (Gourlay *et al.*, 1973), but suggestive sections may be sometimes seen. The purified particles seen after negative staining exhibit no regular subunit structure or arrangement, and some appear empty because of penetration by stain (Figs. 27 and 28); they contain double-stranded DNA (Maniloff *et al.*, 1977a). Their diameters range from 50 to 125 nm (average, 80 nm) (Gourlay *et al.*, 1973). Infection by MVL2 has been shown to be nonlytic (Putzrath and Maniloff, 1977).

A similar virus, named MV-L-pS2-L 172, was isolated independently by Liska (1972). It was not examined by negative staining, but sections of infected colonies (Fig. 29) (Liska and Tkadlicek, 1975) showed particles of 50–90 nm diameter surrounding the host cells. Many particles possessed an irregular, taillike protuberance by which some appeared attached to the cell surface (Fig. 30): Such protuberances are also seen in negative stains of infected MVL2 (Fig. 21) and probably represent the final stage of budding. The width of the unit membrane of the particle is approximately 6.6–8.6 nm. As in MVL2 (Gourlay *et al.*, 1973), recognizable capsids or precursors were not definable within cells infected by MV-Lg-pS2-L 172, but the released virus particles (Fig. 30) demonstrated more uniform interiors then MVL2 (Liska and Tkadlicek, 1975).

## 2. Particles Not Known To Be Enveloped

Intracellular particles suggestive of viruses were first seen by electron microscopy in 1960 in sections of an unidentified mycoplasma (Edwards and Fogh, 1960). They were electron dense and roughly round or slightly prolate and measured 33 by 50 nm in size. Later (Swartzendruber *et al.*, 1967), intracellular round particles of 25–30 nm diameter were identified in sections of a human mycoplasma. Similar particles, 31–35 nm in diameter, were subsequently reported in sections of *Mycoplasma hominis* (Robertson *et al.*, 1972), in which they were seen in linear submembrane

---

FIGURE 21. Particles of MVL2 surrounding, and apparently budding from, a cell of an infected culture of *A. laidlawii* JA-1. Note some "tailed" examples (arrows). Negatively stained with ammonium molybdate. × 51,000; bar = 0.5 μm.

FIGURE 22. Section of preparation similar to that in Fig. 21. Arrows show "tails." × 50,000; bar = 0.5 μm.

FIGURE 23. Free particles of MVL2 with dense centers, among *A. laidlawii* cells in section of infected culture. × 50,000; bar = 0.5 μm. (From Gourlay *et al.*, 1973, with permission of Cambridge University Press.)

FIGURES 24–26. Sectioned individual particles of MVL2, showing dense centers and surrounding membrane. × 125,000; bar = 100 nm. (From Gourlay *et al.*, 1973, with permission of Cambridge University Press.)

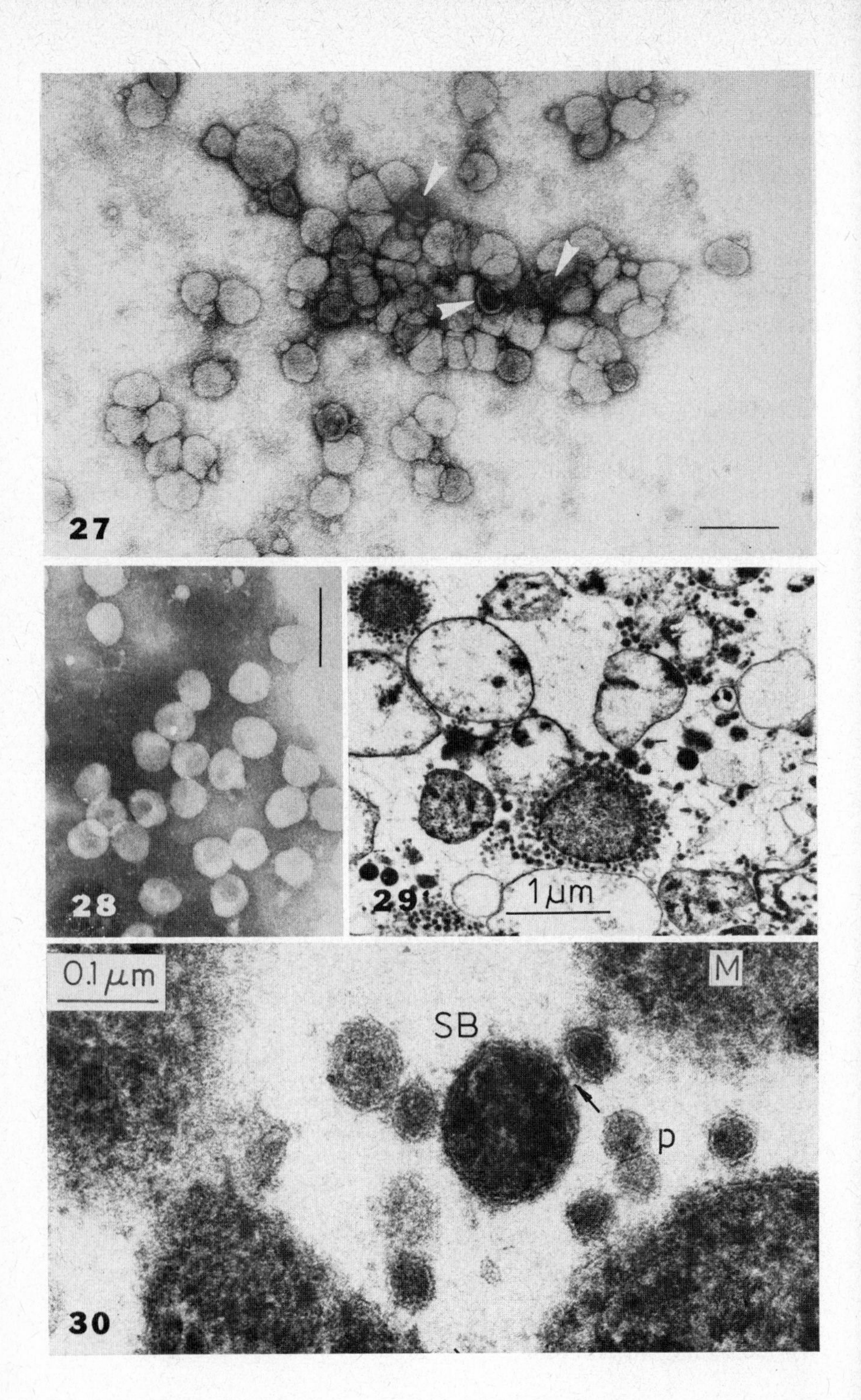
27
28
29
1 μm
0.1 μm
M
SB
p
30

arrays or in hexagonal arrays (Fig. 31). None of these has been propagated or examined by negative staining, but it is probable (because of their density within sectioned cells) that they represent nonenveloped viruses that are bacteriophage-like polyhedrons rather than truly spherical particles.

## C. Polyhedral Viruses

### 1. Short-Tailed Particles (Group L3)

A third virus from *A. laidlawii,* originating uniquely from a spontaneous plaque, was designated Mycoplasmatales virus *laidlawii* 3 (Gourlay and Wyld, 1973). Unlike MVL1 and MVL2, MVL3 is polyhedral in shape (Fig. 32). Like MVL1, it is a naked particle. It possesses a short tail measuring 9 by 25 nm that appears attached to the head by a collar (Fig. 32, insert; Fig. 34). The head is hexagonal in outline and measures 57 nm between opposite planes and 61 nm between opposite vertices. No subunits or other structures were resolved, but a threadlike internal arrangement was suggested and rare elongated heads were seen (Garwes *et al.*, 1975). The viruses, like C3 particles (see Section II, C, 2) may emerge surrounded by host cell membrane (Fig. 33), which is later lost. Naked particles may be seen free or adsorbed to cell surfaces by their tails (Fig. 34). The infection is lytic (unlike MVL1 and MVL2) and, in conformity with results of premature lysis experiments (Liss, 1977), sections reveal mature virus particles within infected cells (Figs. 35–38); emerging virions in membrane buds are also seen (Figs. 36, 39, and 40). Other characteristics reported include a buoyant density of 1.477 g/ml in cesium chloride, the presence of double-stranded DNA, and a capsid composition of five structural polypeptides (Garwes *et al.*, 1975).

### 2. Short-Tailed Particles (Group C3)

Spiroplasmavirus citri 3 (SVC3) was first detected (Cole *et al.*, 1974) in strains of *Spiroplasma citri,* but morphologically identical particles (Fig.

---

FIGURE 27. Purified MVL2, negatively stained with uranyl acetate. Occasional empty, stain-filled particles are seen (arrows). × 112,000; bar = 100 nm.

FIGURE 28. Same as Fig. 27, but negatively stained with ammonium molybdate. × 112,000; bar = 100 nm.

FIGURE 29. Section of colony of *A. laidlawii* S2 infected with MV-Lg-pS2-172. Virus particles are clustered around the producing cells. × 15,000; bar = 1.0 μm. (From Liska and Tkadlecek, 1975, with permission of Academic Press Ltd., London.)

FIGURE 30. Detail of MV-Lg-pS2-172 infection in section, showing *Acholeplasma* cells (M), virus particles (P), and a small cell (small body, SB) to which a virion is attached by a "tail" (arrow). × 160,000; bar = 100 nm. (From Liska and Tkadlecek, 1975, with permission of Academic Press Ltd., London.)

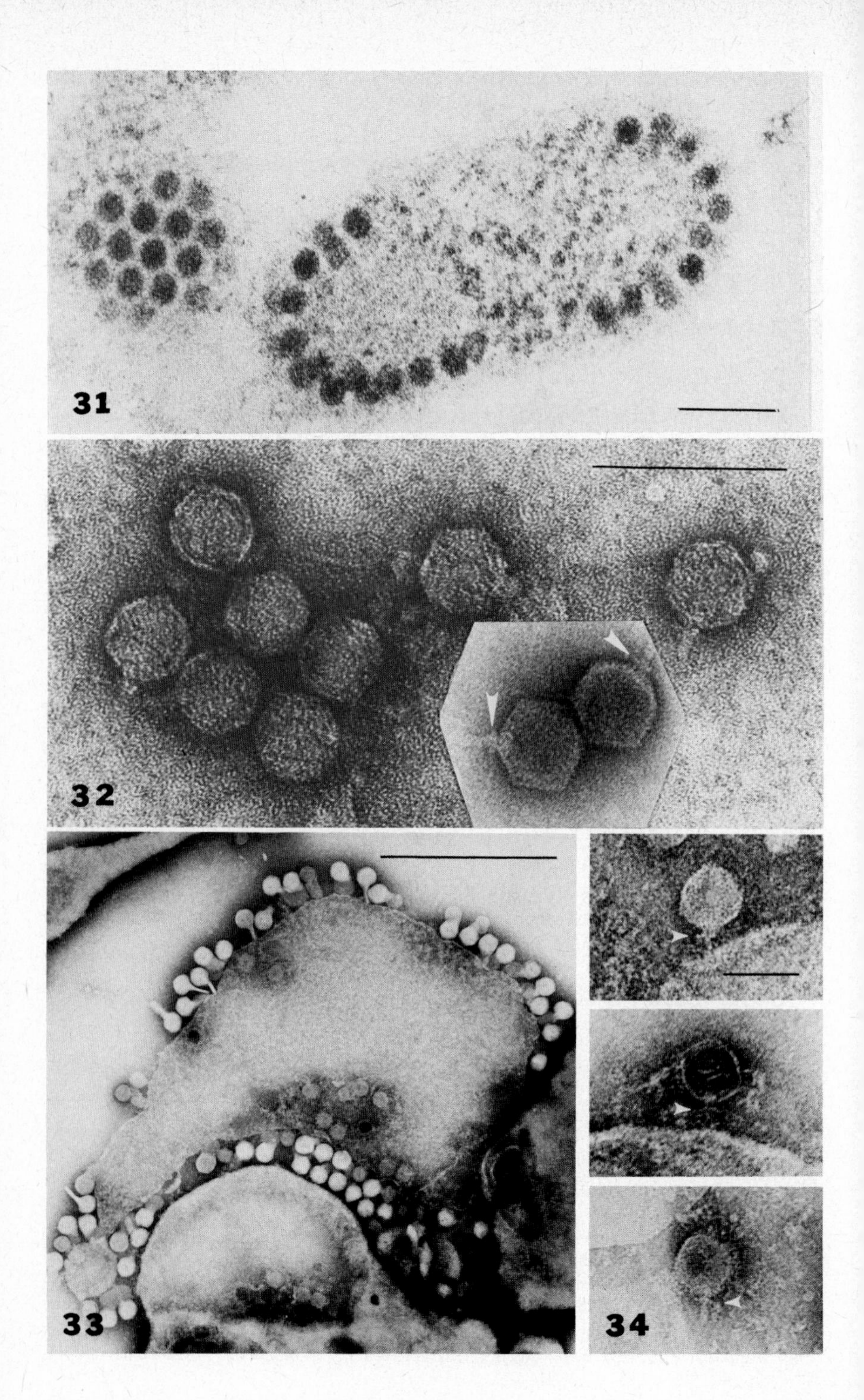
31
32
33
34

28) have since been found in some strains of all spiroplasmas examined (Cole *et al.*, 1976; Cole, 1977, 1978; R. M. Cole, unpublished). These include, in addition to many strains of *S. citri*, some others that have been cultured—including the corn stunt organism (CSO) (Chen and Liao, 1975; Williamson and Whitcomb, 1975), the suckling mouse cataract agent (SMCA) (Tully *et al.*, 1976, 1977), and the tick organism 277F (Brinton and Burgdorfer, 1976)—as well as the uncultured sex-ratio organism (SRO) (Williamson *et al.*, 1977). Though cultivable, other spiroplasmas isolated from cactus (Kondo *et al.*, 1976) and from honeybees (Clark, 1977) have not been reported to contain viruses. In the SRO similar viruses had been previously detected when the host microorganism was thought to be a spirochete (Oishi and Poulson, 1970; Oishi, 1971).

In negatively stained preparations of cultures infected with C3 viruses, several stages of intracellular development of the virus heads can be distinguished (Fig. 41). Some of these, which are different in structure and smaller in diameter than mature heads (Figs. 42 and 43), resemble the proheads that have been described in P22 and other bacteriophages (Lenk *et al.*, 1975). The mature virus particles emerge within buds of membrane (Fig. 44), in which respect they resemble MVL3 and the viruses of the sex-ratio organism (Oishi and Poulson, 1970; Williamson *et al.*, 1977). The membrane is rapidly lost, and naked viruses—both free and adsorbed to other cells of the infected culture (Figs. 45 and 48)—are commonly seen.

Virions of SVC3 are hexagonal in outline but may often appear spherical (Figs. 45–49). Those purified by metrizamide gradients (Cole *et al.*, 1977) are almost entirely intact particles 37–44 by 35–37 nm, with a short tail 13–18 nm in length by 6–7 nm in width and a slightly wider base plate without appendages (Fig. 46) (Cole *et al.*, 1974; Cole, 1977). In cesium chloride gradients. not only is DNA released from many virions (Cole *et al.*, 1977), but also intact tails are freed, thus revealing the base plate, the tail proper, and a proximal disklike portion that is sometimes seen intruding at its attachment site to empty but intact virions (Fig. 47). No collar, as seen in MVL3, is evident. Viruses isolated from both *S. citri* and the

FIGURE 31. Viruslike particles in section of *M. hominis*. × 146,000; bar = 100 nm. (From Robertson *et al.*, 1971, with permission of the National Research Council of Canada.)

FIGURE 32. Purified virions of MVL3, negatively stained with uranyl acetate. × 278,000; bar = 100 nm. (From Garwes *et al.*, 1975, with permission of Cambridge University Press.) Insert: MVL3 negatively stained with ammonium molybdate, at same magnification. A collar (arrows) is visible on these virions.

FIGURE 33. Buds of membrane, some "tailed," surrounding MVL3 emerging from cell of *A. laidlawii* BCL-13, in culture infected at high multiplicity. Negatively stained with ammonium molybdate. × 54,000; bar = 0.5 μm.

FIGURE 34. Examples of MVL3, free or adsorbed, showing collars (arrows). Negatively stained with ammonium molybdate. × 200,000; bar = 50 nm.

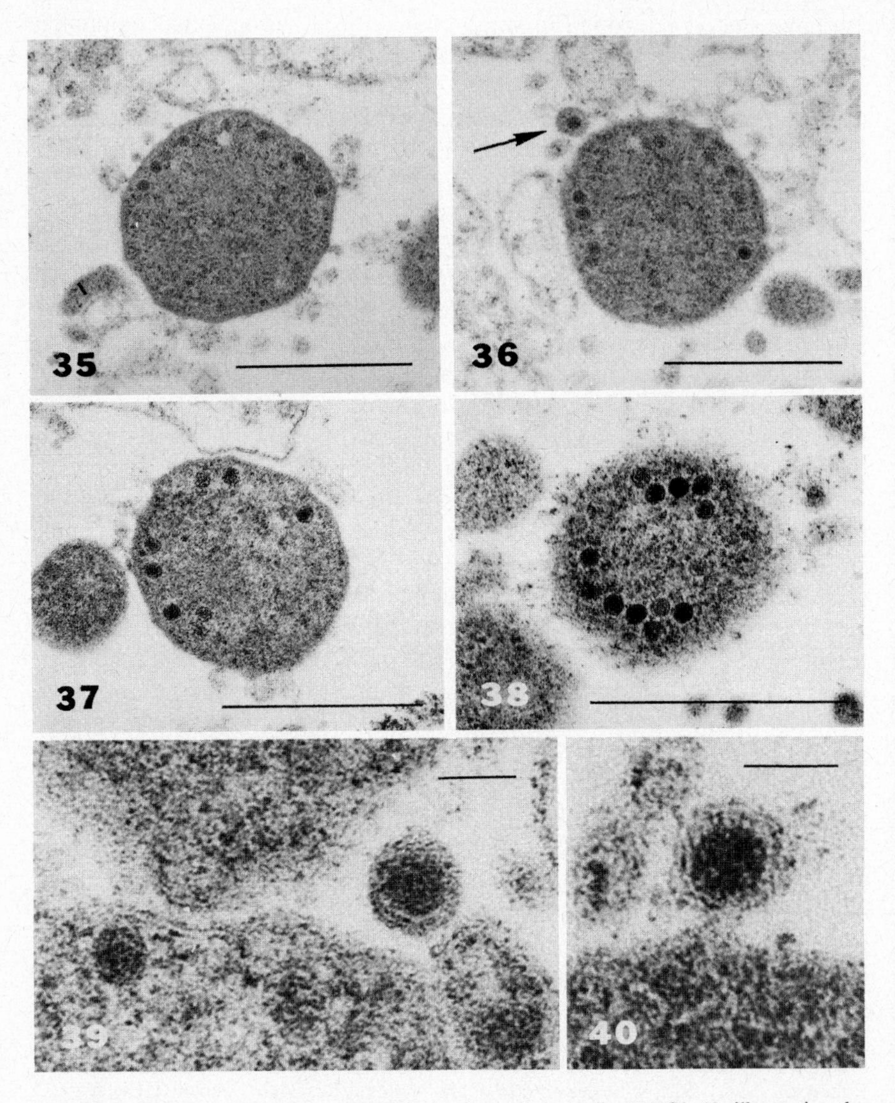


FIGURES 35–38. Intracellular MVL3 in sections of *A. laidlawii* BCL-13, illustrating the usual small numbers of visions seen per cell and one membrane bud (arrow). Magnifications: × 46,800 (Figs. 35 and 36), × 54,600 (Fig. 37), × 70,200 (Fig. 38); bars = 0.5 μm.
FIGURES 39–40. MVL3 virions in membrane buds, in section. × 214,500 (Fig. 39), × 253,500 (Fig. 40); bars = 50 nm.

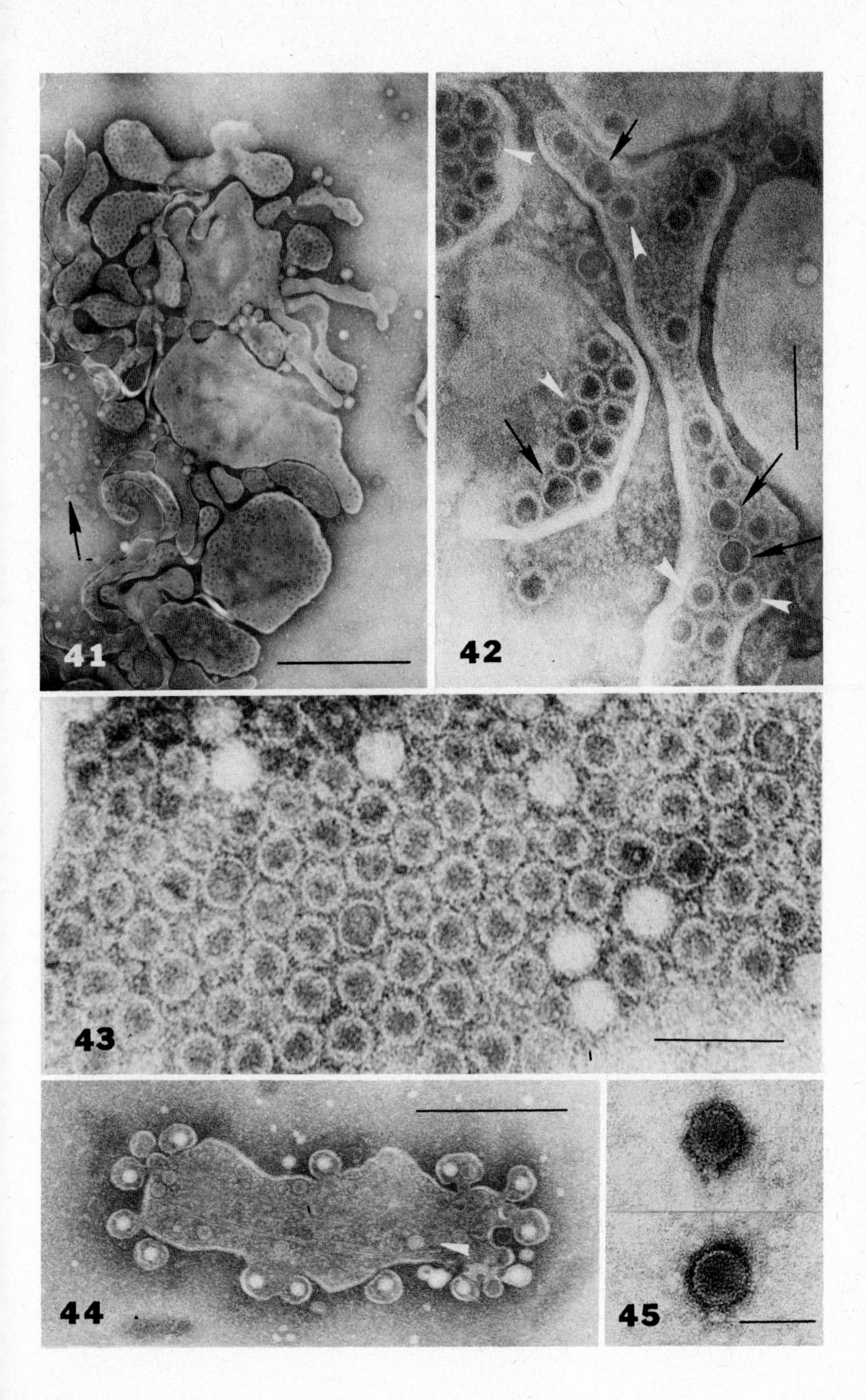
41
42
43
44
45

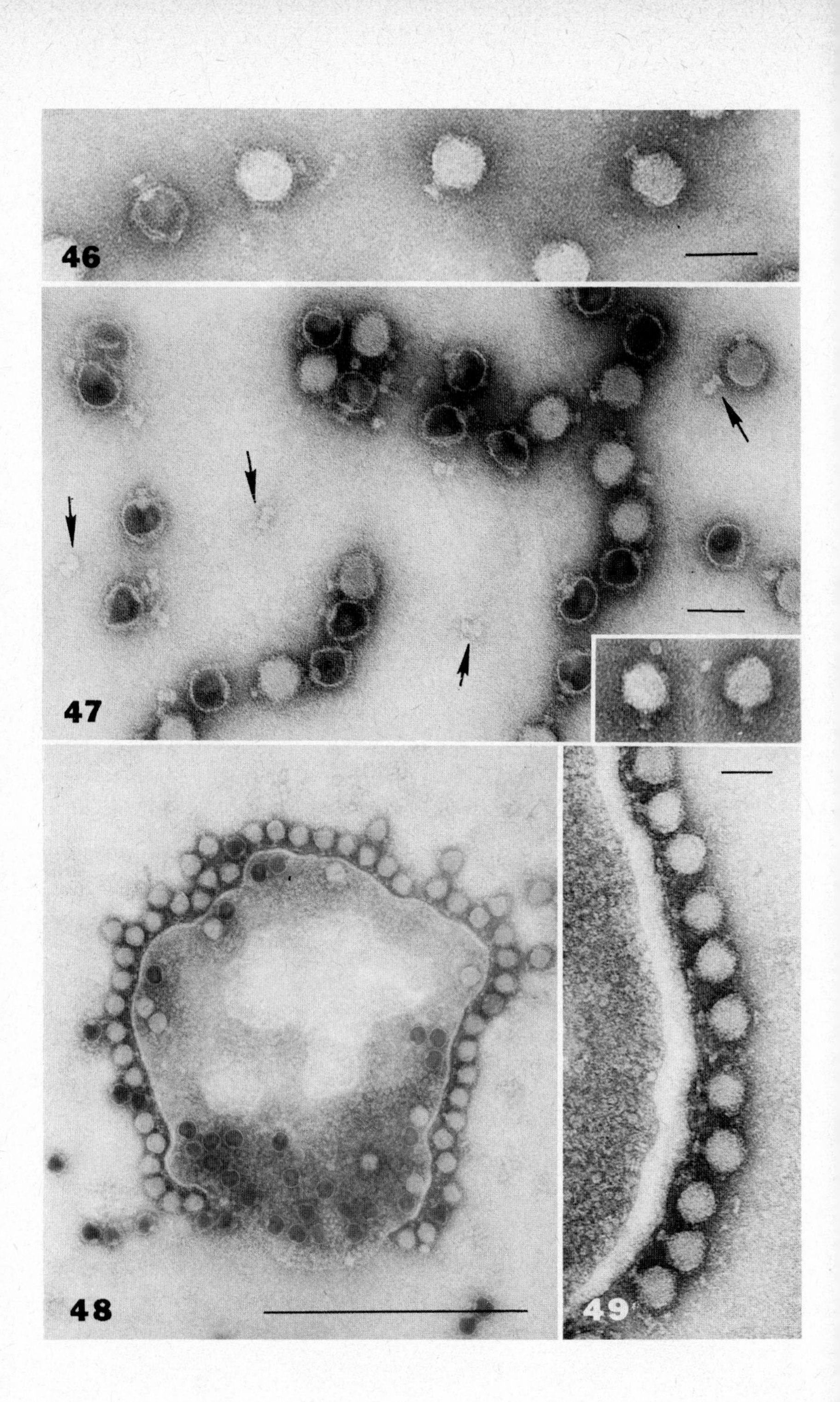
46
47
48
49

SMCA, as well as those seen in other spiroplasmas, are structurally similar. Those from the SRO, (Fig. 47, insert), though originally described as being spherical with a diameter of 50–60 nm (Oishi and Poulson, 1970; Oishi, 1971), are now known to have a head 35–35 by 35–45 nm, to which is attached a short tail 10–12 by 7–9 nm with a base plate (Williamson *et al.*, 1977). These dimensions are similar to those of SVC3 from cultured spiroplasma hosts.

Virions of SVC3 that are released in experimentally or naturally infected cultures commonly adsorb by their base plates to the surface of nearby host cells (Figs. 48 and 49).

Sections of cultures infected with SVC3 show numerous intracellular (Fig. 50) and membrane-enclosed budding virions (Figs. 50 and 51). The number of these seen is probably a function of the multiplicity of infection, with both SVC3 and MVL3.

Viruses of group C3 have been isolated from both the suckling mouse cataract agent and *S. citri* and have been propagated, purified, and par-

---

FIGURE 41. *Spiroplasma citri* 750 infected with SVC3/608, seen at low magnification in early stage. Dark dots are proheads or unfilled heads: some maturing particles are seen in one cell (arrow). Negatively stained with ammonium molybdate. × 34,500; bar = 0.5 μm.

FIGURE 42. Proheads (white arrows) and larger unfilled heads (black arrows) of SVC3 in natural infection of *S. citri* 539. Negatively stained with ammonium molybdate. × 125,200; bar = 100 nm.

FIGURE 43. Mass of proheads and unfilled heads of SVC3, with occasional filled mature particle (white), as extruded from mechanically ruptured host cell. Negatively stained with ammonium molybdate. × 195,000; bar = 100 nm.

FIGURE 44. Membrane buds enclosing emerging virions of SVC3, in naturally infected suckling mouse cataract agent. Negatively stained with ammonium molybdate. × 52,900; bar = 0.5 μm.

FIGURE 45. Virions of SVC3 showing subunits and short tail with base plate. Negatively stained with ammonium molybdate. × 200,000; bar = 50 nm. (From Cole *et al.*, 1974, with permission of Department des Colloques et des publications l'Institut National de la Santé et de la Recherche Medical, Paris.)

FIGURE 46. SVC3/608 virions from metrizamide density gradient band at 1.26 gm/cm$^3$, negatively stained with ammonium molybdate. Virions are largely intact; collars, as seen in MVL3, are absent. × 200,000; bar = 50 nm.

FIGURE 47. SVC3/608 virions from cesium chloride density gradient band at 1.45 gm/cm$^3$, negatively stained with ammonium molybdate. Banding in cesium results in extrusion of DNA from many virions and disrupts some, liberating intact tail structures (arrows). × 161,000; bar = 50 nm. Insert: SVC3-like virus from the sex-ratio organism. Same stain and magnification. (Original material courtesy of D. L. Williamson.)

FIGURE 48. Natural infection of suckling mouse cataract agent, showing SVC3 virions adsorbed to cell producing virions intracellularly. Negatively stained with ammonium molybdate. × 75,000; bar = 0.5 μm. (From Cole, 1977, with permission of Academic Press.)

FIGURE 49. Higher magnification of SVC3 adsorption in natural infection of suckling mouse cataract agent, strain GT-48. Negatively stained with ammonium molybdate. × 160,000; bar = 50 nm.

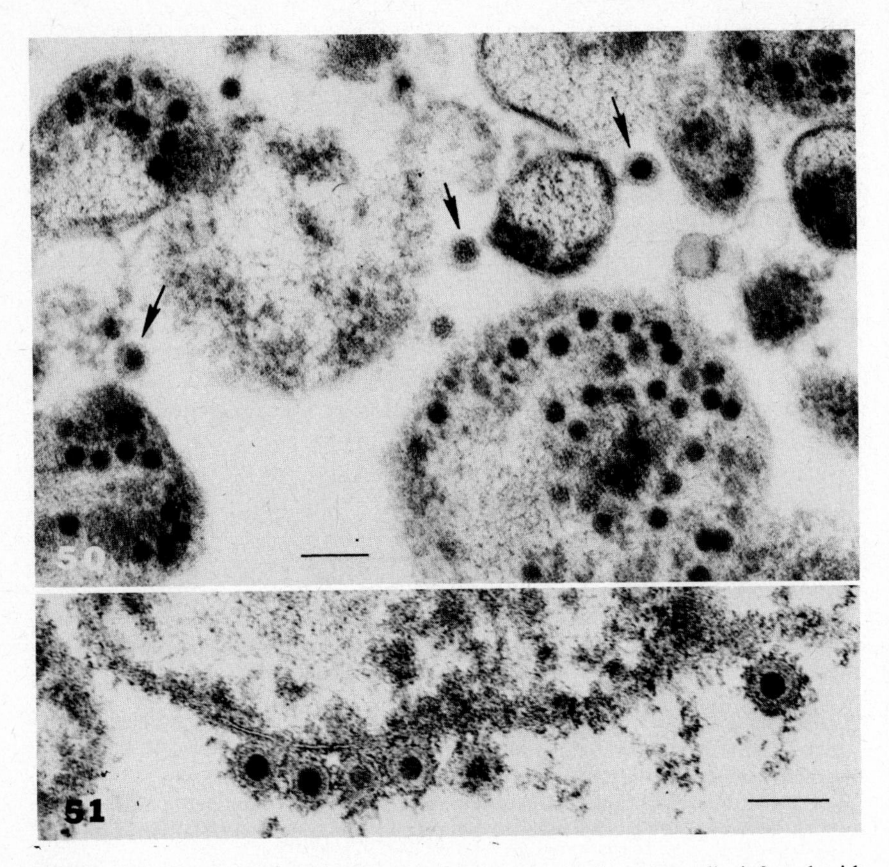


FIGURE 50. Section of suckling mouse cataract agent culture naturally infected with SVC3. Numerous intracellular virions and some emerging in membrane buds (arrows) are present. × 92,000; bar = 100 nm.
FIGURE 51. Detail of SVC3 virions in membrane buds in section of naturally infected *S. citri* 608. × 101,200; bar = 100 nm.

tially characterized. Each contains linear double-stranded DNA of molecular weight $14 \times 10^6$ daltons, and the capsid is composed of five structural proteins (Cole *et al.*, 1977).

### 3. Long-Tailed Particles (Group C2)

The first viruslike particle seen in cultures of *S. citri* was a long-tailed polyhedron characteristic of a type B bacteriophage (Cole *et al.*, 1973a, b). The head is hexagonal in profile and is attached without a collar to the noncontractile tail (Fig. 52), which terminates in a base plate without obvious appendages but nevertheless adsorbs to the outer layer of the host cell (Figs. 53 and 54). No capsid subunits are clearly resolved. Dimensions of the head are 48–51 nm between flat sides and 52–58 nm between vertices. The tail is 78–83 nm long, with a width of 6–8 nm except at the slightly wider base plate (Cole *et al.*, 1974; Cole, 1977). Electron microscopy suggests that the infection is lytic, and that the exit of mature virus particles from the infected cell does not involve a temporary enclosure in host membrane, as in the short-tailed mycoplasma viruses. In negatively stained preparations, heads and tails, empty heads, complete particles, and apparent proheads can all be seen within cells (Fig. 55). Both empty and filled particles are easily distinguished in sections, but tails have not been resolved (Fig. 56).

This particle—named Spiroplasmavirus citri 2 (SVC2) (Cole *et al.*, 1974)—has not been propagated, and most of its properties are therefore unknown. It has been detected in numerous strains of *S. citri* isolated in different parts of the world (Cole *et al.*, 1974; Cole, 1977, 1978) but has not been found in other spiroplasmas examined; therefore, it may represent (unlike SVC1 and SVC3) a virus restricted to one host. It has also been seen in a *S. citri* multiplying in salivary glands of a leafhopper (Townsend *et al.*, 1977).

All information to date indicates that, with one possible exception, mycoplasma viruses are not structurally unique. They are rods or filaments, all of which will probably prove to have helical symmetry, and tailed polyhedrons, which by definition possess cubic symmetry of the head and probably helical symmetry of the tail portions. Whether the polyhedrons are actually octahedrons or icosahedrons has not been clearly determined, and more investigation is also needed to determine the nature of capsomeres, collars, tails, base plates, or possible appendages not as yet detected. What is clear at present is that these viruses are structural counterparts of known bacteriophages; they can be classified, therefore, in one or another morphologic group of bacterial viruses (Bradley, 1971; Ackermann, 1973), although such classification is one of convenience and is only a part of any ultimate scheme (if such is ever achieved; see Lwoff and Tournier, 1971, for example) which must also incorporate information on natural occurrence, host range, nature of nucleic acid, size, and other attributes.

The exception to the above is MVL2. As a roughly spherical membrane-enveloped particle of variable size, which has not been visualized within the host cell before emergence and which apparently contains only double-stranded DNA that is not organized as a nucleocapsid or other internal structure, this simple virus would appear to have no counterpart among known viruses of either prokaryotic or eukaryotic organisms. It will be of interest to confirm its apparently primitive nature,

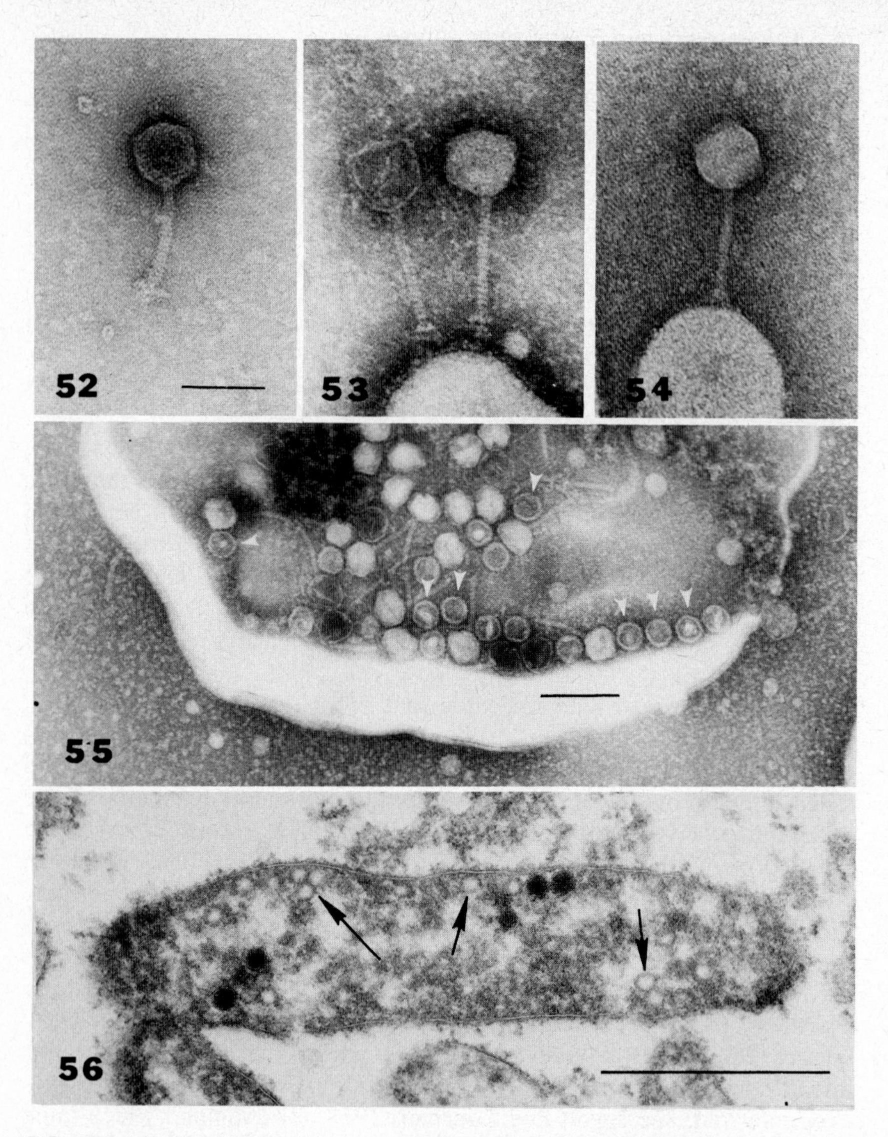


FIGURES 52–54. SVC2 from *S. citri* Morocco (R8A2), free and adsorbed to cell surface. Negatively stained with uranyl acetate (Fig. 52), potassium phosphotungstate (Fig. 53), and ammonium molybdate (Fig. 54). All × 200,000; bar = 50 nm. (Figures 52 and 54 from Cole *et al.*, 1974, with permission Department des Colloques et des Publications of l'Institut National de la Santé et de la Recherche Medicale, Paris; Fig. 53 from Cole *et al.*, 1973a, with permission of American Society for Microbiology.)

FIGURE 55. Intracellular virions, tails, unfilled heads, and proheads (arrows) of SVC2 in *S. citri* Morocco. Negatively stained with potassium phosphotungstate. × 105,000; bar = 100 nm.

FIGURE 56. Section of *S. citri* Morocco naturally infected with SVC2, showing completed heads (black) and empty immature heads or proheads (arrows). × 59,800; bar = 0.5 $\mu$m. (From Cole *et al.* 1973a, with permission of American Society for Microbiology.)

and to examine it, as well as the rodlike viruses and the temporarily enveloped short-tailed polyhedrons, in the light of possible evolutionary pathways among prokaryotes and of the effects of the wall-less state on the nature and life cycle of mycoplasma viruses.

## REFERENCES


Ackermann, H.-W. (1973). *In* "Handbook of Microbiology (A. I. Laskin and H. G. Lechevalier, eds.), 1st ed., Vol. I, pp. 573–607. CRC Press, Inc., West Palm Beach, Florida.

Allen, T. C. (1971). *Virology* **47,** 491–493.

Bradley, D. E. (1971). *In* "Comparative Virology" (K. Maramorosch and E. Kurstak, eds.), pp. 207–253. Academic Press, New York.

Brinton, L. P., and Burgdorfer, W. (1976). *Int. J. Syst. Bacteriol.* **26,** 554–560.

Bruce, J., Gourlay, R. N., Hull, R., and Garwes, D. J. (1972). *J. Gen. Virol.* **16,** 215–221.

Cadilhac, B., and Giannotti, J. (1975). *C. R. Hebd. Seances Acad. Sci., Ser. D* **281,** 539–542.

Chen, T. A., and Liao, C. H. (1975). *Science* **188,** 1015–1017.

Clark, T. B. (1977). *J. Invertebr. Pathol.* **29,** 112.

Cole, R. M. (1977). *In* "The Atlas of Plant and Insect Viruses" (K. Maramorosch, ed.), pp. 451–465. Academic Press, New York.

Cole, R. M. (1978). *In* "Handbook of Microbiology" (A. I. Laskin and H. A. Lechavalier, eds.), 2nd ed., Vol. II. CRC Press, Inc., West Palm Beach, Florida.

Cole, R. M., Tully, J. G., Popkin, T. J., and Bové, J. M. (1973a). *J. Bacteriol.* **115,** 367–386.

Cole, R. M., Tulley, J. G., Popkin, T. J., and Bové, J. M. (1973b). *Ann. N.Y. Acad. Sci.* **225,** 471–493.

Cole, R. M., Tully, J. G., and Popkin, T. J. (1974). *Colloq. Inst. Natl. Sante Rech. Med.* **33,** 125–132.

Cole, R. M., Tully, J. G., Popkin, T. J., and Mitchell, W. O. (1976). *Abstr. Annu. Meet., Am. Soc. Microbiol.* p. 61.

Cole, R. M., Mitchell, W. O., and Garon, C. F. (1977). *Science* **198,** 1262–1263.

Das, J., and Maniloff, J. (1975). *Biochem. Biophys. Res. Commun.* **66,** 599–605.

Edwards, G. A., and Fogh, J. (1960). *J. Bacteriol.* **79,** 267–276.

Garwes, D. J., Pike, B. V., Wyld, S. G., Pocock, D. H., and Gourlay, R. N. (1975). *J. Gen. Virol.* **29,** 11–24.

Gourlay, R. N. (1970). *Nature (London)* **225,** 1165.

Gourlay, R. N. (1971). *J. Gen. Virol.* **12,** 65–67.

Gourlay, R. N. (1972). *Pathog. Mycoplasmas; Ciba Found. Symp., 1972* pp. 145–156.

Gourlay, R. N. (1973). *Ann. N.Y. Acad. Sci.* **225,** 144–148.

Gourlay, R. N. (1974). *Crit. Rev. Microbiol.* **3,** 315–331.

Gourlay, R. N., and Wyld, S. G. (1973). *J. Gen. Virol.* **19,** 279–283.

Gourlay, R. N., Garwes, D. J., Bruce, J., and Wyld, S. G. (1973). *J. Gen. Virol.* **18,** 127–133.

Gourret, J. P., Maillet, P. L., and Gouranton, J. (1973). *J. Gen. Microbiol.* **74,** 241–249.

Kondo, F., McIntosh, A. H., Padhi, S. B., and Maramorosch, K. (1976). *Proc. Soc. Gen. Microbiol.* **3,** Part 4, 155.

Lenk, E., Casjens, S., Weeks, J., and King, J. (1975). *Virology* **68,** 182–199.

Liska, B. (1972). *Stud. Biophys.* **34,** 151–155.

Liska, B., and Tkadlecek, L. (1975). *Folia Microbiol. (Prague)* **20,** 1–7.

Liss, A. (1977). *Virology* **77,** 433–436.

Liss, A., and Maniloff, J. (1971). *Science* **173,** 725–727.

Liss, A., and Maniloff, J. (1972). *Proc. Natl. Acad. Sci. U.S.A.* **69,** 3423–3427.
Liss, A., and Maniloff, J. (1973a). *Biochem. Biophys. Res. Commun.* **51,** 214–218.
Liss, A., and Maniloff, J. (1973b). *Virology* **55,** 118–126.
Lwoff, A., and Tournier, P. (1971). *In* "Comparative Virology" (K. Maramorosch and E. Kurstak, eds.), pp. 1–42. Academic Press, New York.
Maniloff, J. (1972). *Pathog. Mycoplasmas, Ciba Found. Symp., 1972* pp. 156–164.
Maniloff, J., and Das, J. (1975). *In* "DNA Synthesis and Its Regulation" (M. Goulian, P. Hanawalt, and C. F. Fox, eds.), pp. 445–450. Benjamin, Reading, Massachusetts.
Maniloff, J., and Liss, A. (1973). *Ann. N.Y. Acad. Sci.* **225,** 149–158.
Maniloff, J., and Liss, A. (1974). *In* "Viruses, Evolution and Cancer" (E. Kurstak and K. Maramorosch, eds.), pp. 583–604. Academic Press, New York.
Maniloff, J., Das, J., and Christensen, J. R. (1977a). *Adv. Virus Res.* **21,** 343–380.
Maniloff, J., Das, J., and Putzrath, R. M. (1977b). *In* "The Atlas of Plant and Insect Viruses" (K. Maramorosch, ed.), pp. 439–450. Academic Press, New York.
Milne, R. G., Thompson, G. W., and Taylor-Robinson, D. (1972). *Arch. gesamte Virusforsch.* **37,** 378–385.
Oishi, K. (1971) *Genet. Res.* **18,** 45–56.
Oishi, K., and Poulson, D. F. (1970). *Proc. Natl. Acad. Sci. U.S.A.* **76,** 1565–1572.
Ploaie, P. G. (1971). *Rev. Roum. Biol., Ser. Bot.* **16,** 3–6.
Putzrath, R. M., and Maniloff, J. (1977). *J. Virol.* **22.** 308–314.
Robertson, J., Gomersall, M., and Gill, P. (1972). *Can. J. Microbiol.* **18,** 1971–1972.
Swartzendruber, D. C., Clark, J., and Murphy, W. H. (1967). *Bacteriol. Proc.* **67,** 151.
Townsend, E., Markham, P. G., and Plaskitt, K. A. (1977). *Ann. Appl. Biol.* **87,** 307–313.
Tully, J. G., Whitcomb, R. F., Williamson, D. L., and Clark, H. F. (1976). *Nature (London)* **259,** 117–120.
Tully, J. G., Whitcomb, R. F., Clark, H. F., and Williamson, D. L. (1977). *Science* **195,** 892–894.
Williamson, D. L., and Whitcomb, R. F. (1974). *Colloq. Inst. Natl. Sante Rech. Med.* **33,** 283–290.
Williamson, D. L., and Whitcomb, R. F. (1975). *Science* **188,** 1018–1020.
Williamson, D. L., Oishi, K., and Poulson, D. F. (1977). *In* "The Atlas of Plant and Insect Viruses" (K. Maramorosch. ed.), pp. 465–472. Academic Press, New York.

# 15 / MYCOPLASMA AND SPIROPLASMA VIRUSES: MOLECULAR BIOLOGY

*Jack Maniloff, Jyotirmoy Das,[1] Resha M. Putzrath,[2] and Jan A. Nowak*



## I. INTRODUCTION

The isolation of a virus that could infect mycoplasmas was first reported by Gourlay (1970). This isolate was designated MVL1 (Mycoplasmatales virus *laidlawii* 1). Since then more than 50 virus isolates have been reported (Gourlay, 1971, 1972, 1973, 1974; Liss and Maniloff, 1971; Liska, 1972; Gourlay and Wyld, 1973; Phillpotts *et al.*, 1977). These have been classified into three groups (reviewed by Maniloff *et al.*, 1977a): (1) Group 1, consisting of naked bullet-shaped particles; (2) Group 2, consisting of

[1] Present address: Department of Microbiology, Bose Institute, Calcutta 700009, India.
[2] Present address: Department of Physiology, Harvard Medical School, Boston, Massachusetts 02115.



THE MYCOPLASMAS, VOL. 1

ISBN 0-12-078401-7

roughly spherical, enveloped viruses; and (3) Group 3, polyhedral particles with tails. All three groups contain DNA. The general properties of the three mycoplasmavirus groups are summarized in Table I. Most of the isolates are Group 1, eight are Group 2, and one is Group 3.

All Group 1 virus isolates are morphologically identical and serologically related. Some Group 1 isolates have properties, such as host range, ultraviolet or antiserum inactivation rate constants, and one-step growth parameters, different from the original MVL1 isolate and from each other (Liss and Maniloff, 1971; Maniloff and Liss, 1974). Although most mycoplasmaviruses have been isolated from *Acholeplasma laidlawii* strains, some Group 1 isolates have been reported from other *Mycoplasma* and *Acholeplasma* species (Liss and Maniloff, 1971; Gourlay, 1972; Clyde, 1974a,b). This creates a problem in virus nomenclature. For a particular isolate, Liss and Maniloff (1971) suggested following the MV designation by a letter or letters indicating the virus origin (e.g., L for *laidlawii* and Gs for *gallisepticum*) and then an isolate number.

Group 2 viruses have recently been shown to have antigenic and host range differences (Phillpotts *et al.*, 1977). It is not possible to decide how fundamental these differences are, since (as will be discussed below) virus maturation by budding may affect the virion antigenic determinants and host restriction and modification of Group 2 viruses is probably the major factor controlling host range.

All three mycoplasmavirus groups can be assayed as plaque-forming units (PFU), since each plaque is the result of an infection by a single virus. This was demonstrated experimentally for Group 1 and 2 mycoplasmaviruses (Maniloff and Liss, 1974) and can be concluded from the data on Group 3 viruses (Liss, 1977) by calculating the Poisson distribu-

TABLE I. **Properties of Mycoplasmaviruses**[a]

| | Virion morphology | Nucleic acid | | Progeny virus release |
|---|---|---|---|---|
| | | Type | Molecular weight | |
| Group 1 | Naked bullet-shaped particles | Circular single-stranded DNA | $1.5 \times 10^6$ | Nonlytic |
| Group 2 | Roughly spherical enveloped particles | Circular double-stranded DNA | $7.8 \times 10^6$ | Nonlytic |
| Group 3 | Polyhedral particles with short tails | Linear double-stranded DNA | $25.8 \times 10^6$ | Lytic |

[a] References in text.

tion of viruses per cell at different multiplicities of infection (MOI) and comparing this with the measured PFU.

Mycoplasma strains presently used to propagate viruses may also carry viruses. The nature of this presumptive mycoplasmavirus carrier state is not known. However, the carrier state does not appear to interfere with virology studies, since indicator hosts are chosen for which the frequency of spontaneous virus release is extremely low. In addition, treatments that induce Group 2 virus in persistently infected cells do not induce PFU in the *A. laidlawii* strain JA1 indicator host (Putzrath and Maniloff, 1978).

In addition to the three groups of mycoplasmaviruses which have been isolated from *Mycoplasma* and *Acholeplasma* species and propagated on *A. laidlawii* strains, three morphological types of viruslike particles have been observed in spiroplasma cultures (Cole *et al.*, 1974; Cole, 1977). Since it has not been possible to propagate two of these particle types, no biochemical or virological data are available about them. However, the third type has recently been shown to produce plaques on a *Spiroplasma citri* lawn (Cole *et al.*, 1977). Finally, there have been reports of "viruslike particles" in plants and insect tissues carrying "mycoplasma-like organisms" (Ploaie, 1971; Allen, 1972; Gourret *et al.*, 1973; Giannotti *et al.*, 1973). Electron micrographs show bacilliform particles, generally about 25–35 nm by 70–150 nm. These particles do not resemble known mycoplasma- and spiroplasmaviruses, so their relationship to the mycoplasmas is not clear, and since nothing further is known of their biology, they will not be considered here.

## II. MYCOPLASMAVIRUSES

### A. Biophysical and Biochemical Properties

#### 1. Chemical Composition

**a. DNA.** All three mycoplasmavirus groups are DNA viruses. The genome of MVL51 (a Group 1 mycoplasmavirus) is circular, single-stranded DNA. This was first shown by studies on the effect of various specific nucleases on viral DNA infectivity (Liss and Maniloff, 1973a) and has recently been confirmed by electron microscopic observations (Fig. 1). Measurement of double-stranded MVL51 replicative form (RF) DNA by electron microscopy has allowed a more accurate determination of the molecular weight (Nowak *et al.*, 1978; Maniloff *et al.*, 1978) than has been

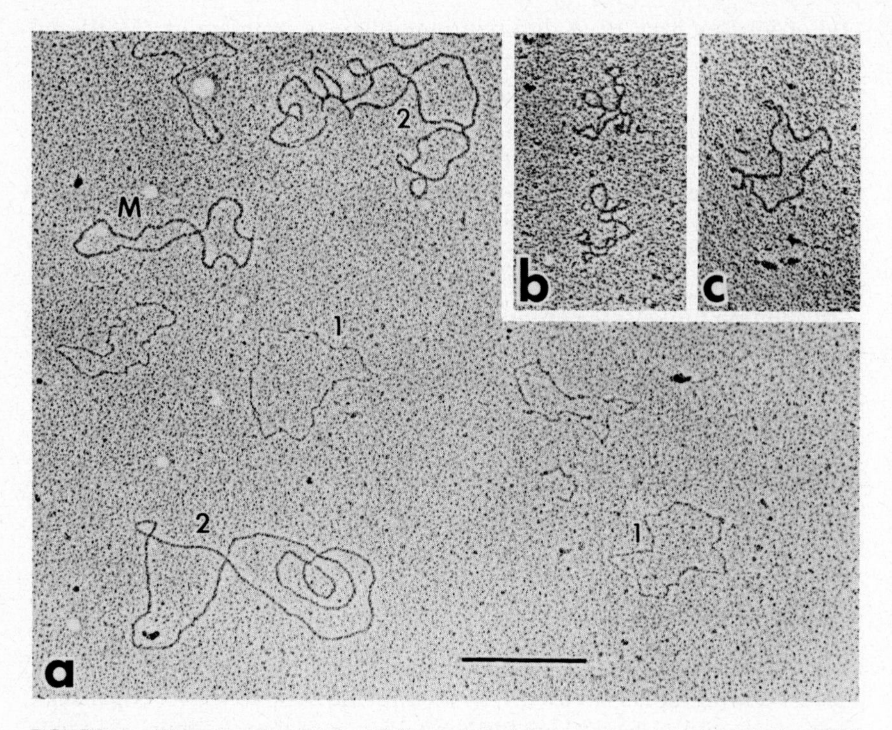


FIGURE 1. Electron micrographs of Group 1 and 2 mycoplasmavirus DNA. (a) DNA from sodium dodecyl sulfate-disrupted Group 1 virions (MVL51) and Group 2 virions (MVL2) was mixed with purified coliphage M13 RF DNA. The following DNAs can be seen: (a) single-stranded circular MVL51 viral DNA (marked 1), MVL2 viral DNA (marked 2), and M13 RF DNA (marked M); (b) MVL51 RFI DNA; (c) MVL51 RFII DNA. Bar denotes 0.5 $\mu$m. (From Maniloff *et al.*, 1978.)

possible by previous sedimentation studies (Liss and Maniloff, 1973a). The MVL51 RF contour length is about 0.71 that of the bacteriophage M13 RF DNA used to calibrate these measurements. When a value of 6370 nucleotides for M13 DNA (Day and Berkowitz, 1977) is used, the MVL51 single-stranded DNA contains about 4520 nucleotides and has a molecular weight (assuming 330 daltons/nucleotide) of about $1.5 \times 10^6$.

Electron microscopic studies have shown that the genome of MVL2 (a Group 2 mycoplasmavirus) is a superhelical molecule of double-stranded DNA (Fig. 1). Measurement of relaxed molecules gives a DNA molecular weight about $7.8 \times 10^6$ (Nowak and Maniloff, 1979; Lombardi and Cole, 1979). Using a different Group 2 mycoplasmavirus isolate, Drasil *et al.* (1978) have found a DNA size of $11 \times 10^6$ daltons. It remains to be seen whether these two different size determinations reflect an actual difference in genome size between Group 2 isolates or differences in calibration procedures in the two laboratories.

Similar electron microscopic studies of MVL3 (the only Group 3 mycoplasmavirus) isolate have revealed that the genome is a molecule of linear double-stranded DNA with a molecular weight of $25.8 \times 10^6$ (Haberer, 1978). Studies of the base composition of MVL3 DNA by different procedures give conflicting GC contents (Garwes *et al.*, 1975).

**b. Proteins.** Purified preparations of a Group 1 mycoplasmavirus, MVL51, were found to contain four proteins of molecular weights 70,000, 53,000, 30,000, and 19,000 (Maniloff and Das, 1975). The approximate stoichiometric ratios of these proteins were 7:1:2:9.

Lombardi and Cole (1978) have reported that purified MVL2 preparations contain six to seven proteins, with molecular weights from 17,000 to 79,000, as determined by sodium dodecyl sulfate polyacrylamide gel electrophoresis. They also noted that their virus preparation contained several other proteins with the same electrophoretic mobility as some cell membrane proteins. Watkins (1977) found that MVL2 has six antigens not found in uninfected cells and observed six gel bands, between 14,500 and 89,000 daltons. Putzrath and Maniloff (1979) have grown MVL2 on *A. laidlawii* strains having different cell membrane protein gel patterns. The purified virions had at least eight protein bands, from 25,000 to 87,000 daltons. Variation in the presence of some of the proteins was seen, so the viral protein distribution depends on which strain the virus had been propagated. Therefore, MVL2 virus preparations contain some cell membrane proteins.

The Group 3 virus, MVL3 was found to contain five proteins with molecular weights 172,000, 81,000, 73,000, 68,000, and 43,000 (Garwes *et al.*, 1975). Based on this published data, we have calculated the stoichiometric ratios of these proteins to be 1:16:2:27:320.

**c. Other components.** Group 2 viruses contain lipids, as first shown by their sensitivity to detergents and organic solvents (Gourlay, 1971), low buoyant density of 1.19 gm/cm$^3$ (Gourlay *et al.*, 1973), and unit membrane of the viral envelope in electron micrographs (Gourlay, 1973). Putzrath and Maniloff (1979) found that the MVL2 viral membrane fatty acid composition and thermal transition temperature are similar to those of the cells on which the viruses are propagated.

Garwes *et al.* (1975) reported that fucose accounted for 1.8% (w/w) of the composition of the Group 3 virus. However, it was not possible to determine whether fucose is present as glycoprotein or is associated with the DNA.

### 2. Antiserum Inactivation

Mycoplasmaviruses have been classified into three groups by examining the ability of specific antiserum to inhibit plaque formation (Gourlay, 1972). Maniloff and Liss (1974) observed a threefold range in inactivation rates among four Group 1 viruses using antiserum against MVL51. This indicates some degree of serological heterogeneity among Group 1 isolates. Using a virus neutralization assay, Phillpotts *et al.* (1977) also found antigenic heterogeneity in a study of six Group 2 isolates. Some of these Group 2 differences may be due to cell antigens either in the viral envelope or as contaminants in the virus preparations, since Group 2 viruses mature by budding from the cell membrane and virus preparations have been shown to contain cell proteins (Lombardi and Cole, 1978; Putzrath and Maniloff, 1979).

### 3. Ultraviolet Inactivation

All three groups of mycoplasmaviruses can be inactivated by ultraviolet (uv) light (reviewed by Maniloff *et al.*, 1977a). The uv doses which give survival of 37% of the viruses are (in order of increasing uv sensitivity): 0.109 $kJ/m^2$ for Group 2, 0.048 $kJ/m^2$ for Group 3, and 0.028 $kJ/m^2$ for Group 1 viruses. Some Group 1 isolates have small but significantly different uv inactivation cross sections, indicating that not all of these isolates are identical.

MVL2, a Group 2 mycoplasmavirus, can be host cell reactivated (Das *et al.*, 1977), a mode of repair of the uv-damaged viral DNA that almost certainly utilizes the *A. laidlawii* excision repair system. Since Group 1 viruses have single-stranded DNA, they cannot be host cell reactivated (Das *et al.*, 1977).

Ultraviolet reactivation of Group 1 and 2 viruses has also been shown (Das *et al.*, 1977). This was measured as the enhanced survival of uv-inactivated virus when the virus was assayed on a lightly uv-irradiated host compared to assay on an unirradiated host (Weigle, 1953). In *Escherichia coli*, ultraviolet reactivation of viral DNA has been proposed to be due to an inducible, error prone repair mechanism (Radman, 1974). This may indicate the existence of such an inducible enzyme system in mycoplasmas.

### 4. Thermal Stability

Group 1 viruses are relatively heat stable. At 60°C and pH 8.0, the surviving fraction of MVL51 is 0.37 after 3 min and 0.03 after 30 min (J. Das and J. Maniloff, unpublished data). Inactivation is exponential only to about 15% survival. Gourlay and Wyld (1972) showed that inactivation is exponential at temperatures above 70°C. Clyde (1974b) reported that Group 1 viruses are resistant to 20 cycles of freezing and thawing.

Group 2 viruses are heat labile (Gourlay, 1971). At 60°C and pH 8.0, the survival of MVL2 is about $4 \times 10^{-4}$ in 5 min and inactivation is exponential (J. Das and J. Maniloff, unpublished data).

Group 3 virus is relatively heat stable at 60°C (Gourlay and Wyld, 1973).

### 5. Detergent and Organic Solvent Inactivation

Group 1 viruses are resistant to relatively high concentrations of detergents. There is no loss of virus titer after treatment with 0.4% Nonidet P40 (Gourlay, 1970), 0.4% Sarkosyl NL97 (Das and Maniloff, 1975), or 0.4% Triton X100 (Liss and Maniloff, 1971). These viruses are also resistant to ether but are sensitive to chloroform (Gourlay, 1970; Gourlay and Wyld, 1972).

Group 2 viruses are sensitive to detergents, ether, and chloroform (Gourlay, 1971), consistent with the fact that these virions are membrane-bound.

Group 3 viruses are resistant to Nonidet P40 and ether (Gourlay and Wyld, 1973).

## B. Biological Properties

### 1. Host Range

Several *Acholeplasma* and *Mycoplasma* species have been tested as indicators of mycoplasmaviruses. Only some *A. laidlawii* strains have been shown to produce plaques with any of the three groups of viruses (reviewed by Maniloff and Liss, 1974).

### 2. Host Modification and Restriction

Many bacteriophages have been shown to exhibit an altered host range as a result of a single cycle of phage growth in a particular host (reviewed by Arber, 1974). This host-induced modification differs from mutational modification of host range, since the property is not heritable and the modification is lost when the phage is grown in some other host.

Host modification and restriction of mycoplasmaviruses has been studied using MVL51 (a Group 1 virus), MVL2 (a Group 2 virus), and *A. laidlawii* strains JA1 and M1305/68 (Maniloff and Das, 1975) (Table II). JA1 neither modifies nor restricts MVL51. M1305/68 slightly restricts MVL51 propagated on JA1 and modifies the virus so that the plating on M1305/68 is improved. However, even the modified virus plates less well on M1305/68 than on JA1. MVL2 showed host controlled modification and restriction in both hosts, giving 1000-fold more plaques on the host in which it had been grown. Host modification and restriction of five other Group 2 mycoplasmavirus isolates was observed by Phillpotts *et al.*

TABLE II. **Relative Number of Virus Plaques on *Acholeplasma laidlawii* Lawns**[a]

| Virus[b] | Relative PFU on *A. laidlawii* | |
|---|---|---|
| | Strain JA1 | Strain 1305 |
| MVL51·JA1 | 1 | 0.2 |
| MVL51·1305 | 1 | 0.6 |
| MVL2·JA1 | 1 | 0.015 |
| MVL2·1305 | 0.011 | 1 |

[a] From Maniloff and Das (1975) and Nowak *et al.* (1976).
[b] The symbols after the dot indicate the last host on which the virus was grown.

(1977). Preliminary data (J. A. Nowak, unpublished observations) show that MVL3 is neither modified nor restricted.

The number of plaques produced on the restricting host depends upon the physiological state of the cells. If the host is irradiated with ultraviolet light before MVL2 infection, there is an increase in plating efficiency on the restricting host (Maniloff *et al.*, 1977b).

### 3. Induction of Interferon

Lombardi and Cole (1978) have found that a Group 2 mycoplasmavirus, but not a Group 1 or Group 3 virus, can induce interferon *in vitro* in sheep peripheral blood lymphocytes. Thermal inactivation of the Group 2 virus reduced the amount of interferon induced, but uv and antiserum inactivation of viral infectivity did not affect its interferon induction.

## C. Replication of Group 1 Viruses

### 1. "One-Step" Growth

Group 1 virus "one-step" growth curves have a short latent period of about 10 min followed by a gradual increase in virus titer over several hours (Liss and Maniloff, 1971, 1973b). This observation led to the original suggestion that viral production was by a nonlytic mechanism. After the rise period the virus titer reaches a plateau. For MVL51 infection in tryptose broth, the plateau is reached about 2 hr after infection and by this time about 150–200 virus particles have been released per infected cell. The one-step growth curves for other Group 1 isolates exhibit the same pattern, but the growth parameters (latent period and amount of progeny released per infected cell) are different for different isolates (Liss and Maniloff, 1971, 1973b).

### 2. Artificial Lysis

The possible formation of intracellular virions during the viral latent period was examined by artifically lysing infected cells (Liss and Maniloff, 1973b). These experiments were carried out like one-step growth experiments, except that each time a sample was removed for PFU assay a parallel sample was removed, lysed with 0.2% Triton X100, and assayed for PFU. During the latent period no infectious intracellular virions could be demonstrated. During the rise period the number of PFU in the lysed sample increased to equal, but never exceed, the number of PFU in the unlysed sample, reflecting the release of the progeny virions. These experiments are consistent with a nonlytic viral infection, with no intracellular pool of completed viruses and virus maturation and release being coincident.

### 3. "Single-Burst" Experiments

Although infection is nonlytic and viruses are released continually, rather than in a burst at cell lysis, the term "burst" is used in discussing virus yield. The number of progeny viruses released from individual MVL51-infected cells was determined by the protocol of Ellis and Delbrück (1939). The average virus yield is 88 per cell at 60 min after infection and 154 per cell at 120 min (Liss and Maniloff, 1973b). These data confirm that each infected cell releases virus continuously at a constant rate.

The "burst size" measured in one-step growth experiments is an average over all infected cells. For Group 1 viruses this ranges from three for the MVL1 isolate to 150–200 for the MVL51 isolate (Liss and Maniloff, 1971, 1973b). Both the virus yield and the growth rate of infected cells vary as a function of MOI (Maniloff and Liss, 1973; Liss and Maniloff, 1973b). At a MOI above 10, virus yield is variable and usually less than that measured at low MOI.

### 4. Growth of Infected Cells

MVL51-infected cells (at a MOI less than 10) produce virus without a loss of cell titer (Liss and Maniloff, 1973b). Above MOI 10, there is a decrease in the number of colony-forming units (Maniloff and Liss, 1973). Infected cells (MOI less than 10) grow slower than uninfected cells: the infected cell doubling time is 160 min, compared to 110 min for uninfected cells (Liss and Maniloff, 1973b). Infected cells also make smaller colonies than uninfected cells: After 3 days of incubation, infected cell colony size is about 0.3 mm and uninfected cell colony size about 0.5 mm (Maniloff and Liss, 1973).

Some early reports on these viruses considered infection to be lytic because plaque formation was assumed to imply lysis of infected cells. Studies of filamentous nonlytic bacterial viruses have shown that their plaques are due to infected cells growing more slowly and making smaller colonies than uninfected ones (Hsu, 1968). Similarly, Group 1 virus-infected mycoplasmas grow more slowly and make smaller colonies (Maniloff and Liss, 1973). Therefore, Group 1 virus plaques are produced by the differential growth rate and size of infected cell colonies, resulting in an uninfected cell lawn appearing more dense than an area of infected cells.

## D. Molecular Details of Group 1 Virus Replication

### 1. Adsorption and Penetration

The first steps of infection by MVL51 are: (1) attachment or adsorption of the virus to the cell, and (2) penetration of viral DNA into the cell. Adsorption is ionic and requires mono- or divalent cations (Fraser and Fleischmann, 1974). Adsorption follows first-order kinetics with a rate constant about $3 \times 10^{-9}$ cm$^2$/min. The theoretical adsorption rate constant ($2\text{–}5 \times 10^{-9}$ cm$^3$/min), calculated assuming single-hit collision kinetics (Fraser and Fleischmann, 1974; Maniloff and Liss, 1974), is in agreement with this experimental value, indicating that nearly every virus–cell collision results in adsorption.

Fraser and Fleischmann (1974) estimated that each cell has about 10 MVL51 adsorption sites. Das and Maniloff (1976a) found that, although infected cells can take up several viral DNA molecules, only two or three parental viral DNA molecules per cell bind to the cell membrane and participate in DNA replication.

The adsorption rate constant is not very temperature dependent over the range of 0° to 42°C, indicating that cell metabolism is probably not required for this process (Fraser and Fleischmann, 1974). Treatment of cells before infection with chloramphenicol or rifampicin had little effect on penetration and binding of viral DNA to the cell membrane sites (Das and Maniloff, 1976c). This indicates that neither viral transcription nor the synthesis of new proteins is required for DNA penetration.

### 2. Replicative Intermediates

The first step of replication involves conversion of parental viral single-stranded DNA to double-stranded replicative forms (Das and Maniloff, 1976a). This step is not affected by pretreatment of cells with chloramphenicol, so it is presumably carried out by preexisting host cell enzymes (Das and Maniloff, 1976c). The replicative forms are RFI, a covalently closed double-stranded circular derivative of the viral DNA, and RFII, a relaxed form of RFI (Fig. 1). The structure of MVL51 RFI and RFII has been determined by ethidium bromide–CsCl equilibrium gradient analysis (Das and Maniloff, 1975) and by electron microscopy (Maniloff *et al.*, 1977a).

During the second stage of replication, parental RF replicates to form a pool of progeny RF molecules (Das and Maniloff, 1975). This requires at least one cellular function, since $REP^-$ cells, which are capable of forming parental RF, are blocked in further replication (Nowak *et al.*, 1976).

In the final phase of replication, RFII molecules serve as precursors for synthesis of progeny viral circular single-stranded DNA (SSI). Growth in nutrient-poor medium allows identification of an intermediate between nascent progeny viral chromosomes (SSI) and mature viruses (Das and Maniloff, 1976b). This intermediate (SSII) is a protein-associated form of SSI and its formation from SSI is inhibited by chloramphenicol. SSII contains two proteins, which are electrophoretically identical to two of the four virion proteins (Maniloff and Das, 1975). These are the 70,000, and 53,000 dalton proteins and, in SSII, have an approximate stoichiometric ratio of 1.4:1.0.

### 3. Steps in DNA Replication

Parental viral DNA becomes associated with the host cell membrane after infection: This association is pronase sensitive (Das and Maniloff, 1976a). The association was demonstrated by infecting cells with $^{32}P$-labeled virus, lysing the cells by freezing and thawing, and sedimenting the lysate through a sucrose gradient over a CsCl shelf. Most parental label sedimented with the cell membrane as a fast sedimenting complex. The remaining label was in slow sedimenting material.

The amount of fast and slow sedimenting parental viral DNA was measured as a function of MOI (Das and Maniloff, 1976a). Membrane-associated parental DNA reached a saturation value and, from this, it was calculated that there are only two or three sites for membrane-associated parental DNA per cell.

Parental viral DNA is not transferred to progeny viruses, as shown by experiments in which cells were infected with $^{32}P$-labeled MVL51. The amount of intracellular $^{32}P$-labeled viral DNA remained constant and no extracellular acid-insoluble $^{32}P$ could be detected (Das and Maniloff, 1976a).

Several different experiments have shown that progeny RF synthesis occurs by symmetric replication at cell membrane sites (Das and Maniloff, 1976a). First, during early infection most nascent DNA is in RF and 80% of nascent DNA is membrane-associated. Second, equilibrium centrifugation analysis of membrane-associated progeny RFII in alkaline CsCl gradients showed nascent DNA in both viral and complementary strands, indicating symmetrical replication of the double-stranded RF. Third, when infected cells were pulsed with [$^3$H]deoxythymidine for short periods during early infection (when RF replication predominates), nascent DNA in slow sedimenting material, relative to fast sedimenting material, increased with increasing pulse length (Maniloff *et al.*, 1977a). This suggests that RF molecules are synthesized at the membrane and then released into the cytoplasm.

Later in infection, while progeny RF replication continues and progeny SSI synthesis increases, most nascent DNA is found as slow sedimenting material. Analysis of slow sedimenting progeny RFII by equilibrium centrifugation in alkaline CsCl gradients showed nascent DNA only in the viral strand (Das and Maniloff, 1975a), indicating that progeny SSI synthesis involves asymmetric replication. This does not rule out some asymmetric synthesis on the membrane.

Electron microscopy of infected cells shows clusters of extracellular progeny viruses at a few membrane sites, indicating that assembly and release occur at a limited number of sites per cell (Liss and Maniloff, 1973b).

### 4. Synthesis of Virus-Specific Proteins

*In vivo* synthesis of MVL51-specific proteins has been studied (Das and Maniloff, 1978). Six virus-specific proteins were identified with molecular weights 70,000, 50,000, 30,000, 19,000, 14,000, and 10,000. The four largest proteins are probably the four virion proteins. The amount of 70,000 dalton protein remains about the same throughout infection, while the pools of the 50,000, 30,000, and 19,000 dalton proteins increase during infection. For the nonstructural proteins, the amount of 14,000 dalton protein is maximal when most viral DNA replication is the RF → RF synthesis, suggesting this protein might have a role in progeny RF synthesis. The amount of 10,000 dalton protein increases late in infection when most viral DNA synthesis is in SSI formation, suggesting this protein might have a role in SSI synthesis.

### 5. Replication in REP$^-$ Cells

*Acholeplasma laidlawii* cells with a REP$^-$ phenotype are able to propagate double-stranded DNA mycoplasmaviruses but not single-stranded DNA mycoplasmaviruses. The block in MVL51 replication in REP$^-$ cells is in the RF → RF step (Nowak *et al.*, 1976). Studies of protein synthesis in infected REP$^-$ cells, at a time when most viral replication should involve RF → RF synthesis, show that the amount of 70,000 dalton protein is the

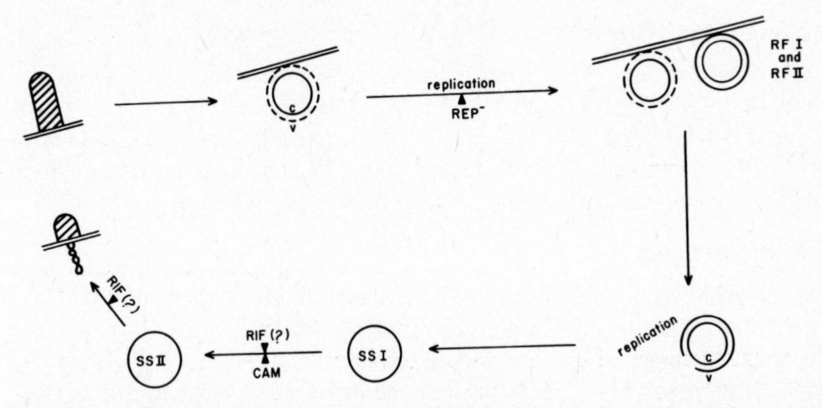


FIGURE 2. Schematic presentation of the replication of Group 1 mycoplasmaviruses. Dotted lines show parental viral DNA strands and continuous lines show progeny DNA strands. Steps blocked by rifampin (RIF), chloramphenicol (CAM), and REP⁻ cell variant are shown. Viral and complementary DNA strands are marked v and c, respectively. The parallel double lines denote the cell membrane. (From Maniloff *et al.*, 1977a.)

same as in wildtype cells (Das and Maniloff, 1978). The amounts of 50,000-, 30,000-, and 19,000-dalton structural proteins are less in REP⁻ cells than in wild-type cells, indicating RF replication may be necessary to stimulate synthesis of these proteins. The amount of 10,000 dalton protein is small in both types of cells at this infection time. However, the 14,000 dalton protein cannot be detected in REP⁻ cells.

### 6. Summary

Figure 2 shows a schematic presentation of the replication of Group 1 mycoplasmaviruses, based on the studies of MVL51 reviewed above. Data on the regulation and role of cellular and viral functions in viral replication are still incomplete.

## E. Replication of Group 2 Viruses

### 1. Adsorption

The first-order adsorption rate constant of MVL2 was experimentally determined to be about $2 \times 10^{-10}$ cm³/min (Putzrath and Maniloff, 1977). The theoretical rate constant (calculated from collision kinetics) is an order of magnitude greater than this experimental result. Therefore, under the experimental conditions investigated, only about 1 in 10 of the potential virus–cell collisions leads to adsorption.

### 2. "One-Step" Growth

The MVL2 "one-step" growth curve has a latent period of about 120 min (118 ± 29 min) followed by an exponential release of progeny virus, lasting 4–6 hr (Putzrath and Maniloff, 1977). During this rise period about 1000 PFU are released per infected cell. The gradual nature of viral release first indicated that MVL2 infection is nonlytic.

### 3. Artificial Lysis

During the latent period of an MVL2 artificial lysis experiment no PFU were found in the lysed sample. The number of PFU in the lysed sample increased during the rise period to equal the number of PFU in the unlysed sample. This reflects the release of progeny virus and is consistent with there being no intracellular pool of completed virus, and virus maturation and release being coincident (Putzrath and Maniloff, 1977).

### 4. Growth of Infected Cells

Since MVL2 infection is nonlytic, infected cells continue to grow. However, the doubling time of infected cells is 4–5 hr, compared to about 1.5 hr for uninfected cells, and infected cells give rise to smaller colonies than uninfected cells (Putzrath and Maniloff, 1977, 1978). This differential growth rate and colony size between infected and uninfected cells explains why these nonlytic viruses are able to produce plaques on indicator lawns.

### 5. Persistent Infection

MVL2 infection of *A. laidlawii* cells leads to the establishment of a persistent infection (Putzrath and Maniloff, 1977, 1978). Persistently infected *A. laidlawii* clones are resistant to superinfection by homologous (Group 2) virus but can be infected by heterologous (Group 1) virus. Although most cells in a persistently infected culture do not plate as infectious centers, they retain the potential to produce virus and transmit this potential as a stable heritable trait. Mitomycin C and ultraviolet light induce an increase in infectious centers in persistently infected cultures.

### 6. Interactions between Viral and Cell Membranes

Since MVL2 is an enveloped virus and matures by budding from the cell membrane (Maniloff *et al.*, 1977c), there must be interactions between viral and cell membranes during both adsorption (and penetration) and maturation. MVL2 infection was found to have an unexpected effect on the osmotic fragility of *A. laidlawii* cells (Putzrath and Maniloff, 1977).

Newly infected cells are more resistant to osmotic shock than uninfected cells: However, after establishment of a persistent infection, the cells become more sensitive and their osmotic fragility is indistinguishable from noninfected cells.

Putzrath and Maniloff (1979) have examined the effect of *A. laidlawii* cell membrane composition on the growth and composition of MVL2 virus. Differences in host cell membrane proteins were found to alter the virion protein composition. Similarly, cells grown in modified medium which altered the cell membrane lipids gave rise to viruses with altered lipid composition: The viral lipid thermal transition temperature was the same as that of the cell membrane. Such alteration in cell lipids allowed studies of viral infection at temperatures below the cell membrane transition temperature. Although cell growth was greatly reduced below the membrane lipid transition temperature, the more ordered state of the cell and viral lipids had no measurable effect on viral adsorption or maturation by budding.

## F. Replication of Group 3 Viruses

### 1. Adsorption

From the adsorption kinetics of MVL3 the first-order virus adsorption rate constant has been found to be $3 \times 10^{-10}$ cm$^3$/min (Haberer, 1978). This is an order of magnitude less than the theoretical rate constant calculated for collision kinetics. Therefore, under the experimental conditions studied, only about 1 in 10 virus–cell collisions leads to adsorption.

### 2. One-Step Growth

MVL3 infection has been studied in *A. laidlawii* strains BCL-13 (Liss, 1977) and M1305/68 (Haberer, 1978). In BCL-13 cells, the one-step growth curve has a 20 min latent period followed by a 50-min rise period, with a burst size of six to seven progeny MVL3 per infected cell. However, the one-step growth curve in M1305/68 cells has a 5 hr latent period, followed by a 3 hr rise period, with a burst size of five virions per infected cell. These data suggest that MVL3 viral infection is lytic.

### 3. Artificial Lysis

These experiments demonstrated the production of intracellular infectious MVL3 virus particles during the latent period, as expected for a lytic infection. In BCL-13 cells, where the virus latent period is 20 min, intracellular infectious progeny virus were found to be present at 10–12 min after infection (Liss, 1977). For M1305/68 cells, where the latent period is 5 hr, infectious intracellular virions were observed 3 hr after infection (Haberer, 1978).

#### 4. Single-Burst Experiments

From these studies of MVL3-infected M1305/68 cells (Haberer, 1978), the average virus yield per infected cell was found to be 5. No increase in PFU could be obtained by longer incubations.

#### 5. Growth of Infected Cells

These experiments confirmed that MVL3 is a lytic virus (Liss, 1977; Haberer, 1978). A loss of viability was observed in infected cultures, with the cell survival fraction being consistent with the interpretation that every cell that received one virion (calculated from the Poisson distribution for the MOI in each experiment) was inactivated as a viable unit.

### G. Transfection

Although DNA-mediated transformation in mycoplasmas has not been demonstrated, DNA isolated from Group 1 viruses is able to transfect mycoplasmas (Liss and Maniloff, 1972, 1974). *Acholeplasma laidlawii* host cells are competent for transfection during the late logarithmic phase of growth. Compared to infection, transfection has a longer latent period and a smaller virus yield. The transfection kinetics indicate that 10–15 min is required for the uptake of viral DNA into a deoxyribonuclease-resistant form.The dose–response curve (plot of the logarithm of the number of transfectants against DNA concentration) for transfection has a slope of two, indicating that an average of two molecules of DNA per cell may be required in order to produce an infection. The maximum transfection efficiency is 3–4 $\times$ $10^5$ viral equivalents of DNA per transfectant.

It has also been found that *Mycoplasma gallisepticum* A5969, which cannot be infected by Group 1 virus, can be transfected by Group 1 viral DNA and produce progeny viruses (Maniloff and Liss, 1973). This is most interesting in view of the requirement for a number of cell functions in Group 1 viral replication (described above).

## III. VIRUSES AND VIRUSLIKE PARTICLES OF SPIROPLASMAS

### A. General Considerations

The spiroplasmas are a group of helical mycoplasmas which have been found in plants and insect vectors associated with a variety of plant diseases (e.g., Davis *et al.*, 1972; Bové *et al.*, 1973; Daniels *et al.*, 1973). The sex-ratio organism (SRO) of *Drosophilia* was originally identified as a spirochete (Poulson and Sakaguchi, 1961) and more recently shown to be a spiroplasma (Williamson and Whitcomb, 1974). In addition, the suckling mouse cataract agent (SMCA), isolated from ticks and originally thought to be a slow virus, has been found to be a spiroplasma (Tully *et al.*, 1977).

Most of the plant spiroplasma isolates examined were found to contain viruslike particles (Cole *et al.*, 1974; Cole, 1977). These studies were done by electron microscopic examination of cell cultures and three different morphological types of particles were described: SVC1 particles are rod-shaped, SVC2 particles are polyhedral with long tails, and SVC3 particles are polyhedral with short tails (see Chapter 14 in this volume).

The eight *Drosophilia* strains carrying the SRO trait that have been studied so far have been found to be infected by viruses (Oishi and Poulson, 1970; Oishi, 1971). Cole (1977) examined one SRO strain and reported both SVC1 and SVC3 particles. The two SMCA strains examined were found to contain SVC3 (Cole, 1977). (See this volume, Chapter 14, and Volume III.)

## B. Biological Properties

Until recently, it had not been possible to propagate viruslike particles associated with spiroplasma cultures (Cole *et al.*, 1974). Varying numbers of particles were observed in electron microscopic preparations of spiroplasma cell cultures. However, particles were not found in every sample nor could they be found in every passage of a culture containing particles in prior and subsequent passages. In general, the relationship of the particles to viruses is circumstantial and depends on their striking morphologic similarity to known viruses. Cole *et al.* (1977) have recently reported the development of a plaque assay for SVC3 isolates (but not SVC1 or SVC2) on an *S. citri* strain. This has enabled large amounts of SVC3 to be obtained for chemical analysis (described below) and should allow the development of spiroplasma virology.

## C. Chemical Composition

The use of an *S. citri* indicator strain allowed Cole *et al.* (1977) to obtain sufficient quantities of SVC3 for chemical analysis. The density of the virus is 1.26 gm/cm$^2$ in metrizamide and 1.45 gm/cm$^2$ in CsCl. The virion genome is linear double-stranded DNA of molecular weight $14 \times 10^6$. Purified virions contained five proteins with molecular weights of 130,000, 110,000, 67,000, 48,000, and 38,000.

## IV. CONCLUSIONS

Although the first mycoplasmavirus was not isolated until 1970 (Gourlay, 1970), since then they have been reported in species of *Mycoplasma, Acholeplasma,* and *Spiroplasma* (described in this chapter). The apparent ubiquity of the mycoplasma- and spiroplasmaviruses may reflect the spread of viruses and/or infectious DNA in the normal ecological situation.

The further development of mycoplasma and spiroplasma virology should allow the evaluation of these viruses as pathogenic determinants. Both the ubiquity of the viruses and considerations of comparative pathology indicate that a viral role in mycoplasma- and spiroplasma-associated disease states needs clarification. In addition, further studies promise to contribute to an understanding of viral replication and virus–cell interactions in systems of plasma membrane-bounded viruses and cells.

Finally, these viruses allow new approaches to studies of the molecular biology of mycoplasmas and spiroplasmas.

## ACKNOWLEDGMENTS

Support for the studies in this laboratory has come from the United States Public Health Service, National Institute of Allergy and Infectious Diseases, Grant AI-10605, and the University of Rochester Biomedical and Environmental Research Project.

## REFERENCES

Allen, T. C. (1972). *Virology* **47,** 491.

Arber, W. (1974). *Prog. Nucleic Acid Res. Mol. Biol.* **14,** 1.

Bové, J. M., Saglio, P., Tully, J. G., Freundt, A. E., Lund, Z., Pillot, J., and Taylor-Robinson, D. (1973). *Ann. N.Y. Acad. Sci.* **225,** 462.

Clyde, W. A. (1974a). *Colloq. Inst. Natl. Sante Rech. Med.* **33,** 109.

Clyde, W. A. (1974b). *Ann. N.Y. Acad. Sci.* **225,** 159.

Cole, R. M. (1977). *In* "Plant and Insect Viruses: An Atlas" (A. J. Dalton, F. Hagenau, and K. Maramorosch, eds.), p. 451. Academic Press, New York.

Cole, R. M., Tully, J. G., and Popkin, T. J. (1974). *Colloq. Inst. Natl. Sante Rech. Med.* **33,** 125.

Cole, R. M., Mitchell, W. O., and Garon, C. F. (1977). *Science* **198,** 1262.

Daniels, M. J., Markham, P. G., Meddins, B. M., Plaskitt, A. K., Townsend, R., and Bar-Joseph, M. (1973). *Nature (London)* **244,** 523.

Das, J., and Maniloff, J. (1975). *Biochem. Biophys. Res. Commun.* **66,** 599.

Das, J., and Maniloff, J. (1976a). *Proc. Natl. Acad. Sci. U.S.A.* **73,** 1489.

Das, J., and Maniloff, J. (1976b). *Microbios* **15,** 127.

Das, J., and Maniloff, J. (1976c). *J. Virol.* **18,** 969.

Das, J., and Maniloff, J. (1978). *Virology* **86,** 186.

Das, J., Howak, J. A., and Maniloff, J. (1977). *J. Bacteriol.* **129,** 1424.
Davis, R. E., Worley, J. F., Whitcomb, R. F., Ishijima, T., and Steere, R. L. (1972). *Science* **176,** 521.
Day, L. A., and Berkowitz, S. A. (1977). *J. Mol. Biol.* **116,** 603.
Drasil, V., Doskar, J., Tkadlecek, L., Bohacek, J., and Koudelka, J. (1978). *Abstr. Fourth Internat. Congr. Virology, The Hague,* p. 174.
Ellis, E. L., and Delbrück, M. (1939). *J. Gen. Physiol.* **22,** 365.
Fraser, D., and Fleischmann, C. (1974). *J. Virol.* **13,** 1067.
Garwes, D. J., Pike, B. V., Wyld, S. G., Pocock, D. H., and Gourlay, R. N. (1975). *J. Gen. Virol.* **29,** 11.
Giannotti, J., Devauchelle, G., Vago, C., and Marchoux, G. (1973). *Ann. Phytopathol.* **5,** 461.
Gourlay, R. N. (1970). *Nature (London)* **225,** 1165.
Gourlay, R. N. (1971). *J. Gen. Virol.* **12.** 65.
Gourlay, R. N. (1972). *Pathog. Mycoplasmas, Ciba Found. Symp., 1972,* p. 145.
Gourlay, R. N. (1973). *Ann. N.Y. Acad. Sci.* **225,** 144.
Gourlay, R. N. (1974). *Crit. Rev. Microbiol.* **3,** 315.
Gourlay, R. N., and Wyld, S. G. (1972). *J. Gen. Virol.* **14,** 15.
Gourlay, R. N., and Wyld, S. G. (1973). *J. Gen. Virol.* **19,** 279.
Gourlay, R. N., Garwes, D. J., Bruce, J., and Wyld, S. G. (1973). *J. Gen. Virol.* **18,** 127.
Gourrett, J. P., Maillet, P. L., and Gouranton, J. (1973). *J. Gen. Microbiol.* **74,** 241.
Haberer, K. (1978). *Abstr. Fourth Internat. Congr. Virology, The Hague,* p. 173.
Hsu, Y. C. (1968). *Bacteriol. Rev.* **32,** 387.
Liska, B. (1972). *Stud. Biophys.* **34,** 151.
Liss, A. (1977). *Virology* **77,** 433.
Liss, A., and Maniloff, J. (1971). *Science* **173,** 725.
Liss, A., and Maniloff, J. (1972). *Proc. Natl. Acad. Sci. U.S.A.* **69,** 3423.
Liss, A., and Maniloff, J. (1973a). *Biochem. Biophys. Res. Commun.* **51,** 214.
Liss, A., and Maniloff, J. (1973b). *Virology* **55,** 118.
Liss, A., and Maniloff, J. (1974). *Microbios* **11,** 107.
Lombardi, P. S., and Cole. B. C. (1978). *Infect. Immun.* **20,** 209.
Lombardi, P. S., and Cole, B. C. (1979). *J. Virol.* (in press).
Maniloff, J., and Das J. (1975). *In* "DNA Synthesis and Its Regulation" (M. Goulian, P. Hanawalt, and C. F. Fox, eds.), p. 445. Benjamin, Reading, Massachusetts.
Maniloff, J., and Liss, A. (1973). *Ann. N.Y. Acad. Sci.* **225,** 149.
Maniloff, J., and Liss, A. (1974). *In* "Virus, Evolution and Cancer" (E. Kurstak and K. Maramorosch, eds.), p. 584. Academic Press, New York.
Maniloff, J., Das, J., and Christensen, J. R. (1977a). *Adv. Virus Res.* **21,** 343.
Maniloff, J., Das, J., and Nowak, J. A. (1977b). *Beltsville Symp. Agric. Res.* **1,** 221.
Maniloff, J., Das, J., and Putzrath, R. M. (1977c). *In* "Insect and Plant Viruses: An Atlas" (A. J. Dalton, F. Hagenau, and K. Maramorosch, eds.), P. 439. Academic Press, New York.
Maniloff, J., Das, J., and Nowak, J. A. (1978). *In* "Single-stranded DNA Phages" (D. T. Denhardt, D. H. Dressler, and D. S. Ray, eds.). Cold Spring Harbor Lab., Cold Spring Harbor, New York (in press).
Nowak, J. A., and Maniloff, J. (1979). *J. Virol.* (in press).
Nowak, J. A., Das, J., and Maniloff, J. (1976). *J. Bacteriol.* **127,** 832.
Nowak, J. A., Maniloff, J., and Das, J. (1978). *FEMS Lett.* **4,** 59.
Oishi, K. (1971). *Genet. Res.* **18,** 45.
Oishi, K., and Poulson, D. F. (1970). *Proc. Natl. Acad. Sci. U.S.A.* **67,** 1565.
Phillpotts, R. J., Patel, K. K. T., and Edward, D. G. (1977). *J. Gen. Virol.* **36,** 211.

Ploaie, P. G. (1971). *Rev. Roum. Biol., Ser. Bot.* **16,** 3.
Poulson, D. F., and Sakaguchi, B. (1961). *Science* **133,** 1489.
Putzrath, R. M., and Maniloff, J. (1977). *J. Virol.* **22,** 308.
Putzrath, R. M., and Maniloff, J. (1978). *J. Virol.* **28,** 254.
Putzrath, R. M., and Maniloff, J. (1979). Submitted for publication.
Radman, M. (1974). *In* "Molecular and Environmental Aspects of Mutagenesis" (L. Prakash *et al.*, ed.), p. 128. Thomas, Springfield, Illinois.
Tully, J. G., Whitcomb, R. F., Clark, H. F., and Williamson, D. L. (1977). *Science* **195,** 892.
Watkins, A. L., (1977). *Annu. Meet. Am. Soc. Microbiol., 77th, 1977, New Orleans* (unpublished).
Weigle, J. J. (1953). *Proc. Natl. Acad. Sci. U.S.A.* **39,** 628.
Williamson, D. L., and Whitcomb, R. F. (1974). *Colloq. Inst. Natl. Sante Rech. Med.* **33,** 283.

# 16 / SPECIAL FEATURES OF THE ACHOLEPLASMAS

*Joseph G. Tully*



## I. INTRODUCTION AND HISTORY

Mycoplasmas currently classified as *Acholeplasma laidlawii* were initially recovered from sewage (Laidlaw and Elford, 1936) and soil or compost (Seiffert, 1937a,b). These organisms differed significantly from other mycoplasmas known at the time in the lack of growth requirement for serum or sterols and in their ability to grow at temperatures as low as 22°C. For over 30 years they were considered to be saprophytes, although no information was available that these wall-free organisms could exist for



THE MYCOPLASMAS, VOL. 1

ISBN 0-12-078401-7

any extended period in such environments. The first evidence of an association between acholeplasmas and animal hosts came in 1950 when a number of strains, subsequently identified as *A. laidlawii,* were recovered from the bovine genital tract (Edward, 1950a). In the intervening years it has been shown that acholeplasmas occur in a wide variety of domesticated animals; this very intimate association probably offers at least one explanation for the entry, persistence, and recovery of acholeplasmas from soil and sewage sites.

New information, which has been developed from attempts to cultivate plant mycoplasmas, offers a number of observations which also suggest that we might have to modify our concepts about acholeplasmas existing as free-living saprophytes. While much of the new information on plant acholeplasmas requires extensive confirmation, it seems probable, on balance, that acholeplasmas do occur in association with plants. Although acholeplasmas can multiply in insects (Whitcomb *et al.*, 1973; Whitcomb and Williamson, 1975), there is little information on mechanisms by which insects might interact with acholeplasmas and plants.

Acholeplasmas have also become important tools in the study of membrane biology. As a result of their relative susceptibility to osmotic lysis, yields of membranes free of other types of intracytoplasmic membranes or cell wall components are readily obtained (see this volume, chapter 8). The acholeplasmas are also of special importance to the cell culturist since they frequently occur as contaminants in vertebrate and invertebrate cell cultures. The persistent problems associated with acholeplasmas in cell cultures arise from their intimate association with animal tissues (see Volume II, Chapter 13).

## II. TAXONOMY

### A. Distinctions at the Family and Genus Level

In the currently accepted taxonomic scheme, members of the order Mycoplasmatales are divided into three families (Mycoplasmataceae, Acholeplasmataceae, and Spiroplasmataceae), based primarily upon differences in their genome size, cellular morphology, and growth requirements for cholesterol. The family Acholeplasmataceae contains a single genus (*Acholeplasma*). The most unique characteristic of the acholeplasmas, and a property which distinguishes them from most other mycoplasmas, is their ability to grow in artificial medium without the addition of animal serum, cholesterol, or other sterols. However, acholeplasmas also possess a number of other molecular, genetic, and biochemical differences which separate them from mycoplasmas classified in the remaining two families (Table I). In addition a recent study of the characteristics of ribosomal ribonucleic acids (rRNAs) obtained from mycoplasmas and acholeplasmas suggests further genetic distinctions between these organisms (Reff *et al.*, 1977).

## B. Subgeneric Classification

Since a requirement for sterol was originally considered to be a fundamental property that distinguished mycoplasmas from true bacteria, the sterol-nonrequiring strains recovered from soil and sewage posed a taxonomic problem for a number of years (Edward, 1967). Resolution of this dilemma became imperative when it was shown that another mycoplasma (*Mycoplasma granularum*) originally thought to require sterols actually belonged to the sterol-nonrequiring group of mycoplasmas (Tully and Razin, 1968). Thus, a proposal was offered to create a new family (Acholeplasmataceae) and genus (*Acholeplasma*) which would give separate taxonomic status to those mycoplasmas (*A. laidlawii* and *A. granularum*) known at the time to have no growth requirements for cholesterol (Edward and Freundt, 1970). This proposal also contained recommended test procedures for establishing the sterol growth requirements of newly named mycoplasmas—tests that would allow assignment of an organism to one of the major families then existing. Shortly thereafter, an acholeplasma isolated from tissue cultures (*A. axanthum*) (Tully and Razin, 1969, 1970) and acholeplasmas from goats (*A. oculi*) (Al-Aubaidi *et al.*, 1973) and cattle (*A. modicum*) (Leach, 1973) were designated as new species.

The current classification scheme for acholeplasmas now includes seven species (see Table II). Differentiation at the species level is presently based on serologic distinctions and a limited number of biochemical or physiologic differences (see Section III). Despite the obvious need for better biologic markers to separate *Acholeplasma* species, nucleic acid hybridization (DNA) tests performed on the first three organisms assigned to this group confirmed the species separation that had been made on the basis of biochemical and serologic properties (McGee *et al.*, 1967; Pollock and Bonner, 1969; Neimark, 1970; E. A. Freundt, personal communication). However, there is need for similar data on some of the recently established *Acholeplasma* species. Indeed, nucleic acid hybridization and other molecular or genetic distinctions provide a substantial and necessary basis for species separation in all mycoplasmas.

TABLE I. **Molecular and Physiologic Distinctions among Families in the Order Mycoplasmatales**

| | Family (genus) | | | |
|---|---|---|---|---|
| | Mycoplasmataceae | | Acholeplasmataceae | Spiroplasmataceae |
| Characteristic | *Mycoplasma* | *Ureaplasma* | *Acholeplasma* | *Spiroplasma* |
| Genome size (daltons)[a] | $5 \times 10^8$ | $5 \times 10^8$ | $1 \times 10^9$ | $1 \times 10^9$ |
| Localization of NADH oxidase activity[b] | Cytoplasm | Cytoplasm | Membrane | Cytoplasm |
| Lactic dehydrogenases specifically activated by fructose 1,6-diphosphate[c] | — | NT[f] | + | NT |
| Biosynthesis of fatty acids from acetate[d] | − | + | + | − |
| Sterol requirement[e] | + | + | − | + |

[a] From Bak *et al.* (1969); Askaa *et al.* (1973); Black *et al.* (1972); Saglio *et al.* (1973).
[b] From Pollack *et al.* (1965); Pollack (1975); Pollack (1978); Kahane *et al.* (1977).
[c] From Neimark (1973).
[d] From Herring and Pollack (1975); Freeman *et al.* (1976); Romano *et al.* (1976); Pollack (1978).
[e] From Razin and Tully (1970); Rottem *et al.* (1971); Saglio *et al.* (1973); Freeman *et al.* (1976).
[f] NT = Not tested.

TABLE II. **Biochemical, Physiologic, and Molecular Properties of *Acholeplasma* Species**

| Species | Type strain | Reference collection number[a] | G + C of DNA (mol %) | Fermentation of: | | Arginine hydrolysis | Esculin hydrolysis | Arbutin hydrolysis | Pigmented carotenoid test | Film and spot reaction |
|---|---|---|---|---|---|---|---|---|---|---|
| | | | | Glucose | Mannose | | | | | |
| *A. axanthum* | S-743 | ATCC 25176<br>NCTC 10138 | 31.3 | + | − | − | + | 4+ (10)[b] | − | − |
| *A. equifetale* | C112 | ATCC 29724 | ND[c] | + | ± | − | ± | − (2) | − | + |
| *A. granularum* | BTS 39 | ATCC 19168<br>NCTC 10128 | 30.5–32.4 | + | − | − | − | − (5) | + | − |
| *A. hippikon* | C1 | ATCC 29725 | ND | + | + | − | − | − (1) | − | + |
| *A. laidlawii* | PG8 | ATCC 23206<br>NCTC 10116 | 31.7–35.7 | + | − (rare pos.) | − | + (few neg.) | ± (10) (4/10 neg.) | + | − |
| *A. modicum* | Squire (PG49) | ATCC 29102<br>NCTC 10134 | 29.3 | + | − | − | − | − (4) | −[d] | − |
| *A. oculi* (formerly *A. oculusi*) | 19L | ATCC 27350 | ND | + | − | − | + | 2+ (7) | + | − |

[a] ATCC = American Type Culture Collection; NCTC = National Collection of Type Cultures (Britain).

[b] From D. L. Rose and J. G. Tully (unpublished). Number of strains tested given in parentheses. The intensity of the reaction is denoted by the number of +, with +4 representing the highest.

[c] ND = Not done.

[d] This species is capable of producing carotenoids but in amounts insufficient to give positive reaction in the test procedure (see text).

### C. Electrophoretic Analysis of Cell Proteins

Polyacrylamide gel electrophoresis of mycoplasma cell proteins, employing the acid gel system first proposed for mycoplasmas (Razin, 1968), also has been used to confirm the assignment of various mycoplasmas to the genus *Acholeplasma* (Tully and Razin, 1968; Tully, 1973; Boden and Kirchhoff, 1977). Most of these studies showed that membranes of acholeplasmas share a number of common proteins (electrophoretic bands) so that adequate evaluation of new strains with this procedure must involve comparison of the candidate strain with several related strains of each recognized species. The use of an alkaline gel system, sodium dodecyl sulfate (SDS) for membrane solubilization, and slab or flat gel electrophoresis probably provides the most optimal conditions for current application of this technique to mycoplasmas (Daniels and Meddins, 1973; Wreghitt *et al.*, 1974).

## III. BIOCHEMICAL AND PHYSIOLOGIC CHARACTERISTICS

### A. Growth Response to Cholesterol

The growth response of mycoplasmas to cholesterol is most accurately determined by assessing growth on a number of serum-free medium preparations to which various concentrations of solubilized cholesterol are added (Razin and Tully, 1970; Edward, 1971). The acholeplasmas usually show no significant growth response with increasing cholesterol levels, whether the response is measured by total cellular yields (protein) in liquid media or by numbers of colony-forming units on solid medium. It is important in these tests to include palmitic acid or Tween 80 in the base medium since some acholeplasmas (e.g., *A. axanthum*) require additional essential fatty acids in the fatty acid-poor base medium before adequate cellular growth can occur (Tully and Razin, 1969). Other factors which may affect the assay and some comments on control preparations have been outlined (Razin and Tully, 1970; Tully, 1973). It should be emphasized that sterol requirements cannot be determined adequately by passage of the organism once or twice on a serum-free agar or broth preparation, because small amounts of sterols are often passed with the initial inoculum or some sterol may occur in other media components employed.

A number of tests for indirect assessment of cholesterol needs have been proposed. Most of these procedures are based upon the observation that sterol-requiring mycoplasmas have a higher concentration of cholesterol in their cell membrane than the sterol-nonrequiring strains and that certain chemical agents (detergents, polyenes, etc.) bind to sterols in the cell membrane. This complex usually results in lysis of the organisms. Thus, the differential sensitivity of membranes of mycoplasmas and acholeplasmas to such compounds as polyanethol sulfonate (Kunze, 1971; Andrews and Kunze, 1972), lysolecithin (Mardh and Taylor-Robinson, 1973; Soltesz and Mardh, 1977), digitonin (Smith and Rothblat, 1960; Freundt *et al.*, 1973), amphotericin (Rottem, 1972), or polyene antibiotics (Grabowski *et al.*, 1976) can provide indirect evidence of cholesterol incorporation into mycoplasma membranes. The digitonin disk plate procedure has been used most extensively and appears to combine reliability with simplicity in distinguishing most acholeplasmas from other mycoplasmas. Sterol-requiring mycoplasmas generally show 4–20 mm zones of growth inhibition around dried disks previously saturated with digitonin (1.5% solution in ethanol), while acholeplasmas are usually resistant to this concentration (Freundt *et al.*, 1973).

### B. Biochemical Activities and Distinctions

The most useful information on biochemical and physiologic properties of established *Acholeplasma* species is summarized in Table II. No single biochemical test has been found that clearly separates these species. All acholeplasmas found to date are glycolytic and none appears to hydrolyze arginine or urea. Although acholeplasmas appear to lack the sugar transport system found in fermentative mycoplasmas (phosphoenol pyruvate-dependent phosphotransferase system), high levels of hexokinase can be demonstrated in soluble fractions of cells (Cirillo and Razin, 1973). It is suggested, at least for *A. laidlawii*, that glucose permeation occurs through a carrier-mediated process, rather than by facilitated diffusion (Read and McElhaney, 1975; Tarshis *et al.*, 1976; this volume, Chapter 12).

The ability to hydrolyze esculin (6,7-dihydrocoumarin 6-glucoside) through a β-D-glucosidase enzyme system occurs in a number of *Acholeplasma* species (Table II) and is apparently absent in sterol-requiring mycoplasmas (Williams and Wittler, 1971; Stipkovits *et al.*, 1973a; Bradbury, 1977). Likewise, hydrolysis of arbutin (hydroquinone β-D-glucopyranoside) has also been proposed as a diagnostic characteristic for certain acholeplasmas, particularly *A. axanthum* (Stipkovits *et al.*, 1973b; Ernø and Stipkovits, 1973). Since only a few *Acholeplasma* strains have been tested the value of the procedure has been questioned. More recently, a comparative study of arbutin hydrolysis was performed on a number of strains of all seven species of *Acholeplasma* (Table II) (D. L. Rose and J. G. Tully, unpublished), using the plate method described originally (Ernø and Stipkovits, 1973). *Acholeplasma axanthum* strains showed very strong (4+) responses in the test, which involved the development of a dark brown to black color throughout an agar medium containing arbutin and ferric citrate. All seven strains of *A. oculi* gave a less intense (2+), but positive, response. Strains of *A. laidlawii* were observed to produce a faint brownish tint to the agar, which was read as a ± reaction, while all other acholeplasmas produced no obvious color changes in the agar after inoculation and incubation.

The occurrence of pigmented carotenoids in mycoplasmas was first noted in strains of *A. laidlawii* (Rothblat and Smith, 1961). In contrast, no evidence could be found that these compounds occurred in mycoplasmas that required sterols for growth. These findings were confirmed in studies directed primarily to understanding the location and function of carotenoids in so-called saprophytic mycoplasmas (Smith, 1963; Razin *et al.*, 1963; Razin and Cleverdon, 1965; Razin and Rottem, 1967). A search for carotenoids in various known *Acholeplasma* species and other unspeciated but putatively distinct acholeplasmas, using a slightly modified detection method (Razin and Cleverdon, 1965) showed that the procedure had merit for separation of sterol-nonrequiring organisms (Tully and Razin, 1968, 1969; Tully, 1973). At present, strains of at least three species (*A. axanthum, A. equifetale,* and *A. hippikon*) appear incapable of producing pigmented carotenoids, although a limited number of strains of the latter two equine acholeplasmas has been tested. In addition, *A. modicum* strains first appeared to be carotenoid-negative. However, chemical analysis, involving a much larger cell mass of this organism, showed the presence of both colored (yellow) and colorless carotenoids (Mayberry *et al.*, 1974). Although these results indicate that the test may be insensitive to very small amounts of carotenoids, the procedure still appears to have value in separation of acholeplasmas.

Finally, the film and spot reaction (Edward, 1950b; Fabricant and Freundt, 1967), which occurs in a number of *Mycoplasma* species, was recently observed in *A. equifetale* and *A. hippikon* (Kirchhoff, 1978). Whether this reaction is characteristic of other strains of these two new species or occurs to any extent in other acholeplasmas is uncertain. A summary of these and other biologic properties of acholeplasmas has recently been tabulated (Tully and Razin, 1977).

## IV. SEROLOGIC RELATIONSHIPS

### A. Evaluation of Serologic Procedures

Like most other mycoplasmas, acholeplasmas seem to possess a limited number of distinctive biologic features. Under such circumstances, serologic analysis assumes great importance in species characterization. Unfortunately, there is no clear consensus about the most suitable serologic method and no one method combines all the requirements of specificity (low cross-reactivity between species), sensitivity (separation of species by small amounts of antibody), or ease of performance. The most useful serologic methods for acholeplasmas, which together meet most of these criteria, include the growth-inhibition (GI), metabolism-inhibition (MI), and fluorescent antibody (FA) tests, all of which rely on antibody attachment to the mycoplasma membrane (Kahane and Razin, 1969; Lemcke, 1973).

The GI test, based upon inhibition of acholeplasma growth around disks saturated with homologous antiserum and performed on an agar medium, is probably the most specific test (Clyde, 1964; Stanbridge and Hayflick, 1967; Dighero *et al.*, 1970). However, the procedure requires highly potent antiserum, which is not always easily prepared. Even with such serum and under optimal conditions, antisera zones of inhibition to homologous strains are rarely larger than 4–5 mm and average only 2–4 mm (Stanbridge and Hayflick, 1967; Ernø and Jurmanova, 1973; Stipkovits, 1973). Also, colony breakthroughs are often observed. Since acholeplasmas grow rapidly, small zones of inhibition may be related to growth that occurs before diffusion of antiserum. Therefore, certain modifications of the test procedure have been suggested to enhance sensitivity, including reduction of the inoculum size (possibly to $10^3$ or $10^4$ colony-forming units per milliliter); reduction or, preferably elimination of the quantity of animal serum in the agar medium; and reduction of the incubation temperature to 25° or 30°C. However, heterologous reactions and nonspecific inhibition by "normal" sera might still not be completely excluded by these modifications so a more potent antiserum is still required to increase the differential between specific and nonspecific reactions.

The MI test, as applied to acholeplasmas, measures the capability of a specific antiserum to inhibit the metabolism of glucose by the organisms (Taylor-Robinson *et al.*, 1966). The inhibitory activity is thought to be related to adsorption of antibody to the mycoplasma cell membrane with a subsequent depression of cell metabolism and growth, frequently leading to lysis of the organism. Although the technical aspects of the procedure are more detailed than the GI test, the test is fairly specific and sensitive for separation of acholeplasmas. Homologous MI titers with antisera to acholeplasmas are generally much lower than those observed with other mycoplasmas and are often in the range of 1:64 to 1:2560 (Stipkovits, 1973; Ernø *et al.*, 1973; Stipkovits and Varga, 1974; Kiska and Gois, 1975). Occasionally, acholeplasmas show heterologous cross-reactions in the MI test but the differences in titer between heterologous and homologous strains are usually sufficient to make species separations. As with the GI test, reductions in the incubation temperature and the serum content of the broth medium usually enhance the specificity of the MI test.

The FA procedure, as recently modified, has become an important and useful test for serologic analysis of all mycoplasmas, particularly for primary identification. The epi-FA test altered the conventional FA technique so that mycoplasma colonies on solid media could be stained directly with fluorescein-conjugated antiserum and then examined microscopically under incident illumination (Del Giudice *et al.*, 1967). While the specificity of the epi-FA test for acholeplasmas is similar to the GI test, its sensitivity is much higher and compares favorably to the MI test. The test has two definite advantages: Conjugated antisera can be titrated and the procedure can detect a mixture of mycoplasma serotypes on agar. Epi-FA end point titers with acholeplasma antisera generally are in the range of 1:80 to 1:640. Antisera possessing high GI antibody levels can usually be used for production of conjugated sera for the epi-FA test, but less potent sera may yield unsatisfactory conjugates. Acholeplasma antiserum for the epi-FA test should be diluted to a concentration that gives strong immunofluorescence with the homologous strain and little or no reaction with heterologous strains. The test has been modified further to conserve the amount of conjugated antiserum needed, to eliminate the necessity of having a number of specific conjugates, or to provide other conveniences (Ertel *et al.*, 1970; Al-Aubaidi and Fabricant, 1971; Rosendal and Black, 1972; Lehmkuhl and Frey, 1974; Bradbury *et al.*, 1976; Ernø, 1977).

One other recent serologic procedure holds promise as a means of distinguishing *Acholeplasma* species but has not received extensive evaluation. This test involves the enzyme-linked, indirect immunoperoxidase reaction, which utilizes specific antiserum, a rabbit immunoglobulin conjugated to horseradish peroxidase, and acholeplasma or mycoplasma colonies on agar (Polak-Vogelzang *et al.*, 1978). The procedure appears to have the specificity of the epi-FA test but eliminates background autofluorescence and the need for special microscopy. The test has been applied recently to detection of mycoplasma antigen in animal tissues and to quantitative detection of mycoplasma antibody (Bruggmann *et al.*, 1977a,b).

Other serologic procedures employed in mycoplasma characterization (complement fixation, growth precipitation, double immunodiffusion, and indirect hemagglutination) do not appear to have any distinct advantages in serologic analysis of acholeplasmas. On the other hand, it should be noted that fully adequate serologic comparisons must be based upon the results of several procedures performed with several different antiserum preparations.

## B. Preparation of Antisera

Several problems confronted the preparation of satisfactory antisera to acholeplasmas. The first difficulty, which also exists with most mycoplasmas, is that serum proteins from media supplemented with animal serum can absorb to mycoplasmas in amounts sufficient to cause the resulting antiserum to contain antibodies to these foreign proteins. Thus, acholeplasma (and mycoplasma) immunizing and test antigens grown in the presence of the same serum will produce excessive nonspecific cross-reactions in a number of serologic procedures, especially in immunodiffusion, complement-fixation, and agglutination tests. This appears to be less of a problem with growth-inhibition, metabolism-inhibition, or immunofluorescence serologic tests. An alternate solution to the problem involves the cultivation of the immunizing antigen in a serum homologous to the host being used for immunization (Bradbury and Jordan, 1972) or the use of agamma serum (horse or calf) for medium supplementation (Kenny, 1969).

However, the major difficulty observed in preparation of antisera to acholeplasmas is the low titer of growth-inhibiting antibody formed, even after intensive immunization. In fact, some rabbits fail to develop any growth-inhibiting antibody to acholeplasmas, regardless of the amount and kind of antigen, use of adjuvants, various immunization routes, hosts, etc. Under optimal conditions, the zones of growth inhibition induced by acholeplasma antisera do not exceed 3–5 mm; such antisera are difficult to use for the separation of species. A possible reason for the refractory response in rabbits and guinea pigs may be the presence of antibody to acholeplasmas in normal sera of these hosts (Kenny, 1969). Dorner *et al.* (1977) recently confirmed this suggestion by showing that normal guinea pig serum contains mycoplasmacidal antibody to membrane phospholipids of *A. laidlawii*. Such antibody may arise from natural infection, or at least from exposure to the organism or to related microbial or plant phospholipids.

To partially counter the various problems, preparation of acholeplasma antiserum should include a number of recommended measures. Normal serum from a number of rabbits or guinea pigs can be prescreened in immunodiffusion tests with candidate antigens and animals exhibiting preexisting antibody excluded from trials. Acholeplasma antigens should be cultivated in serum-free medium or in a medium supplemented with 1–2% bovine serum fraction. Antigens are then suspended in Freund's complete adjuvant and the recommended immunization scheme involves, in sequence, four intramuscular and two intravenous inoculations at biweekly intervals (over a 3-month period). If test bleedings at the end of 3 months do not show development of adequate growth-inhibiting antibody (3–5 mm zones), the immunization trial should be terminated and repeated in other prescreened animals. As a last resort, the potency of antiserum that exhibits low or unsatisfactory growth-inhibiting activity might be enhanced somewhat by concentration techniques (Windsor and Trigwell, 1976).

## V. HABITAT

### A. Ecology and Host Distribution

A summary of the habitat and host distribution of acholeplasmas is given in Table III. More extensive comments and evaluation of the association of acholeplasmas with specific animal hosts can be found in appropriate chapters of Volume II.

As previously noted, the frequent occurrence of acholeplasmas in animals has important implications. In recent years, *Acholeplasma* species have become one of the most prevalent contaminants in tissue cultures, primarily because of their frequent occurrence in commercial bovine serum (Barile and Kern, 1971; Barile *et al.*, 1973). The handling of raw bovine serum heavily contaminated with acholeplasmas can also present other hazards. For example, *A. laidlawii* was recovered from uninoculated commercial mycoplasma broth culture media; in this case, the source of entry apparently was contaminated agamma horse serum (J. G. Tully, unpublished). It was suspected that contamination of working areas with aerosols of acholeplasmas led to the entrance of organisms into the horse serum after processing of the agamma serum. This suspected mechanism may also account for the reported isolation of *A. laidlawii* from commercial dehydrated serum-free tissue culture (Dulbecco basal) medium (Low, 1974).

### B. Problems in Evaluating Pathogenicity

Although acholeplasmas have frequently been isolated from a variety of diseased and healthy animal tissues, there is no clear evidence at present that any of these plays a pathogenic role. However, there are few controlled studies which adequately test the ability of established *Acholeplasma* species to produce disease, either through techniques involving experimental challenge of animal hosts or by the collection of adequate clinical and epidemiologic information. The widespread distribution of acholeplasmas (and mycoplasmas) in animals, as well as the presence of antibody to many of these organisms, not only makes it difficult to induce disease in conventional animals but to interpret the clinical and histologic responses observed. The use of gnotobiotic animals offers definite advantages in experimental studies but is limited by the range of hosts available. Specific pathogen-free (SPF) hosts might be useful for some studies but such animals may not be free of mycoplasmas or acholeplasmas (Stipkovits *et al.*, 1974); their utility, therefore, depends upon the amount of effort expended to define their prechallenge microbial flora. Experimental history of the challenge culture is another important variable, since faulty information arises from the use of uncloned or mixed cultures or from a challenge with attenuated (avirulent) strains arising from prolonged *in vitro* passage. Thus, difficulties in providing completely controlled conditions in experimental investigations of possible pathogenicity of *A. axanthum* (Stipkovits *et al.*, 1974) and *A. oculi* (Al-Aubaidi *et al.*, 1973) still leave major questions unanswered. Such problems have also resulted in conflicting epidemiologic and challenge data on the role of *A. laidlawii* in bovine mastitis (Erfle and Brunner, 1977; Jasper, 1977).

## C. Acholeplasmas in Plants

The recovery of acholeplasmas from plants and plant materials has been reported with increasing frequency the past few years. *Acholeplasma laidlawii* has been isolated from clover with phyllody disease (Horne and Taylor-Robinson, 1972; Spaar *et al.*, 1974), from *Rudbeckia purpurea* (black-eyed susans) (Spaar *et al.*, 1974), and from periwinkle (Colleno *et al.*, 1972). The apparent occurrence of *A. laidlawii* within the sieve tubes of *Vinca rosea* was recently noted (de Leeuw, 1977). While identification involved direct staining of histologic sections of plant material with *A. laidlawii* antiserum, no control studies were reported. *Acholeplasma axanthum* has been recovered frequently from diseased (lethal yellowing) coconut palms (S. Eden-Green, personal communication). Unidentified acholeplasmas were also isolated from *Vinca rosea* (Spaar *et al.*, 1974). Several other investigators have isolated acholeplasmas from plants but have not published their findings.

Although most of these observations and reports lack an adequate set of controls needed to completely exclude entrance of exogenous acholeplasmas (from animal contact with plant material, from bovine or other animal serum used in cultivation, etc.), the total impact of this preliminary, but cumulative, evidence suggests that acholeplasmas inhabit plants. Thus, although there is no definitive evidence at present that acholeplasmas play any role in plant disease, the possibility should not be summarily dismissed. Perhaps the role of acholeplasmas as plant patho-

TABLE III. **Habitat and Experimental Pathogenicity of the Acholeplasmas**

| Species | Host and tissue distribution | Experimental pathology | Key references |
|---|---|---|---|
| *A. axanthum* | Bovine: nasal cavity, lymph nodes kidney, and serum<br>Porcine: lung, peribronchial lymph nodes (also sewage effluent)<br>Equine: oral cavity<br>Avian: goose embryos | Intranasal challenge of specific pathogen-free piglets induced mild clinical symptoms and gross lesions (and histologic changes) in lungs; organism also pathogenic for goose and chicken embryos | Tully (1973), Stipkovits *et al.* (1973b), 1974, 1975a,b), Barile *et al.* (1973), Ogata *et al.* (1974), Kisary *et al.* (1976), Orning *et al.* (1978) |
| *A. equifetale* | Equine: Nasopharynx and trachea<br>Avian: Trachea and cloaca (chicken) | Not studied | Kirchhoff (1974), Allam and Lemcke (1975), Kirchhoff (1978), Bradbury (1978) |
| *A. granularum* | Porcine: nasal cavity, lung, and feces<br>Equine: conjunctivae and nasopharynx | No multiplication in embryonated hen's eggs; aerosol challenge of SPF pigs produced no clinical or histologic evidence of disease | Switzer (1969), Taylor-Robinson and Dinter (1968), Goiš *et al.* (1969), Roberts and Goiš (1970), Jericho *et al.* (1971), Bannerman and Nicolet (1971), Ross (1973), Ogata *et al.* (1974), Roberts and Little (1976), Lemcke (Volume II, Chapter 5) |
| *A. hippikon* | Equine: nasopharynx | Not studied | Kirchoff (1974, 1978) |

| | | | | |
|---|---|---|---|---|
| *A. laidlawii* | Avian:<br>Bovine:<br>Caprine:<br>Canine:<br>Equine:<br>Feline:<br>Murine:<br>Ovine:<br>Porcine:<br>Primates: | from the oral cavity and respiratory and/or genital tract secretions of most hosts; also from the eye, lymph nodes, semen, and serum | Conflicting results recorded on occurrence of mastitis in cattle given experimental challenge in udder; some strains pathogenic for chick embryos | Edward (1950a), Leach (1967), Taylor-Robinson and Dinter (1968); Goiš *et al.* (1969), Gourlay and Thomas (1969), Tan and Miles (1972), Rosendal and Laber (1973), Barile *et al.* (1973), Leach (1973), Langford (1974), Kirchhoff (1974), Koshimizu and Ogata (1974), Hill (1974), Ogata *et al.* (1974), Stipkovits *et al.* (1975a), Allam and Lemcke (1975), Ernø (1975), Kisary *et al.* (1976), Erfle and Brunner (1977), Jasper (1977) |
| *A. modicum* | Bovine:<br>Porcine: | lungs, thoracic fluids, lymph nodes, and semen<br>nasal secretions | Not studied | Langer and Carmichael (1963), Leach (1967, 1973), Bokori *et al.* (1971), Stipkovits (1973), Stipkovits *et al.* (1975a), Ernø (1975) |
| *A. oculi* | Caprine:<br>Porcine:<br>Equine:<br>Guinea pig: | eye (conjunctivitis)<br>nasal secretions<br>nasopharynx, lung, spinal fluid, joint, semen<br>external genitalia | Intravenous inoculation of conventional goats produced signs of pneumonia and death within 6 days; conjunctival inoculation of conventional goats produced mild conjunctivitis | Al-Aubaidi *et al.* (1973), Allam *et al.*, (1973); Allam and Lemcke (1975), Kuksa Goiš (1975), E. V. Langford, unpublished, R. M. Lemcke (Volume II, Chapter 5) |

gens will be as difficult to study as it has been in animal hosts. The role of insects in the natural dissemination and plant pathogenicity of acholeplasmas also remains to be determined. However, experimental inoculation of acholeplasmas into various insects, including leafhoppers known to be vectors of plant mycoplasmas, showed multiplication and retention of the organisms for considerable time periods (up to 24 days) (Whitcomb *et al.*, 1973; Whitcomb and Williamson, 1975; S. Eden-Green, personal communication), so the potential exists for an important interaction.

## VI. FUTURE OUTLOOK

There is need for further information on the basic biologic and serologic distinctions among acholeplasmas. Nothing has been said in this review about the problem of virus infections of acholeplasmas since this topic is considered in depth in Chapters 14 and 15 of this volume. However, in view of the frequent occurrence of nonlytic viruses, and their maturation from membrane locations, it is possible that virus infections might be responsible for some biologic and antigenic alterations within the acholeplasmas. Because serologic differences are a major taxonomic tool in acholeplasmas, it would seem advisable to delineate the role of viruses in changing the antigenic structure of these organisms.

Although the conditions required to establish the etiologic role of acholeplasmas in disease are complicated by host and microbial factors, it is evident that only when these details are emphasized and met will conclusive information be derived on the pathogenicity of these organisms to plants and animals.

## REFERENCES

Al-Aubaidi, J. M., and Fabricant, J. (1971). *Cornell Vet.* **61,** 519–542.

Al-Aubaidi, J. M., Dardiri, A. H., Muscoplatt, C. C., and McCauley, E. H. (1973). *Cornell Vet.* **63,** 117–129.

Allam, N. M., and Lemcke, R. M. (1975). *J. Hyg.* **74,** 385–408.

Allam, N. M., Powell, D. G., Andrews, B. E., and Lemcke, R. M. (1973). *Vet. Rec.* **93,** 402.

Andrews, B. E., and Kunze, M. (1972). *Med. Microbiol. Immunol.* **157,** 175.

Askaa, G., Christiansen, C., and Ernø, H. (1973). *J. Gen. Microbiol.* **75,** 283–286.

Bak, A. L., Black, F. T., Christiansen, C., and Freundt, E. A. (1969) *Nature (London)* **224,** 1209–1210.

Bannerman, E. S. N., and Nicolet, J. (1971). *Schweiz. Arch. Tierheilkd.* **113,** 697–710.

Barile, M. F., and Kern, J. (1971). *Proc. Soc. Exp. Biol. Med.* **138,** 432–437.

Barile, M. F., Hopps, H. E., Grabowski, M., Riggs, D. B., and Del Giudice, R. A. (1973). *Ann. N.Y. Acad. Sci.* **225,** 251–264.

Black, F. T., Christiansen, C., and Askaa, G. (1972). *Int. J. Syst. Bacteriol.* **22,** 241–242.
Boden, K., and Kirchhoff, H. (1977). *Zentralbl. Bakteriol., Hyg., Parasitenkd., Infektionskr. Abt. 1: Orig.* **237,** 342–350.
Bokori, J., Horvath, Z., Stipkovits, L., and Molnar, L. (1971). *Acta Vet. Acad. Sci. Hung.* **21,** 61–73.
Bradbury, J. M. (1977). *J. Clin. Microbiol.* **5,** 531–534.
Bradbury, J. M. (1978). *Vet. Record* **102,** 316.
Bradbury, J. M., and Jordan, F. T. W. (1972). *J. Hyg.* **70,** 267–278.
Bradbury, J. M., Oriel, C. A., and Jordan, F. T. W. (1976). *J. Clin, Microbiol.* **3,** 449–452.
Bruggmann, S., Keller, H., Bertschinger, H. U., and Engberg, B. (1977a). *Vet. Rec.* **101,** 109–111.
Bruggmann, S., Engberg, B., and Ehrensperger, F. (1977b). *Vet. Rec.* **101,** 137.
Cirillo, V. P., and Razin, S. (1973). *J. Bacteriol.* **113,** 212–217.
Clyde, W. A., Jr. (1964). *J. Immunol.* **92,** 958–965.
Colleno, A., LeNormand, M., and Maillet, P. (1972). *Pathog. Mycoplasmas, Ciba Found. Symp., 1972* Discussion p. 220.
Daniels, M. J., and Meddins, B. M. (1973). *J. Gen. Microbiol.* **76,** 239–242.
de Leeuw, G. T. N. (1977). *Nat. Tech.* **45,** 74–89.
Del Giudice, R. A., Robillard, N. F., and Carski, T. R. (1967). *J. Bacteriol.* **93,** 1205–1209.
Dighero, M. W., Bradstreet, C. M. P., and Andrews, B. E. (1970). *J. Appl. Bacteriol.* **33,** 750–757.
Dorner, I., Brunner, H., Schiefer, H-G., Loos, M., and Wellensiek, H.-J. (1977). *Infect. Immun.* **18,** 1–7.
Edward, D. G. ff. (1950a). *J. Gen. Microbiol.* **4,** 4–15.
Edward, D. G. ff. (1950b). *J. Gen. Microbiol.* **4,** 311–320.
Edward, D. G. ff. (1967). *Ann. N.Y. Acad. Sci.* **143,** 7–8.
Edward, D. G. ff. (1971). *J. Gen. Microbiol.* **69,** 205–210.
Edward, D. G. ff., and Freundt, E. A. (1970). *J. Gen. Microbiol.* **62,** 1–2.
Erfle, V., and Brunner, A. (1977). *Berl. Muench. Tieraerztl. Wochenschr.* **90,** 28–34.
Ernø, H. (1975). *Acta Vet. Scand.* **16,** 321–323.
Ernø, H. (1977). *Acta Vet. Scand* **18,** 176–186.
Ernø, H., and Jurmanova, K. (1973). *Acta Vet. Scand.* **14,** 524–537.
Ernø, H., and Stipkovits, L. (1973). *Acta Vet. Scand.* **14,** 450–463.
Ernø, H., Jurmanova, K., and Leach, R. H. (1973). *Acta Vet. Scand.* **14,** 511–523.
Ertel, P. Y., Ertel, I. J., Somerson, N. L., and Pollack, J. D. (1970). *Proc. Soc. Exp. Biol. Med.* **134,** 441–446.
Fabricant, J., and Freundt, E. A. (1967). *Ann. N.Y. Acad. Sci.* **143,** 50–58.
Freeman, B. A., Sisenstein, R., McManus, T. T., Woodward, J. E., Lee, I. M., and Mudd, J. B. (1976). *J. Bacteriol.* **125,** 946–954.
Freundt, E. A., Andrews, B. E., Ernø, H., Kunze, M., and Black, F. T. (1973). *Zentralbl. Bakteriol., Hyg. Parasitenkd., Infektionskr. Abt. 1: Orig.* **225,** 104–112.
Goiš, M., Černý, M., Rozkošný. V., and Sovadina, M. (1969). *Zentralbl. Veterinaermed., Reihe B* **16,** 253–265.
Gourlay, R. N., and Thomas, L. H. (1969). *Vet. Rec.* **84,** 416–417.
Grabowski, M. W., Rottem, S., and Barile, M. F. (1976). *J. Clin. Microbiol.* **3,** 110–112.
Herring, P. K., and Pollack, J. D. (1975). *Int. J. Syst. Bacteriol.* **24,** 73–78.
Hill, A. (1974). *Vet. Rec.* **94,** 385.
Horne, R. W., and Taylor-Robinson, D. (1973). *Pathog. Mycoplasmas, Ciba Found. Symp., 1972* Discussion, p. 59.
Jasper, D. E. (1977). *J. Am. Vet. Med. Assoc.* **170,** 1167–1172.

Jericho, K. W. F., Austwick, P. K. C., Hodges, R. T., and Dixon, J. B. (1971). *J. Comp. Pathol.* **81,** 13–21.
Kahane, I., and Razin, S. (1969). *J. Bacteriol.* **100,** 187–194.
Kahane, I., Greenstein, S., and Razin, S. (1977). *J. Gen. Microbiol.* **101,** 173–176.
Kenny, G. E. (1969). *J. Bacteriol.* **98,** 1044–1055.
Kirchhoff, H. (1974. *Zentralbl. Veterinaermed., Reihe B* **21,** 207–210.
Kirchhoff, H. (1978). *Int. J. Syst. Bacteriol.* **28,** 76–81.
Kisary, J., El-Ebeedy, A. A., and Stipkovits, L. (1976). *Avian Pathol.* **5,** 15–20.
Koshimizu, K., and Ogata, M. (1974). *Jpn. J. Vet. Sci.* **36,** 391–406.
Kuksa, F., and Goiš, M. (1975). *In vitro CSSR (Czech.)* **4,** 42–51.
Kunze, M. (1971). *Zentralbl. Bakteriol, Parasitenkd., Infektionskr. Hyg., Abt. 1: Orig.* **216,** 501–505.
Laidlaw, P. P., and Elford, W. J. (1936). *Proc. R. Soc. London, Ser. B* **120,** 292–303.
Langer, P. H., and Carmichael, L. E. (1963). *Proc. U.S. Livestock Sanitary Assoc.* pp. 129–137.
Langford, E. V. (1974). *J. Wildl. Dis.* **10,** 420–422.
Leach, R. H. (1967). *Ann. N.Y. Acad. Sci.* **143,** 305–316.
Leach, R. H. (1973). *J. Gen. Microbiol.* **75,** 135–153.
Lehmkuhl, H. D., and Frey, M. L. (1974). *Appl. Microbiol.* **27,** 1170–1171.
Lemcke, R. M. (1973). *Ann. N.Y. Acad. Sci.* **225,** 46–53.
Low, I. E. (1974). *Appl. Microbiol.* **27,** 1046–1052.
McGee, Z. A., Rogul, M., and Wittler, R. G. (1967). *Ann. N.Y. Acad. Sci.* **143,** 21–30.
Mardh, P.-A., and Taylor-Robinson, D. (1973). *Med. Microbiol. Immunol.* **158,** 219–226.
Mayberry, W. R., Smith, P. F., and Langworthy, T. A. (1974). *J. Bacteriol.* **118,** 898–904.
Neimark, H. C. (1970). *J. Gen. Microbiol.* **63,** 249–263.
Neimark, H. C. (1973). *Ann. N.Y. Acad. Sci.* **225,** 14–21.
Ogata, M., Watabe, J., and Koshimizu, K. (1974). *Jpn. J. Vet. Sci.* **36,** 43–51.
Orning, A. P., Ross, R. F., and Barile, M. F. (1978). *Am. J. Vet. Res.* **39,** 1169–1174.
Polak-Vogelzang, A. A., Hagenaars, R., and Nagel, J. (1978). *J. Gen. Microbiol.* **106,** 241–249.
Pollack, J. D. (1975). *Int. J. Syst. Bacteriol.* **25,** 108–113.
Pollack, J. D. (1978). *Int. J. Syst. Bacteriol.* **28,** 425–426.
Pollack, J. D., Razin, S., and Cleverdon, R. C. (1965). *J. Bacteriol.* **90,** 617–622.
Pollock, M. E., and Bonner, S. V. (1969). *Bacteriol. Proc.* p. 32.
Razin, S. (1968). *J. Bacteriol.* **96,** 687–694.
Razin, S., and Cleverdon, R. C. (1965). *J. Gen. Microbiol.* **41,** 409–415.
Razin, S., and Rottem, S. (1967). *J. Bacteriol.* **93,** 1181–1182.
Razin, S., and Tully, J. G. (1970). *J. Bacteriol.* **102,** 306–310.
Razin, S., Argaman, M., and Avigan, J. (1963). *J. Gen. Microbiol.* **33,** 477–487.
Read, B. D., and McElhaney, R. N. (1975). *J. Bacteriol.* **123,** 47–55.
Reff, M. E., Stanbridge, E. J., and Schneider, E. L. (1977). *Int. J. Syst. Bacteriol.* **27,** 185–193.
Roberts, D. H., and Gois, M. (1970). *Vet. Rec.* **87,** 214–215.
Roberts, D. H., and Little, T. W. A. (1976). *Vet. Rec.* **99,** 13.
Romano, N., Rottem, S., and Razin, S. (1976). *J. Bacteriol.* **128,** 170–173.
Rosendal, S., and Black, F. T. (1972). *Acta Pathol. Microbiol. Scand., Ser. B* **80,** 615–622.
Rosendal, S., and Laber, G. (1973). *Zentralbl. Bakteriol., Parasitenkd., Infektionskr. Hyg., Abt. 1: Orig.* **225,** 346–349.
Ross, R. F. (1973). *Ann. N.Y. Acad. Sci.* **225,** 347–368.
Rothblat, G. H., and Smith, P. F. (1961). *J. Bacteriol.* **82,** 479–491.
Rottem, S. (1972). *Appl. Microbiol.* **23,** 659–660.

Rottem, S., Pfendt, E. A., and Hayflick, L. (1971). *J. Bacteriol.* **105,** 323–330.
Saglio, P., L'Hospital, M., LaFlèche, D., Dupont, G., Bové, J. M., Tully, J. G., and Freundt, E. A. (1973). *Int. J. Syst. Bacteriol.* **23,** 191–204.
Seiffert, G. (1937a). *Zentralbl. Bakteriol., Parasitenkd., Infektionskr. Hyg., Abt. 1: Orig.* **139,** 337–342.
Seiffert, G. (1937b). *Zentralbl. Bakteriol., Parasitenkd., Infektionskr. Hyg., Abt. 1: Orig.* **140,** 168–172.
Smith, P. F. (1963). *J. Gen. Microbiol.* **32,** 307–319.
Smith, P. F., and Rothblat, G. H. (1960). *J. Bacteriol.* **80,** 842–850.
Soltesz, L. V., and Mardh, P-A. (1977). *Acta Pathol. Microbiol. Scand., Sect. B* **85,** 255–261.
Spaar, D., Kleinhempel, H., Muller, H. M., Stanarius, A., and Schimmel, D. (1974). *Colloq. Inst. Nat. Sante Rech. Med.* **33,** 207–214.
Stanbridge, E. J., and Hayflick, L. (1967). *J. Bacteriol.* **93,** 1392–1396.
Stipkovits, L. (1973). *Acta Vet. Acad. Sci. Hung.* **23,** 315–323.
Stipkovits, L., and Varga, L. (1974). *Acta Vet. Acad. Sci. Hung.* **24,** 139–149.
Stipkovits, L., Schimmel, D., and Varga, L. (1973a). *Acta Vet. Acad. Sci. Hung.* **23,** 307–313.
Stipkovits, L., Varga, L., and Schimmel, D. (1973b). *Acta Vet. Acad. Sci. Hung.* **23,** 361–368.
Stipkovits, L., Romváry, J., Nagy, Z., Bodon, L., and Varga, L. (1974). *J. Hyg.* **72,** 289–296.
Stipkovits, L., Bodon, L., Romváry, J., and Varga, L. (1975a). *Acta Microbiol. Acad. Sci. Hung.* **22,** 45–51.
Stipkovits, L., El-Ebeedy, A. A., Kisary, J., and Varga, L. (1975b). *Avian Pathol.* **4,** 35–43.
Switzer, W. P. (1969). *In* "The Mycoplasmatales and L-Phase of Bacteria" (L. Hayflick, ed.), pp. 607–619. Appleton, New York.
Tan, R. J. S., and Miles, J. A. R. (1972). *Br. Vet. J.* **128,** 87–90.
Tarshis, M. A., Bekkouzjin, A. G., Ladygina, G., and Panchenko, L. F. (1976). *J. Bacteriol.* **125,** 1–7.
Taylor-Robinson, D., and Dinter, Z. (1968). *J. Gen. Microbiol.* **53,** 221–229.
Taylor-Robinson, D., Purcell, R. H., Wong, D. C., and Chanock, R. M. (1966). *J. Hyg.* **64,** 91–104.
Tully, J. G. (1973). *Ann. N.Y. Acad. Sci.* **225,** 74–93.
Tully, J. G., and Razin, S. (1968). *J. Bacteriol.* **95,** 1504–1512.
Tully, J. G., and Razin, S. (1969). *J. Bacteriol.* **98,** 970–978.
Tully, J. G., and Razin, S. (1970). *J. Bacteriol.* **103,** 751–754.
Tully, J. G., and Razin, S. (1977). *In* "Handbook of Microbiology" (A. I. Laskin and H. Lechevalier, eds.), 2nd ed., Vol. 1, pp. 405–459. Chem. Rubber Publ. Co., Cleveland, Ohio.
Whitcomb, R. F., and Williamson, D. L. (1975). *Ann. N.Y. Acad. Sci.* **266,** 260–275.
Whitcomb, R. F., Tully, J. G., Bové, J. M., and Saglio, P. (1973). *Science* **182,** 1251–1253.
Williams, C. O., and Wittler, R. G. (1971). *Int. J. Syst. Bacteriol.* **21,** 73–77.
Windsor, G. D., and Trigwell, J. A. (1976). *Res. Vet. Sci.* **20,** 221–222.
Wreghitt, T. G., Windsor, G. D., and Butler, M. (1974). *Appl. Microbiol.* **28,** 530–533.

# 17 / SPECIAL FEATURES OF UREAPLASMAS

*M. C. Shepard and G. K. Masover*





THE MYCOPLASMAS, VOL. 1

ISBN 0-12-078401-7

## I. INTRODUCTION

*Ureaplasma urealyticum* (Shepard *et al.*, 1974) has long been associated with primary and recurrent nongonococcal urethritis (NGU) in the human male (Shepard, 1954, 1956). Its possible causative role in this disease was first suggested by Shepard (1959, 1966, 1970, 1973b, 1974; Shepard *et al.*, 1964; Shepard and Calvy, 1965). However, its causative role in NGU has been controversial (McCormack *et al.*, 1973; Taylor-Robinson, 1977). New significance was recently given to the etiologic role of *U. urealyticum* in nongonococcal urethritis. Its pathogenicity for the human genital tract has now been established and Koch's postulates fulfilled by the experimental intraurethral inoculation of two human volunteers with two different strains of *U. urealyticum* isolated from men with NGU (Taylor-Robinson *et al.*, 1977). Thus, *Ureaplasma* urethritis has now been produced experimentally in man, and *U. urealyticum* has become the second member of the Mycoplasmatales to be established as a disease-producing human pathogen (the first being *Mycoplasma pneumoniae,* the cause of primary atypical pneumonia). Ureaplasmas and mycoplasmas in human genitourinary infections are covered in Volume II.

## II. HISTORICAL BACKGROUND

*Ureaplasma urealyticum* was formerly known by such trivial names as "T-strain of Mycoplasmas," "T strains," and "T mycoplasmas." The organism was first isolated from male patients with primary and recurrent nongonococcal urethritis. The colony size of *U. urealyticum* was exceedingly small (7–15 μm diameter), and the colonies were referred to as "tiny-form PPLO colonies." The minute size and characteristic structure of these colonies of *U. urealyticum* were first described and illustrated in two photomicrographs by Shepard (1954). It was later established that these minute colonies were those of a living, self-replicating agent and a previously undescribed member of the *Mycoplasma* group. Two years following their discovery in 1954, the organisms were described in greater detail and their distinctive morphologic characteristics were illustrated (Shepard, 1956).

Progress in the study of *U. urealyticum* was extremely slow and laborious. Although excellent outgrowth of the organism (in number of colonies) occurred in *primary* cultures of urethral exudates, relatively poor growth was obtained in attempts at subsequent serial passages on the same type of agar medium (Shepard, 1960). These early difficulties were the result of (1) a nutritionally inadequate agar culture medium of high alkalinity (standard for the then-known mycoplasmas) and (2) the failure of the medium to furnish required, unknown substances or growth factors present in the original clinical exudate. The native protein enrichment for these early media was generally human ascitic fluid which varied considerably in growth-promoting ability from lot to lot. It was not until enrichment was changed to horse serum that progress in the subcultivation, maintenance and study of *U. urealyticum* was achieved (Shepard, 1967). Shepard and Lunceford (1965) found that the optimal reaction for multiplication and colony growth of *U. urealyticum* was pH 6.0 ± 0.5 and not pH 7.8–8.0 as formerly employed in accordance with accepted methodology for the then-known mycoplasmas. This finding significantly improved the growth of *U. urealyticum* in agar cultures (larger colonies) and in fluid cultures (higher titers).

Cultivation of *U. urealyticum* in *fluid* media presented added difficulties. Standard practice called for transfers to be made 48-hourly, but such cultures experienced a progressive decline in titers and early death. Ford (1962) significantly observed that the organisms reached maximal titers after only 16 hr of incubation. Knowledge of the rapid growth of *U. urealyticum* in fluid medium established reliable procedures for maintenance of strains in broth cultures. Maximal titers, however, seldom reached more than $10^5$–$10^7$ colony-forming units per milliliter (CFU/ml), in contrast to generally high titers of $10^9$ CFU/ml obtained with many of the classical, large-colony mycoplasmas, e.g., *Mycoplasma hominis*.

The discovery of a urease enzyme system in *U. urealyticum,* and the ability of the organism to hydrolyze urea with the production of ammonia (Shepard, 1966; Purcell *et al.*, 1966; Ford and MacDonald, 1967; Shepard and Lunceford, 1967) was a significant milestone in the understanding of the biology of this organism. This distinguishing property justified the proposal of a new genus, *Ureaplasma,* for the T mycoplasmas, and an official binomial, *Ureaplasma urealyticum,* in the family Mycoplasmataceae (Shepard *et al.*, 1974). Diagnostic methods for detection and identification of *U. urealyticum* subsequently were developed, based upon the demonstration of urease (Shepard, 1973a). Purcell *et al.* (1966, 1967) employed a urea-containing broth in a metabolism inhibition color test for measurement of antibody to *U. urealyticum*. Shepard and Lunceford (1970a) introduced a ureas color test medium (U9) as a diagnostic aid in the identification of *U. urealyticum* in clinical specimens. This medium was also useful in quantifying cultures and studying the quantitative relationship of *U. urealyticum* to the clinical course of nongonococcal urethritis in the male (Shepard, 1973b, 1974). A urease color test fluid medium known as "B broth" was recently reported which employs bromthymol blue indicator instead of phenol red (Robertson, 1978). Robertson's "B broth" is a sensitive indicator medium for detection of *U. urealyticum* in clinical specimens and promotes rapid growth without rapid loss of viability.

Postive identification of *U. urealyticum* in primary *agar* cultures was first made possible by development of a direct spot test for urease in colonies on standard agars (Shepard and Howard, 1970; Shepard, 1973a). The biochemical principle of the direct test for urease was utilized in a differential agar medium (A7) (Shepard and Lunceford, 1976). The A7 differential medium is an important aid in the identification of *U. urealyticum* in primary agar cultures of clinical specimens in the diagnostic laboratory. *Ureaplasma* colonies from humans and lower animals are deep brown by transmitted light under the low power of the microscope, either by the direct spot test for urease or on the A7 differential agar medium, due to deposition of a manganese reaction product (manganese dioxide). Colonies of the classical, large-colony *Mycoplasma* and *Acholeplasma* species, as well as L colonies of *Proteus* species, are unreactive to the direct spot test for urease or on the A7 differential agar.

Much has been learned in recent years concerning techniques for isolation, identification, and cultivation of *U. urealyticum,* and the general biology of the organism. Newer information has recently been reported concerning ultrastructure and certain biochemical characteristics. These and other findings which emphasize the special features of the ureaplasmas are discussed in the following sections and summarized in Table I.

## III. MORPHOLOGY AND ULTRASTRUCTURE

### A. *Ureaplasma* Organisms

#### 1. In Fluid Cultures

The morphology of human and animal strains of ureaplasmas in young broth cultures during logarithmic growth is basically similar to that of

TABLE I. **Biologic Properties of Ureaplasmas**

| Property | Characteristic |
|---|---|
| Morphology | Basically small, spherical organisms averaging 330 nm in diameter (size range: 100–850 nm); also pleomorphic, short, bacillary elements with one end pointed seen in association with infected epithelial cells in clinical exudates; elongated elements sometimes seen |
| Agar colonies | Minute colonies 15–30 $\mu$m in diameter with irregular border; larger colonies with "cauliflower head" morphology; "fried egg" colonies on buffered agar; deep brown colonies with direct spot test for urease in *Ureaplasma* colonies on standard agars and on differential agar medium A7 |
| Preferred atmosphere | Depends upon medium employed; on standard *Ureaplasma* agars of pH 6.0, generally 5–15% $CO_2$ in air or nitrogen; good growth anaerobically with $H_2 + CO_2$; poor growth aerobically; high levels of $CO_2$ should be used with caution |
| Optimal pH for growth | $6.0 \pm 0.5$ |
| Genome size (daltons) | $4.1–4.8 \times 10^8$ |
| DNA guanine + cytosine (buoyant density; mol %) | 26.9–29.8 |
| Optimal temperature | 35°–37°C |
| Action on carbohydrates | Not fermented (hexokinase negative) |
| Cholesterol required | + |
| Digitonin sensitivity | + |
| Aminopeptidase | + |
| Esterase | + |
| $\alpha$-Glycerophosphate dehydrogenase | + |
| L-Histidine ammonia-lyase | + |
| Malate dehydrogenase | + |
| Phosphatase | + |
| Urease | + |
| Alanine dehydrogenase | − |
| Arginine deiminase | − |
| Catalase | − |
| Glutamate dehydrogenase | − |
| Lactic dehydrogenase (LDH) | − |
| NAD-dependent L-(+)-LDH | − |
| NAD-independent L-(+)-LDH | − |
| NAD-independent L-(−)-LDH | − |
| Proteolytic activity | + |

*(continued)*

TABLE I (*Continued*)

| Property | Characteristic |
|---|---|
| Hemolysis of guinea pig erythrocytes | + (beta) |
| Hemadsorption of guinea pig erythrocytes | + (human serotype III) |
| Sensitivity to erythromycin | + |
| Sensitivity to thallium acetate | + |
| Sensitivity to lincomycin | − |
| Tetrazolium reduction (aerobic and anaerobic) | − |
| Antigenicity | At least eight serotypes among human strains; at least eight serotypes among bovine strains, which are distinct from human strains. |
| Habitat | The genitourinary tract of man and lower animals, the respiratory tract of animals (cattle), and the pharynx and rectum of man |
| Pathogenicity | Produces ureaplasmal urethritis in man; pathogenic in some animal species and is associated with reproductive failure in man and animals (cats) |

other mycoplasmas and consists of dense, round to ovoid elements approximately 330 nm in diameter, with range of approximately 100–850 nm, and occasionally up to 1000 nm (Taylor-Robinson *et al.*, 1968; Rottem *et al.*, 1971; Black *et al.*, 1972a; Black, 1973a; Whitescarver and Furness, 1975). As with other mycoplasmas, short bacillary elements, branching filaments, and other pleomorphic forms often can be observed, depending upon the composition of the culture medium, pH, method of fixation, and method of examination (Ford and MacDonald, 1963; Shepard, 1967). The usual method of examination of broth cultures is by means of Giemsa-stained smears (Shepard, 1957). More recently, a hematoxylin–Giemsa stain was developed (Shepard, 1977) to demonstrate *U. urealyticum* in clinical materials and from broth cultures (Fig. 1). By this method, the organisms appeared as discrete, round or ovoid elements, singly, in pairs, triads, tetrads, short chains of three to five elements, and various-sized aggregates. No filamentous elements were seen. The organisms are gram-negative, although they stain weakly by this method (Shepard, 1969). The basic mode of multiplication of *U. urealyticum* appears to be a simple budding process, and two or more buds may appear simultaneously (Shepard, 1969; Whitescarver and Furness, 1975). Such a budding process from multiple buds could explain the ramifying, polydirectional manner which results in growth flocs or aggregates of varying sizes

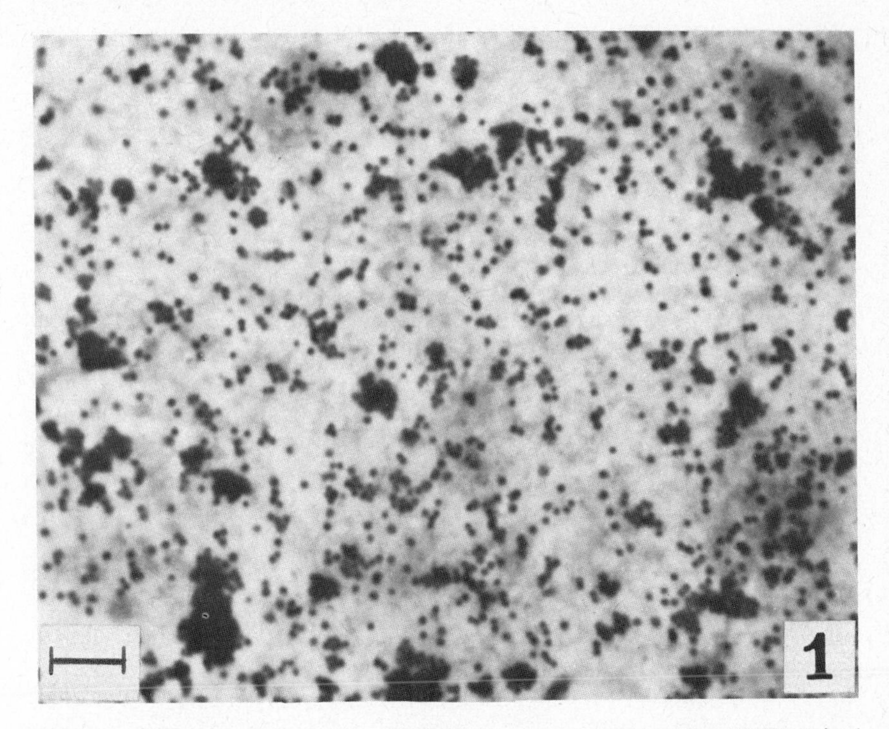


FIGURE 1. *Ureaplasma urealyticum* from a 17-hr culture in liquid medium U17B, stained by the hematoxylin–Giemsa method of Shepard. Twice-washed organisms concentrated 80-fold, suspended in Tyrode's buffer containing 0.8% formaldehyde fixative. Scale line = 5 μm. (From Shepard, 1977, reprinted by permission of the copyright owner.)

in broth cultures, or colonies in agar cultures. Replication by binary fission may also exist in ureaplasmas. In broth cultures, in addition to development of single organisms, pairs, etc., *U. urealyticum* has a tendency to multiply in loosely bound flocs or aggregates of organisms within a gelatinous matrix (Shepard, 1969). The organisms are basically "mononucleate" (i.e., with a single genome), but the "nucleus" replicates before a bud is sufficiently mature to be separated, thus resulting in a population in broth cultures of both "mono- and binucleate" organisms (Furness, 1975).

Examination by electron microscopy of broth-grown organisms produces excellent images and furnishes information not possible to obtain by conventional stained smears examined by light microscopy. The method of fixation is very important in order to preserve original morphology, and fixation while the organisms are in fluid suspension (prefixation) generally gives most reliable results (Maniloff *et al.*, 1965; Anderson and Barile, 1965; Lemcke, 1972). Improper fixation may result in loss of original form with resultant flattening and distortion, producing organisms that are unrealistically large and deceptive in morphology.

Scanning electron microscopy was employed by Klainer and Pollack (1973) to examine the surface structure of mycoplasmas, including two strains of *U. urealyticum*. In addition to round elements, elongated, rod-shaped *helical* elements were observed which measured 250 × 2,000 nm in preparations made from one strain. Helical elements were not observed in preparations made in the same manner of any of the classical, large-colony mycoplasmas or acholeplasmas examined. This is the first report of helical elements in cultures of *U. urealyticum*.

Phase-contrast microscopy was employed by Razin *et al.* (1977a) to study the morphology of broth-grown *U. urealyticum*. The usual morphologic elements were observed, and coccoidal organisms seemed to predominate. Most organisms appeared singly or in pairs. Long filaments and long chains of coccoid elements, common in classical mycoplasma cultures, were not observed.

The ultrastructure of *U. urealyticum* grown in broth cultures, as observed by electron microscopy of embedded, ultrathin sections (and preparations subjected to negative-staining procedures), consists of mostly rounded to ovoid organisms. The organisms ranged in size from 120 to 1000 nm in diameter (Williams, 1967; Taylor-Robinson *et al.*, 1968; Rottem *et al.*, 1971; Black *et al.*, 1972a; Black, 1974; Whitescarver and Furness, 1975). In addition, however, filamentous organisms were observed (Rottem *et al.*, 1971; Black *et al.*, 1972a) which reached a length of 2000 nm and had a width of 50–300 nm. Unlike Black *et al.* (1972a) and Rottem *et al.* (1971), filamentous elements were never observed by Whitescarver and Furness (1975) (although the same strains of *U. urealyticum* were used in two of the studies) and they concluded that the organism is normally spherical in shape. *Ureaplasma urealyticum* is bounded by a single trilaminar membrane approximately 10 nm thick consisting of piluslike structures radiating from the surface (Black *et al.*, 1972a; Whitescarver and Furness, 1975). Confirmation of the gelatinous matrix around *U. urealyticum* (Shepard, 1967) was reported by Whitescarver and Furness (1975), employing the pseudoreplica technique. Further confirmation of a capsular, extramembranous layer was reported by Robertson and Smook (1976) who found cytochemical evidence of extramembranous carbohydrates. By means of ruthenium red staining of thin-sectioned organisms examined by electron microscopy, a matlike, electron-dense layer 20–30 nm thick was observed. Application of the concanavalin A–iron dextran stain indicated that the outer layer contained glucosyl residues. The ultrastructure of *U. urealyticum* also has been shown to consist of randomly distributed ribosomes (occasionally arranged in regular geometric patterns) and vacuoles (Black *et al.*, 1972a). The ribosomes measured 12–15 nm, and nuclear fibroids 7.5–9 nm (Whitescarver and Furness, 1975).

### 2. In Agar Colonies

In colonies of *U. urealyticum* the individual organisms appear as minute, coccoidal particles growing in loosely bound masses downward into the agar gel. They are best visualized in wet-stained agar block preparations stained with Dienes' stain (Shepard, 1956). Black and Vinther (1977) provided excellent insight (Fig. 7) into the morphology and ultrastructure of *U. urealyticum* in agar colonies grown for 3, 5, and 10 days, thin-sectioned, and examined by electron microscopy. The arrangement of organisms within the colonies was less dense and less stratified than that observed in most other classical *Mycoplasma* colonies, confirming the earlier observations of Shepard (1956). Small, dense, pleomorphic organisms were located mainly along the colony border toward the agar and exhibited ribosome arrangements within the organisms and hairlike extramembranous surfaces. Thus, the ultrastructure of individual organisms in agar-grown colonies of *U. urealyticum* (Black and Vinther, 1977) is similar to that of ureaplasmas grown in fluid media.

### 3. In Clinical Exudates

Large numbers of *U. urealyticum* organisms may occur free in clinical exudates and urine specimens from male patients with *Ureaplasma*-associated urethritis. However, the organisms are difficult to recognize free in such exudates stained by Giemsa or similar types of staining procedures. The organisms are most easily recognized in association with infected epithelial cells scraped from the urethral mucosa (Shepard, 1957, 1966). The method of collection, fixation, and staining of smears prepared from urethral epithelial scrapings was also described (Shepard, 1977). The morphology of cell-associated *U. urealyticum* consists of round to ovoid organisms with variable degrees of pleomorphism, occurring mostly singly, but often in one or more aggregates within or on the surface of infected epithelial cells (Shepard, 1957). Whether the organisms are intracellular or primarily attached to the cell membrane is controversial. An important characteristic of cell-associated *U. urealyticum* in stained urethral exudates is the frequent occurrence of pleomorphic, short, bacillary, usually single organisms (or foci of single organisms) with *monopointed ends*. Shepard *et al.* (1974) and Shepard (1977) suggested that the pointed ends may represent small buds resulting from unequal division (as seen in Giemsa- or hematoxylin–Giemsa-stained smears). This theory is supported by the illustration shown in Fig. 2 of *U. urealyticum* examined by thin-section electron microscopy. The general appearance of *U. urealyticum* associated with an infected urethral epithelial cell in clinical exudate is shown in Fig. 3.

### B. *Ureaplasma* Colonies

The morphology of *U. urealyticum* colonies on unbuffered standard agars of pH 6.0 is generally that of a small (20–30 $\mu$m), somewhat refractile circular colony with an irregular border, growing downward into the medium (Shepard, 1956, 1957, 1959), as illustrated in Fig. 4. Although colony morphology of ureaplasmas is generally uniform on standard agars, colony size can be extremely variable. Under crowded conditions, colony size may be reduced to 7–10 $\mu$m and become nearly unrecognizable, as illustrated by Shepard and Lunceford (1976). Unusually small colonies of *U. urealyticum* can be mistaken for chemical deposits in primary cultures of urine specimens from certain NGU patients and can be overlooked (Shepard and Lunceford, 1975). Their formation is not

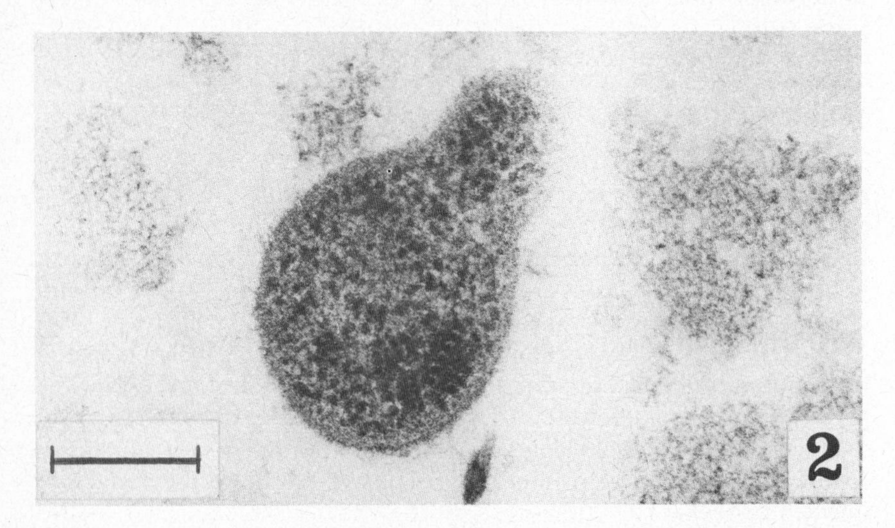


FIGURE 2. *Ureaplasma urealyticum* in a thin section, stained and counterstained in uranyl acetate/lead citrate and lightly carbon coated. Preparation illustrates a budding organism near the end of the replication cycle. Such an organism observed in clinical material (urethral exudate) stained by Giemsa or hematoxylin–Giemsa and examined by light microscopy would be interpreted as a monopointed organism and is a characteristic feature commonly seen in clinical exudates. Scale line = 0.25 $\mu$m. (From Whitescarver and Furness, 1975, reprinted with the permission of the copyright owner.)

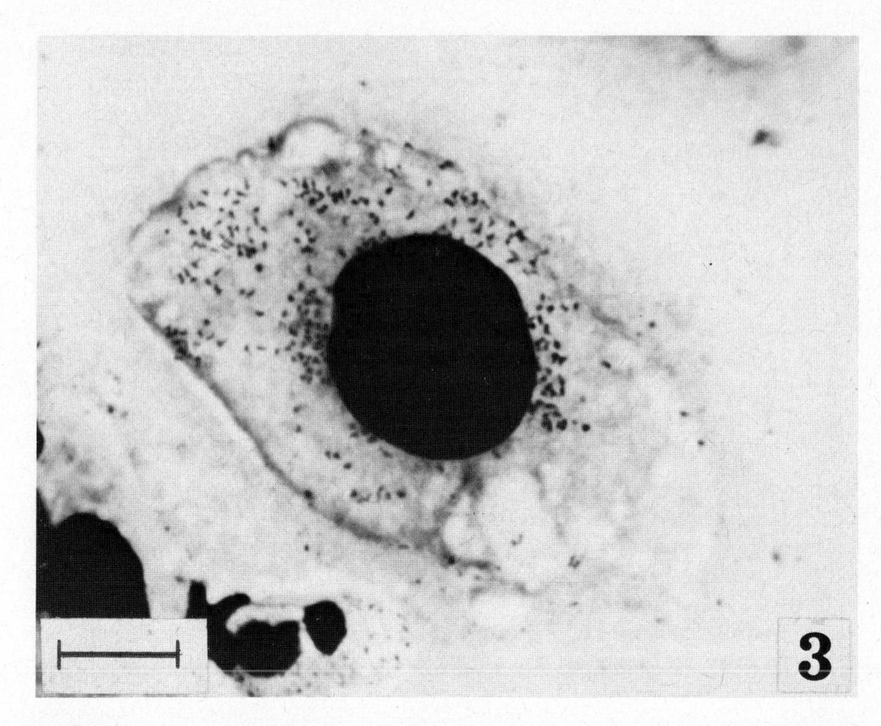


FIGURE 3. *Ureaplasma urealyticum*-infected epithelial cell from the anterior urethra of a male patient with nongonococcal urethritis. Urethral exudate was collected by epithelial scraping. Smear was fixed by methanol and stained for 3 hr in 1:50 Giemsa stain (pH 7.0). The characteristic morphology of cell-associated *U. urealyticum* in clinical exudates consists primarily of pleomorphic, short elements, frequently with monopointed ends. Scale line = 10 $\mu$m. Agar cultures from this same NGU patient grew out large numbers of *U. urealyticum* colonies in pure culture. × 1700.

completely understood but is believed to be associated with high urea excretion levels and/or elevated levels of divalent cations in the urines of certain NGU patients. In uncrowded conditions of cultivation, colonies may attain diameters of 30 to over 50 $\mu$m, depending on strain differences, agar volume, and inoculum size (Lee *et al.*, 1974). Larger colonies often become multilobate and look like "cauliflower head" colonies (Shepard *et al.*, 1974). Under special conditions of cultivation on unbuffered agar media of pH 6.0, such as increased carbon dioxide to 20% (Ford and MacDonald, 1963; Shepard, 1967); increased agar hardness, volume, and cation content (magnesium) (Furness, 1973b; Lee *et al.*, 1974); or incorporation of buffers such as L-histidine (Ajello and Romano, 1975; Romano *et al.*, 1975), HEPES buffer (Manchee and Taylor-Robinson, 1969; Furness, 1973b; Lee *et al.* 1974), or phosphates (Windsor *et al.*, 1975;

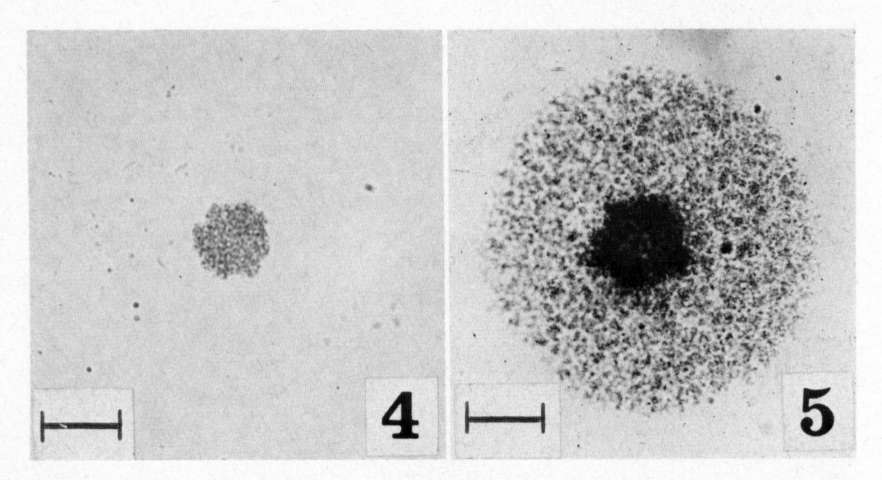


FIGURE 4. A characteristic colony of *Ureaplasma urealyticum* growing on standard unbuffered agar of pH 6.0. Colony size is approximately 25 μm in diameter. Strain K42 in second serial passage following primary isolation from a male patient with nongonococcal urethritis. Scale line = 25 μm. × 430.

FIGURE 5. A colony of *Ureaplasma urealyticum* showing "fried egg" morphology, growing on standard unbuffered agar of pH 6.0. The medium was additionally supplemented with L-cysteine to 0.01%. The development of large colony size and "fried egg" morphology was the result of the property of this particular strain and not a response to incorporation of special buffers or increased carbon dioxide. From an Agfachrome color transparency. Scale line = 25 μm. × 400.

Romano *et al.*, 1975), *U. urealyticum* colonies produce surface growth and assume the "fried egg" morphology of classical *Mycoplasma* species (Fig. 5). Razin *et al.* (1977b) recommended a pH 6.0 agar medium supplemented with 0.01 *M* urea and 0.01 *M* putrescine, incubated under a gaseous environment of 100% carbon dioxide. On this medium colonies of a laboratory-adapted strain of *U. urealyticum* were large (235 μm with 5% horse serum enrichment) and of "fried egg" morphology. A differential agar medium (A7) containing added urea and a manganese salt was developed for the identification of *U. urealyticum* in primary cultures of clinical materials (Shepard and Lunceford, 1976). Colonies of the organism on the A7 medium grow as deep brown colored colonies (Fig. 6) as seen by direct transmitted light under the low power of the microscope. The deep brown color (white by indirect, oblique illumination) is due to a manganese reaction product. The medium is specific for colonies of *Ureaplasma,* both human and lower animal strains, and gives positive identification.

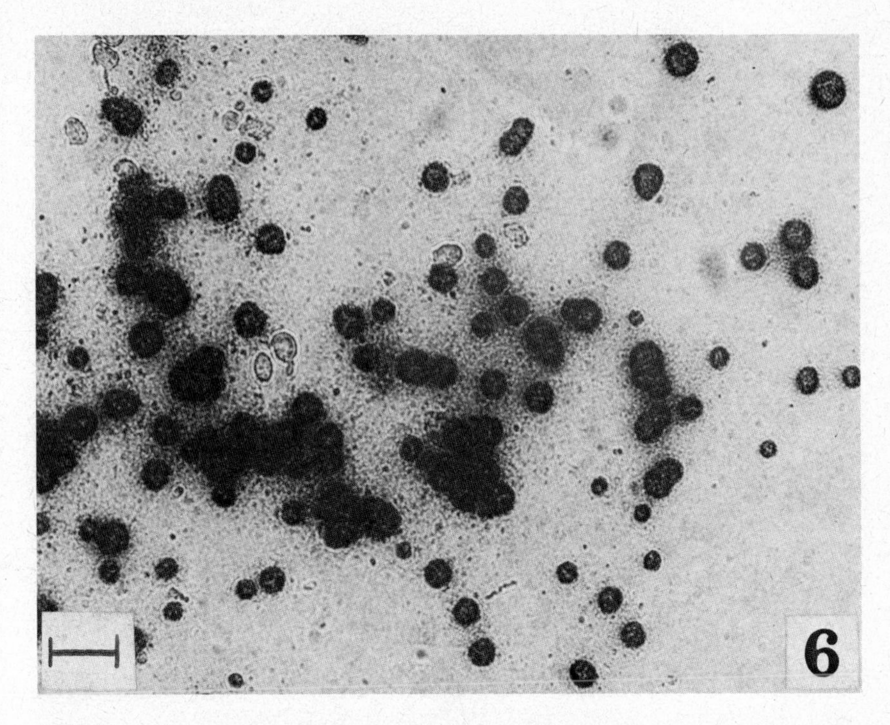


FIGURE 6. Characteristic appearance of colonies of *Ureaplasma urealyticum* growing on differential agar medium A7 of Shepard and Lunceford. The colonies are deep golden-brown to brown due to deposition of a manganese reaction product within and on the surface of the colonies, as a result of urea hydrolysis and ammonia production by the organism. Both human and lower animal strains of *Ureaplasma* elicit this reaction on differential agar medium A7 (and by the direct spot test for urease in colonies of ureaplasmas on *standard* agar media). Scale line = 50 $\mu$m. × 180.

## IV. EFFECTS OF PHYSICAL AND CHEMICAL ENVIRONMENT

### A. Heat

The optimal temperature for multiplication and growth of *U. urealyticum* is 35°–37°C (Shepard, 1956; Ford, 1962; Black, 1973a; Shepard *et al.*, 1974; Furness, 1975). Growth has been reported to occur at 40°C (Black, 1973a; Furness, 1975), but multiplication was adversely affected at this temperature, and up to a four fold reduction in titer may occur. The lag phase of many strains is increased, and mean generation times of 10 different strains of *U. urealyticum* were significantly increased

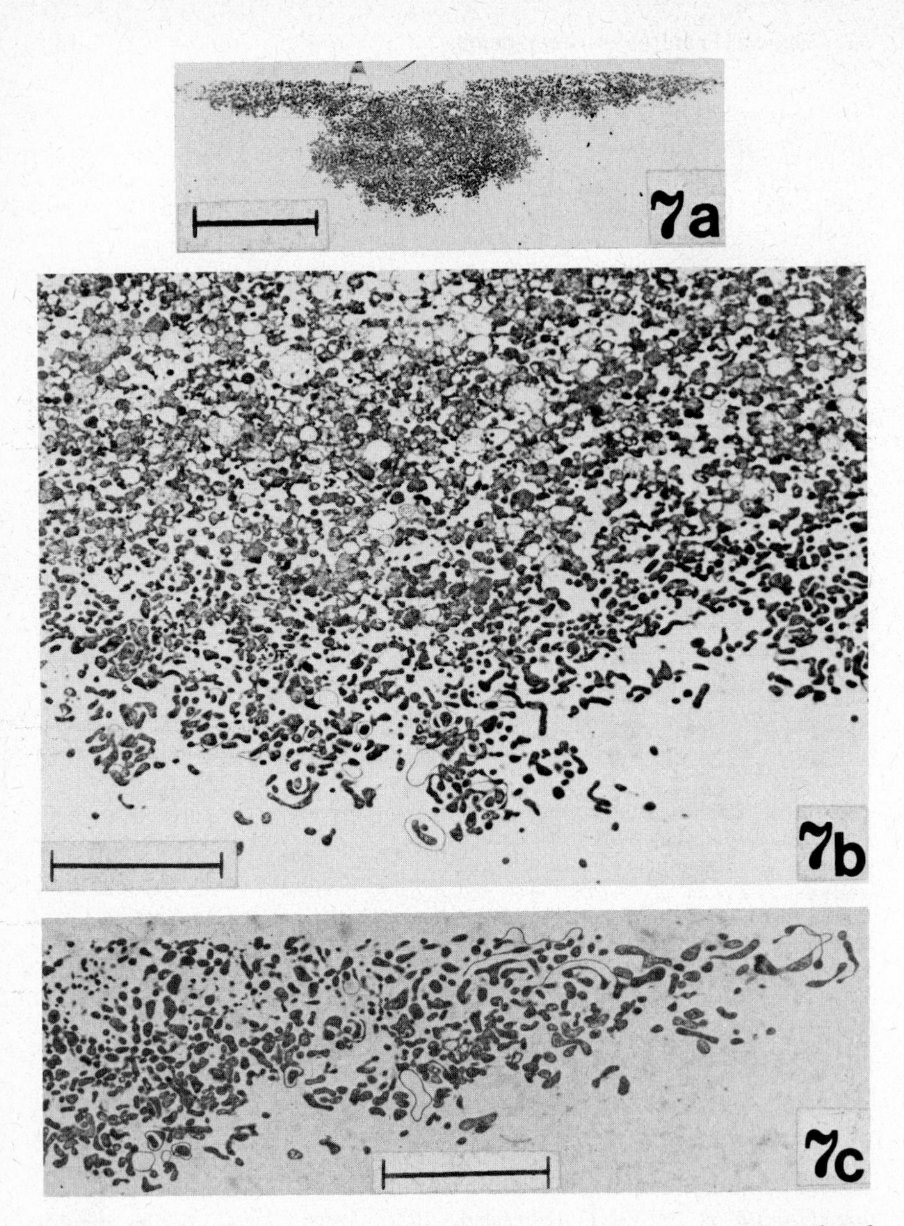


FIGURE 7. Ultrastructure of an agar-grown colony of *U. urealyticum*. (a) Cross section of a 3-day-old agar colony, showing a relatively thin zone of peripheral surface growth and a deeper, more dense central core of growth with a characteristically irregular border (× 672). Scale line = 25 $\mu$m. (b) Higher magnification of the deeper central growth of the colony shown in (a). The outer edge of the colony consists of dense organisms of variable morphology, whereas organisms in the center of the deeper growth are larger, ovoid, and less dense (× 4650). Scale line = 5 $\mu$m. (c) Higher magnification of peripheral *surface* growth of colony shown above in (a), illustrating organisms of quite variable morphology (× 4500). Scale line = 5 $\mu$m. (From Black and Vinther, 1977, reprinted with the permission of the copyright owner.)

by an elevation in temperature of only 3°C from 37°C to 40°C (Furness, 1975). No growth occurs at 42°C or 45°C (Black, 1973a; Furness, 1975). Multiplication occurs at slower rates at temperatures of 30°C and 27°C, but these reduced temperatures of incubation are useful in serial subcultivation of the organisms in fluid medium (30°C) at 24-hr intervals (Ford and MacDonald, 1967) and in the performance of the growth inhibition test (Black, 1973c). On solid medium, growth of *U. urealyticum* at 20°C is variable, and some strains fail to grow at this temperature; in liquid medium, no growth of eight serotype strains could be detected (Black, 1973a). *Ureaplasma urealyticum* is susceptible to heat inactivation at temperatures of 56°C and 60°C and the effectiveness of inactivation is a function of time. Ten different strains all were inactivated within 2.5 min at 56°C (Furness, 1975), and two strains were inactivated within 5 min at 56°C (Taylor-Robinson *et al.*, 1968). Heat inactivation of eight serotype strains at 60°C for 10 min was successful only for serotypes I, III, and IV, but treatment at the same temperature for 30 min inactivated all eight strains (Black, 1973a). The thermal death time for *U. urealyticum* probably lies between 10 and 30 min at a temperature of 60°C and varies with the strain, suspending medium, and other factors.

## B. Cold

Broth cultures of *U. urealyticum* remain viable up to 16 days at 4°C and for 90 days at −20°C (Ford, 1962). It is best not to retain broth cultures in the refrigerator longer than 2 weeks before transfer to fresh medium. Viability of *U. urealyticum* in broth cultures of pH 6.0 (without added urea) is retained up to 5 days at room temperature (Shepard *et al.*, 1974). Thus, freshly inoculated (but not incubated) broth cultures can be shipped over time periods of 5 days or more without serious loss of viability. Young (16–20 hr) broth cultures of *U. urealyticum* in screw-capped tubes or vials, further sealed with plastic tape, can be reliably stored in the frozen state at −60° to −85°C for periods of 1 year or more (Black, 1973a), Shepard *et al.*, 1974). Storage in the frozen state at −20°C has been recommended (Black, 1973a), but in the experience of one of us (MCS) storage at −20°C has been unreliable and is not recommended. Preservation by freeze drying (lyophilization) has proved to be reliable and convenient for long-term storage and for shipping (Black, 1973a; Shepard *et al.*, 1974).

## C. Ultraviolet Radiation and Sonic Energy

Ten different strains of *U. urealyticum* were examined for susceptibility to ultraviolet (uv) irradiation by Furness (1975) and all were found to behave in a similar manner. Most organisms in single cell suspensions were killed at an exponential rate, and thereafter the remaining organisms were inactivated more slowly. The biphasic curves suggested that the suspensions contained both "mononucleate" and "binucleate" (i.e., with single and double genome) organisms and, in this respect, were similar to those of *Mycoplasma mycoides* subsp. *mycoides* (Furness and DeMaggio, 1973). *Ureaplasma urealyticum* is considered basically "mononucleate," and the occurrence of "binucleate" organisms in single-cell suspensions suggests that the nucleus replicates before a bud or daughter cell is sufficiently mature to be separated from the parent cell by sonication (Furness, 1975). Sonic energy has proved to be useful in dispersing aggregated organisms in concentrated suspensions prepared from centrifuged broth cultures for use as antigens or for lyophilization (M. C. Shepard, unpublished observations). Ureaplasmas may be exponentially killed after 3–8 min sonication, depending upon the method of applying sonic energy, power level used, suspending medium, and strain differences. Furness (1975) observed that sonication disrupted multicellular aggregates of organisms into single cells within 2 min and that stationary phase organisms died more rapidly than exponential-phase organisms.

### D. Optimal Reaction (pH)

In contrast to the reaction preferred by the majority of classical *Mycoplasma* species (pH 7.6–8.0), all of the human and lower animal strains of *U. urealyticum* prefer an acid reaction of pH 6.0 ± 0.5 in liquid and solid media (Shepard and Lunceford, 1965). Many of the early problems in the cultivation of *U. urealyticum* prior to 1965 were related to an unfavorable (alkaline) reaction of the media.

### E. Gaseous Requirement

Standard fluid cultures of pH 6.0 without buffers are generally incubated in screw-capped or stoppered vessels without carbon dioxide addition. On standard *agar* media of pH 6.0 without added buffers, a gaseous mixture of 5–15% carbon dioxide in air or nitrogen (obtained by three cycles of evacuation and gas mixture replacement) provides best growth (Shepard *et al.*, 1974). Higher concentrations of carbon dioxide in such gaseous mixture (20% or more) favor development of thin zones of surface growth (Shepard, 1969) and colonies may assume "fried egg" morphology. Satisfactory growth can also be achieved by incubation of agar cultures by the modified Fortner method (Shepard, 1967). Incubation in a candle extinction jar is less satisfactory.

On standard agar media of pH 6.0 *with the addition of various buffers,* further supplemented with 0.01 *M* urea and 0.01 *M* putrescine dihydrochloride, high colony numbers and large colony size were achieved with a laboratory-adapted strain of *U. urealyticum* by incubation in 100% carbon dioxide (Razin *et al.*, 1977a). Large colony size of "fried egg" morphology developed under 100% carbon dioxide atmosphere, and isolated colonies exceeded 200 μm in diameter. However, the mass of *Ureaplasma* organisms in the surface zone of growth was much smaller than that in classical *Mycoplasma* colonies of the same size, and the surface growth was much thinner, making their recognition more difficult (Shepard, 1967; Razin *et al.*, 1977a). Masover *et al.* (1977a) studied the separate effects of urea, carbon dioxide, ammonia, and pH on growth. Carbon dioxide by gas bubbling technique acted as a superior buffer in liquid cultures in the pH range of 5.7–6.8 which is optimal for ureaplasmas. However, longer incubation times were sometimes required to achieve peak titers, which did not exceed $5 \times 10^7$ 50% Color Changing Units/ml (CCU/ml). For standard unbuffered agars, 5–15% carbon dioxide levels are recommended, as stated earlier. For differential agar media, such as the A7 medium of Shepard and Lunceford (1976), a relatively low carbon dioxide concentration is recommended, and higher levels are inhibitory. For example, excellent growth of *U. urealyticum* in primary cultures on the A7 medium is achieved by incubation in the BBL GasPak Anaerobic System, where a small level of carbon dioxide (in addition to hydrogen) is provided. Levels of 100% carbon dioxide must be employed with caution, particularly with media designed for primary isolation of ureaplasmas from humans and from lower animals. The gaseous environment must be carefully tailored to fit the medium, and the concentration of added urea and other supplements and the concentration and type of buffers may become critical factors determining the optimal carbon dioxide level required.

## V. NUTRITIONAL FACTORS IMPORTANT FOR GROWTH

### A. Major Factors

#### 1. Mammalian Serum

Media for isolation and cultivation of *U. urealyticum* require enrichment with mammalian serum. Horse serum, in concentrations of from 10 to 20%, is perhaps best for this purpose for reasons of cost and nutritional value. The known nutritional factors for ureaplasmas furnished by unheated, normal horse serum are *urea* (25–40 mg/100 ml, with an average value around 30 mg/100 ml) and *cholesterol* (approximately 70 mg/100 ml). Essentially no multiplication of ureaplasmas occurs in the absence of serum or in the presence of properly *dialyzed* serum for enrichment (Masover *et al.*, 1974; Shepard *et al.*, 1974). Ultrafiltrates of horse serum (as well as the serum from which the ultrafiltrate was obtained) support growth of ureaplasmas (Shepard, 1967). There are undoubtedly unknown factors in serum which are required or are beneficial to growth of ureaplasmas. A tripeptide isolated from human serum which stimulates cell growth (Pickart and Thaler, 1973) might be stimulatory for growth of *U. urealyticum*. The tripeptide (glycyl-L-histidyl-lysine acetate) was incorporated in a urease color test medium ("B broth") by Robertson (1978). Rarely encountered lots of horse serum contain an unknown factor(s) capable of supporting high titers ($10^{10}$ CCU/ml) of animal ureaplasmas in liquid media (Gourlay and Thomas, 1970). One of us (MCS) examined samples from 22 consecutive, different commercial lots of horse serum over a period of 2 years and failed to discover serum yielding such unusual performance.

### 2. Urea

Urea is the substrate which differentiates the ureaplasmas from all other members of the Mycoplasmatales. Yet, approximately 13 years after its discovery by Shepard (1966) and independently by Purcell *et al.* (1966) as a substrate of *U. urealyticum,* we know almost nothing about the role of urea in the physiology and growth of the organism. The requirement for urea is controversial but of fundamental biologic importance. The fate of urea nitrogen is unknown in the metabolism of the ureaplasmas, as it is for a wide variety of unrelated organisms that hydrolyze urea. It was hypothesized (Masover *et al.*, 1977c) that the intracellular ammonium ion (from urea nitrogen) plays a role in proton elimination or acid–base balance, which might be coupled to an energy-producing ion gradient and/or transport mechanisms (Harold, 1972). The fate of urea carbon, however, is known. When $^{14}$C-labeled urea was metabolized by ureaplasmas, approximately 95% of the radioactivity was recovered as $^{14}CO_2$, indicating that urea carbon is not utilized (Ford *et al.*, 1970). Urease activity is shared by a wide variety of unrelated microorganisms of which the ureaplasmas are the smallest. The question of urea requirement is even more provocative in the light of the finding by Delisle (1977) that *U. urealyticum* may possess up to three different ureases. What is the function of urease such that an organism with as few as 600–700 genes in its entire genome finds it of value to produce three of them?

The ureaplasmas are unique among the Mollicutes with respect to their ability to hydrolyze urea with the production of ammonia and carbon dioxide (Shepard, 1967; Shepard and Lunceford, 1967; Ford and MacDonald, 1967). Arginine is not hydrolyzed and the ureaplasmas are arginine deiminase-negative (Woodson *et al.*, 1965). The first demonstration of the hydrolysis of urea in the laboratory of one of us (MCS) was accomplished November 2, 1964 employing a serum-free, partially defined fluid medium (Stuart *et al.*, 1945) which was readjusted to pH 6.0. Subsequent early versions of serum-free fluid media for studies of ureaplasma urease activity included semidefined medium U-4 (Shepard and Lunceford, 1967) and serum-free urea medium U-7A, in which the "Boston T-strain" was first isolated and identified as a *Ureaplasma*. Subsequently, low levels of horse serum (4–5%) were incorporated to increase sensitivity. Treatment of horse serum with commercial urease (followed by readjustment of reaction to pH 6.0) rendered the serum completely inactive as an enrichment for ureaplasmas (Shepard and Lunceford, 1967). Similarly, specially dialyzed horse serum (residual urea = 3 $\mu$g/ml) failed to support growth of *U. urealyticum* (Shepard, 1967; Shepard and Lunceford, 1967; Shepard *et al.*, 1974). There are, however, other factors in serum that are important for multiplication, growth, and urease activity of ureaplasmas, and restoration of urea alone in media supplemented with specially dialyzed horse serum may not support the above activities without additional supplementation with some of these other factors. Masover and Hayflick (1973, 1974) and Masover *et al.* (1976a) studied the replication of *U. urealyticum* in media prepared from dialyzed components containing small amounts of added urea and found that the least amount of urea which allowed growth was 10 $\mu$g/ml (Masover *et al.*, 1974). Furness and Coles (1975) found the minimal urea requirement for growth to be between 2.5 and 10 $\mu$g/ml. Certain polyamines have been shown by Masover and Hayflick (1973, 1974) to exhibit interesting growth-promoting abilities in both dialyzed and nondialyzed media. Putrescine alone did not support growth of ureaplasmas, but the *combination* of urea and putrescine did support growth of a laboratory-adapted strain (Razin *et al.*, 1977a,b). These authors recommended the routine incorporation of both 0.01 *M* putrescine and 0.01 *M* added urea in agar media for cultivation of *U. urealyticum*. Kenny and Cartwright (1977) studied the effect of urea on growth of *U. urealyticum* and reported that without urea, growth did not occur. Growth was limited not only by urea concentration but also by the buffer capacity of the medium [also emphasized by Masover, *et al.* (1977a) and Razin *et al.* (1977a)]. The least amount of urea which supported growth was 0.032 m*M* (approximately 2.0 $\mu$g/ml) and the maximum yield of organisms ($8 \times 10^7$ CFU/ml) was observed at a concentration of 32 m*M* urea. However, the maximum growth rate of *U.*

*urealyticum* was observed at 3.2 m*M* urea, a value close to the $K_m$ determined for urease activity using whole cells (Masover *et al.*, 1976a). At lower concentrations, the growth rate was limited by urea concentrations. Kenny and Cartwright (1977) also reported generation times ranging from 8.0 hr at 0.032 m*M* to 1.6 hr at 3.2 m*M* urea. The rate of *urea hydrolysis* was not considered in this study, although it has been shown that cultures of $10^5$–$10^6$ CCU/ml of *U. urealyticum* will hydrolyze repeated additions of 10 m*M* urea rapidly (Masover *et al.*, 1977a). The rate of *ammonium ion* accumulation in ureaplasma cultures was shown to be independent of the growth rate and of the initial urea concentration above 0.025%. However, the *quantity* of ammonium ion which accumulated depended upon the initial urea concentration (Masover *et al.*, 1977b).

### 3. Lipids (Sterols and Fatty Acids)

A requirement for sterols has been demonstrated for ureaplasmas (Rottem *et al.*, 1971). This requirement is a major consideration in the taxonomy of the class Mollicutes, and the requirement for sterols is a unique feature of all species in the genus *Mycoplasma*. Ureaplasmas were very sensitive to digitonin, amphotericin B, and progesterone, and this sensitivity (together with a relatively high content of cholesterol found in the organisms) indicated a requirement for sterols. The sterol requirement could be met by cholesterol and by β-sitosterol. The importance of serum in the growth of ureaplasmas, in addition to supplying urea, is that it also satisfies the requirement for native protein and cholesterol (Rottem *et al.*, 1971). The requirement for sterol for all eight *Ureaplasma* serotypes was confirmed by Black (1973a). Rottem *et al.* (1971) also examined the possible requirement for fatty acids by ureaplasmas. The addition of saturated fatty acids (myristic, palmitic, or stearic acid) and unsaturated fatty acids (e.g., oleic) had no effect on growth, indicating that ureaplasmas do not require exogenous fatty acids for growth. In an earlier study of the lipid composition of a strain of *Ureaplasma,* Romano *et al.* (1972) reported the neutral lipid fraction to be characterized by a large amount of free fatty acids, and the phospholipid fraction by a predominant quantity of phosphatidic acid. Cholesterol was likewise found, and the ratio of cholesteryl esters to cholesterol was estimated to be about 1 : 4. In addition to phosphatidyl ethanolamine, an unidentified polar lipid (thought to be a diamino hydroxy compound containing adjacent fatty acid ester and *N*-acyl groups) was found. The occurrence of these two compounds distinguishes the lipids of this strain of *U. urealyticum* from other species of *Mycoplasma*. It was recently demonstrated (Romano *et al.*, 1976) that a strain of *U. urealyticum* (serotype 6) incorporated radioactivity into its lipids from 1-[$^{14}$C]acetate in growth medium. About 80% of the label was associated with saturated fatty acids; the remaining was found in the unsaturated fatty acid fraction. In the saturated fatty acid fraction, the label was present in the peaks of *palmitate, myristate,* and *stearate,* whereas in the unsaturated methyl ester fraction most of the radioactivity was found in the peak of palmitoleate. This *Ureaplasma* strain was capable of *de novo* synthesis of both saturated and unsaturated fatty acids from acetate, and in this respect differed from all other strains of *Mycoplasma* and *Acholeplasma* investigated to date.

## B. Other Factors

### 1. Yeast Extract

Evidence for the value of yeast extract in media for isolation and cultivation of ureaplasmas from man and lower animals is conflicting. In the laboratory of one of us (MCS) yeast extract was omitted from the agar medium for primary isolation of ureaplasmas from clinical materials because it provided no obvious benefit (Shepard, 1969). Similarly, Furness and Trocola (1977) found that the growth curves of four strains of ureaplasmas did not differ significantly in broth with and without yeast extract. Nevertheless, yeast extract was reinstated in media for ureaplasmas by one of us (MCS) to avoid confusion of *M. hominis* colonies growing without "fried egg" morphology with those of *U. urealyticum.* Yeast extract is required by *M. hominis* for vigorous growth and development of characteristic, well-developed surface growth zones. The usual 10% level of supplementation of yeast extract in media is partially inhibitory to many strains of ureaplasmas in primary cultures. As a result, Shepard and Howard (1970) and Shepard and Lunceford (1975) reduced the yeast extract concentration to 1.0% in agar media. A similar recommendation was voiced by Sueltmann *et al.* (1971), who observed that 1.0% yeast extract supported good growth in liquid medium and frequently was superior to 10% supplementation.

### 2. Amino Acids, Vitamins, and Cofactors

Very little is known about the nutritional requirements of the ureaplasmas for amino acids, vitamins, and cofactors. Ajello and Romano (1975) and Romano *et al.* (1975) found that the addition of L-histidine to growth medium prolonged the stationary phase and survival of a strain of *U. urealyticum.* A study performed with $^{14}$C-labeled urea demonstrated that L-histidine was acting as a buffer and, like HEPES and phosphate buffers (Romano *et al.,* 1975), retarded the rise in pH due to hydrolysis of urea in the medium. In a more recent study by Ajello *et al.* (1977), cell-free extracts from a strain of *U. urealyticum* deaminated L-histidine to urocanate and ammonia through an enzymatic reaction carried out by L-histidine ammonia-lyase, an enzyme also referred to as histidine-$\alpha$-deaminase. This is the first report of possible utilization of an amino acid by a strain of *U. urealyticum*. The deamination of L-histidine required the presence of a thiol such as reduced glutathione or 2-mercaptoethanol, and no enzymatic activity was detected in the absence of thiol. The liberation of ammonia as a result of deamination of L-histidine might account for some of the extra ammonia measured in urea broth cultures above that which would be expected from the total hydrolysis of the urea present in dialyzed broth medium (Masover *et al.*, 1977b).

L-Cysteine (0.01%) was found to yield 10-fold higher titers in fluid media, improved growth of colonies in solid media, and extended usefulness (up to 7 days) of the direct spot test for urease (M. C. Shepard and C. D. Lunceford, unpublished findings). L-Cysteine and some other thiols (2-mercaptoethanol, dithiothreitol, reduced glutathione, and thioglycolate) were the only thiols incorporated in media that improved performance or resulted in a modest increase in titer in fluid medium. L-Cysteine probably acts as a ureaplasma urease protector against inhibition by heavy-metal ions, and also as a reducing agent. The precise mechanism of its action is not known. L-Cysteine was therefore incorporated in all subsequent media for isolation and identification of ureaplasmas (Shepard and Lunceford, 1970a, 1976; Shepard, 1977). Tull *et al.* (1975) confirmed the usefulness of thiols in improving growth of ureaplasmas on differential agar A6 (Shepard and Lunceford, 1970b) and found 0.5% sodium thioglycolate as well as L-cysteine to enhance growth, but not methionine, cystine, or reduced glutathione. There is so far no evidence that L-cysteine is incorporated for metabolic purposes by ureaplasmas. The amino acids of the urea cycle (arginine, ornithine, and citrulline) are not hydrolyzed by ureaplasmas (Shepard, 1977). Furness and Trocola (1977) observed that three strains of *U. urealyticum* grew only in broth containing horse serum (10%) and died in broth supplemented with serum fraction A in place of horse serum. The same authors reported that a strain of *U. urealyticum* capable of growing well in broth supplemented with serum fraction A was stimulated by a combination of three amino acids, arginine, methionine, and cystine. However, neither amino acid alone stimulated growth.

M. C. Shepard and C. D. Lunceford (unpublished findings) observed that the poor growth-promoting ability of overheated (56°C) horse serum inadvertently used as serum enrichment for ureaplasmas was due to inactivation of pantothenic acid. The performance of such sera was restored to full activity by the addition of calcium pantothenate (10–40 μg/ml), suggesting that ureaplasmas may require pantothenate (normally supplied by unheated normal horse serum).

In 1971 M. C. Shepard and C. D. Lunceford (unpublished findings) observed that a differential agar medium (A6) supplemented with *dialyzed* horse serum (20%) and urea alone supported poor growth of seven ureaplasma strains in clinical exudates and one strain failed to grow. However, when "other additives," in the form of a commercial, chemically defined mixture of cofactors, vitamins, and amino acids (CVA Enrichment, GIBCO Diagnostics) *plus urea* were added as a supplement, abundant growth of five of the same eight strains occurred. On the basis of these studies, supplementation of standard media of pH 6.0 with 10–20% unheated horse serum, 0.02–0.05% added urea, 0.01% L-cysteine HCl, and 0.5% CVA Enrichment provided a superior fluid or agar medium for the primary isolation of ureaplasmas and the cultivation of especially fastidious ureaplasma strains from canine, bovine, and human sources. Media containing these supplements have been reported (Shepard and Lunceford, 1976).

## VI. BIOCHEMICAL PROPERTIES

### A. Enzymes

#### 1. Urease

The occurrence of urease in ureaplasmas and the influence of urea on growth of ureaplasmas was first reported in detail by Shepard and Lunceford (1967) and by Ford and MacDonald (1967). The optimal reaction for maximal urease activity was found to be pH 6.0 ± 0.5, which is also the optimal reaction for multiplication and growth of ureaplasmas (Shepard and Lunceford, 1965). The optimal initial concentration of urea for maximal urease activity (urea hydrolysis) in a urea broth of pH 6.0 was reported by Shepard and Lunceford (1967) to be 60 mg per 100 ml (0.01 *M*); a maximal titer of $1.2 \times 10^6$ CFU/ml was also observed at 0.01 *M* initial urea concentration. However, these values depend upon the strain of ureaplasma used and the experimental conditions employed. Masover *et al.* (1976a) examined the urea-hydrolyzing activity of a strain of ureaplasma using whole organisms and cell-free enzyme (urease) preparations and found the optimal urea concentration to be approximately 33.6 mg/100 ml (0.0056 *M*). The pH optimum was observed to be between 5.0 and 6.0, confirming the observation made by Shepard and Lunceford (1967).

However, ureaplasma urease activity in cell-free lysates may decline steeply at pH values higher than 8.0 (Swanberg *et al.*, 1978). Shepard and Lunceford (1967) reported that at high pH values of 8.0 or more, ureaplasma *viability* also declines sharply. Ford (1962) stated that a pH of 8.0 was lethal for ureaplasmas. *Ureaplasma* urease (used in the broad sense) is inhibited by heavy-metal ions such as 0.001 *M* mercuric chloride (Shepard and Howard, 1970). Hydroxamic acids (sorbyl-, benzoyl-, and 3-aminobenzoyl-) were first reported by Ford (1972) to inhibit growth and urease activity of ureaplasmas. Sorbyl hydroxamic acid (0.0001 *M*) inhibited the release of $^{14}CO_2$ from $^{14}C$ urea by washed organisms in a short incubation time of 4 hr. Ford (1973) concluded that urease activity of ureaplasmas was a direct requirement for multiplication of the organisms. Masover *et al.* (1974) observed that a 0.00018 *M* concentration of acetohydroxamic acid (AHA) did not inhibit laboratory-adapted strain No. 960 of *Ureaplasma urealyticum* or its ability to hydrolyze urea. The organisms multiplied at a slower rate in the presence of AHA than in its absence. The effect of acetohydroxamic acid on cell-free soluble urease preparations from *U. urealyticum* was examined by the same authors (Masover *et al.*, 1976a). Soluble urease activity failed to be inhibited by 0.0001 *M* AHA, whereas increasing amounts of AHA (to 0.005 *M*) caused decreasing urease activity. M. C. Shepard (unpublished findings) found that the direct (spot) test for urease in agar colonies of *U. urealyticum* (Shepard and Howard, 1970) was completely nullified by pretreatment of the colonies with 0.166 *M* acetohydroxamic acid solution (followed by 12 washings with distilled water before application of the test). It was further observed that AHA in a concentration of 80 $\mu$g/ml (0.0011 *M*) incorporated in a urease color test medium (Shepard and Lunceford, 1970a) was effective in completely blocking urease activity in this medium. A monovalent organic mercury compound, *p*-chloromercuribenzoic acid, selectively blocks sulfhydryl groups, forming stable mercaptides. Urease activity of two strains of *U. urealyticum* in a urease color test medium (Shepard and Lunceford, 1970a) was completely inhibited by 0.00025 *M* sodium *p*-chloromercuribenzoate (M. C. Shepard, unpublished findings).

The localization of urease in *U. urealyticum* was first reported by Masover *et al.* (1976a) and independently by Vinther (1976) and also by Masover *et al.* (1977c) in cell-free, soluble fractions of the organisms. Urease activity was confined to the cytoplasm and was not membrane-bound and was reported to be a constitutive enzyme, not adaptive (Masover *et al.*, 1977c). Digitonin lysates lost urease activity within 2–3 weeks when stored at −20°C; however, most of the activity remained for at least 5 weeks when stored at −70°C (Masover *et al.*, 1976a). Masover *et al.* (1977c) noted that significant urease activity could also be detected in nonviable organisms. Shepard and Howard (1970) observed that colonies of *U. urealyticum* that were in *excess* of 48 hr old (at 37°C) often failed to yield a positive test for urease by the direct (spot) test, suggesting that older colonies contained nonviable organisms which had lost enzyme activity.

Cytochemical methods were employed by Vinther (1976) to localize urease activity in two different strains of *U. urealyticum* to supplement chemical measurements of cell-free fractions. Urease activity was confined to the cytoplasmic, soluble fractions of the organisms (Fig. 8) and confirmed the findings of Masover *et al.* (1976a, 1977c). In addition, the cytochemical method obviated the need for cell fractionation. The method introduced by Shepard and Howard (1970) for light microscopic identification of *U. urealyticum* colonies (employing a manganous chloride reagent) was adapted to electron microscopy. The method is based on the precipitation of manganese dioxide, which is electron opaque, at the site of ammonia release as a result of urea hydrolysis by ureaplasma urease. The results obtained by this cytochemical method nicely supported the previous findings of cytoplasmic sites of the enzyme.

Delisle (1977) examined cytoplasmic fractions of recent isolates of

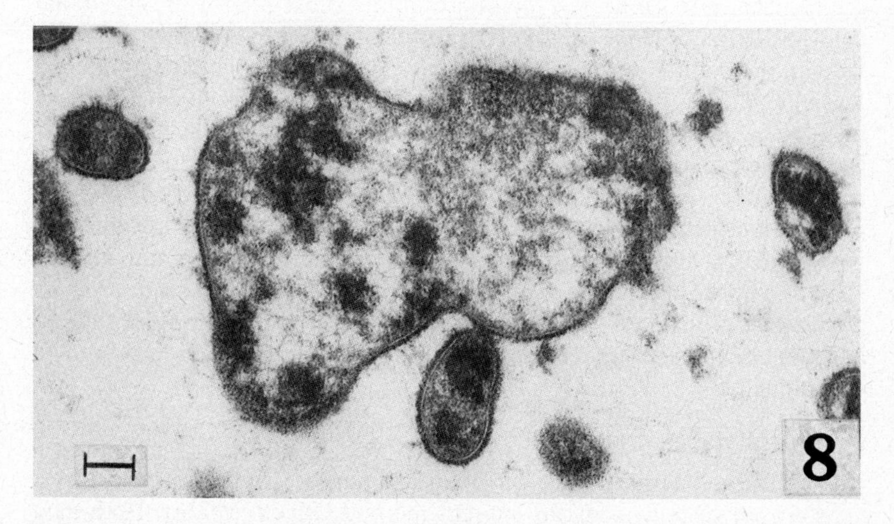


FIGURE 8. Cytochemical localization of urease in *U. urealyticum*. Thin section of organisms treated with an adaptation for electron microscopy of the direct test for urease of Shepard and Howard (1970), using a urea–manganese reagent. Electron-dense regions in the cytoplasm of the large cell and in two adjacent smaller cells are $MnO_2$ reaction product which is produced by the urea–manganese reagent in the presence of urease. Scale line = 0.1 μm. × 70,000. (From Vinther, 1976, reprinted with permission of the copyright owner.)

untyped strains of *U. urealyticum* and discovered that ureaplasmas contain not one, but three stable urease isoenzyme forms which could be distinguished by zymogram banding patterns. This technique has been used successfully with a variety of enzymes from bacterial and fungal species. Active zymogram banding was also observed with malate dehydrogenase, α-glycerophosphate dehydrogenase and esterase, but clear, repeatable differences in zymogram banding were found only with urease. These results represent the first demonstration of the complexity of urease activity in the genus *Ureaplasma*. Enzyme banding patterns were *not* detected after polyacrylamide gel electrophoresis of soluble cytoplasmic fractions with the following enzymes: alanine and glutamate dehydrogenase, lactic dehydrogenase (LDH), nicotinamide adenine dinucleotide (NAD)-dependent L-(+)-LDH, NAD-independent L-(+)-LDH, or NAD-independent L-(−)-LDH (Delisle, 1977).

### 2. L-Histidine Ammonia-Lyase

L-Histidine added to liquid media prolongs the stationary phase and survival of *U. urealyticum* and increases colony size and manner of superficial growth ("fried egg" morphology) in agar cultures (Ajello and Romano, 1975; Romano *et al.*, 1975). Its function was that of a buffer to retard rise in pH. However, the same authors subsequently observed that cell-free extracts from *U. urealyticum* deaminated L-histidine to urocanate and ammonia through the enzymatic action of L-histidine ammonia-lyase (E.C. 4.3.1.3). The enzyme is also referred to as histidase or histidine-α-deaminase (Ajello *et al.*, 1977). Activity of the enzyme required the presence of thiols, e.g., 2-mercaptoethanol. Statistically significant increases of enzyme activity were demonstrated in crude enzyme preparations obtained from *U. urealyticum* in fluid medium supplemented with 16 m*M* L-histidine. Although L-histidine ammonia-lyase activity was shown to occur in cell-free extracts, whether intact ureaplasma *organisms* metabolize L-histidine has not been determined, and the role of the enzyme in *Ureaplasma* metabolism and its physiologic function remain to be determined.

### 3. Phosphatases and ATPase

Phosphatase activity was demonstrated in 44 different strains of *U. urealyticum* tested, using an indirect method (Black, 1973b). Both acid and alkaline phosphatases were found. Phosphatase activity was demonstrated in all of eight *U. urealyticum* serotypes studied (Black, 1973a). Thus, 52 different strains of the organism were tested, and all contained phosphatase. Masover *et al.* (1977c) reported that adenosine triphosphatase in *U. urealyticum* is membrane-associated, as it is with other mycoplasmas. Delisle (1977) showed that cytoplasmic fractions of the organism had neither acid nor alkaline phosphatase, indicating that the enzymes were membrane-associated.

### 4. Aminopeptidase

Proteolytic activity is a well-known property of some *Mycoplasma* species, and aminopeptidases have been detected in the membranes of *Acholeplasma laidlawii* and *Mycoplasma fermentans*. The eight serotype strains of *U. urealyticum* were examined by Vinther and Black (1974) and all were found to possess aminopeptidase activity. The ureaplasmas released alanine from L-alanyl-L-alanyl-L-alanine, and serotypes VII and VIII released free amino acids from all of the unsubstituted peptides studied and from leucine amide. Negative findings were obtained after incubation with N-blocked peptides. Watanabe *et al.* (1973) tested three strains of *U. urealyticum* for their ability to attack horse serum proteins, and all three strains showed proteolytic activity. Egg yolk protein was similarly attacked, but weakly. The ureaplasmas showed no proteolytic activity against skimmed milk or gelatin. However, Black (1973a) reported negative tests for lecithinase activity (employing egg yolk emulsion) and negative tests for serum digestion (employing the method of Aluotto *et al.*, 1970). The eight serotype *U. urealyticum* strains were used as test organisms.

### 5. Catalase and Other Enzymatic Activities

The ureaplasmas are catalase-negative (Shepard *et al.*, 1974) which is in agreement with findings for some other mycoplasmas, e.g., *A. laidlawii, Mycoplasma agalactiae, Mycoplasma arthritidis, M. mycoides* subsp. *mycoides,* and *M. pneumoniae* (Low *et al.*, 1968). Ureaplasmas are hexokinase-negative and dextrose and other carbohydrates are not fermented as determined by acid production, nor is aesculin metabolized. Arginine deiminase is not produced (Woodson *et al.*, 1965; Black, 1973a; Shepard *et al.*, 1974). Neither methylene blue nor tetrazolium are reduced, aerobically or anaerobically (Black, 1973a). Delisle (1977) examined cytoplasmic fractions of *U. urealyticum* by polyacrylamide gel electrophoresis zymogram banding patterns and found the preparations negative for alanine dehydrogenase, glutamate dehydrogenase, and lactate dehydrogenase; but *active* zymogram patterns were found with malate dehydrogenase (single band), $\alpha$-glycerophosphate dehydrogenase (single band), esterase (one to four bands) and urease (two to three bands, as discussed previously in this section).

### B. Genome Size and Base Composition of *Ureaplasma* DNA

Bak *et al.* (1969) showed that the genome sizes of the mycoplasmas varied in accordance with the requirement for sterols. The Acholeplasmataceae have genome sizes of about $9 \times 10^8$ daltons, whereas the Mycoplasmataceae have genome sizes from $4 \times 10^8$ to $5 \times 10^8$ daltons. These same authors determined the genome size of two human *Ureaplasma* strains and calculated the molecular weights to be $4.7 \times 10^8$ and $4.4 \times 10^8$ daltons, which is in good agreement with the values shown above for the sterol-requiring members of the Mycoplasmatales. The ureaplasmas also require sterols (Rottem *et al.*, 1971). The genome sizes of all of the eight serotypes of *U. urealyticum* were found to range from $4.1 \times 10^8$ to $4.8 \times 10^8$ daltons, thus indicating that the ureaplasmas are a homogeneous group with respect to their genome size, and gave further support for placing the ureaplasmas in the sterol-requiring family Mycoplasmataceae (Black *et al.*, 1972b).

The melting profiles ($T_m$) of purified DNA from seven human ureaplasma strains (serotypes I through VII) were found to be similar, ranging from 80.65° to 81.00°C, with an average value of 80.78°C (Bak and Black, 1968). The DNA base composition of these same seven serotype strains, plus *U. urealyticum* serotype VIII (Black *et al.*, 1972b), in mol % guanine + cytosine (G + C) ranged from 27.7 to 28.5, suggesting a degree of genetic homogeneity among the human ureaplasmas. The DNA base composition of five bovine ureaplasmas was found to be 29.0–29.8% G + C (Howard *et al.*, 1974). It was suggested from this that the human and bovine strains represent different species or subspecies of *Ureaplasma*. It should be noted that this low G + C content is similar to that found in other members of the family Mycoplasmataceae (26–32% G + C) (McGee *et al.*, 1967).

### C. Cell Proteins

In addition to genetic homogeneity of the human ureaplasmas with respect to similar genome size and DNA base composition, similarities in cellular protein values may also be taken to indicate genetic relatedness, since the protein composition of the organism is dictated by the genome (Razin and Rottem, 1967). The electrophoretic patterns of 12 human *U. urealyticum* strains were strikingly similar, and six strains exhibited almost identical patterns (Razin *et al.*, 1970). These findings further suggest that the human ureaplasmas are genetically closely related and support their classification in a single species. Taylor-Robinson *et al.* (1971) provided further evidence in support of genetic homogeneity of the ureaplasmas by electrophoretic examination of the proteins of five strains of human and animal origin. The 11 bands observed with the ureaplasma strains were closely similar and could be regarded as belonging to a single closely related group.

## D. Membrane Composition

Mycoplasma membranes consist mainly of proteins and lipids, as do other biological membranes. The protein comprises about two-thirds of the mass of the mycoplasma membrane, the balance being almost entirely lipid. Carbohydrates are usually found in biomembranes as components of glycolipids and glycoproteins, and this is the case with mycoplasma membranes also. Two rather unusual carbohydrate polymers are associated with mycoplasma membranes (Razin, 1975). One is a galactan associated with *M. mycoides* subsp. *mycoides,* covering the exterior cell surface forming a slime layer (Gourlay and Thrower, 1968). In the case of ureaplasmas, Robertson and Smook (1976) reported cytochemical evidence for extramembranous carbohydrates. An extramembrane layer of polyanions was demonstrated by ruthenium red technique and found to contain glucosyl-like residues. Purified membranes from seven strains of *U. urealyticum* were reported by Whitescarver *et al.* (1975) to contain between 1 and 7% carbohydrates, and mannose, galactose, and glucose were identified. The same ureaplasma strains were examined for amino acid and protein composition by Whitescarver *et al.* (1976). The amino acid composition was similar to that of other biological membranes.

Membranes of *U. urealyticum* contain cholesterol (Rottem *et al.*, 1971), as do the membranes of other members of the sterol-requiring family Mycoplasmataceae and the spiroplasmas. As would be expected, such cells are susceptible to lysis by digitonin (Masover *et al.*, 1977c). The membrane of the ureaplasmas has other qualities which appear to be unique among the Mollicutes. For example, the fatty acid composition varies from other species in that a larger proportion of free fatty acids and phosphatidic acid are present in ureaplasmas, and phosphatidyl ethanolamine and a diamino hydroxy polar lipid have also been identified in the ureaplasmas (Romano *et al.*, 1972). Adenosine triphosphatase activity was found in ureaplasma membranes (Masover *et al.*, 1977c) as it is in other mycoplasmas.

## E. Hemolysin

Ureaplasmas of human and animal origin produce a soluble hemolysin which is capable of lysing erythrocytes of various animal species

(Shepard, 1967; Manchee and Taylor-Robinson, 1970; Black, 1973a). Guinea pig erythrocytes suspended in an agar overlay were lysed by colonies of *U. urealyticum* which produced a β-hemolysin (Shepard, 1967). No hemolysis was seen with horse, calf, ox, swine, rabbit, or chicken erythrocytes under the same experimental conditions. Sobeslavsky and Chanock (1968) failed to detect hemolysis of guinea pig erythrocytes by colonies of three *Ureaplasma* strains of human origin and suggested that this may have been due to the small size of the colonies. Shepard (1969) observed that hemolysis of guinea pig erythrocytes failed to occur unless *freshly prepared* agar medium was employed, suggesting that a labile factor in fresh medium (or horse serum enrichment) is required by the organism to produce the soluble β-hemolysin. Manchee and Taylor-Robinson (1970) reported that ureaplasma colonies of simian, canine, and bovine origin hemolyzed both homologous and guinea pig erythrocytes. Four human ureaplasma strains failed to hemolyze human erythrocytes and gave variable results with guinea pig erythrocytes. The hemolysis observed was of the $\alpha'$ type and was considered to be a peroxide, since they could inhibit hemolysis of erythrocytes by a canine ureaplasma with catalase. Conversely, all of the eight human ureaplasma serotypes produced a soluble β-hemolysin that completely lysed guinea pig and rabbit erythrocytes, both aerobically and anaerobically (Black, 1973a). Under the same conditions, human erythrocytes were unaffected and peroxide production was not demonstrated by the inhibition of hemolysis by catalase or directly by using the benzidine blood–agar plate method. Black (1973a) introduced a solid medium containing 1.5% guinea pig, rabbit, or human erythrocytes, as well as an erythrocyte–broth medium to demonstrate hemolysis by *U. urealyticum* strains.

### F. Hemadsorption

Solid medium containing HEPES buffer was used to demonstrate hemadsorption of human, guinea pig, and fowl erythrocytes by eight different human and animal strains of ureaplasma. Only colonies of a simian (squirrel monkey throat) ureaplasma adsorbed the erythrocytes tested, and the attachment was tenuous (Manchee and Taylor-Robinson, 1969). Subsequently, Black (1973a) examined eight serotype strains of *U. urealyticum* for their ability to adsorb guinea pig, rabbit, and human erythrocytes (using a modified Shepard medium). Hemadsorption was demonstrated only with guinea pig erythrocytes, and only by colonies of serotype III *U. urealyticum*. This is the only biologic test so far described that is capable of distinguishing between serotypes of human ureaplasmas.

### G. Tissue Culture Cell Adsorption

Agar colonies of six different human ureaplasmas adsorbed EDTA-dispersed tissue culture-grown HeLa cells from suspensions, colonies of some *Ureaplasma* strains more vigorously than others. Attachment of the cells was tenacious and resisted vigorous washing (Manchee and Taylor-Robinson, 1969). Two animal ureaplasmas, a simian (throat) and a canine (semen) strain, failed to adsorb HeLa cells under the same conditions. The specificity of the HeLa cell adsorption reaction was suggested by inhibition of cell attachment by specific antiserum (D. Taylor-Robinson, personal communication).

### H. Possible Toxic Products

The possible elaboration of a toxin or toxic product of some type by some or all strains of *U. urealyticum* is controversial but of fundamental biologic importance. The explanation of the well-known behavior of ureaplasmas in fluid media—the abrupt cessation of growth once the culture has attained a titer of $10^6$–$10^8$ CCU/ml—is unknown. The observations of Razin *et al.* (1977b) appear to rule out the possibility that this is due to the exhaustion of urea. The possibility that the rise in pH due to liberated ammonia is causing the low titer was ruled out by their findings that under carbon dioxide this does not happen, and yet the titer remains low. A possible explanation of this behavior of ureaplasmas in fluid cultures was suggested by Furness (1973a) who proposed that the abrupt cessation of growth is caused by the accumulation of a *toxic factor*. This factor is catalase-resistant, thermostable, and dialyzable and appeared to be responsible for the failure of spent cultures to support growth of ureaplasmas, even when supplemented with fresh serum and urea. All ureaplasma strains tested produced a toxic factor that inactivated not only the strain producing it but several other strains, suggesting that production of toxic substances is a characteristic of the ureaplasmas. The identity of this toxic factor is unknown. However, it must be emphasized that any experiment which produces higher titers or prolonged stationary growth of ureaplasmas in liquid cultures (Brighton *et al.*, 1967; Hendley and Allred, 1972; Windsor and Trigwell, 1976; Robertson, 1978) argues against a "toxic factor" in such cultures.

Another aspect of the problem of ureaplasma toxicity concerns the cytotoxic and cytopathic effect of growth of ureaplasmas in tissue cell cultures, for example, see Section IX. Now that ureaplasmas have been demonstrated to be pathogenic for animals (Gourlay, 1974) and to be pathogenic for man (Taylor-Robinson *et al.*, 1977), the possibility of ureaplasma toxicity *in vivo* has assumed new significance. One of the major products elaborated by ureaplasmas during active hydrolysis of urea is *ammonia*. Ammonia has been shown to be toxic for cells and tissues, both *in vitro* and *in vivo* (MacLaren, 1969; Visek, 1970; LeVeen *et al.*, 1978). Since *U. urealyticum* is an actively ureolytic organism, large quantities of ammonia (as *ammonium ion* at physiologic pH) may be produced within small areas of infection at the cellular level—on the urethral mucosa or within infected mucous glands and ducts of Littré, for example. It is possible that such strong local accumulations of ammonium ion on mucous membranes may act as a toxic factor (competing for hydrogen ions?) and contribute to the pathogenic potential of the ureaplasmas.

## VII. SENSITIVITY TO ANTIBIOTICS AND ANTIMICROBIAL AGENTS

### A. Sensitivity to Antibiotics

Human and animal ureaplasmas are susceptible to the inhibitory action of the following antibiotics, listed in order of approximate decreasing effectiveness *in vitro*: doxycycline, minocycline, declomycin, tetracycline, erythromycin, chlortetracycline, oxytetracycline, chloramphenicol, streptomycin, spectinomycin, spiromycin, kanamycin, and gentamycin (Ford, 1962; Shepard *et al.*, 1966; Taylor-Robinson *et al.*, 1968; Braun *et al.*, 1970a; Black, 1973a; Spaepen *et al.*, 1976; Spaepen and Kundsin, 1977). This ranking is not necessarily the same in all laboratories conducting antibiotic sensitivity studies, and the results depend upon the method used (agar disk, broth, or broth disk, etc.), culture medium, pH, and other factors. In general, the clinical effectiveness agrees with the results of antibiotic susceptibility testing. An increasing number of ureaplasma strains isolated from patients are showing resistance to tetracycline and its derivatives (Ford and Smith, 1974; Hofstetter *et al.*, 1976; Spaepen *et al.*, 1976). Seven (13%) of 54 *U. urealyticum* isolates tested were resistant to all five of the following antibiotics: tetracycline, minocycline, doxycycline, demeclocycline, and erythromycin (Spaepen and Kundsin, 1977). R. B. Kundsin (personal communication, 1977) indicated that chloramphenicol, streptomycin, and gentamycin showed promising usefulness against ureaplasmas, with susceptibilities of 84%, 72%, and 42% of strains tested, respectively.

### B. Resistance to Antibiotics and Antimicrobial Agents

Ureaplasmas are generally unaffected by the following agents: penicillin (including the semisynthetic penicillins, e.g., ampicillin), sulfonamides and rifampin (Shepard *et al.*, 1974), cephaloridine (Csonka *et al.*, 1967; Braun *et al.*, 1970a), aurothiomalate (Taylor-Robinson *et al.*, 1968; Ford, 1972), and lincomycin (Shipley *et al.*, 1968; Csonka and Corse, 1970; Braun *et al.*, 1970a). Lincomycin has been found useful in the recovery and purification of ureaplasmas from mixed cultures containing both ureaplasmas and *M. hominis,* since the latter is inhibited by lincomycin, whereas ureaplasmas are not (Braun *et al.*, 1970b; Shepard *et al.*, 1974). Trimethoprim lactate in a concentration (5 $\mu$g/ml) that completely inhibits most strains of *Proteus* in an agar system of pH 6.0 is inactive against ureaplasmas (Shepard *et al.*, 1974).

### C. Selective Growth Inhibition by Antimicrobial Agents

The following agents exhibit *selective* growth inhibition of ureaplasmas but in general are inactive against most classical mycoplasmas, e.g., *M. hominis:* thallium acetate, erythromycin, 5-iodo-2′-deoxyuridine, hydroxyurea, acetohydroxamic acid, and hydroxamic acid derivatives. The classical mycoplasmas are largely unaffected by these agents at concentrations which are completely inhibitory to ureaplasmas, as discussed more fully by Shepard *et al.* (1974) and Shepard (1977). It is emphasized that growth and urease inhibition by hydroxamic acids is incompletely understood, and the mechanisms of inhibition remain to be clarified.

## VIII. SEROLOGY

### A. Human Ureaplasmas

The human ureaplasmas are antigenically distinct from all other recognized mycoplasmas and consist of a heterogeneous group, presently of eight serotypes (Black, 1970, 1973c), or 11 serotypes (Lin *et al.*, 1972). The validity of the eight serotypes proposed by Black (1970, 1973c) was established by comparative serologic testing, employing four different serologic methods: a metabolism inhibition test, a growth inhibition test, an indirect immunofluorescence test, and an indirect hemagglutination test (Black, 1970). The serotype strains examined by Black were originally isolated and serologically studied by Ford (1966, 1967). The growth inhibition test was subsequently modified for application to the serologic study of ureaplasma serotypes (Black, 1973c) and found specific and well-suited for the classification and identification of human ureaplasmas. Potent antisera are required, however, and the reactivity of antisera is significantly affected by the composition of the growth medium used to grow the antigens (Masover *et al.*, 1975). An indirect immunofluorescence technique employing *unfixed* colonies of ureaplasmas (Rosendal and Black, 1972) was found by Black and Krogsgaard-Jensen (1974) to possess the same degree of specificity as the growth inhibition test. The growth inhibition method of Black (1973c) was used by Piot (1976) to serotype ureaplasma strains isolated from patients with nongonococcal urethritis, with gonorrhea, and from two control groups. He found no difference in distribution of serotypes among 45 isolates from these groups, and no serotype could be correlated with disease symptoms. The same investigator (Piot, 1977) later serotyped the eight type-strains plus 39 genital isolates and compared the growth inhibition test (Black, 1973c) with the indirect immunofluorescence test of Rosendal and Black (1972), using unfixed ureaplasma colonies. The advantage of the latter test is its ability to detect *mixed infections* of more than one serotype of *U. urealyticum* on primary isolation plates. The significance of this statement can be appreciated by the detection of mixed primary cultures in as many as 36% of isolates examined by the immunofluorescence method (Piot, 1977).

Lin *et al.* (1972), employing a complement-dependent mycoplasmacidal test (Lin and Kass, 1970), identified 11 different serotypes of *U. urealyticum*. Antisera were prepared against ureaplasmas that had been isolated from the genital tracts of pregnant women and from infants. Cross-reactivity of these sera showed the presence of five serogroups on the basis of shared common antigens, and of 11 serotypes within these groups. The ureaplasmas came from a variety of sources and all could be typed using the available 14 sera. No single type seemed to be associated with nongonococcal urethritis, pregnancy, or infancy. The method of Lin *et al.* (1972) appeared to be specific and sensitive. Urea was omitted from the reaction mixture, since the ammonia produced by hydrolysis of the urea by the organisms presumably inactivated the C4 component of the guinea pig complement. However, Masover *et al.* (1975) showed that it was unlikely that ammonia produced from urea had any effect on complement-dependent killing of ureaplasmas. The serotypic heterogeneity of human genital ureaplasma isolates was reemphasized by Lin and Kass (1973), further indicating that more than one serotype may coexist in the same specimen. A comparative review of the eight serotypes proposed by Black (1970, 1973c) and of the 11 serotypes proposed by Lin *et al.* (1972) should be undertaken with the purpose of reconciling serotypic differences and establishing a new standardization of serotypes among the ureaplasmas.

### B. Animal Ureaplasmas

*Bovine* strains, like the *U. urealyticum* strains from humans, are also heterogeneous (Taylor-Robinson *et al.*, 1969: Howard and Gourlay, 1972). The first bovine ureaplasma strains isolated in the United States ("San Angelo T-strains") were found to be unrelated serologically to two human ureaplasmas and similarly unrelated to two bovine ureaplasma strains of urinary tract origin (Livingston, 1972). Eight bovine ureaplasmas were examined serologically by three different test procedures by Howard and Gourlay (1973). They were serologically heterogeneous and were distinct from human ureaplasmas as well as a caprine, a simian, and a canine strain of *Ureaplasma.* Ureaplasmas isolated from sheep and goats (Livingston and Gauer, 1975) were similarly unrelated antigenically to human strains, ten bovine strains, and three bovine strains isolated in England. At least eight serotypes of *bovine* ureaplasmas were isolated from the respiratory and reproductive tracts of cattle in the United States by Livingston and Gauer (1974), employing the immune inactivation technique of Lin and Kass (1970) and Lin *et al.* (1972). The relationship of the eight bovine serotypes isolated in the United States by Livingston and Gauer (1974) to the eight bovine serotypes identified in Europe by Howard and Gourlay (1973) is unknown. As a group, the pulmonic isolates appeared more closely related than did the isolates from the genitourinary tract. More than one serotype was occasionally isolated from a single animal site (Livingston and Gauer, 1974) and the possible inadequacy of the cloning procedure may have contributed to lack of pure serotypes in every instance. The same difficulty arises in the cloning of ureaplasma isolates from human sources as well. The limit–dilution method of cloning should be discontinued in favor of the filtration–dilution method recommended by the Subcommittee on the Taxonomy of Mycoplasmatales (1972), using the smallest pore size (200–450 nm) possible. Proper, repeated cloning gives a high degree of assurance that the strain of *Ureaplasma*, for example, is pure; but it does not necessarily guarantee that it is pure.

## IX. *Ureaplasma*–HOST CELL INTERACTIONS

Shepard (1957) noted a close association of ureaplasmas and host epithelial cells in clinical exudates (collected by urethral scrapings) shortly after the organisms were first recognized. Despite the widespread problem of mycoplasma contamination of tissue-cultured cells, there is but one report of a *Ureaplasma* contaminant in a cell culture (Sethi, 1972). The reason why more reports of *Ureaplasma* cell culture contamination have not appeared is not understood. As early as 1963 an established HeLa-S3 cell culture was experimentally infected with strain K71 *U. urealyticum* in the laboratory of one of us (MCS). A cytopathogenic effect (CPE) was observed 48 hr postinoculation. Previous attempts to infect HeLa and McCoy cell cultures with this ureaplasma strain had failed. The success in establishing this cell culture infection may have been attributed to the repeated (24 hourly) incorporation of urea in the growth and maintenance media (M.C. Shepard, unpublished observations). However, urea was employed only occasionally in cell culture fluids by Mazzali and Taylor-Robinson (1971), and the usefulness of urea incorporation remains to be more fully studied. The latter authors succeeded in establishing ureaplasma infection in three different cell lines (L132, HeLa, and Vero), but the infections were of relatively short duration. Mazzali and Taylor-Robinson (1971) treated ureaplasma-infected cell cultures with streptomycin and concluded that the surviving ureaplasmas were intracellular. Masover *et al.* (1976b) succeeded in infecting a WI-38 line of normal human embryonic lung fibroblasts with *U. urealyticum* and observed reduced rates of multiplication of infected cells and reduced plating efficiency (Fig. 9) in addition to morphologic changes usually associated with mycoplasma infection of animal cells *in vitro*. The cytotoxic effect was sensitive to aureomycin but not to penicillin. It was not related to depletion of amino acids or nucleic acid precursors in the cell culture fluid, but it did require that the host cells be growing. Masover *et al.* (1976b) observed that the ability of ureaplasmas to hydrolyze urea is lost or not expressed after association with the cells, a finding that confirms an earlier, similar observation made by M. C. Shepard (unpublished observations). The organism produced cytopathic effects but appeared to have been altered in some unknown way by its association with the cells. Does the cell supply some factor(s) *in vivo* that is supplied by urea *in vitro*? Persistent infection of these cells could be established, however, but required pretreatment of the ureaplasmas with trypsin (Masover *et al.*, 1977d). This finding is important as it relates to the possible mechanisms involved in mycoplasma–host cell interactions in general, and suggests that a close membrane-to-membrane association is required for persistent infection to occur.

Organ cultures have likewise been shown to be susceptible to experimental infection with ureaplasmas. Taylor-Robinson and Carney (1974) observed that both genital and oral ureaplasmas multiplied and often

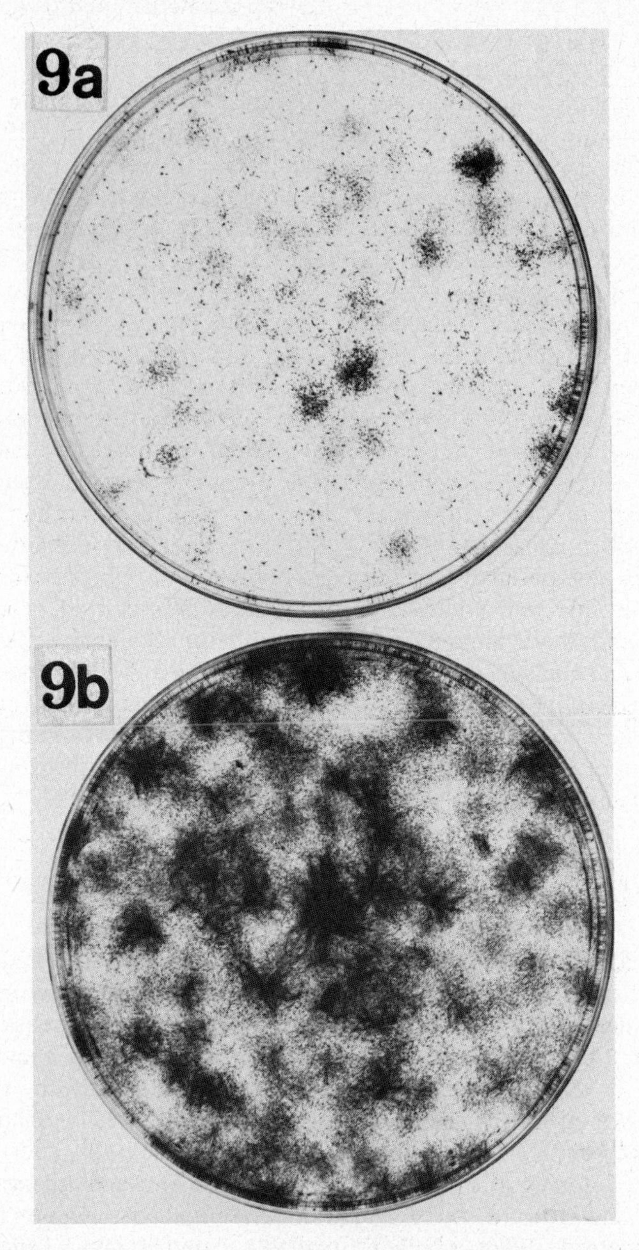


FIGURE 9. Effect of *Ureaplasma urealyticum* infection of WI-38 tissue culture cells on plating efficiency. Culture dishes were seeded with 1000 cells per dish in 5 ml of complete cell culture medium. (a) The upper dish was seeded with 0.1 ml of log phase *U. urealyticum* broth culture (multiplicity of infection = three organisms per cell). (b) Uninoculated control WI-38 cells which received 0.1 ml of sterile ureaplasma medium. Culture medium was changed on days 5, 10, and 15, and cells were fixed and stained with Giemsa stain on day 20. (From Masover *et al.*, 1977d, reprinted by permission of the copyright owner.)

persisted in fallopian tube organ cultures. No loss of ciliary activity or other apparent damage was seen in ureaplasma-infected organ cultures. Quite opposite findings were reported by Stalheim *et al.* (1976), who studied the growth and effects of ureaplasmas in bovine oviductal organ cultures. *Ureaplasma* strains from the human genital tract and the genital and respiratory tracts of cattle all succeeded in growing in experimentally inoculated uterine tube organ cultures, stopping ciliary activity and causing histologic lesions. Titers of $10^8$–$10^9$ CCU/ml were attained, and 24–144 hr after inoculation ciliostasis was complete. This was followed by collapse and sloughing of the cilia, bulging and vacuolization of secretory and ciliated cells, and finally disorganization, necrosis, and desquamation of the epithelium. It is interesting that ciliostasis also occurred after additions of nonviable ureaplasmas or washed, disrupted organisms. This strongly suggests the production of a cytotoxin by ureaplasmas. The relationship of this cytotoxin to the toxic product described in *broth* cultures of ureaplasmas (Furness, 1973a) is presently unknown. However, it is also possible that the cilia stopping effect produced by ureaplasmas in infected oviductal organ cultures was caused by ammonia ($NH_4^+$), and that ammonia and the cytotoxin (Stalheim *et al.,* 1976, Stalheim and Gallagher, 1977) are one and the same. The pathogenicity of human and bovine ureaplasmas for oviductal epithelium should encourage further studies on the role of ureaplasmas in reproductive failure. A further discussion of *Mycoplasma*–cell culture interactions is offered in Volume II, Chapter 13.

## X. LABORATORY IDENTIFICATION

Ureaplasmas of human and animal origin are detected and reliably identified in primary and secondary cultures of clinical specimens and other materials by the demonstration of *Ureaplasma* urease activity (Shepard, 1973a). Three methods which make use of this characteristic property of ureaplasmas are: (1) a urease color test broth (Ford and MacDonald, 1967; Taylor-Robinson *et al*., 1968, 1971; Shepard and Lunceford, 1970a, 1976; Robertson, 1978); (2) a direct (spot) test for urease in colonies of ureaplasma growing out on standard agar media (Shepard and Howard, 1970); and (3) a differential agar medium (Shepard and Lunceford, 1976) which is available commercially. Detection of ureaplasma urease activity in urease color test media and recognition of urease activity in agar cultures are described and illustrated in detail in the literature cited. Fastidious strains of *U. urealyticum* exist which may be difficult to isolate in primary cultures.

## XI. TAXONOMY

The ureaplasmas of human and lower animal origin are distinguished by their ability to hydrolyze urea to ammonia and carbon dioxide and by their possession of a urease enzyme system. This singular property significantly sets the ureaplasmas apart from all other members of the order Mycoplasmatales. On the basis of this property, a separate genus (*Ureaplasma*) in the family Mycoplasmataceae was proposed, containing a single species, *Ureaplasma urealyticum* (Shepard *et al.*, 1974; Approved Lists of Bacterial Names submitted by the Ad Hoc Committee of the Judicial Commission of the ICSB, 1976). The species presently consists of at least eight serotypes (Black, 1970, 1973c) and possibly 11 (Lin *et al.*, 1972). The type strain is *U. urealyticum* 960-(CX8), serotype VIII, ATCC No. 27618. Lower animal strains of *U. urealyticum* appear to be serologically distinct from human strains.

## XII. FUTURE CONSIDERATIONS

Although we have learned much about the biology of the ureaplasmas over the last two decades, there are nevertheless wide gaps in our knowledge of the organism and many important questions remain to be answered. We are particularly ignorant in the area of ureaplasma physiology and metabolism. For example, we still do not know the energy source of the organism, nor do we understand the role of urea and of ammonia nitrogen in the metabolism of the ureaplasmas. We have just begun to understand the importance of lipids in the physiology of ureaplasmas, and the biosynthesis of fatty acids. These are fundamental questions. What is the nature and function of the extramembranous layer surrounding the organism? Progress in antigenic and biochemical studies of human and lower animal ureaplasmas is hampered by our inability to grow the organisms to high titers in liquid cultures. In spite of this inability, prolonged stationary growth of the organisms in liquid culture can nevertheless be used to advantage. The first use of a continuous culture system in a closed apparatus employing the chemostat principle for the cultivation of *U. urealyticum* was by Brighton *et al.* (1967). Viable counts of $10^5$–$10^6$ CFU/ml were achieved over a period of 216 hr (9 days) by means of such a system. Recently, one of us (G. K. Masover, unpublished findings) utilized continuous culture of a laboratory-adapted strain of *U. urealyticum* in a chemostat system. Continuous production of *Ureaplasma* organisms to a titer of $10^7$ CCU/ml over a period of >300 hr (12.5

days) was accomplished, employing a flow rate of fresh medium of 35 ml/hr in a 350-ml vessel (dilution rate = 0.1). Incubation temperature was 37°C, and reaction of the culture was maintained at between pH 6.9 and 7.0. Multiplication of ureaplasmas under these experimental conditions was unaffected by urea concentrations from 5.0 to 20.0 m*M* (0.03–0.12%). Such yields of organisms should begin to provide acceptable cell mass for further biochemical and antigenic studies of the ureaplasmas.

Now that *U. urealyticum* has been proved to be a human pathogen and a cause of urethritis in humans (Taylor-Robinson *et al.*, 1977) other questions become more pressing, such as *Ureaplasma*–host cell interactions, demonstration and localization of *Ureaplasma* organisms in the human genital tract in patients with urethritis, and the possible relationships of serotypes to pathogenicity. Are certain serotypes avirulent and carried in the normal genitourinary tract without producing symptoms? Can they produce symptoms under certain conditions of microbial or other stress? Are certain serotypes of more pathogenic potential than other serotypes? These and many more questions are still seeking answers to contribute to a more full understanding of the fundamental biology of the ureaplasmas and their role as human and lower animal pathogens.

## REFERENCES

Ad Hoc Committee of the Judicial Commission of the ICSB (1976). *Int. J. Syst. Bacteriol.* **26,** 563–599.

Ajello, F., and Romano, N. (1975). *Appl. Microbiol.* **8,** 183–187.

Ajello, F., Romano, N., and Massenti, M. F. (1977). *Bol. Ist. Sieroter. Milan.* **56,** 343–350.

Aluotto, B. B., Wittler, R. G., Williams, C. O., and Faber, J. E. (1970). *Int. J. Syst. Bacteriol.* **20,** 35–58.

Anderson, D. R., and Barile, M. F. (1965). *J. Bacteriol.* **90,** 180–192.

Bak, A. L., and Black, F. T. (1968). *Nature (London)* **219,** 1044–1045.

Bak, A. L., Black, F. T., and Christiansen, C. (1969). *Nature (London)* **224,** 1209–1210.

Black, F. T. (1970). *Proc. Int. Congr. Infect. Dis., 5th 1970,* Vol. 1, pp. 407–411.

Black, F. T. (1973a). *Ann. N.Y. Acad. Sci.* **225,** 131–143.

Black, F. T. (1973b). *Int. J. Syst. Bacteriol.* **23,** 65–66.

Black, F. T. (1973c). *Appl. Microbiol.* **25,** 528–533.

Black, F. T. (1974). Doctorate dissertation, pp. 1–74. Inst. Med. Microbiol., University of Aarhus, Aarhus, Denmark.

Black, F. T., and Krogsgaard-Jensen, A. (1974). *Acta Pathol. Microbiol. Scand., Sect. B* **82,** 345–353.

Black, F. T., and Vinther, O. (1977). *Acta Pathol. Microbiol. Scand., Sect. B* **85,** 281–285.

Black, F. T., Birch-Andersen, A., and Freundt, E. A. (1972a). *J. Bacteriol.* **111,** 254–259.

Black, F. T., Christiansen, C., and Askaa, G. (1972b). *Int. J. Syst. Bacteriol.* **22,** 241–242.

Braun, P., Klein, J. O., and Kass, E. H. (1970a). *Appl. Microbiol.* **19,** 62–70.

Braun, P., Klein, J. O., Lee, Y.-H., and Kass, E. H. (1970b). *J. Infect. Dis.* **121,** 391–400.

Brighton, W. D., Windsor, G. D., Andrews, B. E., and Williams, R. E. O. (1967). *Mon. Bull. Minist. Health Public Health Lab. Serv. (G.B.)* **26,** 154–158.

Csonka, G. W., and Corse, J. (1970). *Br. J. Vener. Dis.* **46,** 203–204.

Csonka, G. W., Williams, R. E. O., and Corse, J. (1967). *Ann. N.Y. Acad. Sci.* **143,** 194–198.

Delisle, G. J. (1977). *J. Bacteriol.* **130,** 1390–1392.

Ford, D. K. (1962). *J. Bacteriol.* **84,** 1028–1034.

Ford, D. K. (1966). *Arthritis Rheum.* **9,** 503–504.

Ford, D. K. (1967). *Ann. N.Y. Acad. Sci.* **143,** 501–504.

Ford, D. K. (1972). *Antimicrob. Agents & Chemother.* **2,** 340–343.

Ford, D. K. (1973). *J. Infect. Dis.* **127,** Suppl., S82–S83.

Ford, D. K., and MacDonald, J. (1963). *J. Bacteriol.* **85,** 649–653.

Ford, D. K., and MacDonald, J. (1967). *J. Bacteriol.* **93,** 1509–1512.

Ford, D. K., and Smith, J. R. (1974). *Br. J. Vener. Dis.* **50,** 373–374.

Ford, D. K., McCandlish, K. L., and Gronlund, A. F. (1970). *J. Bacteriol.* **102,** 605–606.

Furness, G. (1973a). *J. Infect. Dis.* **127,** 9–16.

Furness, G. (1973b). *J. Infect. Dis.* **128,** 703–709.

Furness, G. (1975). *J. Infect. Dis.* **132,** 592–596.

Furness, G., and Coles, S. (1975). *Proc. Soc. Exp. Biol. Med.* **150,** 807–809.

Furness, G., and DeMaggio, M. (1973). *J. Infect. Dis.* **127,** 563–566.

Furness, G., and Trocola, M. (1977). *J. Infect. Dis.* **135,** 507–511.

Gourlay, R. N. (1974). *Colloq. Inst. Nat. Sante Rech. Med. (INSERM)* **33,** 365–374.

Gourlay, R. N., and Thomas, L. H. (1970). *J. Comp. Pathol.* **80,** 585–594.

Gourlay, R. N., and Thrower, K. J. (1968). *J. Gen. Microbiol.* **54,** 155.

Harold, F. M. (1972). *Bacteriol. Rev.* **36,** 172–230.

Hendley, J. O., and Allred, E. N. (1972). *Infect. Immun.* **5,** 164–168.

Hofstetter, A., Blenk, H., and Rangoonwala, R. (1976). *Muench. Med. Wochenschr.* **118,** 49–50.

Howard, C. J., and Gourlay, R. N. (1972). *Br. Vet. J.* **128,** 37–41.

Howard, C. J., and Gourlay, R. N. (1973). *J. Gen. Microbiol.* **79,** 129–134.

Howard, C. J., Gourlay, R. N., Garwes, D. J., Pocock, D. H., and Collins, J. (1974). *Int. J. Syst. Bacteriol.* **24,** 373–374.

Kenny, G. E., and Cartwright, F. D. (1977). *J. Bacteriol.* **132,** 144–150.

Klainer, A. S., and Pollack, J. D. (1973). *Ann. N.Y. Acad. Sci.* **225,** 236–245.

Lee, Y.-H., Donner, A., Bailey, P. E., Alpert, S., and McCormack, W. M. (1974). *J. Lab. Clin. Med.* **84,** 766–770.

Lemcke, R. (1972). *J. Bacteriol.* **110,** 1154–1162.

LeVeen, E. G., Falk, G., Moon, I., Mazzapica, N., and LeVeen, H. H. (1978). *Am. J. Surg.* **135,** 53–56.

Lin, J.-S., and Kass, E. H. (1970). *J. Infect. Dis.* **122,** 93–95.

Lin, J.-S., and Kass, E. H. (1973). *Infect. Immun.* **7,** 499–500.

Lin, J.-S. L., Kendrick, M. I., and Kass, E. H. (1972). *J. Infect. Dis.* **126,** 658–663.

Livingston, C. W., Jr. (1972). *Am. J. Vet. Res.* **33,** 1925–1929.

Livingston, C. W., Jr., and Gauer, B. B. (1974). *Am. J. Vet. Res.* **35,** 1469–1471.

Livingston, C. W., Jr., and Gauer, B. B. (1975). *Am. J. Vet. Res.* **36,** 313–314.

Low, I. E., Eaton, M. D., and Proctor, P. (1968). *J. Bacteriol.* **95,** 1425–1430.

McCormack, W. M., Braun, P., Lee, Y.-H., Klein, J. O., and Kass, E. H. (1973). *N. Engl. J. Med.* **288,** 78–89.

McGee, Z. A., Rogul, M., and Wittler, R. G. (1967). *Ann. N.Y. Acad. Sci.* **143,** 21–30.
MacLaren, D. M. (1969). *J. Pathol.* **97,** 43–49.
Manchee, R. J., and Taylor-Robinson, D. (1969). *J. Bacteriol.* **100,** 78–85.
Manchee, R. J., and Taylor-Robinson, D. (1970). *J. Med. Microbiol.* **3,** 539–546.
Maniloff, J., Morowitz, H. J., and Barrnett, R. J. (1965). *J. Bacteriol.* **90,** 193–204.
Masover, G. K., and Hayflick, L. (1973). *Ann. N.Y. Acad. Sci.* **225,** 118–130.
Masover, G. K., and Hayflick, L. (1974). *J. Bacteriol.* **118,** 46–52.
Masover, G. K., Benson, J. R, and Hayflick, L. (1974). *J. Bacteriol.* **117,** 765–774.
Masover, G. K., Mischak, R. P., and Hayflick, L. (1975). *Infect. Immun.* **11,** 530–539.
Masover, G. K., Sawyer, J. E., and Hayflick, L. (1976a). *J. Bacteriol.* **125,** 581–587.
Masover, G. K., Namba, M., and Hayflick, L. (1976b). *Exp. Cell Res.* **99,** 363–374.
Masover, G. K., Razin, S., and Hayflick, L. (1977a). *J. Bacteriol.* **130,** 292–296.
Masover, G. K., Catlin, J., and Hayflick, L. (1977b). *J. Gen. Microbiol.* **98,** 587–593.
Masover, G. K., Razin, S., and Hayflick, L. (1977c). *J. Bacteriol.* **130,** 297–302.
Masover, G. K., Palant, M., Zerrudo, Z., and Hayflick, L. (1977d). *In* "Nongonococcal Urethritis and Related Infections" (D. Hobson and K. K. Holmes, eds.), pp. 364–369. Am. Soc. Microbiol., Washington, D.C.
Masover, G. K., Perez, R., and Matin, A. (1978). *Infect. Immun.* **23,** 172–174.
Mazzali, R., and Taylor-Robinson, D. (1971). *J. Med. Microbiol.* **4,** 125–138.
Pickart, L., and Thaler, M. M. (1973). *Nature (London) New Biol.* **243,** 85–87.
Piot, P. (1976). *Br. J. Vener. Dis.* **52,** 266–268.
Piot, P. (1977). *Br. J. Vener. Dis.* **53,** 186–189.
Purcell, R. H., Taylor-Robinson, D., Wong, D., and Chanock, R. M. (1966). *J. Bacteriol.* **92,** 6–12.
Purcell, R. H., Wong, D., Chanock, R. M., Taylor-Robinson, D., Canchola, J., and Valdesuso, J. (1967). *Ann. N.Y. Acad. Sci.* **143,** 664–675.
Razin, S. (1975). *Prog. Surf. Membr. Sci.* **9,** 257–312.
Razin, S., and Rottem, S. (1967). *J. Bacteriol.* **94,** 1807–1810.
Razin, S., Valdesuso, J., Purcell, R. H., and Chanock, R. M. (1970). *J. Bacteriol.* **103,** 702–706.
Razin, S., Masover, G. K., Palant, M., and Hayflick, L. (1977a). *J. Bacteriol.* **130,** 464–471.
Razin, S., Masover, G. K., and Hayflick, L. (1977b). *In* "Nongonococcal Urethritis and Related Infections" (D. Hobson and K. K. Holmes, eds.), pp. 358–363. Am. Soc. Microbiol., Washington, D.C.
Robertson, J. (1978). *J. Clin. Microbiol.* **7,** 127–132.
Robertson, J., and Smook, E. (1976). *J. Bacteriol.* **128,** 658–660.
Romano, N., Smith, P. F., and Mayberry, W. R. (1972). *J. Bacteriol.* **109,** 565–569.
Romano, N., Ajello, F., Massenti, M. F., and Scarlata, G. (1975). *Boll. Ist. Sieroter. Milan.* **54,** 292–295.
Romano, N., Rottem, S., and Razin, S. (1976). *J. Bacteriol.* **128,** 170–173.
Rosendal, S., and Black, F. T. (1972). *Acta Pathol. Microbiol. Scand., Sect. B* **80,** 615–622.
Rottem, S., Pfendt, E. A., and Hayflick, L. (1971). *J. Bacteriol.* **105,** 523–530.
Sethi, K. K. (1972). *Zentralbl. Backteriol., Parasitenkd., Infektionskr. Hyg., Abt. I: Orig., Reihe A:* **219,** 550.
Shepard, M. C. (1954). *Am. J. Syph., Gonorrhea, Vener. Dis.* **38,** 113–124.
Shepard, M. C. (1956). *J. Bacteriol.* **71,** 363–369.
Shepard, M. C. (1957). *J. Bacteriol.* **73,** 162–171.
Shepard, M. C. (1959). *Urol. Int.* **9,** 252–257.
Shepard, M. C. (1960). *Ann. N.Y. Acad. Sci.* **79,** 397–402.
Shepard, M. C. (1966). *Health Lab. Sci.* **3,** 163–169.

Shepard, M. C. (1967). *Ann. N.Y. Acad. Sci.* **143,** 505–514.
Shepard, M. C. (1969). *In* "The Mycoplasmatales and the L-Phase of Bacteria" (L. Hayflick, ed.), pp. 49–65. Appleton, New York.
Shepard, M. C. (1970). *J. Am. Med. Assoc.* **211,** 1335–1340.
Shepard, M. C. (1973a). *J. Infect. Dis.* **127,** Suppl., S22–S25.
Shepard, M. C. (1973b). *Proc. Int. Vener. Dis. Symp., 2nd, 1972,* pp. 83–87.
Shepard, M. C. (1974). *Collog. Inst. Natl. Sante Rech. Med. (INSERM)* **33,** 375–380.
Shepard, M. C. (1977). *In* "Nongonococcal Urethritis and Related Infections" (D. Hobson and K. K. Holmes, eds.), pp. 345–357. Am. Soc. Microbiol., Washington, D.C.
Shepard, M. C., and Calvy, G. L. (1965). *N. Engl. J. Med.* **272,** 848–851.
Shepard, M. C., and Howard, D. R. (1970). *Ann. N.Y. Acad. Sci.* **174,** 809–819.
Shepard, M. C., and Lunceford, C. D. (1965). *J. Bacteriol.* **89,** 265–270.
Shepard, M. C., and Lunceford, C. D. (1967). *J. Bacteriol.* **93,** 1513–1520.
Shepard, M. C., and Lunceford, C. D. (1970a). *Appl. Microbiol.* **20,** 539–543.
Shepard, M. C., and Lunceford, C. D. (1970b). *Bacteriol. Proc.* p. 83.
Shepard, M. C., and Lunceford, C. D. (1975). *J. Clin. Microbiol.* **2,** 456–458.
Shepard, M. C., and Lunceford, C. D. (1976). *J. Clin. Microbiol.* **3,** 613–625.
Shepard, M. C., and Lunceford, C. D. (1978). *J. Clin. Microbiol.* **8,** 566–574.
Shepard, M. C., Alexander, C. E., Jr., Lunceford, C. D., and Campbell, P. E. (1964). *J. Am. Med. Assoc.* **188,** 729–735.
Shepard, M. C., Lunceford, C. D., and Baker, R. L. (1966). *Br. J. Vener. Dis.* **42,** 21–24.
Shepard, M. C., Lunceford, C. D., Ford, D. K., Purcell, R. H., Taylor-Robinson, D., Razin, S., and Black, F. T. (1974). *Int. J. Syst. Bacteriol.* **24,** 160–171.
Shipley, A. S., Bowman, J., and O'Connor, J. J. (1968). *Med. J. Aust.* **1,** 794–796.
Sobeslavsky, O., and Chanock, R. M. (1968). *Proc. Soc. Exp. Biol. Med.* **129,** 531–535.
Spaepen, M. S., and Kundsin, R. B. (1977). *Antimicrob. Agents & Chemother.* **11,** 267–270.
Spaepen, M. S., Kundsin, R. B., and Horne, H. B. (1976). *Antimicrob. Agents & Chemother.* **9,** 1012–1018.
Stalheim and Gallagher (1977). *Infect. Immun.* **15,** 995–996.
Stalheim, O. H. V., Proctor, J. S., and Gallagher, J. E. (1976). *Infect. Immun.* **13,** 915–925.
Stuart, C. A., Stratum, E. V., and Rustigian, R. (1945). *J. Bacteriol.* **49,** 437–444.
Subcommittee on the Taxonomy of Mycoplasmatales (1972). *Int. J. Syst. Bacteriol.* **22,** 184–188.
Sueltmann, S., Allen, V., Inhorn, S. L., and Benforado, J. M. (1971). *Health Lab. Sci.* **8,** 62–66.
Swanberg, S. L., Masover, G. K., and Hayflick, L. (1978). *J. Gen. Microbiol.* **108,** 221–225.
Taylor-Robinson, D. (1977). *In* "Nongonococcal Urethritis and Related Infections" (D. Hobson and K. K. Holmes, eds.), pp. 30–37. Am. Soc. Microbiol., Washington, D.C.
Taylor-Robinson, D., and Carney, F. E., Jr. (1974). *Br. J. Vener. Dis.* **50,** 212–216.
Taylor-Robinson, D., Williams, M. H., and Haig, D. A. (1968). *J. Gen. Microbiol.* **54,** 33–46.
Taylor-Robinson, D., Thomas, M., and Dawson, P. L. (1969). *J. Med. Microbiol.* **2,** 527–533.
Taylor-Robinson, D., Martin-Bourgon, C., Watanabe, T., and Addey, J. P. (1971). *J. Gen. Microbiol.* **68,** 97–107.
Taylor-Robinson, D., Csonka, G. W., and Prentice, M. J. (1977). *Q. J. Med.* [N.S.] **46,** 309–326.
Tull, A. H., Blair, E. B., Fishman, D. L., and Heatley, G. J. (1975). *J. Clin. Microbiol.* **1,** 234–236.
Vinther, O. (1976). *Acta Pathol. Microbiol. Scand., Sect. B* **84,** 217–224.

Vinther, O., and Black, F. T. (1974). *Acta Pathol. Microbiol. Scand., Sect. B* **82,** 917–918.

Visek, W. J. (1970). *Agric. Sci. Rev.* (Coop. State Res. Serv., U.S. Dep. Agric.) **8,** 9–23.

Watanabe, T., Mishima, K., and Horikawa, T. (1973). *Jpn. J. Microbiol.* **17,** 151–153.

Whitescarver, J., and Furness, G. (1975). *J. Med. Microbiol.* **8,** 349–355.

Whitescarver, J., Castillo, F., and Furness, G. (1975). *Proc. Soc. Exp. Biol. Med.* **150,** 20–22.

Whitescarver, J., Trocola, M., Campana, T., Marks, R., and Furness, G. (1976). *Proc. Soc. Exp. Biol. Med.* **151,** 68–71.

Williams, M. H. (1967). *Ann. N.Y. Acad. Sci.* **143,** 397–400.

Windsor, G. D., and Trigwell, J. A. (1976). *J. Med. Microbiol.* **9,** 101–103.

Windsor, G. D., Edward, D. G. ff., and Trigwell, J. A. (1975). *J. Med. Microbiol.* **8,** 183–187.

Woodson, B. A., McCarty, K. S., and Shepard, M. C. (1965). *Arch. Biochem. Biophys.* **109,** 364–371.

# 18 / SPECIAL FEATURES OF THERMOPLASMAS

*Thomas A. Langworthy*



## I. INTRODUCTION

*Thermoplasma acidophilum,* the thermophilic, acidophilic associate member of the Mollicutes, not only survives in the combined extremes of high temperature and low pH but requires both for growth and reproduction. As a cell wall-less, free-living saprophyte in a hot acid environment, *Thermoplasma* is unique among organisms. Obvious interest has arisen in its ability to cope with such a harsh environment. So far, studies indicate that the special structural and physiologic features are as unique as the existence of *Thermoplasma* might imply.



THE MYCOPLASMAS, VOL. 1

ISBN 0-12-078401-7

## II. OCCURRENCE

The first notion of the existence of *Thermoplasma* came when it was isolated by Darland *et al.* (1970) from self-heating coal refuse piles in southern Indiana. Such a strange habitat is man-made, generated from waste materials in the coal recovery process. Acid production results from the oxidation of pyritic materials, while the self-heating process provides high temperatures which may be seen as steam emission areas on the surface of the piles. Initially three isolates were obtained from regions of a single pile where, by field measurements, temperatures ranged from 32° to 80°C and pH from 1 to 5. Since the transient nature of self-heating coal refuse piles appeared unlikely as a primary habitat, an exhaustive search was undertaken of other natural hot acid environments, principally the acid hot springs and sulfatara soils in the Yellowstone National Park (Belly *et al.*, 1973). This source seemed likely since other thermoacidophiles such as *Sulfolobus acidocaldarius* (Brock *et al.*, 1972) and *Bacillus acidocaldarius* (Darland and Brock, 1971) had been found there. The search, however, failed in over 700 isolation attempts. On the other hand, further extensive samplings of piles in southern Indiana, as well as western Pennsylvania, resulted in 113 isolations in 486 attempts from 20 of 30 individual piles examined. The isolates could be differentiated by immunofluorescence and immunodiffusion analysis into five antigenic groups, indicating serologic diversity among thermoplasmas (Belly *et al.*, 1973; Bohlool and Brock, 1974). Self-heating coal refuse piles are, albeit unlikely, still the only known habitat of thermoplasmas.

## III. PHYSIOLOGIC ASPECTS

### A. Morphology

*Thermoplasma* possesses a typical mycoplasmal ultrastructure and colonial morphology. Light and phase microscopy reveal pleomorphic forms ranging in size from 0.1 to 0.2 $\mu$m, as well as larger forms and typical budding and filamentous characteristics (Belly *et al.*, 1973). Electron micrographs of thin-sectioned cells (Darland *et al.*, 1970) reveal nuclear material dispersed through the cytoplasm which is surrounded by only a trilaminar membrane approximately 5–10 nm thick (Fig. 1). The lack of a cell wall was confirmed by cellular insensitivity to vancomycin (penicillin is degraded under the acidic conditions), sensitivity to novobiocin, lysis by sodium lauryl sulfate, absence of hexosamines, and the ability to pass a

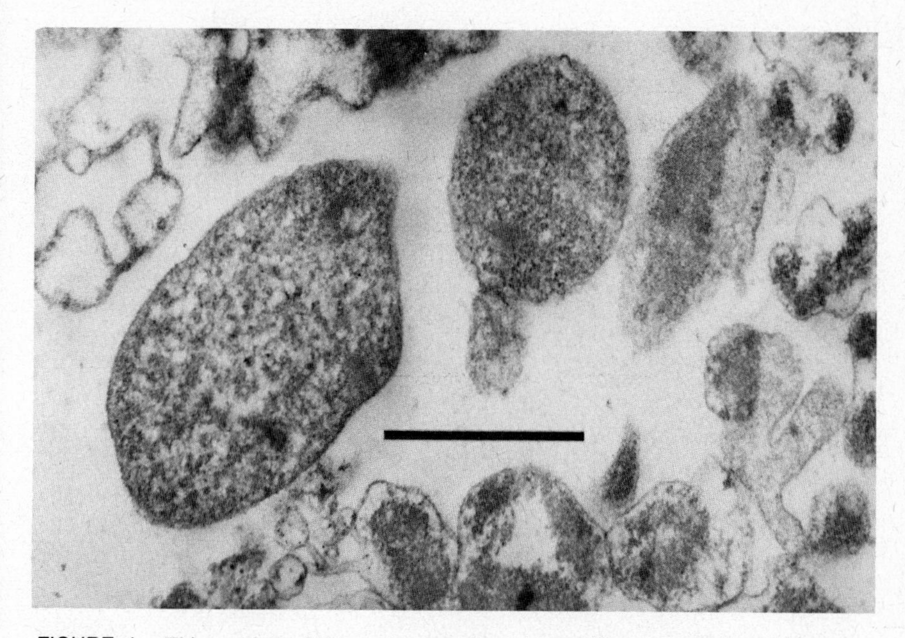

FIGURE 1. Thin section of *Thermoplasma*. Bar, 1 μm. (Courtesy of K. J. Mayberry-Carson.)

0.45 μm but not 0.22 μm filter. Freeze-fractured cells do demonstrate a peculiarity which may be related to the thermoplasmal membrane structure. Typically, cells are observed in cross-fracture through the cytoplasm and membrane. Normally observed inner and outer membrane layers are not found (P. Ververgaert, personal communication).

Colonial morphology presents special problems because of the acidic media and high temperature, e.g., dehydration and/or hydrolysis of the agar. This problem is surmounted (Belly *et al.*, 1973) by mixing double-strength liquid media and Ionagar previously cooled to 45°C. Dehydration is prevented by incubation in moist, humidified containers. After incubation small colonies arise which have the characteristic umbonate (fried egg) appearance (Fig. 2). Closer examination by scanning electron microscopy (Mayberry-Carson *et al.*, 1974a) demonstrates cells growing on the agar surface ranging in size from 0.5 to 1.9 μm, singly and in clumps, with the occasional appearance of large bodies (5 μm in diameter) similar to those observed in *Acholeplasma laidlawii* (Klainer and Pollack, 1973). The characteristic imbricate surface texture of cells lacking cell walls is also observed (Fig. 3).

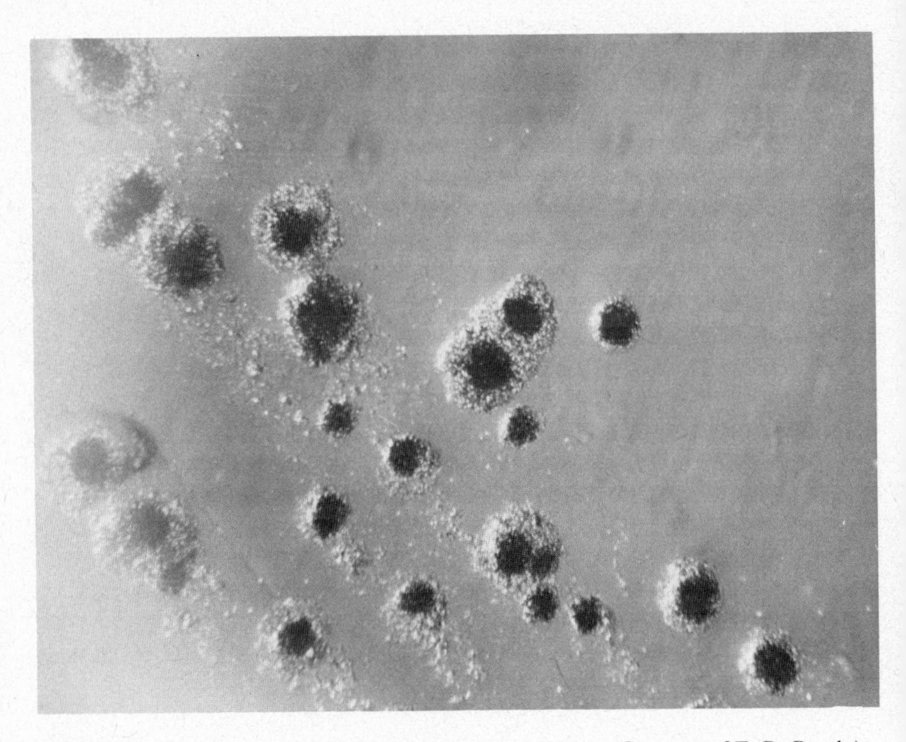

FIGURE 2. *Thermoplasma* colonies grown on agar medium. (Courtesy of T. D. Brock.)

## B. Growth

### 1. Conditions for Growth

*Thermoplasma* is an obligate thermoacidophile which requires pH 2 and 59°C for optimum growth. During the course of growth, no change is observed in the pH of the culture medium. The growth range spans the limits of pH 0.5–4.0 and 40°–62°C, though growth rates are markedly reduced at the extremes (Belly *et al.*, 1973; Smith *et al.*, 1973). *Thermoplasma* is an aerobe possessing *c*- and *o*-type cytochromes but lacking either the *a*- or *b*-types (Belly *et al.*, 1973) and contains the naphthoquinone vitamin $K_2$-7 (Langworthy *et al.*, 1972), suggesting a complete respiratory chain. *Thermoplasma* is a heterotroph which is best cultured in a liquid medium composed of inorganic salts, adjusted to pH 2 with sulfuric acid, followed by supplementation with 0.1% yeast extract and 1% glucose (Belly *et al.*, 1973) (Table I).

Besides temperature and pH, growth rates and cell yields are sensitive to a variety of other parameters (Smith *et al.*, 1973). Since the amount of

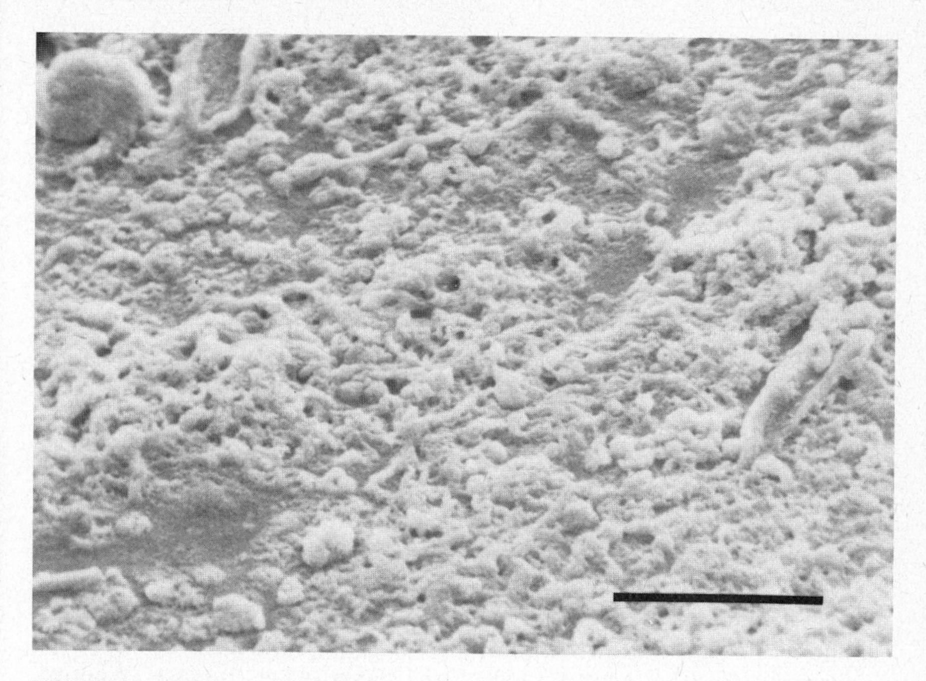

FIGURE 3. Scanning electron micrograph of *Thermoplasma* cells on agar surface. Note the imbricate texture characteristic of cells without cell walls. A larger body is apparent in the upper left field. Bar, 10 $\mu$m. (Courtesy of I. L. Roth and K. J. Mayberry-Carson.)

dissolved oxygen is limiting at 59°C, cell yields and growth rates are greatly improved by increasing the surface exposure of the medium, by shaking cultures, and more so by sparging. For some unexplainable reason, total cell yields decrease with inoculum sizes of less than 5%. Cell growth is also variable, depending upon the particular lot of yeast extract used to support growth.

Growth curves (measurements at $OD_{540}$) of shaken cultures started

TABLE I. **Growth Characteristics of *Thermoplasma*[a]**

| | |
|---|---|
| Optimum temperature | 59°C |
| Optimum pH | 2.0 |
| Temperature range | 45–62°C |
| pH range | 0.5–4.0 |
| Generation time (approx.) | 4–5 hr |
| Oxygen | Strict aerobe |
| Nutrition | Requires yeast extract |
| Nutritional tolerance[b] | <0.25% |

[a] Data compiled from Belly *et al.* (1973) and Smith *et al.* (1973).

[b] Concentrations of yeast extract greater than 0.25% become inhibitory to growth.

from a 5% inoculum under optimal conditions demonstrate a lag phase of about 20 hr, with the exponential phase extending to 50–60 hr at which point the stationary phase is reached (Smith *et al.*, 1973). Viable cell numbers, estimated by the most probable number technique necessitated by the difficulties in plating out cells for viable counts, were found to increase with the optical density to the beginning of the stationary phase reaching about $1 \times 10^9$ to $1 \times 10^{10}$ cells per milliliter. As cultures aged, there was a drastic loss of viability, though not as drastic as the reduction in optical density. The estimated generation time for *Thermoplasma* from the study of Smith *et al.* (1973) was 3.5–5 hr, which correlates with the initial report of about 4 hr calculated from Arrhenius plots (Darland *et al.*, 1970).

## 2. Nutrition

The nutritional demands of *Thermoplasma* appear to be the least complex among the Mollicutes, involving only inorganic salts, 1% glucose, and 0.1% yeast extract. Of the strains initially tested by Belly *et al.* (1973), all grew on yeast extract but not on casamino acids, peptone, or glucose when supplied alone as the sole carbon source. In yeast extract-limited cultures (0.025%), growth could apparently be stimulated by replacing glucose with either sucrose, galactose, mannose, or fructose. Whereas yeast extract concentrations below 0.025% were growth limiting, *Thermoplasma* could not tolerate concentrations above 0.25%. Higher concentrations became growth inhibitory.

The specificity and absolute requirement for yeast extract for growth and reproduction became rapidly apparent. Of several hundred compounds tested as substitutes for yeast extract by Smith and associates (1975) none supported growth with the exception of a few peptones, ferredoxins, and aged glutathione, each of which only gave a meager growth response. This prompted an attempt to characterize the component(s) in yeast extract necessary for growth (Smith *et al.*, 1975). Fractionation of yeast extract by ethanol precipitation followed by aqueous-phenol extraction, trypsin digestion, and finally gel permeation chromatography resulted in only about a 21-fold increase in specific growth-promoting activity. Analyses of this fraction (Table II) suggests the growth factor to be a polypeptide of molecular weight approximating 1000, containing eight to ten amino acids and having only one free amino group per molecule. The polypeptide, bound irreversibly to charcoal and strong cation-exchange resins, was not retarded by anion-exchange or weak cation-exchange resins, was dialyzable, and avidly bound cations. It was composed largely of basic and dicarboxylic amino acids, which,

TABLE II. **Composition of "Growth Factor" from Yeast Extract**[a]

| Amino acids | Mol % |
|---|---|
| Asp, Glu, $NH_3$ | 12–16 |
| Gly, Pro, Ser, Ala, Leu, Lys | 5–10 |
| Val, Thr, Ile, Arg, Phe, Try, His, Met | 1–4 |
| Cys | <1 |
| Total basic amino acids: 27% | |
| Total acidic amino acids: 9% | |
| Estimated molecular weight: ~1000 | |
| One free amino group/per molecule | |

[a] Data from Smith *et al.* (1975).

coupled with the aforementioned properties, suggests a protamine- or histonelike molecule. An array of known protamines, histones, or polyamines, however, could not support growth. It was concluded that the polypeptide(s) might function in providing essential amino acids in a permeable form in the extremely acidic environment. Alternatively, in view of the large quantities required for growth, perhaps it functions as a scavenger for some trace metals or possibly as protection of the cellular surface from $H^+$ ions.

### 3. Maintenance

*Thermoplasma* is best carried in the laboratory as actively growing cultures by continuous transfer every 2–3 days. Cultures do retain viability if refrigerated for periods up to 1 month. For longer term storage, cultures may sometimes be recovered from the frozen state. Viability is lost upon lyophylization of cells, either directly from the growth medium or after washing at pH 5.5. Viability appears to be retained if 48 hr cultures (5 ml) are concentrated at room temperature followed by addition of a 10% solution of yeast extract (0.1 ml) to the cell pellet prior to lyophilization (T. A. Langworthy, preliminary observation).

## C. Intracellular pH

Intracellular ionic conditions influence both activity and stability of metabolic processes. In *Thermoplasma,* internal $H^+$ concentration is of particular interest as certainly such acid-labile molecules as ATP and

DNA would be degraded if the internal and external pH values were in equilibrium.

Hsung and Haug (1975) estimated internal pH values by measuring the passive distribution of the weak acid 5,5-dimethyl-2,4-oxazolidinedione (DMO) within cells. Since the p*K* value of DMO is 6.1 at 56°C, most molecules are nonionized at pH 2–4. At pH values above the p*K* value some molecules will tend to become ionized. After the internal and external water volumes are determined, intracellular pH can be calculated from the ratio of the concentrations of the internal ionized form to that of the nonionized form. Using this approach, it was concluded that the intracellular pH of *Thermoplasma* was 6.4–6.9. Measurement of cell lysates after sonication gave values of pH 6.3–6.8 which seemed in agreement. The pH optimum of cytoplasmic malate dehydrogenase also had a broad optimum near pH 8, again implying an internal pH near neutrality. Furthermore, cells boiled for 5 hr at 100°C or cells treated with the metabolic inhibitors 2,4-dinitrophenol, iodoacetate, or sodium azide still possessed intracellular pH values near neutrality (6.4–6.7). This study suggested the pH gradient across the membrane to be maintained by structural features of the membrane rather than by an active $H^+$ ion extrusion mechanism. In a later study, the potential across the membrane was estimated at 120 mV (Hsung and Haug, 1977).

In a second study, Searcy (1976a) determined intracellular pH and potassium concentration as well. Advantage was taken of the feature of *Thermoplasma* by which cells lyse at pH values near neutrality. Cell suspensions were titrated to neutrality with base and then back titrated with acid. Intersection of the two titration curves at pH 5.4 ± 0.02 was taken as the intracellular pH. Measurement of cells ruptured in the French press gave comparable values (pH 5.5 ± 0.1). In contrast to the results of Hsung and Haug, pH values of boiled cells were in equilibrium with external pH values. Searcy has suggested that absorption of DMO to the membrane may have resulted in overestimating the values reported by Hsung and Haug. This also raises the question of the stability of the metabolic inhibitors at acid pH. In either case, *Thermoplasma* possesses a pH gradient of 3.5–4.5 units across its membrane which is exposed to the hot acid environment. An internal pH value nearer 5.4 is probably conceptually more satisfying in view of the fact that *Thermoplasma* begins to lose viability and cellular integrity at pH values between 5.0 and 6.0.

In contrast to intracellular $H^+$ ions which are not in equilibrium with the environment, $K^+$ ions are in close osmotic equilibrium (Searcy, 1976a). Internal $K^+$ concentrations may be as low as 17 m*M*, which appears to be the lowest intracellular concentration recorded.

## D. The Genome

The genome size of *Thermoplasma* approximates $1 \times 10^9$ daltons and has a guanine plus cytosine (G + C) content of 46% determined by both buoyant density (Christiansen *et al.*, 1975) and by chromatography of bases after hydrolysis (Searcy and Doyle, 1975). The genome size of *Thermoplasma* resembles that of the family Acholeplasmataceae within the order Mycoplasmatales. Original reports of a lower G + C content near 25% (Darland *et al.*, 1970; Belly *et al.*, 1973) were apparently in error due likely to depurination of the DNA by acid in the older cultures used for analysis (Searcy and Doyle, 1975).

A special feature of isolated thermoplasmal DNA is its association with an acid-soluble, basic protein (Searcy, 1975). Several of its properties resemble those of eukaryotic histones; e.g., it is acid-soluble, is positively charged at neutral pH, and dissociates from the DNA at high ionic strength. It migrates at the same rate as histone IV(F2al) from calf thymus by polyacrylamide gel electrophoresis. The molecule is rich in basic amino acids (23%) and has an unusually high content (16–20%) of the amides of acidic amino acids (Table III). Searcy (1975) initially suggested this basic protein may function in the thermal stabilization of the DNA or possibly protect it against depurination at the relatively high internal pH of 5.5. Further experimentation, however (Searcy, 1976b), revealed it provided neither function, eliminating these possibilities. It still remains to be shown whether this protein is in fact associated with the DNA intracellularly or if it is actually acquired during the isolation procedure. In this regard, it is of interest to note the striking similarities between its

TABLE III. **Composition of "Histonelike" Protein from *Thermoplasma*[a]**

| Amino acids | Mol % |
|---|---|
| Glu, Lys, $NH_3$ | 10–20 |
| Val, Ala, Ser, Ile, Arg, Gly, Asp, Thr | 6–9 |
| Phe, Leu, Pro, Try | 1–5 |
| Trp, Met, Cys | <1 |
| Total basic amino acids: 23% | |
| Total acidic amino acids: 20% | |
| Estimated molecular weight: ~10,000 | |

[a] Data from Searcy (1975).

composition (Table III) and the composition of the partially characterized growth factor from yeast extract (Table II). Preliminary observations (P. F. Smith, personal communication) indicate that the acid-soluble proteins from *Thermoplasma* whole cells, from the isolated DNA, and from yeast chromatin all support growth. Might the "growth factor" and histonelike protein be one and the same?

## IV. MEMBRANE-SURFACE PROPERTIES

### A. Cellular Stability

#### 1. Physical Properties

In contrast to most mycoplasmas, *Thermoplasma* is exceedingly resistant to a variety of physical and chemical agents, e.g., mechanical disruption, sonic oscillation, nonionic detergents, ethylenediamintetraacetic acid (EDTA), primary alcohols, digitonin, pronase, and trypsin (Belly and Brock, 1972; Smith *et al.*, 1973). Cells are osmotically stable upon suspension in distilled water and are thermally stable at 100°C for 30 min, in contrast to prepared protoplasts or many mycoplasmas which lyse under these conditions. *Thermoplasma* is lysed by increasing the ionic strength higher than 0.2 and by cationic and anionic detergents, though concentrations are about eight times greater than normally required for lysis of mycoplasmas. The extreme rigidity of cells is borne out in their characteristic cross-fracture (see Section III, A) and by electron spin resonance studies (Smith *et al.*, 1974). The mobility of spin labels in the membrane is even less than that reported for the extreme halophile *Halobacterium cutirubrum*, suggesting that the thermoplasmal membrane may be the most rigid yet known.

Perhaps one of the more interesting features of *Thermoplasma*, in view of its resistance to many physical treatments, is the organism's specific requirement for $H^+$ ions for the maintenance of cellular integrity (Belly and Brock, 1972; Smith *et al.*, 1973). Cells remain stable and viable between pH 1 and 5. At pH values greater than 6, lysis ensues with the concomitant loss of viability, protein, and uv-absorbing material. Lysis is essentially instantaneous at pH values above 7. Imagine, a cell lysed by neutrality! This phenomenon appears to be independent of cellular concentration or temperature. The requirement for $H^+$ ions is specific, as other monovalent or divalent cations, or sucrose in concentrations up to 20%, cannot substitute for $H^+$ or protect against lysis. Following lysis by high pH, Smith *et al.* (1973) were unable to recover membranes by sedimentation at 100,000 *g* for 4 hr, suggesting solubilization at high pH values. Membranous-like vesicles were found to reassemble, however, upon dialysis of lysates against deionized water (pH 5.5). Because of the extreme resistance to cellular breakage by mechanical means under acidic conditions and disaggregation at high pH values, Smith and associates were able to obtain native membranes by adjustment of cell suspensions to pH 5 and moderate ionic strength (0.05) prior to breakage by sonication. Still, about 50% of the cells were resistant to breakage. In a much later report, detailing the discovery of most of the already reported features of *Thermoplasma*, Ruwart and Haug (1975) were able to recover membranous material from high-pH lysates by the judicious manipulation of cellular protein concentrations.

### 2. Role of Surface Charge

The relationship of electrical charges on the membrane surface to the $H^+$ requirement for cell stability was studied by Smith *et al.* (1973). Although the ratio of free —COOH groups to free —$NH_2$ groups (4:1) on the surface of *Thermoplasma* was the same as that for *Acholeplasma laidlawii* B, the total number of these groups per unit of protein was less than half the number found for *A. laidlawii*. The effect of altering the surface charges was examined (Fig. 4). Normally, cells or membranes are stable at low pH and dissociate at high pH. By reacting membranes with glycine methyl ester, which removes the potential negative charge of free —COOH groups, membranes became stable over the entire pH range. Reaction of membranes with ethylene diamine, which not only removes the potential negative charge of the —COOH groups but adds a potential positive charge in the form of an —$NH_2$ group, reversed membrane stability. Membranes became stable at high pH and disaggregated at low pH. These observations suggest that the surface of *Thermoplasma* is quite hydrophobic. The $H^+$ requirement for cellular stability may be in part ascribed to the necessity for protonation of the free —COOH groups on the surface. This requirement is met at low pH. At pH values near neutrality, —COOH groups become ionized (—$COO^-$) with the resultant repulsion of negative charges likely responsible for destabilization of the membrane.

## B. Membrane Composition

The direct exposure of the thermoplasmal membrane to the hot acid environment has directed attention to its structural features. Gross chemical analysis reveals a composition of approximately 60% protein, 25% lipid, and 10% carbohydrate.

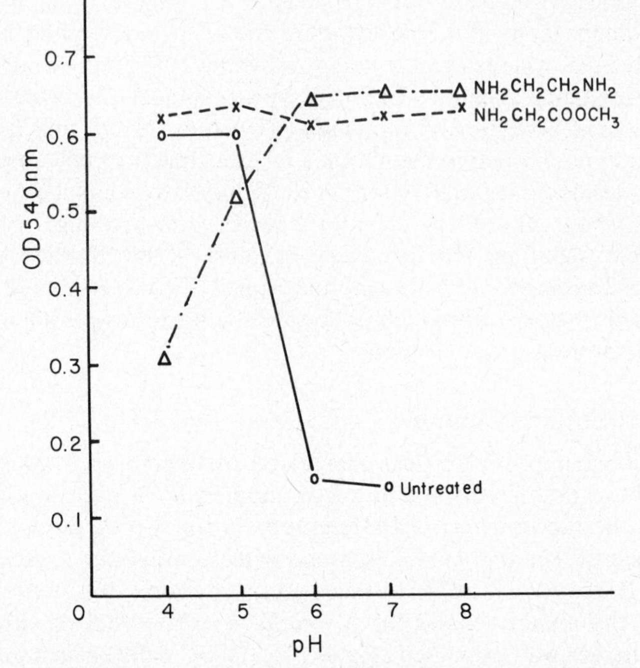


FIGURE 4. Effect of charge alteration on stability of *Thermoplasma* membranes to pH. ○——○, untreated; ×----×, treated with glycine methyl ester; △-·-·-△, treated with ethylene diamine. Buffers: pH 4–5, acetate buffers; pH 6–8, phosphate buffers. (From data of Smith *et al.*, 1973.)

### 1. Proteins

The protein spectrum of membranes appears as a heterogeneous mixture with molecular weights generally less than 80,000 when examined by polyacrylamide gel electrophoresis. The amino acid composition and distribution of membrane proteins are not unlike that of other mycoplasmas. No outstanding feature of the membrane proteins is as yet readily apparent (Smith *et al.*, 1973; Ruwart and Haug, 1975).

### 2. Dialkyl Diglycerol Tetraethers

As might not be entirely unexpected, the membrane lipids of *Thermoplasma* contain ether lipids rather than typical fatty acid-derived glyceride residues. The assembly of these ethers, however, is unique among organisms. They are comprised structurally of two *sn*-2,3-glycerol molecules bridged through ether linkages by two fully saturated, isopranoid branched, $C_{40}$-terminal diols (Langworthy, 1977) as shown in Fig. 5. The

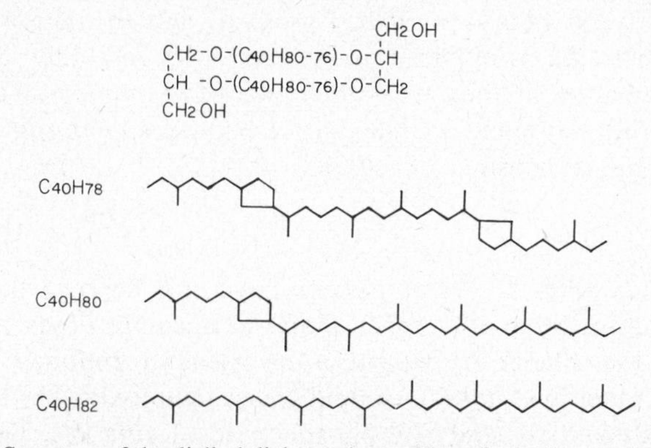


FIGURE 5. Structure of the dialkyl diglycerol tetraethers from *Thermoplasma* (Langworthy, 1977). The cis or trans configuration of the hydroxyl groups has not been established. Identical pairs of $C_{40}$ alkyl chains may be either $C_{40}H_{82}$ (acyclic), $C_{40}H_{80}$ (monocyclic), or $C_{40}H_{78}$ (bicyclic). (Structure of hydrocarbon chains proposed by de Rosa *et al.*, 1977.)

alkyl chains may be either $C_{40}H_{82}$ (acyclic), $C_{40}H_{80}$ (monocyclic), or $C_{40}H_{78}$ (bicyclic). Cyclization within the hydrocarbons appears to be in the form of pentacyclic rings (de Rosa *et al.*, 1977). The tetraethers contain an identical pair of either of the aforementioned hydrocarbons. In *Thermoplasma* the acyclic hydrocarbon-containing tetraether species predominates (65%), with lesser amounts of the monocyclic (33%) and bicyclic (2%) components (de Rosa *et al.*, 1976). The tetraethers possess a structural resemblance to two molecules of di-*O*-phytanylglycerol, as found in the extreme halophiles (Kates *et al.*, 1965), which have been joined by head covalent linkage at the terminal ends of the $C_{20}$ alkyl chains. Biosynthesis of the tetraethers is not known although [$^{14}C$] mevalonate incorporation into the *O*-alkyl chains indicates the mevalonate pathway is operative. Tetraethers account for about 15% of the membrane by weight.

Elucidation of the tetraether assembly has not been without some controversy. Originally the ethers in *Thermoplasma* were thought to be di-*O*-alkyl glycerol ethers possessing two $C_{40}$ alkyl chains linked to glycerol (Langworthy *et al.*, 1972). Further studies (de Rosa *et al.*, 1974, 1976) revealed the presence of equimolar proportions of glycerol and $C_{40}$-terminal alkyl diols in the structure, which suggested a *sn*-2,3-glycerol combined through ether linkages by a single $C_{40}$ alkyl chain forming a 40-carbon macrocyclic loop. The diglycerol tetraether assembly was finally established (Langworthy, 1977) by demonstrating the existence of not one but two free primary hydroxyl functions in the ether and by analysis of its monoacetate and monomethoxy derivatives. The ether was shown to contain two glycerols, two alkyl chains, and six oxygens (four in ether linkages and two as hydroxyl groups), and to have a molecular weight of 1300–1292 as required by the tetraether assembly. Though the tetraethers are unusual, they are not unique to *Thermoplasma* but are also found in another thermoacidophile, *Sulfolobus acidocaldarius* (Langworthy *et al.*, 1974; de Rosa *et al.*, 1976).

### 3. Complex Lipids

The complex lipids of *Thermoplasma* have only been partially characterized (Langworthy *et al.*, 1972). Total extractable lipids account for about 3% of the cellular dry weight or, as already mentioned, about 25% of the membrane. Distribution of lipid classes approximates 17% neutral lipids, 25% glycolipids, and 57% phospholipids (Table IV). Though all of the glycolipids and phospholipids were originally reported to contain glycerol diethers, they are now known to possess diglycerol tetraether residues.

Only vitamin $K_{2\text{-}7}$ and free diglycerol tetraethers have been identified out of about nine components in the neutral lipid fraction. The appearance of a sterol, which does not become radiolabeled during growth, likely arises from the medium. Similarly the small amounts of fatty acids in *Thermoplasma* reported by Ruwart and Haug (1975) may have derived from the same source as incorporation of radioactivity was not examined.

Among the six glycolipid species none has been identified with certainty, although the two major components appear to be monoglycosyl and diglycosyl diglycerol tetraethers.

Typical phosphatides are absent in the phospholipid fraction. Instead, lipid phosphorous is accounted for by glycerolphosphoryl radical-derived phosphoglycolipids. The major component, representing about 80% of the

TABLE IV. **Lipid Composition of *Thermoplasma***[a]

| Lipid class | Percent total lipid | Partially characterized components |
|---|---|---|
| Neutral lipids | 17.5 | Vitamin $K_2$-7, diglycerol tetraethers |
| Glycolipids | 25.6 | Monoglycosyl and diglycosyl diglycerol tetraethers |
| Phospholipids | 56.9 | Glycerolphosphoryl glycosyl diglycerol tetraether, $NH_2$-containing phosphoglycolipids |

[a] Data from Langworthy *et al.* (1972).

phospholipids and nearly 50% of the total lipid in the organism, was tentatively postulated to be a glycerolphosphoryl monoglycosyl diglycerol tetraether. The carbohydrate was unusual. Along with most of the other carbohydrates in the lipids, it has not been identified. Four other minor phosphoglycolipids appeared to possess up to three carbohydrate residues along with free amino groups, possibly as amino sugars. Specifically sought phosphonate, sulfonate, or sulfate residues were absent. Though phosphoglycolipids occur along with typical phosphatides in other mycoplasmas (Chapter 9) all of the phospholipids in *Thermoplasma* are present as carbohydrate-containing phosphoglycolipids.

Presently, only general features of the complex lipids are known. The occurrence of diglycerol tetraethers serving as the hydrophobic moieties in the glycolipids and phosphoglycolipids provides interesting possibilities for their structural assembly. Diglycerol tetraethers possess two hydroxyl groups (Fig. 5) for the attachment of polar residues. In the glycolipids, sugars might be attached asymmetrically to only one side of the tetraether or symmetrically to both sides. For phosphoglycolipids even more possibilities exist, e.g., linkage of glycerolphosphoryl glycosyl residues to one or both hydroxyls or a carbohydrate and glycerol phosphate attached to opposite sides of the molecule. Elucidation of which complex lipid structures exist in *Thermoplasma* awaits future studies.

## 4. Lipopolysaccharide

In addition to some contribution by glycolipids and phosphoglycolipids, most of the carbohydrate associated with the thermoplasmal membrane arises from a new type of diglycerol tetraether-derived lipopolysaccharide (LPS). The polymer is comprised of 24 mannosyl units and one glucosyl unit terminating in a diglycerol tetraether (Mayberry-Carson *et al.*, 1974b). The tentative structure is postulated as [man*p*-(1 $\xrightarrow{\alpha}$ 2)-man*p*-(1 $\xrightarrow{\alpha}$ 2)-man*p*-(1 $\xrightarrow{\alpha}$ 3)]$_8$-glc*p*-diglycerol tetraether, with a monomeric molecular weight of approximately 5300. The polymer is chemically distinct from gram-negative bacterial LPS in that it resembles a glycolipid possessing an extended oligosaccharide chain. Morphologic examination by transmission electron microscopy of aggregates in aqueous solution (MW $>$ 1,200,000) reveals ribbonlike structures about 5 nm in width similar to gram-negative bacterial LPS (Mayberry-Carson *et al.*, 1975). The LPS which is readily extractable from whole cells or membranes in nearly the pure state by hot aqueous-phenol, represents about 3% of the cellular dry weight. The isolated polymer is stable at pH 2 and 59°C for periods up to 1 week. Its location on the cellular surface has not yet been established, though it demonstrates antigenic specificity when reacted with antisera prepared against membranes (Sugiyama *et al.*, 1974). The isolation of an

LPS from *Thermoplasma* has lead to the finding of, heretofore unsuspected, similar diacylglycerol-derived neutral and amino sugar-containing polymers among several of *Thermoplasma*'s mesophilic counterparts (Smith *et al.*, 1976; Chapter 9).

## V. SPECIAL FEATURES AND THERMOACIDOPHILY

Since its discovery, studies on *Thermoplasma* have revealed several features in particular by which the organism seems to have ensured its survival. For the most part, direct experimental data on the role of these properties in thermoacidophily is lacking, yet certain features might be speculated upon.

Thermoplasma seems to have taken advantage of its acidic environment by requiring $H^+$ ions for maintenance of cellular integrity. The necessity of $H^+$ ions for protonation of surface acidic groups likely explains the organism's obligatory requirement for low pH. An analogous feature is shared by the extreme halophiles which specifically require $Na^+$ ions for cellular integrity (Larsen, 1967).

The existence of a pH gradient of 3.5–4.5 units across the membrane which is directly exposed to hot acid implies an unusual membrane assembly. The occurrence of diglycerol tetraethers suggests this may indeed be the case. By virtue of the extension of the $C_{40}$ alkyl chains across the membrane in covalent linkage to glycerol on either side, a biologically functional lipid monolayer is generated rather than a typical lipid bilayer normally formed by the hydrophobic interaction of separate and opposite glyceride residues. The diglycerol tetraethers may therefore impart rigidity and stability to the exposed membrane by, in effect, holding the lipid domain together in the hot acid environment. Under these conditions an exposed normal bilayer membrane assembly might be subject to disorientation. Although influenced by the nature of membrane protein and carbohydrate, electron spin resonance studies indicate an exceedingly rigid membrane, while typical cross-fracture is observed in freeze-etched cells. Both of these features would appear to correlate with the presence of a membrane monolayer. In addition to the extremely hydrophobic membrane surface, the presence of ether linkages in the diglycerol tetraethers would seem related to acid resistance. The long $C_{40}$ alkyl chains likely provide appropriate membrane fluidity at the high temperatures. The purpose of pentacyclic rings in the alkyl chains, thereby regulating the degree of rotational freedom, has been suggested to also influence hydrocarbon fluidity (de Rosa *et al.*, 1977).

Unfortunately, the complex lipid structures have yet to be identified.

Since the diglycerol tetraethers possess a symmetrical assembly, membrane asymmetry will be provided by the individual complex lipid structures by virtue of attachment of polar residues to either the inner or the outer hydroxyl groups of the tetraethers. The preponderance of carbohydrate-containing complex lipids in *Thermoplasma* may also contribute to membrane stability, since high glycolipid content appears characteristic of several extremely thermophilic bacteria (Ray *et al.*, 1971; Oshima and Yamakawa, 1974; Langworthy *et al.*, 1976).

Finally, the seemingly restricted habitat of *Thermoplasma* may bear some relationship to its nutritional demands. Smith *et al.* (1975) observed that the growth-promoting oligopeptide(s) from yeast extract is concentrated and tightly bound by activated charcoal. When the charcoal–oligopeptide complex was supplied to cultures lacking yeast extract, *Thermoplasma* was capable of growth. Though tightly bound, the growth factors appear to be released in sufficient quantities by the hot acid environment. Perhaps the required factors are present and similarly concentrated in naturally occurring carboniferous areas, thereby restricting *Thermoplasma*'s natural habitat to self heating coal refuse piles. It appears fortunate, indeed, that the growth substance(s) or facsimile is present in at least yeast extract or perhaps we may never have been confronted by thermoplasmas.

## VI. FUTURE OUTLOOK

So far, attempts to understand the biology of *Thermoplasma* have centered on selected aspects of the organism, such as nutritional requirements, nucleic acids, and, principally, membrane structure. Still, only the basic features of these particular areas have been elucidated. The membrane has received most attention because of its exposure to the hot acid environment. Its study has certainly been rewarding in partially assessing the biochemical basis of thermoacidophily. However, nearly all aspects of *Thermoplasma*'s physiology remain unknown, especially in terms of internal biochemical constitution and function. Features of intermediary metabolism, proteins, protein synthesis, enzymology, respiration, etc., await investigation. Of special interest should be the area of membrane transport in relationship to the large pH gradient maintained across the membrane. Much work remains for our understanding of *Thermoplasma*'s existence. Future studies should not only aid in establishing the biochemical basis of thermoacidophily but, because of *Thermoplasma*'s anomolous existence, may provide valuable insight into mechanisms of basic cellular functions which may not otherwise be apparent under normal conditions for life.

## VII. CONCLUSION

Studies on *Thermoplasma* have not yet failed to reveal new and interesting biologic features, as well as stimulating one's prowess in designing laboratory studies. Interest should continue in this organism which combines high temperature and low pH as a requirement for its existence. As such, *Thermoplasma* represents not only a unique mycoplasma but an extension of our terrestrial and extraterrestrial yardstick for life. Certainly many more new and special features remain to be uncovered which should aid in explaining *Thermoplasma*'s existence and perhaps its evolutionary past.

## REFERENCES

Belly, R. T., and Brock, T. D. (1972). *J. Gen. Microbiol.* **73,** 465–469.

Belly, R. T., Bohlool, B. B., and Brock, T. D. (1973). *Ann. N.Y. Acad. Sci.* **225,** 94–107.

Bohlool, B. B., and Brock, T. D. (1974). *Infect. Immun.* **10,** 280–281.

Brock, T. D., Brock, K. M., Belly, R. T., and Weiss, R. L. (1972). *Arch. Mikrobiol.* **84,** 54–68.

Christiansen, C., Freundt, E. A., and Black, F. T. (1975). *Int. J. Syst. Bacteriol.* **25,** 99–101.

Darland, G., and Brock, T. D. (1971). *J. Gen. Microbiol.* **67,** 9–15.

Darland, G., Brock, T. D., Samsonoff, W., and Conti, S. F. (1970). *Science* **170,** 1416–1418.

de Rosa, M., Gambacorta, A., Minale, L., and Bu'Lock, J. D. (1974). *J. Chem. Soc., Chem. Commun.* pp. 543–544.

de Rosa, M., Gambacorta, A., and Bu'Lock, J. D. (1976). *Phytochemistry* **15,** 143–145.

de Rosa, M., de Rosa, S., Gambacorta, A., and Bu'Lock, J. D. (1977). *J. Chem. Soc., Chem. Commun.* pp. 514–515.

Hsung, J. C., and Haug, A. (1975). *Biochim. Biophys. Acta* **389,** 477–482.

Hsung, J. C., and Haug, A. (1977). *FEBS. Lett.* **73,** 47–50.

Kates, M., Yengoyan, L. S., and Sastry, P. S. (1965). *Biochim. Biophys. Acta* **98,** 252–268.

Klainer, A. S., and Pollack, J. D. (1973). *Ann. N.Y. Acad. Sci.* **255,** 236–245.

Langworthy, T. A. (1977). *Biochim. Biophys. Acta* **487,** 37–50.

Langworthy, T. A., Smith, P. F., and Mayberry, W. R. (1972). *J. Bacteriol.* **112,** 1193–1200.

Langworthy, T. A., Mayberry, W. R., and Smith, P. F. (1974). *J. Bacteriol.* **119,** 106–116.

Langworthy, T. A., Mayberry, W. R., and Smith, P. F. (1976). *Biochim. Biophys. Acta* **431,** 550–569.

Larsen, H. (1967). *Adv. Microbiol. Physiol.* **1,** 97–132.

Mayberry-Carson, K. J., Roth, I. L., Harris, J. L., and Smith, P. F. (1974a). *J. Bacteriol.* **120,** 1472–1475.

Mayberry-Carson, K. J., Langworthy, T. A., Mayberry, W. R., and Smith, P. F. (1974b). *Biochim. Biophys. Acta* **360,** 217–229.

Mayberry-Carson, K. J., Roth, I. L., and Smith, P. F. (1975). *J. Bacteriol.* **121,** 700–703.

Oshima, M., and Yamakawa, T. (1974). *Biochemistry*. **13,** 1140–1146.
Ray, P. H., White, D. C., and Brock, T. D. (1971). *J. Bacteriol.* **108,** 227–235.
Ruwart, M. J., and Haug, A. (1975). *Biochemistry* **14,** 860–866.
Searcy, D. G. (1975). *Biochim. Biophys. Acta* **395,** 535–547.
Searcy, D. G. (1976a). *Biochim. Biophys. Acta* **451,** 278–286.
Searcy, D. G. (1976b). *In* "Molecular Mechanisms in the Control of Gene Expression" (D. P. Nierlich, W. J. Rutter, and C. F. Fox, eds.), pp. 51–57. Academic Press, New York.
Searcy, D. G., and Doyle, E. K. (1975). *Int. J. Syst. Bacteriol.* **25,** 286–289.
Smith, G. G., Ruwart, M. J., and Haug, A. (1974). *FEBS Lett.* **45,** 96–98.
Smith, P. F., Langworthy, T. A., Mayberry, W. R., and Hougland, A. E. (1973). *J. Bacteriol.* **116,** 1019–1028.
Smith, P. F., Langworthy, T. A., and Smith, M. R. (1975). *J. Bacteriol.* **124,** 884–892.
Smith, P. F., Langworthy, T. A., and Mayberry, W. R. (1976). *J. Bacteriol.* **125,** 916–922.
Sugiyama, T., Smith, P. F., Langworthy, T. A., and Mayberry, W. R. (1974). *Infect. Immun.* **10,** 1273–1279.

# 19 / SPECIAL FEATURES OF ANAEROPLASMAS

*I. M. Robinson*



## I. INTRODUCTION

In 1966, Hungate described an obligately anaerobic rumen microorganism that lysed bacterial cells. Subsequently, the organism was characterized by Robinson and Hungate (1973) as a mycoplasma. Various heat-killed, gram-negative cells, including *Escherichia coli,* were lysed by an extracellular enzyme that attacked the peptidoglycan layer of the cell wall; gram-positive cells (*Bacillus megaterium*) were not attacked (Robinson and Hungate, 1973). Another characteristic of these mycoplasmas was their ability to hydrolyze casein (skim milk).



THE MYCOPLASMAS, VOL. 1

ISBN 0-12-078401-7

Similar strains of anaerobic mycoplasma were isolated from the ruminal contents of both sheep and cattle by Robinson and Allison (1975). In addition, they reported the presence of nonbacteriolytic, anaerobic mycoplasmas that were not able to digest bacterial cells (Robinson *et al.*, 1975). Prins and van Den Vorstenbosch (1975), at the University of Utrecht, isolated cytoclastic anaerobic mycoplasmas from rumen contents of cattle ($2.4 \times 10^3$/ml of rumen fluid). The organisms grew rapidly on living or autoclaved gram-negative bacteria, causing lysis, but they did not grow on gram-positive cells. The ability of Prins' isolates to lyse living bacteria is interesting because previous isolates of anaerobic rumen mycoplasmas (Robinson and Allison, 1975; Robinson and Hungate, 1973) lysed only nonviable gram-negative cells. C. J. Smith and R. B. Hespell, at the University of Illinois (personal communication), found high concentrations of anaerobic mycoplasmas with the ability to lyse rumen bacteria in rumen contents of cattle fed high-grain diets. Isolated strains were highly proteolytic and nonfermentative, and in this latter property they differed from mycoplasmas previously isolated from the rumen.

These studies show that anaerobic mycoplasmas exist in the rumen contents of animals fed a variety of diets and located in diverse geographic regions. The anaerobic nature of these mycoplasmas and the lytic activity possessed by some strains distinguish them from other mycoplasmas that have been described.

Neither lytic nor nonlytic mycoplasma-like organisms were detected in cultures from cecal material of rabbits, hamsters, horses, pigs, or turkeys or the rumen content of a deer when dilutions ($10^3$) were tested (Robinson *et al.*, 1975). The methods and media used were the same as those used for studies with rumen contents. Anaerobic mycoplasmas appear to be ubiquitous in the rumens of cattle and sheep. Further studies will probably reveal a much larger, heterogeneous group of anaerobic mycoplasmas than is indicated at present.

## II. NUMBERS FOUND IN THE RUMEN

Table I shows the counts of anaerobic mycoplasmas in samples of rumen contents collected at different times from fistulated cattle and sheep. Lytic mycoplasmas, able to hydrolyze both killed *E. coli* cells and casein (skim milk), were present at from $10^5$ to $10^7$ viable units per gram of rumen contents. The mean of 10 samples of ruminal contents from four cows and three sheep was $7.6 \times 10^6$/gm. These results are similar to cultural counts reported by Robinson and Hungate (1973). Nonlytic mycoplasmas were consistently present at higher concentrations ($10^7$–

TABLE I. **Colony Counts[a] of Anaerobic Mycoplasmas from the Rumens of Cattle and Sheep**

| Experiment | Animal source[b] | Lytic (CFU/gm) | Nonlytic (CFU/gm) |
|---|---|---|---|
| 1 | Cow 1 | $2.0 \times 10^6$ | —[c] |
| 2 | Cow 1 | $1.4 \times 10^6$ | —[c] |
| 3 | Cow 1 | $2.0 \times 10^7$ | —[c] |
| 4 | Cow 1 | $8.0 \times 10^6$ | $4.0 \times 10^7$ |
| 5 | Cow 2 | $2.0 \times 10^7$ | $6.2 \times 10^7$ |
| 6 | Cow 3 | $3.5 \times 10^6$ | $4.6 \times 10^7$ |
| 7 | Cow 4 | $5.4 \times 10^6$ | $2.7 \times 10^7$ |
| 8 | Sheep 1 | $2.1 \times 10^5$ | $4.0 \times 10^7$ |
| 9 | Sheep 1 | $2.6 \times 10^5$ | $1.7 \times 10^7$ |
| 10 | Sheep 2 | $8.2 \times 10^6$ | $3.2 \times 10^7$ |

[a] Colony-forming units (CFU) per gram of rumen contents.
[b] Rumen samples collected at various times over a period of 18 months.
[c] Nonlytic CFU observed but not counted.

$10^8$/gm of ruminal contents) than lytic mycoplasmas (Robinson *et al.*, 1975).

## III. CLASSIFICATION

The anaerobic nature of these mycoplasmas is a unique, stable characteristic and is the basis for establishing the genus *Anaeroplasma* (Robinson *et al.*, 1975). The requirement for anaerobic conditions for growth is in contrast to that in other mycoplasmas that require reduced oxygen tension for initial isolation but lose this requirement upon repeated transfer (Vandemark, 1969). Plasmalogens (alk-l-enyl-glyceryl ether), which are found in various anaerobic bacteria but not in aerobic bacteria, are major components of polar lipids from anaeroplasmas (Langworthy *et al.*, 1975). The bacteriolytic capability possessed by some of the anaeroplasmas has not been reported for other mycoplasmas.

*Anaeroplasma* cells (16- to 18-hr-old cultures) are coccoid, about 500 nm in diameter, gram negative, and nonmotile. In wet mount preparations examined by phase microscopy, single cells, clumps, dumbbell forms, and clusters of 2 to 10 coccoid forms joined by short filaments are seen. In electron micrographs of negatively stained preparations (cells from 24-hr-old cultures), pleomorphic forms are observed; these include filamentous, budding, and bleblike structures. Electron micrographs of thin sections show cells bounded by a trilaminar membrane with no distinguishable cell wall (Fig. 1).

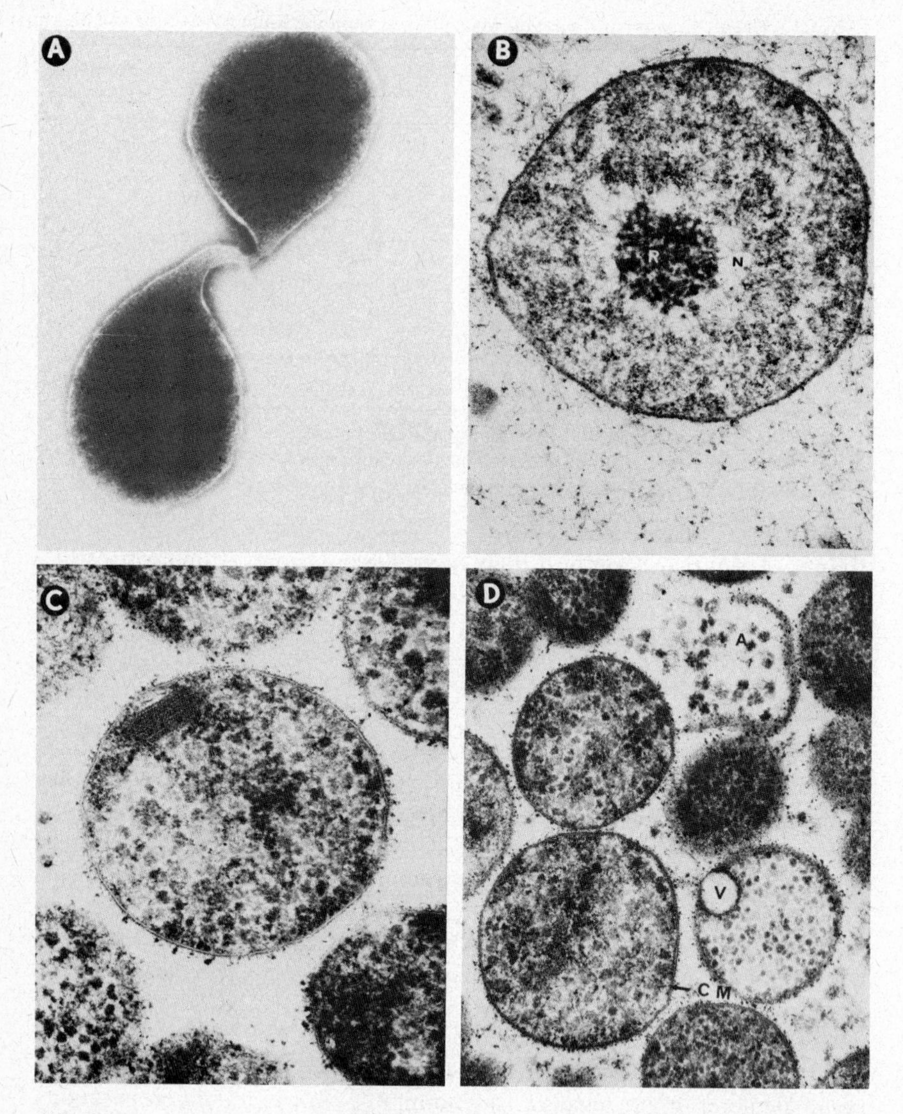


FIGURE 1. (A) Two mycoplasma cells with a tubular membranous-like connection (negative stain preparations) (× 121,000). (B) Anaerobic mycoplasma (strain 7LA); 24-hr culture. Centrally located in the cell is a less dense region within which there are delicate strands of nuclear material (N). A cluster of ribosomes (R) can also be seen (× 120,000). (C) Anaerobic mycoplasma (strain 7LA); 24-hr culture. A corncob-like pattern of electron-dense material is readily visible (× 150,000). (D) Anaerobic mycoplasma (strain 6-1); 24-hr culture. An older cell contains an empty, membrane-bound vesicle (V). An autolyzed cell (A) is also present. The trilaminal cytoplasmic membrane (CM) is readily visible (× 67,250).

Starch is fermented to produce various acids (generally acetic, formic, propionic, lactic, and succinic), ethanol, and gases (primarily $CO_2$ but some strains also produce $H_2$). Strains vary in their ability to ferment other carbohydrates. Bacteriolytic and nonbacteriolytic, as well as sterol-dependent and sterol-independent strains of anaerobic mycoplasmas, have been described.

*Anaeroplasma abactoclasticum* is the type species of *Anaeroplasma*. Strain 6-1 (ATCC 27879), the type strain of *A. abactoclasticum,* requires cholesterol for growth but lacks the extracellular bacterioclastic and proteolytic enzymes that characterize the lytic species. The guanine plus cytosine (G + C) content of deoxyribonucleic acid from strain 6-1 is 29.3 mol %. Nonbacteriolytic strains 161, 162, and 163 do not require cholesterol for growth (Table II), and the G + C contents of strains 161 and 162 are 40.3 and 40.5 mol %, respectively. Because mycoplasmas are placed in separate families based upon sterol requirements, these sterol-independent, nonbacteriolytic strains cannot be placed with sterol-dependent strains in *Anaeroplasma*. Further study is required to clarify their taxonomic status. Strain 161 has been deposited in the American Type Culture Collection as ATCC 27880.

*Anaeroplasma bactoclasticum* (Robinson and Hungate) comb. nov. includes anaerobic mycoplasmas that have proteolytic and bacteriolytic enzymes and require cholesterol for growth. The type strain, JR (ATCC 27112), was transferred from *Acholeplasma bactoclasticum* to the genus *Anaeroplasma* when Robinson and Allison (1975) discovered that cholesterol was required for growth. *Anaeroplasma bactoclasticum* strains

TABLE II. **Differential Characteristics of Anaerobic Mycoplasma Strains**[a]

| Serovar | Strain | Bacteriolytic | Requirement for cholesterol | Guanine plus cytosine (mol %) |
|---|---|---|---|---|
| 1 | JR | + | + | 33.7 |
| 2 | 5LA | + | + | ND[b] |
| | 5LB | + | + | ND |
| | 5LC | + | + | ND |
| | 7LA | + | + | 32.5 |
| 3 | 6-1 | − | + | 29.3 |
| | 171 | − | + | 29.5 |
| 4 | 161 | − | − | 40.2 |
| | 162 | − | − | 40.3 |
| | 163 | − | − | ND |

[a] Cultural, biophysical, and biochemical properties (Robinson and Allison, 1975; Robinson *et al.*, 1975).

[b] ND = not done.

7LA, 5LA, 5LB, and 5LC are similar to the type strain but comprise a single serologic group that differs from the type strain and from anaerobic nonbacterioclastic mycoplasmas. The G + C content of strain JR was 33.7 mol % and that of strain 7LA was 32.5 mol % (Robinson and Allison, 1975).

## IV. ANAEROBIC CULTURE TECHNIQUE

The roll tube anaerobic culture technique of Hungate (1950) has usually been used to culture anaerobic mycoplasmas (Robinson and Allison, 1975; Robinson *et al.*, 1975; Robinson, 1972; Robinson and Hungate, 1973). C. J. Smith and R. B. Hespell, at the University of Illinois (personal communication), however, used an agar-overlay plating technique in an anaerobic hood. Anaerobic mycoplasmas grew only in prereduced medium maintained in a system for exclusion of oxygen. If resazurin in a test medium became oxidized, the organisms failed to grow. The inhibitory effect of oxygen on growth was not altered by repeated subculturing of these organisms.

### A. Media

Table III shows the composition of media used to isolate, grow, and maintain anaerobic mycoplasmas. Media were autoclaved and maintained under anaerobic conditions in rubber-stoppered flasks or tubes with a gas phase of $O_2$-free $CO_2$. Cysteine hydrochloride (0.05%) was added to produce a low oxidation–reduction potential, and resazurin (0.0001%) was added as an indicator of anaerobiosis. Primary isolation medium (PIM) was similar to medium 98-5 that was previously used to culture rumen bacteria (Bryant and Robinson, 1961), except that it contained autoclaved *E. coli* cells (0.5%, w/v) and benzylpenicillic acid (1000 U/ml) was added. Clarified rumen fluid broth (CRFB) medium was similar to PIM, except that Trypticase (0.2%) and yeast extract (0.05%) were added, and the percentages of glucose, cellobiose, and starch were increased (0.2% of each instead of 0.05%); agar, benzylpenicillic acid, and *E. coli* cells were deleted. This medium supported growth of all strains of anaerobic mycoplasmas studied. Cultures were stored at −40°C in CRFB. Modified medium 10 (MM-10) was medium 10 of Caldwell and Bryant (1966), except that increased concentrations of glucose, cellobiose, and starch (0.2% instead of 0.05%) were used and sodium sulfide and agar were deleted. Various modifications of MM-10 were used in nutritional studies. Medium D was used in some experiments to enumerate and isolate lytic

TABLE III. **Composition of Media for Anaerobic Mycoplasmas**

| Component | Percentage in medium | | | |
|---|---|---|---|---|
| | PIM | CRFB | MM-10 | D |
| Clarified rumen fluid[a] (v/v) | 40 | 40 | — | — |
| Glucose (w/v) | 0.05 | 0.2 | 0.2 | — |
| Cellobiose (w/v) | 0.05 | 0.2 | 0.2 | — |
| Starch (w/v) | 0.05 | 0.2 | 0.2 | 0.2 |
| Minerals[b] (v/v) | 3.75 | 3.75 | 3.75 | 3.75 |
| Trypticase (w/v) | 0.2 | 0.2 | 0.2 | 0.2[c] |
| Yeast extract (w/v) | 0.1 | 0.1 | 0.1 | 0.1[d] |
| Volatile fatty acids[e] (v/v) | — | — | 0.31 | 0.31 |
| Resazurin (w/v) | 0.0001 | 0.0001 | 0.0001 | 0.0001 |
| Hemin (w/v) | — | — | 0.0001 | — |
| Lipopolysaccharide[f] (w/v) | — | — | 0.025 | 0.025[g] |
| Cholesterol[h] (w/v) | — | — | 2.0 | 2.0 |
| Autoclaved *E. coli* cells[i] (w/v) | 0.5 | — | — | — |
| $Na_2CO_3$ (w/v) | 0.4 | 0.4 | 0.4 | 0.4 |
| Cysteine HCL (w/v) | 0.05 | 0.05 | 0.05 | 0.05 |
| Agar (w/v) | 1.5 | — | — | — |
| Benzylpenicillic acid[j] (w/v) | 0.0006 | — | — | — |

[a] Rumen contents strained through several layers of cheesecloth, autoclaved, and clarified by centrifugation.

[b] Minerals (final concentrations) $K_2HPO_4$, $1.7 \times 10^{-3}$ *M;* $KH_2PO_4$, $1.3 \times 10^{-3}$ *M;* NaCl, $7.6 \times 10^{-4}$ *M;* $(NH_4)_2SO_4$, $3.4 \times 10^{-3}$ *M;* $CaCl_2$, $4.1 \times 10^{-4}$ *M;* and $MgSO_4 \cdot 7H_2O$, $3.8 \times 10^{-4}$ *M*.

[c] In some experiments Trypticase was replaced by amino acids (0.02%, w/v, of each): L-alanine, L-arginine, L-aspartic, L-asparagine, L-glutamic, L-glutamine, L-glycine, L-histidine, L-isoleucine, L-leucine, L-lysine, L-methionine, L-phenylalanine, L-proline, L-serine, L-threonine, L-tryptophan, L-tyrosine, L-valine.

[d] In some experiments, yeast extract was replaced by vitamins (final concentrations) 0.2 mg/100 ml each of thiamine–HCl, Ca panthothenate, nicotinamide, riboflavin, and pyridoxal; 0.01 mg/100 ml of *p*-aminobenzoic acid; 0.005 mg/100 ml each of biotin, folic acid, and DL-thioctic acid; 0.002 mg/100 ml of $B_{12}$.

[e] Final concentrations (millimolar) of volatile fatty acids in the medium were: acetic, 29; propionic, 8.0; *n*-butyric, 4.3; isobutyric, 1.1; *n*-valeric, isovaleric, and DL-$\alpha$-methylbutyric, 0.9 each.

[f] Lipopolysaccharide (LPS); lipopolysaccharide, Boivin.

[g] In some experiments LPS was replaced by phosphatidyl choline (soybean); final concentration was 0.05%.

[h] Cholesterol: 20 mg of cholesterol in 1 ml of ethanol, made to 20 ml with water (final concentration) (20 $\mu$g/ml).

[i] Autoclaved *E. coli* cells (see preparation in text).

[j] Final concentration, 1000 U/ml.

and nonlytic anaerobic mycoplasmas by deleting the lipid source, either lipopolysaccharide, Boivin (LPS-B) or phosphatidyl choline, and then adding *E. coli* cells, agar, and benzylpenicillic acid, as described for PIM. When Trypticase, yeast extract, and LPS-B of medium D were replaced by amino acids, vitamins, and phosphatidyl choline, growth of anaerobic mycoplasma strains was achieved in a completely defined medium (Robinson *et al.*, 1975).

### B. Isolating and Culturing

To isolate anaerobic mycoplasmas, 1 gm of each sample was placed in a sterile, rubber-stoppered tube (18 by 150 mm) containing 9 ml of sterile anaerobic mineral dilution solution. This suspension was serially diluted in tubes of anaerobic dilution solution with anaerobic methods. Volumes (0.1 ml) of appropriate dilutions were inoculated into tubes of melted PIM held in a water bath at 45°C, and the tubes were rolled in an ice-water bath equipped with a motorized roller. Inoculated roll tubes were incubated at 37°C and periodically examined visually for the presence of colonies surrounded by clear zones. The clear zones developed around colonies that possessed lytic activity caused by the partial digestion of suspended *E. coli* cells. Discrete zones often could be seen within 48 hr. Roll tubes were also examined with a stereoscopic microscope for the presence of mycoplasma-like colonies that did not produce clear zones.

The colonial characteristics of lytic and nonlytic mycoplasmas were similar. Subsurface colonies were golden, irregular, and often multilobed. Surface colonies had a dense center and a translucent periphery, presenting the so-called fried egg appearance. The average diameter of surface colonies was about 1 mm. Pure cultures were picked as individual colonies with a Pasteur pipette and subcultured in CRFB medium. Cell suspensions in liquid medium were serially diluted in PIM, and colonies were reisolated from roll tubes of the highest dilutions showing growth after incubation at 37°C for 48 hr.

When benzylpenicillic acid was deleted from the isolation medium, mycoplasma-like CFU could still be detected. Isolates obtained from lytic and nonlytic colonies, when subcultured in medium devoid of benzylpenicillic acid, did not revert to a bacterial form. The lytic or nonlytic properties of these cultures were a stable characteristic and did not change with subsequent subculture (Robinson *et al.*, 1975).

## V. NUTRITION, PHYSIOLOGY, AND METABOLISM

The anaerobic mycoplasma strains (Table II) all grew in a medium containing mineral salts, rumen fluid, and soluble starch (CRFB). The optimum level of rumen fluid was about 40% for strain 6-1, while 60% was optimum for strain 7LA. Robinson and Hungate (1973) used 33% clarified rumen fluid for their strain of anaerobic mycoplasma. A typical bacterial growth curve was observed when strain 6-1 was grown in CRFB medium. Growth was most rapid in the first 6–8 hr. The generation time was about 90 min. Although growth, as measured by numbers of CFU, indicated that the culture was past the exponential growth phase before turbidity had reached maximal values, turbidity measurements were usually used as an index of growth because they were more convenient. In some media, turbidity often dropped rapidly shortly after the culture had reached maximum absorbance. Thus, absorbance was measured frequently during the exponential growth period. This phenomenon was inconsistent, however, and could not be related to a particular culture condition. Electron-microscopic examination of negatively stained preparations of the lysed material did not reveal structures resembling phage. Growth was also measured by following incorporation of [$^3$H]thymidine (Byfield and Scherbaum, 1966) in some studies (Robinson and Rhoades, 1977).

The optimum temperature for growth of *Anaeroplasma* was 37°C and the optimum pH was 6.5–7.0; these growth conditions are also the optimum for most anaerobic organisms found in the rumen. The osmotic pressure in bovine rumen fluid is about 7.8 atm (Hungate, 1942) and is close to the 5–6 atm estimated to be the internal osmotic pressure of *Acholeplasma laidlawii* (Spears and Provost, 1967). Colonies of anaerobic mycoplasmas failed to develop in PIM containing 0.2% thallium acetate. Thallium acetate has been used as a bacterial inhibitor for the isolation of most mycoplasmas other than ureaplasmas (Morton and Lecce, 1953). Growth of strain 6-1 was inhibited when bacitracin (0.7 $\mu$mol/ml), streptomycin (0.36 $\mu$mol/ml), or D-cycloserine (5 $\mu$mol/ml) was added to CRFB medium. The fact that D-cycloserine inhibited growth was unexpected and without explanation. These materials inhibited anaeroplasmas at concentrations much lower than the minimal inhibitory levels reported for many other mycoplasmas (Ward *et al.*, 1958). The addition of 0.5 mg of benzylpenicillic acid per milliliter (1000 U/ml) to the medium did not inhibit growth of any of the strains tested.

Soluble starch supported growth of all *Anaeroplasma* strains studied. For optimum growth the level was 0.2%. Other carbohydrates had variable effects on growth. Maltose supported growth of strain 6-1 slightly but had very little effect on growth of strains 7LA and JR. Only strain JR used galactose, and only strain 7LA used glucose. Arabinose, fructose, glycerol, lactose, mannose, raffinose, salicin, sucrose, and xylose did not support growth of any strains.

Fermentation products from galactose by strain JR (ATCC 27112) were acetic, formic, lactic, and propionic acids; ethanol; and $CO_2$ and $H_2$

(Robinson and Hungate, 1973). The distribution of $^{14}C$ in fermentation products from strain 6-1 grown in medium containing uniformly $^{14}C$-labeled starch was acetate, 36.6%; formate, 19.0%; lactate, 17.2%; $CO_2$, 8.2%; ethanol, 4.8%; succinate, 1.1%; and several unidentified products, 5.6% (Robinson *et al.*, 1975).

Unlike most mycoplasmas, the *Anaeroplasma* do not require nucleic acids for growth (medium D and MM-10). In this respect, *Anaeroplasma* resembles most rumen bacteria (Smith and Smith, 1977).

The requirements of *Anaeroplasma* for minerals, vitamins, and nitrogen have not been defined. Studies showed that strain 6-1 grew when trypticase and yeast extract in medium D was replaced by amino acids and vitamins. In these experiments, ammonium sulfate was deleted.

The effects of cholesterol and LPS-B on growth of anaerobic mycoplasmas in a medium free of rumen fluid are given in Table IV. None of the mycoplasmas grew in the basal medium alone or in the basal medium plus cholesterol. A factor in LPS-B was essential for growth of all strains; cholesterol was required for all strains except nonlytic strains 161, 162, and 163.

The concentration of cholesterol required for one-half maximum growth of cholesterol-dependent strains was 5 $\mu$g/ml. A high concentra-

TABLE IV. **Growth of *Anaeroplasma* Strains in Medium Free of Rumen Fluid**

| | Growth (OD at 600 nm × 100)[a] | |
|---|---|---|
| | Basal[b] + LPS[c] | Basal + LPS + cholesterol[d] |
| Sterol-dependent strains | | |
| JR | 0 | 8 |
| 7LA | 0 | 15 |
| 5LA | 0 | 51 |
| 5LB | 0 | 53 |
| 5LC | 0 | 52 |
| 6-1 | 0 | 69 |
| Sterol-independent strains | | |
| 161 | 18 | 18 |
| 162 | 22 | 28 |
| 163 | 20 | 25 |

[a] Maximum reading within 48 hr.

[b] Basal medium: starch, Trypticase, vitamins, volatile fatty acids, minerals, $Na_2CO_3$, cysteine (Robinson *et al.*, 1975). None of the strains grew in basal alone or in basal plus cholesterol.

[c] Lipopolysaccharide (Boivin), 0.25 mg/ml medium.

[d] Cholesterol (20 $\mu$g/ml).

tion of cholesterol (100 μg/ml) was inhibitory. Growth of strains which did not require cholesterol was only slightly improved by adding cholesterol to the culture medium. Growth of cholesterol-dependent strains was inhibited by low concentrations of digitonin (20 μg/ml), whereas growth of sterol-independent strains was not affected by 200 μg of digitonin per milliliter and only slightly inhibited by 500 μg/ml. These data support our observation (Robinson and Allison, 1975; Robinson *et al.*, 1975) on cholesterol dependence for growth because mycoplasmas that depend upon cholesterol are sensitive to low concentrations of digitonin (Razin and Shafer, 1969).

A growth response to LPS by *Anaeroplasma* could be shown even when the culture medium (CRFB) contained rumen fluid (Robinson *et al.*, 1975; Robinson, 1972). Growth increased when CRFB medium was supplemented with increased amounts of LPS (0.01–1.0 mg/ml). Concentrations above 1.0 mg/ml interfered with absorbance measurements and were not tested. Strain 6-1 failed to grow when the LPS-B, which was added to MM-10, was hydrolyzed at pH 10 for 60 min at 100°C. The LPS growth factor was extracted into chloroform–methanol after mild acid hydrolysis (0.1 *N* formic acid at 100°C for 60 min). Strain 6-1 did not grow when the aqueous fraction was added to medium D. Mild acid hydrolysis splits off the lipid A component of LPS-B. Lipid A, after extraction into chloroform–methanol or other lipid solvents, supported growth of strain 6-1.

The lipid requirement of strain 6-1 could also be met by compounds other than LPS. Strain 6-1 grew in medium D when LPS was replaced by phosphatidyl choline, phosphatidyl glycerol, diphosphatidyl glycerol, phosphatidyl ethanolamine, lysophosphatidyl choline, and phosphatidic acids (listed in order in which they stimulated growth). Growth, as measured by absorbance, was proportional to the concentration of phosphatidyl choline at 0.5 and 1.0 mg/ml. Glycerolphosphoryl choline, glycerolphosphoryl ethanolamine, glycerolphosphoryl serine, and glycerol phosphate would not effectively replace LPS. Other strains of *Anaeroplasma* showed a similar growth response to the lipids. Like LPS-B, hydrolysis of phosphatidyl choline and phosphatidyl ethanolamine at pH 10 for 60 min at 100°C destroyed their ability to promote growth; this indicates that the intact phospholipid was needed for growth.

Other substances examined for their ability to supply the lipid growth factor requirement of *Anaeroplasma* were monoolein, diolein, and triolein, and lauric, myristic, myristoleic, palmitic, palmitoleic, stearic, oleic, linoleic, linolenic, arachidic, *cis*-11-arachidic, and arachidonic acids. None of these, when assayed alone or in various combinations with glycerol and choline, would replace intact phospholipid.

Esterified fatty acids apparently were the essential component of the polar lipids needed for growth because glycerolphosphoryl choline, glycerolphosphoryl ethanolamine, glycerolphosphoryl serine, and glycerol phosphate would not support growth. Furthermore, the requirement appears to be specific for unsaturated rather than saturated fatty acids. Strain 6-1 grew in medium D in which either oleic or linoleic acid was esterified to glycerolphosphoryl choline but not when lauric, myristic, palmitic, or stearic acid was esterified. This lipid requirement differs from that reported for *Acholeplasma laidlawii* (Razin and Rottem, 1963) or anaerobically grown *Saccharomyces cerevisiae* (Altherthum and Rose, 1973), which use unsaturated free fatty acids.

Experiments with two *Anaeroplasma* strains, sterol-dependent strain 6-1 and sterol-independent strain 161, showed that an insignificant amount of 2-[$^{14}$C] acetate was incorporated into cellular constituents during growth; this suggests that these organisms are unable to synthesize long-chain fatty acids from acetate. All strains of *Anaeroplasma* appear to require phospholipids, and either LPS-B or glycerolphosphoryl compounds with esterified oleic or linoleic acids meet this need.

## VI. LYTIC FACTOR

The lytic factors diffused from colonies grown in agar and also were active in liquid cultures. Suspensions of autoclaved rumen bacteria, *E. coli, Salmonella typhimurium,* and *Spirillum serpens,* were lysed by an ammonium sulfate precipitate prepared from a culture supernatant of *A. bactoclasticum*. Murein sacculi of *E. coli* were lysed by a similar preparation, but freeze-dried *Micrococcus lysodeikticus* cells were not lysed (Robinson and Hungate, 1973). Milk clearing is also caused by a diffusible enzyme(s).

Robinson (1972) hypothesized that the cell lytic and proteolytic activities were performed by the same enzyme. Both cell lysis and proteolysis (measured as skim milk clearing) were reversibly inhibited by exposure to air, and the activity was restored by reducing agents (mercaptoethanol or dithiothreitol). Both activities were enriched to the same degree in ammonium sulfate precipitates and eluted as a single peak on gel filtration (Sephadex 100). Both lytic activities exhibited the same response to temperature in phosphate buffer (most active at 55°C, inactive at 65°C), with greater activity at pH 6.8 than at 5.7, 7.6, or 8.0.

Heat and lipid extraction of susceptible cells facilitated lysis by culture supernatant of *A. bactoclasticum* (Robinson, 1972). These results were consistent with the failure to demonstrate lipase activity; thus, disruption of the cell envelope may be required for the enzyme to gain access to the peptidoglycan in the absence of lipase activity.

## VII. SEROLOGY

On the basis of agglutination, gel precipitin, and growth inhibition tests, 10 anaerobic mycoplasma strains (Table II) were separated into four serovars compatible with group separations based upon cultural, biochemical, and biophysical properties (Robinson and Rhoades, 1977). Also, hyperimmune sera prepared against representative strains from each of the four serovars failed to cross-react with antigens of the following aerobic bovine mycoplasmas or acholeplasmas: (B) PG-11; (C) PG-10; (D) FX-1; (E) B5P; (F) B-38; (G) RLPG-51; (H) B12PA; (I) MC-1; (J) RPG-47; (K) B-74; (L) B-144; and (M) B-44 (Robinson and Rhoades, 1977). These findings support our contention that anaerobic mycoplasmas should be placed in a taxonomic group that is apart from *Mycoplasma* and *Acholeplasma*.

## VIII. SIGNIFICANCE IN THE RUMEN

The ecologic role of these organisms in the rumen has not been determined. Although the concentrations of these organisms in the rumen are usually low when compared with the concentrations of other bacteria, anaerobic mycoplasma contribute to the pool of microbial fermentation products. The turnover of microbial protoplasma in the rumen is probably significant and has been estimated to be as high as 20% (Smith and Smith, 1977). The bacteriolytic anaerobic mycoplasmas, along with rumen protozoa and bacteriophage, probably contribute to this turnover.

## IX. CONCLUSION

Both sterol-dependent and sterol-independent, obligately anaerobic mycoplasmas have been characterized. They have cultural, biochemical, and serologic properties that distinguish them from other mycoplasmas. The anaerobic conditions required for growth and the physiologic characteristics of these mycoplasmas suggest that they are well adapted to grow in the rumen.

Strains of *Anaeroplasma* have been grown on media containing only chemically defined constituents. All strains appear to require phospholipid. Chemically synthesized phosphatidyl choline meets this requirement only if the esterified fatty acids are unsaturated. Plasmalogen (alk-l-enyl glyceryl ether) is a major constituent of the polar lipids of *Anaeroplasma* strains tested.

Several unclassified anaerobic mycoplasma strains share some biologic properties with *A. abactoclasticum* but they are not serologically related, they do not require cholesterol for growth, and their G + C content is significantly higher.

Anaerobic mycoplasmas appear to be a heterogeneous group that have been found only in the rumen of cattle and sheep. Each new group of anaerobic mycoplasmas that has been isolated had one or more different properties. Thus, apparently more biotypes exist in nature. Whether their distribution is limited to the rumens of cattle and sheep remains to be determined.

## REFERENCES


Alterthum, F., and Rose, A. H. (1973). *J. Gen. Microbiol.* **77,** 371–382.

Bryant, M. P., and Robinson, I. M. (1961). *J. Dairy Sci.* **44,** 1446–1456.

Byfield, J. E., and Scherbaum, O. H. (1966). *Anal. Biochem.* **17,** 434–443.

Caldwell, D. R., and Bryant, M. P. (1966). *Appl. Microbiol.* **14,** 794–801.

Hungate, R. E. (1942). *Biol. Bull. (Woods Hole, Mass.)* **83,** 303–319.

Hungate, R. E. (1950). *Bacteriol. Rev.* **14,** 1–49.

Hungate, R. E. (1966). "The Rumen and Its Microbes." Academic Press, New York.

Langworthy, T. A., Mayberry, W. R., Smith, P. F., and Robinson, I. M. (1975). *J. Bacteriol.* **122,** 785–787.

Morton, H. E., and Lecce, J. G. (1953). *J. Bacteriol.* **66,** 646–649.

Prins, R. A., and van Den Vorstenbosch, C. J. A. H. V. (1975). *Landbouwhogesch. Wageningen, Misc. Pap.,* **11,** 15–24.

Razin, S., and Rottem, S. (1963). *J. Gen. Microbiol.* **33,** 459–470.

Razin, S., and Shafer, Z. (1969). *J. Gen. Microbiol.* **58,** 327–339.

Robinson, I. M., and Allison, M. J. (1975). *Int. J. Syst. Bacteriol.* **25,** 182–186.

Robinson, I. M., and Rhoades, K. R. (1977). *Int. J. Syst. Bacteriol.* **27,** 200–203.

Robinson, I. M., Allison, M. J., and Hartman, P. A. (1975) *Int. J. Syst. Bacteriol.* **25,** 173–181.

Robinson, J. P. (1972). *Diss. Abstr. Int. B* **32,** 7196B.

Robinson, J. P., and Hungate, R. E. (1973). *Int. Syst. Bacteriol.* **23,** 171–181.

Smith, R. C., and Smith, R. H. (1977). *Br. J. Nutr.* **37,** 388–394.

Spears, D. M., and Provost, J. (1967). *Can. J. Microbiol.* **13,** 213–225.

Vandemark, P. J. (1969). *In* "The Mycoplasmatales and the L-phase Bacteria," (L. Hayflick, ed.), pp. 491–501. Appleton, New York.

Ward, J. R., Madoff, S., and Dienes, L. (1958). *Proc. Soc. Exp. Biol. Med.* **97,** 132–135.

# SUBJECT INDEX

H



I

N



O



P

## Q

**U**

**V**